LONDON MATHEMATICAL SOCIETY LECTURE NOTE SERIES

Managing Editor: Professor N.J. Hitchin, Mathematics Institute,
University of Oxford, 24–29 St Giles, Oxford OX1 3TG, United Kingdom

The titles below are available from booksellers, or, in case of difficulty, from Cambridge University Press.

46 *p*-adic Analysis: a short course on recent work, N. KOBLITZ
59 Applicable differential geometry, M. CRAMPIN & F.A.E. PIRANI
66 Several complex variables and complex manifolds II, M.J. FIELD
86 Topological topics, I.M. JAMES (ed)
87 Surveys in set theory, A.R.D. MATHIAS (ed)
88 FPF ring theory, C. FAITH & S. PAGE
89 An F-space sampler, N.J. KALTON, N.T. PECK & J.W. ROBERTS
90 Polytopes and symmetry, S.A. ROBERTSON
92 Representations of rings over skew fields, A.H. SCHOFIELD
93 Aspects of topology, I.M. JAMES & E.H. KRONHEIMER (eds)
96 Diophantine equations over function fields, R.C. MASON
97 Varieties of constructive mathematics, D.S. BRIDGES & F. RICHMAN
98 Localization in Noetherian rings, A.V. JATEGAONKAR
99 Methods of differential geometry in algebraic topology, M. KAROUBI & C. LERUSTE
100 Stopping time techniques for analysts and probabilists, L. EGGHE
104 Elliptic structures on 3-manifolds, C.B. THOMAS
105 A local spectral theory for closed operators, I. ERDELYI & WANG SHENGWANG
107 Compactification of Siegel moduli schemes, C.-L. CHAI
109 Diophantine analysis, J. LOXTON & A. VAN DER POORTEN (eds)
113 Lectures on the asymptotic theory of ideals, D. REES
114 Lectures on Bochner-Riesz means, K.M. DAVIS & Y.-C. CHANG
116 Representations of algebras, P.J. WEBB (ed)
119 Triangulated categories in the representation theory of finite-dimensional algebras, D. HAPPEL
121 Proceedings of *Groups - St Andrews 1985*, E. ROBERTSON & C. CAMPBELL (eds)
128 Descriptive set theory and the structure of sets of uniqueness, A.S. KECHRIS & A. LOUVEAU
130 Model theory and modules, M. PREST
131 Algebraic, extremal & metric combinatorics, M.-M. DEZA, P. FRANKL & I.G. ROSENBERG (eds)
132 Whitehead groups of finite groups, ROBERT OLIVER
133 Linear algebraic monoids, MOHAN S. PUTCHA
134 Number theory and dynamical systems, M. DODSON & J. VICKERS (eds)
137 Analysis at Urbana, I, E. BERKSON, T. PECK, & J. UHL (eds)
138 Analysis at Urbana, II, E. BERKSON, T. PECK, & J. UHL (eds)
139 Advances in homotopy theory, S. SALAMON, B. STEER & W. SUTHERLAND (eds)
140 Geometric aspects of Banach spaces, E.M. PEINADOR & A. RODES (eds)
141 Surveys in combinatorics 1989, J. SIEMONS (ed)
144 Introduction to uniform spaces, I.M. JAMES
146 Cohen-Macaulay modules over Cohen-Macaulay rings, Y. YOSHINO
148 Helices and vector bundles, A.N. RUDAKOV *et al*
149 Solitons, nonlinear evolution equations and inverse scattering, M. ABLOWITZ & P. CLARKSON
150 Geometry of low-dimensional manifolds 1, S. DONALDSON & C.B. THOMAS (eds)
151 Geometry of low-dimensional manifolds 2, S. DONALDSON & C.B. THOMAS (eds)
152 Oligomorphic permutation groups, P. CAMERON
153 L-functions and arithmetic, J. COATES & M.J. TAYLOR (eds)
155 Classification theories of polarized varieties, TAKAO FUJITA
156 Twistors in mathematics and physics, T.N. BAILEY & R.J. BASTON (eds)
158 Geometry of Banach spaces, P.F.X. MÜLLER & W. SCHACHERMAYER (eds)
159 Groups St Andrews 1989 volume 1, C.M. CAMPBELL & E.F. ROBERTSON (eds)
160 Groups St Andrews 1989 volume 2, C.M. CAMPBELL & E.F. ROBERTSON (eds)
161 Lectures on block theory, BURKHARD KÜLSHAMMER
162 Harmonic analysis and representation theory, A. FIGA-TALAMANCA & C. NEBBIA
163 Topics in varieties of group representations, S.M. VOVSI
164 Quasi-symmetric designs, M.S. SHRIKANDE & S.S. SANE
166 Surveys in combinatorics, 1991, A.D. KEEDWELL (ed)
168 Representations of algebras, H. TACHIKAWA & S. BRENNER (eds)
169 Boolean function complexity, M.S. PATERSON (ed)
170 Manifolds with singularities and the Adams-Novikov spectral sequence, B. BOTVINNIK
171 Squares, A.R. RAJWADE
172 Algebraic varieties, GEORGE R. KEMPF
173 Discrete groups and geometry, W.J. HARVEY & C. MACLACHLAN (eds)
174 Lectures on mechanics, J.E. MARSDEN
175 Adams memorial symposium on algebraic topology 1, N. RAY & G. WALKER (eds)
176 Adams memorial symposium on algebraic topology 2, N. RAY & G. WALKER (eds)
177 Applications of categories in computer science, M. FOURMAN, P. JOHNSTONE & A. PITTS (eds)
178 Lower K- and L-theory, A. RANICKI
179 Complex projective geometry, G. ELLINGSRUD *et al*
180 Lectures on ergodic theory and Pesin theory on compact manifolds, M. POLLICOTT
181 Geometric group theory I, G.A. NIBLO & M.A. ROLLER (eds)
182 Geometric group theory II, G.A. NIBLO & M.A. ROLLER (eds)

183 Shintani zeta functions, A. YUKIE
184 Arithmetical functions, W. SCHWARZ & J. SPILKER
185 Representations of solvable groups, O. MANZ & T.R. WOLF
186 Complexity: knots, colourings and counting, D.J.A. WELSH
187 Surveys in combinatorics, 1993, K. WALKER (ed)
188 Local analysis for the odd order theorem, H. BENDER & G. GLAUBERMAN
189 Locally presentable and accessible categories, J. ADAMEK & J. ROSICKY
190 Polynomial invariants of finite groups, D.J. BENSON
191 Finite geometry and combinatorics, F. DE CLERCK *et al*
192 Symplectic geometry, D. SALAMON (ed)
194 Independent random variables and rearrangement invariant spaces, M. BRAVERMAN
195 Arithmetic of blowup algebras, WOLMER VASCONCELOS
196 Microlocal analysis for differential operators, A. GRIGIS & J. SJÖSTRAND
197 Two-dimensional homotopy and combinatorial group theory, C. HOG-ANGELONI *et al*
198 The algebraic characterization of geometric 4-manifolds, J.A. HILLMAN
199 Invariant potential theory in the unit ball of $\mathbf{C}^n$, MANFRED STOLL
200 The Grothendieck theory of dessins d'enfant, L. SCHNEPS (ed)
201 Singularities, JEAN-PAUL BRASSELET (ed)
202 The technique of pseudodifferential operators, H.O. CORDES
203 Hochschild cohomology of von Neumann algebras, A. SINCLAIR & R. SMITH
204 Combinatorial and geometric group theory, A.J. DUNCAN, N.D. GILBERT & J. HOWIE (eds)
205 Ergodic theory and its connections with harmonic analysis, K. PETERSEN & I. SALAMA (eds)
207 Groups of Lie type and their geometries, W.M. KANTOR & L. DI MARTINO (eds)
208 Vector bundles in algebraic geometry, N.J. HITCHIN, P. NEWSTEAD & W.M. OXBURY (eds)
209 Arithmetic of diagonal hypersurfaces over finite fields, F.Q. GOUVÊA & N. YUI
210 Hilbert C*-modules, E.C. LANCE
211 Groups 93 Galway / St Andrews I, C.M. CAMPBELL *et al* (eds)
212 Groups 93 Galway / St Andrews II, C.M. CAMPBELL *et al* (eds)
214 Generalised Euler-Jacobi inversion formula and asymptotics beyond all orders, V. KOWALENKO *et al*
215 Number theory 1992–93, S. DAVID (ed)
216 Stochastic partial differential equations, A. ETHERIDGE (ed)
217 Quadratic forms with applications to algebraic geometry and topology, A. PFISTER
218 Surveys in combinatorics, 1995, PETER ROWLINSON (ed)
220 Algebraic set theory, A. JOYAL & I. MOERDIJK
221 Harmonic approximation, S.J. GARDINER
222 Advances in linear logic, J.-Y. GIRARD, Y. LAFONT & L. REGNIER (eds)
223 Analytic semigroups and semilinear initial boundary value problems, KAZUAKI TAIRA
224 Computability, enumerability, unsolvability, S.B. COOPER, T.A. SLAMAN & S.S. WAINER (eds)
225 A mathematical introduction to string theory, S. ALBEVERIO, J. JOST, S. PAYCHA, S. SCARLATTI
226 Novikov conjectures, index theorems and rigidity I, S. FERRY, A. RANICKI & J. ROSENBERG (eds)
227 Novikov conjectures, index theorems and rigidity II, S. FERRY, A. RANICKI & J. ROSENBERG (eds)
228 Ergodic theory of $\mathbf{Z}^d$ actions, M. POLLICOTT & K. SCHMIDT (eds)
229 Ergodicity for infinite dimensional systems, G. DA PRATO & J. ZABCZYK
230 Prolegomena to a middlebrow arithmetic of curves of genus 2, J.W.S. CASSELS & E.V. FLYNN
231 Semigroup theory and its applications, K.H. HOFMANN & M.W. MISLOVE (eds)
232 The descriptive set theory of Polish group actions, H. BECKER & A.S. KECHRIS
233 Finite fields and applications, S. COHEN & H. NIEDERREITER (eds)
234 Introduction to subfactors, V. JONES & V.S. SUNDER
235 Number theory 1993–94, S. DAVID (ed)
236 The James forest, H. FETTER & B. GAMBOA DE BUEN
237 Sieve methods, exponential sums, and their applications in number theory, G.R.H. GREAVES *et al*
238 Representation theory and algebraic geometry, A. MARTSINKOVSKY & G. TODOROV (eds)
239 Clifford algebras and spinors, P. LOUNESTO
240 Stable groups, FRANK O. WAGNER
241 Surveys in combinatorics, 1997, R.A. BAILEY (ed)
242 Geometric Galois actions I, L. SCHNEPS & P. LOCHAK (eds)
243 Geometric Galois actions II, L. SCHNEPS & P. LOCHAK (eds)
244 Model theory of groups and automorphism groups, D. EVANS (ed)
245 Geometry, combinatorial designs and related structures, J.W.P. HIRSCHFELD *et al*
246 *p*-Automorphisms of finite *p*-groups, E.I. KHUKHRO
247 Analytic number theory, Y. MOTOHASHI (ed)
248 Tame topology and o-minimal structures, LOU VAN DEN DRIES
249 The atlas of finite groups: ten years on, ROBERT CURTIS & ROBERT WILSON (eds)
250 Characters and blocks of finite groups, G. NAVARRO
251 Gröbner bases and applications, B. BUCHBERGER & F. WINKLER (eds)
252 Geometry and cohomology in group theory, P. KROPHOLLER, G. NIBLO, R. STÖHR (eds)
253 The *q*-Schur algebra, S. DONKIN
254 Galois representations in arithmetic algebraic geometry, A.J. SCHOLL & R.L. TAYLOR (eds)
255 Symmetries and integrability of difference equations, P.A. CLARKSON & F.W. NIJHOFF (eds)
257 An introduction to noncommutative differential geometry and its physical applications, 2ed, J. MADORE
258 Sets and Proofs, S.B. COOPER & J.K. TRUSS (eds)
259 Models and Computability, S.B. COOPER & J.K. TRUSS (eds)
260 Groups St Andrews 1997 in Bath, I, C.M. CAMPBELL *et al* (eds)
261 Groups St Andrews 1997 in Bath, II, C.M. CAMPBELL *et al* (eds)
264 Recent trends in algebraic geometry, K. HULEK, F. CATANESE, C. PETERS & M. REID (eds)

London Mathematical Society Lecture Note Series. 258

Sets and Proofs

Invited papers from Logic Colloquium '97 - European Meeting of the Association for Symbolic Logic, Leeds, July 1997

Edited by

S. Barry Cooper
University of Leeds

John K. Truss
University of Leeds

CAMBRIDGE
UNIVERSITY PRESS

PUBLISHED BY THE PRESS SYNDICATE OF THE UNIVERSITY OF CAMBRIDGE
The Pitt Building, Trumpington Street, Cambridge CB2 1RP, United Kingdom

CAMBRIDGE UNIVERSITY PRESS
The Edinburgh Building, Cambridge, CB2 2RU, UK http://www.cup.cam.ac.uk
40 West 20th Street, New York, NY 10011-4211, USA http://www.cup.org
10 Stamford Road, Oakleigh, Melbourne 3166, Australia

First published 1999

Printed in the United Kingdom at the University Press, Cambridge

A catalogue record for this book is available from the British Library

ISBN 0 521 63549 7 paperback

Contents

Preface

Basic science, and within that pure mathematics, has a unique ability to surprise and change our view of the world we live in. But more often than not, its fundamental 'relevance' has emerged in ways impossible to have anticipated. As has often been remarked, that of the best basic science (say of non-Euclidean geometry, or of Hilbert spaces, or of the universal Turing machine) is independent of limited views of potential applicability.

Logic Colloquium '97, held in Leeds, England, 6th – 13th July, 1997, set out to reflect all that was best in contemporary logic, and **Models and Computability** and **Sets and Proofs** comprise two volumes of refereed articles, mainly based on the invited talks given at that meeting. Thanks to the programme committee (its other members being George Boolos, Sam Buss, Wilfrid Hodges, Martin Hyland, Alistair Lachlan, Alain Louveau, Yiannis Moschovakis, Leszeck Pacholski, Helmut Schwichtenberg, Ted Slaman and Hugh Woodin) and the special sessions organisers (Klaus Ambos-Spies, Sy Friedman, Wilfrid Hodges, Gerhard Jaeger, Steffen Lempp, Anand Pillay and Helmut Schwichtenberg), the editors have been able to call on a rich and distinguished array of authors. It is of great regret that one of our programme committee members was not able to see the success to which he had substantially contributed, and the **British Logic Colloquium Lecture**, given by Paul Benacerraf, took the form of a tribute to his memory. It would be difficult for us to improve on the introduction to Professor Benacerraf's article (on p.27 of this volume) provided by the following extract from the comments received from the referee (necessarily anonymous):

> 'Ever since Paul Benacerraf published "What numbers could not be" (1965) and "Mathematical truth" (1973), his views have been seminal in the development of philosophy of mathematics, and for this reason one can expect that any paper by him that revisits the issues he first discussed in those papers (see footnote 1) will be of immediate interest for the subject. Also, the present paper is written as a very personal tribute to George Boolos, a student of Paul Benacerraf, whose early death, at the age of 55, deprived philosophy of mathematics of one of its leading – and one may also say – best loved contributors, and the Association of Symbolic Logic of its serving President, so it is very particularly fitting that this paper should be published in the proceedings of a meeting of the ASL.'

Logic Colloquium '97 was also the first such conference in Britain since the death of Robin Gandy (the first president of the British Logic Colloquium) on 20 November, 1995, and **Models and Computability**, the companion volume, is dedicated to his memory. As observed by Andrew Hodges[1]:

> 'Robin Gandy's death on 20 November 1995 has ended the strongest living link with Alan Turing, with whom he was all of intimate friend, student and colleague. He inherited all Turing's mathematical books and papers; and thereafter also carried forward part of Turing's intellectual tradition; more precisely he took on the subject that Turing lost interest in, by becoming a pre-eminent British figure in the revival and renewal of mathematical logic.'

Appropriately, Gerald Sacks, Robin's long-time friend and occasional research collaborator, gave the **Robin Gandy Lecture** (eloquently introduced by Joe Shoenfield). The original invitation to Professor Sacks had suggested a theme of "Computability Theory – The First Sixty Years", perhaps with Robin's 1988 article on "The Confluence of Ideas in 1936" in mind. But Gerald, undoubtedly grand but never grandiose, responded in his own, very personal, way, and the paper based on his lecture (on p.367 of **Models and Computability**) provides, among other things, fascinating background to the development of such landmarks of contemporary computability theory as the Sacks Density Theorem.

An element of arbitrariness in the allocation of topics and particular papers between the two volumes has been unavoidable. The prominence of effective model theory as a conference topic was one determining factor in the particular distribution adopted. Other decisions, such as the inclusion of the two papers from the 'Philosophy of Proof' part of the conference programme in **Models and Computability**, were more practical in origin. We hope that the overall benefits of convenience to the reader will be sufficient compensation.

Together, we hope that **Sets and Proofs** and **Models and Computability** will provide readers with with a comprehensive guide to the current state of mathematical logic, and while not pretending to the definitiveness of a handbook, perhaps communicating more of the excitement of a subject in flight. All the authors are leaders in their fields, some articles pushing forward the technical boundaries of the subject, others providing readable and authoritative overviews of particular important topics. (All the contributors have been encouraged to include a good introduction, putting their work in context.) A number of papers can be expected to become classics, essential to any good library (individual or institutional).

In any project of this magnitude, it is impossible to thank all those who have helped. Special thanks are due to all the authors (and to the small number who tried, and failed to deliver!), and to the host of referees who coped with tight deadlines without complaint. On the technical side, we would like to thank Margaret Williams, Audrey Landford, Tim Hainsworth, Frank Drake, David Knapp, Zarina Akhtar, Eric Cole, Kevin McEvoy, and Ben Salzberg and Benjamin Thoma at Blue

[1]At http://www.turing.org.uk/turing/scrapbook/robin.html.

Sky Research. Finally, thanks to Rebecca Mikulin and Roger Astley at Cambridge University Press for their advice and inexhaustible (it seemed) patience.

All royalties accruing from the sale of **Models and Computability** and **Sets and Proofs** go directly to the British Logic Colloquium.

We dedicate this volume to the memory of George Boolos.

S. Barry Cooper
John K. Truss
Leeds, November 1998

An Introduction To Finitary Analyses Of Proof Figures*

Toshiyasu Arai†
Faculty of Integrated Arts and Sciences
Hiroshima University
Higashi-Hiroshima, 739-8521 Japan
e-mail: arai@mis.hiroshima-u.ac.jp

Abstract

In this paper we expound the approach to proof theory due to Gentzen-Takeuti: the finitary analysis of proof figures using ordinal diagrams. As an example we take up theories of strength measured by the Howard ordinal. First we recall Takeuti's consistency proof for a subsystem BI of second order arithmetic in a retrospective analysis of the genesis of ordinal diagrams. Second a theory of Π_2-reflecting ordinals is analysed in the spirit of Gentzen-Takeuti. This paper is intended to be an introduction to an ongoing project of the author's, in which a proof theory for theories of recursively large ordinals is developed.

G. Gentzen published his new version of a consistency proof for first order number theory in 1938 [26]. He had already given two consistency proofs [24] and [25]. The first used a *constructive but rather abstract notion of functionals.* In the second he had first introduced *transfinite ordinals* in proof theory. Although he formulated the result as a consistency proof, his interest seems to involve a divergence from Hilbert's program. Concerning this development G. Kreisel [28] p. 262 wrote:

*This is a part of a résumé for a talk given at the WORKSHOP ON PROOF THEORY, ORDINAL ANALYSIS AND THEIR APPLICATIONS May 29th-31st, 1997 at the University of Münster, Germany. I would like to thank W. Pohlers and A. Weiermann for hospitality during my visit to Münster.

†I would like to thank my colleague, Takao Shohoji for his generous support to my visit to Leeds. I would also like to thank the anonymous referee for his or her helpful suggestions, and the editor, J. K. Truss for his generous treatment.

> ..., by introducing a *quantitative ordinal measure* he (=Gentzen) forces us to pay attention to *combinatorial complexity*[1] and thereby makes it at least more difficult for us to slip into an abstract reading.

It seems that the purpose of the third 'Neue Fassung' is to make a lucid exposure of this combinatorial complexity which Gentzen discovered in *finite proof figures* of number theory.

G. Takeuti [44], [45] followed this idea and developed a proof theory of subsystems of second order arithmetic including the Π^1_1-Comprehension Axiom, $\Pi^1_1 - CA$.

We are following in the steps of Gentzen and Takeuti, and are now developing a proof theory for theories of recursively large ordinals. This paper is intended to be an introduction to this ongoing project [4], [5], [6],[7],[10], [11], [12], [13], [14] and [15]. We expound the approach to proof theory due to Gentzen-Takeuti: the finitary analysis of proof figures using ordinal diagrams. As an example we take up theories of strength measured by the Howard ordinal. First we recall Takeuti's consistency proof for a subsystem BI of second order arithmetic in a retrospective analysis of a genesis of ordinal diagrams. Second a theory of Π_2-reflecting ordinals is analysed in the spirit of Gentzen-Takeuti. Only an outline is given. For fuller details see [12].

For more on the aims and another approach to the proof theory of strong theories, see W. Pohlers [36] and M. Rathjen [37] and [38].

1 Proof theory à la Gentzen-Takeuti

Proof theory à la Gentzen-Takeuti [26], [46] proceeds as follows;

(G1) Let P be a proof whose endsequent Γ has a restricted form, e.g., an arithmetical sequent. Define a reduction procedure r which rewrites such a proof P to yield other proofs $\{r(P,n) : n \in I\}$ of sequents Γ_n provided that P has not yet reduced to a certain canonical form.

For example when we want to show that the arithmetical provable sequent Γ is true, the sequents Γ_n are chosen so that Γ is true iff every Γ_n $(n \in I)$ is true. Also if P is in an irreducible form, then the endsequent is seen to be true outright.

(G2) From the structure of the proof P, we abstract a structure related to this procedure r and throw away any irrelevant residue. This gives a finite figure $o(P)$.

[1] my italics

Following G. Takeuti [44] we call the figure $o(P)$ the *ordinal diagram* (o.d.), and write the set of ordinal diagrams as $\mathcal{O}$.

(G3) Define a relation $<$ on $\mathcal{O}$ by $o(r(P, n)) < o(P)$ for any $n \in I$.

(G4) Show that the relation $<$ on $\mathcal{O}$ is well founded.
Usually $<$ is a linear ordering and hence $(\mathcal{O}, <)$ is a notation system for ordinals.

When the endsequent of a proof P is an arithmetical sequent, we in fact construct a cut-free ω-proof of the sequent whose height is less than or equal to (the order type of) the o.d. $o(P)$ attached to P.

O.d.s are constructed so that each constructor for o.d.s reflects a reduction step on proofs. For example, in Takeuti's cut elimination procedure, a constructor $d_\Omega : \alpha \mapsto d_\Omega\alpha$ corresponds to an inference rule *substitution rule*, which is introduced in eliminating a Σ^1_1-cut formula, *cf.* subsection 3.2. Also the constructor reflects the Collapsing-Bounding lemma 6.2 in the proof theory of ordinals, *cf.* subsection 6.1.

2 A system $O(\Omega)$ of ordinal diagrams

In this section we define a system $O(\Omega)$ of ordinal diagrams. $O(\Omega)$ is equivalent to Takeuti's system $O(2,1)$ and is essentially the same as Schütte's notation system $\Sigma(1)$ in [41] or $\Sigma^*(1)$ in Levitz-Schütte [30]. The Howard ordinal is denoted by the o.d. $d_\Omega\varepsilon_{\Omega+1}$.

The notation system $O(\Omega)$ is introduced in [1] for expounding a proof theory of a subsystem BI of second order arithmetic due to G. Takeuti. Then the system is used for a proof theory of a theory of Π_2-reflecting ordinals in [3]. $O(\Omega)\,|\,\alpha = \{\beta \in O(\Omega) : \beta < \alpha\}$ is wellfor each $\alpha < d_\Omega\varepsilon_{\Omega+1}$.

Let $0, \Omega, +, \omega$(exponential with base ω) and d be distinct symbols. Each element (called an *ordinal diagram*) in the set $O(\Omega)$ is a finite sequence of these symbols.

$0, \Omega$ are *atomic diagrams*, and *constructors* in the system $O(\Omega)$ are $+$, $\alpha \mapsto \omega^\alpha$ and $d_\Omega : \alpha \mapsto d_\Omega\alpha$. Each diagram of the form $d_\Omega\alpha$ and Ω is defined to be an ε-number:

$$\beta < d_\Omega\alpha \Rightarrow \omega^\beta < d_\Omega\alpha$$

Digression. Let us explain where the operation $d_\Omega : \alpha \mapsto d_\Omega\alpha$ came from. In [23] Buchholz and Schütte defined a *partially* defined collapsing function $\psi_\sigma\alpha$ for each (recursively) regular ordinal σ. And then a *totally* defined collapsing function $d_\sigma\alpha$ is defined from $\psi_\sigma\alpha$.[2] Using these totally defined collapsing

[2] For more on partially and totally defined collapsing functions see Buchholz [16], [17], Pfeiffer [32], [33], [34] and Weiermann [47].

functions Schütte [42] obtains an upper bound for the proof-theoretic ordinal of a subsystem $\Pi^1_2 - Sep + BI$ of second order arithmetic which is equivalent to $\Delta^1_2 - CA + BI$.

In examining the relation $d_\sigma\alpha < d_\sigma\beta$ we noticed that this has some connections with Takeuti's o.d.s. Specifically $d_{\Omega_i}\alpha < d_{\Omega_i}\beta$ is akin to $\alpha <_i \beta$ in the sense of Takeuti.[3]

Therefore we decided to choose the simplest, brutal way[4]: Regard $d_\sigma\alpha$ itself as an expression, the symbol d followed by sequences σ and α of symbols. We soon found out that this approach is consonant with the spirit of Gentzen-Takeuti. The constructor $d_\sigma : \alpha \mapsto d_\sigma\alpha$[5] represents faithfully a reduction step on proofs. In this way a system $O(1; I)$ of o.d.s is constructed and using this a cut elimination theorem for a second order logic calculus SBL is given in [2], which is a logical counterpart of the subsystem $\Pi^1_2 - Sep + BI$ of second order arithmetic. *End of Digression*

The *natural sum* or commutative sum $\alpha\#\beta$ is defined as usual. The order relations between ε-numbers are defined as follows.

1. $d_\Omega\alpha < \Omega$

2. $d_\Omega\alpha < d_\Omega\beta$ holds if one of the following conditions is fulfilled. [6]

   (a) $d_\Omega\alpha \leq K_\Omega\beta (\Leftrightarrow_{df} \exists\delta \in K_\Omega\beta(d_\Omega\alpha \leq \delta))$

   (b) $K_\Omega\alpha < d_\Omega\beta (\Leftrightarrow_{df} \forall\gamma \in K_\Omega\alpha(\gamma < d_\Omega\beta))\ \&\ \alpha < \beta$

3. $K_\Omega\alpha$ denotes the finite set of maximal subdiagrams of α which have the form $d_\Omega\gamma$, i.e., $K_\Omega\alpha$ consists of the ε-numbers below Ω which are needed for the unique representation of α in Cantor normal form:

   (a) $K_\Omega 0 = K_\Omega\Omega = \emptyset$

   (b) $K_\Omega(\alpha_1 + \cdots + \alpha_n) = \bigcup\{K_\Omega\alpha_i : 1 \leq i \leq n\}$

   (c) $K_\Omega\omega^\alpha = K_\Omega\alpha$

   (d) $K_\Omega d_\Omega\alpha = \{d_\Omega\alpha\}$

Then we have the following facts.

---

[3]But this is presumably not equivalent to Takeuti's. We do not know the exact relationships. Connections with Takeuti's (or Kino's) ordinal diagrams and other notation systems were already established in H. Levitz [29], Levitz-Schütte [30], Pfeiffer [35] and Buchholz [17].

[4]This has also been done by H. Pfeiffer [34].

[5]α in $d_\sigma\alpha$ is not restricted to the case $\alpha \geq \sigma$ differing from the totally defined collapsing functions in [23].

[6]Note that the definition of $<$ for ε-numbers depends on the definition of $<$ for o.d.s which are not ε-numbers.

(< 1) $d_\Omega \alpha < \Omega$

(< 2) $K_\Omega \alpha < d_\Omega \alpha$

(< 3) $K_\Omega \alpha \leq \alpha$

(< 4) $\beta < \Omega \,\&\, K_\Omega \beta < d_\Omega \alpha \;\Rightarrow\; \beta < d_\Omega \alpha$

An *essentially less than* or a *collapsibly less than relation* $\alpha \ll \beta$ is defined by

$$\alpha \ll \beta \Leftrightarrow K_\Omega \alpha < d_\Omega \beta \,\&\, \alpha < \beta$$

and we set

$$\alpha \underline{\ll} \beta \Leftrightarrow \alpha \ll \beta \text{ or } \alpha = \beta$$

Then we have $d_\Omega \alpha \ll d_\Omega \beta \Leftrightarrow d_\Omega \alpha < d_\Omega \beta$ and

(preservation) $\alpha \ll \beta \Rightarrow d_\Omega \alpha \ll d_\Omega \beta \,\&\, \alpha \# \gamma \ll \beta \# \gamma \,\&\, \omega^\alpha \ll \omega^\beta$

($\ll \#$) $\alpha \underline{\ll} \alpha \# \beta$

($\ll \omega$) $\alpha \underline{\ll} \omega^\alpha$

The sytem $O(\Omega)$ is nothing but the notation system defined in [39], which is constructed in set theory. Put

$$k_\Omega \alpha = \max(K_\Omega \alpha \cup \{0\}) \,\&\, \Omega = \omega_1 \text{ (the least uncountable ordinal)}$$

Define sets $D(\alpha)$ and ordinals $d_\Omega \alpha$ by simultaneous recursion on ordinals α as follows:

1. $\{\Omega\} \cup (k_\Omega \alpha + 1) \subseteq D(\alpha)$
2. $D(\alpha)$ is closed under $+, \omega^\beta$
3. $\delta \in D(\alpha) \cap \alpha \Rightarrow d_\Omega \delta \in D(\alpha)$
4. $d_\Omega \alpha = \min\{\xi : \xi \notin D(\alpha)\}$

Then we see for $\alpha, \beta \in D(\varepsilon_{\Omega+1})$

1. $d_\Omega \alpha < \Omega = \omega_1$
2. $d_\Omega \beta \leq K_\Omega \alpha \Rightarrow d_\Omega \beta < d_\Omega \alpha$
3. $\alpha < \beta \,\&\, K_\Omega \alpha < d_\Omega \beta \Rightarrow d_\Omega \alpha < d_\Omega \beta$
4. $d_\Omega \alpha = d_\Omega \beta \Rightarrow \alpha = \beta$
5. $d_\Omega \alpha = D(\alpha) \cap \Omega$

Thus this gives a set-theoretic interpretation $D(\varepsilon_{\Omega+1})$ of the system $O(\Omega)$.

3 A retrospective analysis of the genesis of ordinal diagrams

In this section we first recall the proof of cut elimination for a second order logic calculus LBI in [43]. The proof uses transfinite induction up to ε_0. Transfinite induction up to ε_0 is needed to show the cut elimination theorem for LBI since it implies finitistically the 1-consistency of the first order arithmetic PA. This is seen from relativization in [46]. Next through reexamining the proof of cut elimination for the logic calculus LBI we give a consistency proof of a subsystem BI of second order arithmetic. The consistency proof is based on transfinite induction up to the Howard ordinal $d_\Omega \varepsilon_{\Omega+1}$. This reexamination leads us to a genesis of ordinal diagrams.

The material in this section comes from [1].

3.1 Cut elimination for a second order logic calculus LBI

In this subsection we recall the proof of cut elimination for a second order logic calculus LBI in [43]. The proof uses transfinite induction up to ε_0.

LBI is formulated in Tait's logic calculus, i.e., one-sided sequent calculus and $\Gamma, \Delta, \ldots$ denote *sequents*, i.e., finite sets of formulae.

For simplicity we assume that the language of LBI is obtained from a first order language without unary predicate constants by adding unary predicate variables $X, Y, \ldots$. Specifically a prime formula takes the form $X(t)$ or $\neg X(t)$ for a term t. First we introduce a predicative second order logic calculus LK_2. In LK_2 introduction rules of second order quantifiers are:

$$\frac{\Gamma, F(X)}{\Gamma, \forall X F(X)} (\forall^2) \quad \text{and} \quad \frac{\Gamma, F(Y)}{\Gamma, \exists X F(X)} (\exists^2)$$

Definition 3.1 *1. Π^1-formula and Σ^1-formula:* A formula is said to be a *Π^1-formula* [*Σ^1-formula*] if it contains no second order existential quantifier [no second order universal quantifier respectively].

2. A second order logic calculus LBI: A logic calculus LBI is obtained from LK_2 by adding the following inference rule (BI):

$$\frac{\Gamma, F(V)}{\Gamma, \exists X F(X)} (BI)$$

where $\exists X F(X)$ is a Σ^1-formula and V is an arbitrary *abstract*[7] $\{x\}A(x)$, viz. the *auxiliary formula* $F(V)$ of the inference denotes the result of replacing each formula $X(t)$ in $F(X)$ by $A(t)$.

[7] If V is a second order variable, then the inference rule is regarded as an $(\exists^2)$

The following theorem is proved by ε_0-induction.

Theorem 3.1 (G. Takeuti [43]) *If P is proof in LBI ending with a Π^1-sequent Π, then Π is cut-free provable in LK_2.*

Remark. This is easily seen model-theoretically: from Gödel's completeness theorem and the soundness of the calculus LBI for full models in which the second order part of the model is the power set of the first order universe.

From this theorem using the Joker translation in Päppinghaus [31] we get finitistically

Corollary 3.1 *If a sequent is provable in LBI, then it is cut-free provable in the full impredicative second order calculus G^1LC in [46].*

Now we prove Theorem 3.1. Our proof is a modification of proofs in [26] and [43].

Definition 3.2 *The grade $gr(A)$ of a formula A and the height $h(\Gamma; P)$ of (an occurrence of) a sequent Γ in a proof P.*

1. The *grade* $gr(A)$ of a formula A is defined so that $gr(\neg A) = gr(A)$ and by counting the number of occurrences of logical symbols *except that we set $gr(A) = 0$ for a Σ^1-formula A.*

2. The *height* $h(\Gamma; P)$ of (an occurrence of) a sequent Γ in a proof P is defined to be the maximal grade of cut formulae and auxiliary formulae of inferences (BI) occurring below Γ. For example at a cut rule

$$\frac{\Gamma, A \quad \neg A, \Lambda}{\Gamma, \Lambda}\ (cut)$$

we set $h(\Gamma, A; P) = h(\neg A, \Lambda; P) = \max\{gr(A), h(\Gamma, \Lambda; P)\}$ and at an inference (BI)

$$\frac{\Gamma, F(V)}{\Gamma, \exists X F(X)}\ (BI)$$

$h(\Gamma, F(V); P) = \max\{gr(F(V)), h(\Gamma, \exists X F(X); P)\}$.

Let P be a proof in LBI, Γ (an occurrence of) a sequent and J (an occurrence of) an inference rule in P. We assign ordinals $o(\Gamma; P), o(J; P) < \varepsilon_0$ inductively as follows:

1. $o(\Gamma; P) = 1$ if Γ is an initial sequent $\Lambda, \neg A, A$ for an arbitrary formula A.

2. At a cut rule the definition is as in [26], viz. the lowersequent receives a tower $\omega_h(\alpha_0\#\alpha_1)$ of ω with the difference $h = h(\Gamma, \neg A; P) - h(\Gamma, \Lambda; P)$ of the heights of upper and lower sequents and the natural sum $o(J; P) = \alpha_0\#\alpha_1$ of ordinals attached to uppersequents.

$$\frac{\Gamma, \neg A :\ \alpha_0 \quad A, \Lambda :\ \alpha_1}{\Gamma, \Lambda :\ \omega_h(\alpha_0\#\alpha_1)}\ (cut)\ J :\ \alpha_0\#\alpha_1$$

3. At a logical inference except (BI) the lowersequent receives the ordinal $\alpha + 1$ with the natural sum α of ordinals attached to uppersequents.

4. At a (BI), set

$$\frac{\Gamma, F(V) :\ \alpha}{\Gamma, \exists X F(X) :\ \omega_h(\omega\#\alpha)}\ (BI)\ J :\ \omega\#\alpha$$

with the difference $h = h(\Gamma, F(V); P) - h(\Gamma, \exists X F(X); P)$ of heights.

5. Finally set

$$o(P) = o(\Gamma_{end}; P)$$

with the endsequent Γ_{end} of P.

Theorem 3.1 is proved by transfinite induction on $o(P) < \varepsilon_0$. Consider the crucial case where P has the following shape.

$$\dfrac{\dfrac{\dfrac{\vdots}{\Gamma_0, \neg F(X) :\ \beta_0}}{\dfrac{\Gamma_0, \forall X \neg F(X) :\ \beta_0 + 1}{\vdots \atop \Gamma, \forall X \neg F(X) :\ \beta}}\ (\forall^2) \qquad \dfrac{\dfrac{\vdots}{F(V), \Lambda_0 :\ \alpha_0}}{\dfrac{\exists X F(X), \Lambda_0 :\ \omega_h(\omega\#\alpha_0)}{\vdots \atop \exists X F(X), \Lambda :\ \alpha}}\ (BI) :\ \omega\#\alpha_0}{\Gamma, \Lambda :\ \beta\#\alpha \atop {\vdots \atop \Pi}}\ (cut) \qquad P$$

where $\exists X F(X)$ is a Σ^1-formula and $\Pi, \neg F(X)$ is a Π^1-sequent.

First the inference rule $(\forall^2)$ is replaced by a (*weakening*) to yield the following $P_0(X)$.

$$\dfrac{\dfrac{\dfrac{\vdots}{\Gamma_0, \neg F(X) :\ \beta_0}}{\dfrac{\Gamma_0, \neg F(X), \forall X \neg F(X) :\ \beta_0}{\vdots \atop \Gamma, \neg F(X), \forall X \neg F(X) :\ \beta_1}}\ (weakening) \qquad {\vdots \atop \exists X F(X), \Lambda :\ \alpha}}{\Gamma, \Lambda, \neg F(X) :\ \beta_1\#\alpha \atop {\vdots \atop \Pi, \neg F(X)}} \qquad P_0(X)$$

$P_0(X)$ is simpler than P since the inference $(\forall^2)$ vanishes. This reflects to the associated o.d.s so that

$$o(\Gamma_0, \neg F(X), \forall X \neg F(X); P_0(X)) = \beta_0 < \beta_0 + 1 = o(\Gamma_0, \forall X \neg F(X); P)$$

and

$$o(\Gamma, \neg F(X), \forall X \neg F(X); P_0(X)) = \beta_1 < \beta = o(\Gamma, \forall X \neg F(X); P)$$

Therefore $o(P_0(X)) < o(P)$. By IH pick a cut-free proof $P_0^{cf}(X)$ of $\Pi, \neg F(X)$ and let P' be the following.

$$\begin{array}{c}
\begin{array}{ccc}
 & & \vdots\; P_0^{cf}(V) \qquad\qquad \vdots \\
 & & \dfrac{\dfrac{\Pi, \neg F(V) : n \qquad F(V), \Lambda_0 : \alpha_0}{\Pi, \Lambda_0 : \omega_h(n\#\alpha_0)}\,(cut) : n\#\alpha_0}{\exists X F(X), \Lambda_0, \Pi : \omega_h(n\#\alpha_0)}\,(weakening) \\
\vdots & & \vdots \\
\end{array} \\
\dfrac{\Gamma, \forall X \neg F(X) : \beta \qquad \exists X F(X), \Lambda, \Pi : \alpha_1}{\Gamma, \Lambda, \Pi : \beta\#\alpha_1} \\
\vdots \\
\Pi \qquad\qquad\qquad\qquad\qquad\qquad P'
\end{array}$$

where $P_0^{cf}(V)$ denotes the result of substituting the abstract V for the variable X in $P_0^{cf}(X)$.

Since $P_0^{cf}(X)$ is cut-free proof and the endsequent is a Π^1-sequent $\Pi, \neg F(X)$, no second order existential quantifier occurs in it. In particular no introduction rule for a second order existential quantifier occurs in it. Hence both $P_0^{cf}(V)$ and P' are proofs in LBI and the o.d.

$$o(P_0^{cf}(X)) = o(\Pi, \neg F(X); P_0^{cf}(X)) = o(\Pi, \neg F(V); P')$$

is a finite ordinal $n < \omega$.

Therefore we have

$$o(\exists X F(X), \Lambda_0, \Pi; P') = \omega_h(n\#\alpha_0) < \omega_h(\omega\#\alpha_0) = o(\exists X F(X), \Lambda_0; P)$$

Observe here that we counted the grade $gr(F(V))$ of an auxiliary formula $F(V)$ of an inference (BI) as part of the height of a sequent since a (cut) with the cut formula $F(V)$ is expected. Hence

$$o(\exists X F(X), \Lambda, \Pi; P') = \alpha_1 < \alpha = o(\exists X F(X), \Lambda; P) \text{ and } o(P') < o(P)$$

This proves Theorem 3.1 by ε_0-induction.

3.2 Genesis of ordinal diagrams

In this subsection we reexamine the proof of the cut elimination theorem in the previous subsection 3.1 and give a consistency proof of a subsystem BI of second order arithmetic. The consistency proof uses transfinite induction up to the Howard ordinal $d_\Omega \varepsilon_{\Omega+1}$. BI is obtained from LBI (in the language of second order arithmetic) by adding axioms for first order constants and the mathematical induction axiom schema:

$$(ind)\ \Gamma, F(0) \wedge \forall x(F(x) \to F(x+1)) \to F(t)$$

This reexamination leads us to a genesis of ordinal diagrams.

In the previous subsection 3.1 we first rewrote P to yield $P_0(X)$, second we got a cut-free proof $P_0^{cf}(X)$ by IH and finally we obtained P' from $P_0^{cf}(V)$ and P. Here an invisible operation is involved to yield another proof $P_0^{cf}(X)$ from a proof $P_0(X)$, i.e., we cannot grasp what $P_0^{cf}(X)$ looks like. Now let us denote this operation as a 'higher order inference rule'. Let (sub) (*substitution inference*) denote the following 'inference rule':

$$P_0^{cf}(V) = \left\{ \begin{array}{c} \vdots\ P_0(X) \\ \dfrac{\Gamma(X)}{\Gamma(V)}\ (sub) \end{array} \right.$$

where the uppersequent $\Gamma(X)$ is a Π^1-sequent.

Temporarily we regard the result $P_0^{cf}(V)$ of applying the 'inference rule' (sub) to $P_0(X)$ as a proof obtained from $P_0(X)$ by eliminating cuts to yield $P_0^{cf}(X)$ and then substituting an abstract V for the variable X. According to the cut elimination procedure in the previous subsection, the cut elimination procedure performed by the 'higher order inference rule' $(sub) : P_0(X) \mapsto P_0^{cf}(V)$ immitates the procedure eliminating cuts in $P_0(X)$ to yield $P_0^{cf}(X)$ as if the abstract V were the prime abstract X.

Keeping this in mind we express P' as follows (where $P_0(X)$ denotes the proof displayed in subsection 3.1):

$$\begin{array}{c}
\dfrac{\dfrac{\dfrac{\dfrac{\vdots\ P_0(X) \quad}{\Pi, \neg F(X)}}{\Pi, \neg F(V)}\ (sub) \qquad \dfrac{\vdots}{F(V), \Lambda_0}}{\dfrac{\Pi, \Lambda_0}{\exists X F(X), \Lambda_0, \Pi}} \\
\vdots \\
\dfrac{\Gamma, \forall X \neg F(X) \qquad \exists X F(X), \Lambda, \Pi}{\Gamma, \Lambda, \Pi} \\
\vdots \\
\Pi \qquad\qquad r(P)
\end{array}$$

Under the interpretation of the 'inference rule' (*sub*) this figure certainly denotes a proof in LBI. But the proof does not stand before our eyes as the figure itself. Therefore in what follows we count the 'higher order inference rule' (*sub*) as an official inference rule in an extension of LBI. Specifically if P is a proof in (the extended) LBI with a Π^1-endsequent $\Gamma(X)$ and we add a line and a sequent $\Gamma(V)$ under P to yield another figure P_1, then we count the figure P_1 as a proof in LBI.

Since we have added a new inference rule (*sub*) and o.d.s attached to proofs should reflect transformations on the associated proofs, we introduce a new constructor $d_\Omega : \alpha \mapsto d_\Omega\alpha$ of o.d.s. Attached o.d.s are changed as follows.

1. For simplicity we assume that every proof ends with a vacuous (*sub*).

$$\frac{\begin{array}{c}\vdots\\ \Pi\end{array}}{\Pi}\ (sub)$$

2. Let Γ be a sequent in a proof which is not the endsequent. The *resolvent* of Γ is the uppersequent of the uppermost substitution (*sub*) below Γ.

3. The height $h(\Gamma; P)$ of a sequent Γ in a proof P is redefined to be the maximal grade of cut formulae and auxiliary formulae of inferences (BI) occurring between Γ and the resolvent of Γ. In particular $h(\Gamma; P) = 0$ for any uppersequent Γ of a (*sub*). This change of heights is consonant with the interpretation of the 'higher order inference rule' (*sub*). That is, we eliminate cuts in the upper part of a (*sub*) apart from remaining parts.

4. Replace the addition ω at a (BI) by Ω:

$$\frac{\Gamma, F(V) : \alpha}{\Gamma, \exists X F(X) : \omega_h(\Omega\#\alpha)}\ (BI)\, J : \Omega\#\alpha$$

by the difference $h = h(\Gamma, F(V); P) - h(\Gamma, \exists X F(X); P)$ of newly redefined heights.

5. Set

$$\frac{\Gamma(X) : \alpha}{\Gamma(V) : d_\Omega\alpha}\ (sub)$$

In the reduction step for (BI) below, (BI) is replaced by a (*cut*) in such a way that the last inference of one of the subproofs is a (*sub*). So we add Ω at a (BI) and require (< 1) $d_\Omega\alpha < \Omega$ for any α.

6. The whole proof P receives an o.d. $o(P) = o(\Gamma_{end}; P)$ of the form $d_\Omega\gamma$ since the last inference rule is a (*sub*).

7. When we treat the subsystem BI of second order arithmetic an induction axiom receives ω: $o(\Gamma, F(0) \wedge \forall x(F(x) \to F(x+1)) \to F(t); P) = \omega$

Let us examine how this works for a consistency proof of the subsystem BI of second order arithmetic, which uses induction on the o.d. $o(P) \in O(\Omega) \mid d_{\Omega}\varepsilon_{\Omega+1}$ attached to a proof P of the empty sequent. Note that the proof in the previous subsection does not work for BI: even if the endsequent is empty and $\neg F(X)$ is a Π^1-formula, any proof $P_0^{cf}(X)$ in a normal form does not enjoy the subformula property because of the presence of the induction axiom and hence an inference rule (BI) may occur in it.

Theorem 3.2 *The consistency of BI follows finitistically from the fact that the system $(O(\Omega) \mid d_{\Omega}\varepsilon_{\Omega+1}, <)$ is a well ordering.*

Consider again the crucial case, where P has the following shape.

$$
\dfrac{\dfrac{\dfrac{\vdots}{\Gamma_0, \neg F_0(X) : \beta_0}}{\dfrac{\Gamma_0, \forall X \neg F_0(X) : \beta_0 + 1}{\vdots \atop \Gamma, \forall X \neg F(X) : \beta}}\,(\forall^2) \qquad \dfrac{\dfrac{\vdots}{F(V), \Lambda_0 : \alpha_0}}{\dfrac{\exists X F(X), \Lambda_0 : \omega_h(\Omega \# \alpha_0)}{\vdots \atop \exists X F(X), \Lambda : \alpha}}\,(BI) : \Omega\#\alpha_0}{\dfrac{\Gamma, \Lambda : \beta \# \alpha}{\vdots \atop \dfrac{\Pi : \gamma}{\Pi' : d_\Omega \gamma}\,(sub) \atop \vdots}}\,(cut)
$$

P

where Π denotes the resolvent of the sequent $\exists X F(X), \Lambda_0$. Note that it may be the case that $F_0 \not\equiv F$, since F_0 may be affected by (sub)s. Since any uppersequent of a (sub) has to be a Π^1-sequent, no (sub) occurs between $\exists X F(X), \Lambda_0$ and $\exists X F(X), \Lambda$ and hence the resolvent Π of the sequent $\exists X F(X), \Lambda_0$ is below the (cut).

Let $r(P)$ be the following proof:

$$\vdots$$
$$\frac{\Gamma_0, \neg F_0(X) : \beta_0}{\Gamma_0, \neg F_0(X), \forall X \neg F_0(X) : \beta_0}$$
$$\vdots \qquad\qquad \vdots$$
$$\frac{\Gamma, \neg F(X), \forall X F(X) : \beta_1 \qquad \exists X F(X), \Lambda : \alpha}{\Gamma, \Lambda, \neg F(X) : \beta_1 \# \alpha}$$
$$\vdots$$
$$\frac{\dfrac{\Pi, \neg F(X) : \gamma_1}{\Pi, \neg F(V) : d_\Omega \gamma_1}\,(sub) \qquad \begin{matrix}\vdots \\ F(V), \Lambda_0 : \alpha_0\end{matrix}}{\dfrac{\Pi, \Lambda_0 : \omega_h(d_\Omega \gamma_1 \# \alpha_0)}{\exists X F(X), \Lambda_0, \Pi : \omega_h(d_\Omega \gamma_1 \# \alpha_0)}} : d_\Omega \gamma_1 \# \alpha_0$$
$$\vdots \qquad\qquad \vdots$$
$$\frac{\Gamma, \forall X \neg F(X) : \beta \qquad \exists X F(X), \Lambda, \Pi : \alpha_2}{\Gamma, \Lambda, \Pi : \beta \# \alpha_2}$$
$$\vdots$$
$$\frac{\Pi : \gamma_2}{\Pi' : d_\Omega \gamma_2}\,(sub)$$
$$\vdots$$

$r(P)$

Then $o(r(P)) < o(P)$ is seen as follows. First from $\beta_0 \ll \beta_0 + 1$ we see $\beta_1 \ll \beta$ and hence $\gamma_1 \ll \gamma$, *cf.* (**preservation**), ($\ll \#$) and ($\ll \omega$) in section 2. Therefore

$$d_\Omega \gamma_1 < d_\Omega \gamma \tag{1}$$

Also by (< 1) $d_\Omega \gamma_1 < \Omega$ we have $d_\Omega \gamma_1 \# \alpha_0 < Ome \# \alpha_0$. Note that no ($sub$) occurs between $\exists X F(X), \Lambda_0$ and Π by the definition of the resolvent. In other words no collapsing operation d_Ω is applied when we go through the branch from $\exists X F(X), \Lambda_0$ to Π downwards. Therefore we get

$$\gamma_2 < \gamma \,\&\, K_\Omega \gamma_2 \subseteq K_\Omega \gamma \cup \{d_\Omega \gamma_1\} \tag{2}$$

By (1), (2) and (< 2) we conclude $d_\Omega \gamma_2 < d_\Omega \gamma$ by the definition. From this we see $o(r(P)) < o(P)$.

We have proved that BI is consistent by transfinite induction up to the Howard ordinal $d_\Omega \varepsilon_{\Omega+1}$.

Remark.

1. Reduction steps by Gentzen [26] and Takeuti [45] are closely related to cut-elimination for infinitary derivations using the $\Omega_{\mu+1}$-rule which was

first introduced by Buchholz [18]. In particular the reduction step using substitution inferences which was initiated by Takeuti is a forerunner of the extended Hydra game in [19]. This vague observation was implicitly stated in [8]. Recently Buchholz [21], [22] gave a precise explanation of the Gentzen-Takeuti reduction steps in terms of cut-elimination for infinitary derivations: the Gentzen-Takeuti reduction steps and ordinal assignment is derived from infinitary proof theory.

2. Obviously a slight modification of the consistency proof for BI yields a cut-elimination proof of the logic calculus LBI, Theorem 3.1. In each reduction step the size of the reduct $r(P)$ is at most twice of one of P. This reflects the size (norm) of o.d.s attached to proofs. This observation leads us to a theorem in [9] which says that the number of reduction steps needed in reaching to a cut-free proof (in the worst cases) is bounded by a slow growing functions indexed by o.d.s in $O(\Omega) \mid d_\Omega \varepsilon_{\Omega+1}$ and defined by a norm bounding transfinite recursion.

   Functions defined by a norm bounding transfinite recursion are utilized in a proof theory by Weiermann [48]. Slow growing functions defined by a hereditarily norm bounding transfinite recursion are introduced by Buchholz [20]. We [9] are inspired by these works [48] and [20].

4 A base theory T_0

Everything in this and the next section is fully expounded in [12], so we do not give full details.

In this section we define a base theory T_0 of ordinals. Let $\mathcal{L}_0$ denote the first order language whose constants are; =(equals), <(less than), 0(zero), 1(one), +(plus), ·(times), j(pairing),$()_0$, $()_1$(projections, i.e., inverses to j).

For each Δ_0 formula $\mathcal{A}(X, a, b)$ with a binary predicate X in $\mathcal{L}_0 \cup \{X\}$ we introduce a binary predicate constant $R^\mathcal{A}$ and a ternary one $R^\mathcal{A}_<$ by transfinite recursion on ordinals a:

$$b \in R^\mathcal{A}_a \Leftrightarrow_{df} R^\mathcal{A}(a,b) \Leftrightarrow \mathcal{A}(R^\mathcal{A}_{<a}, a, b)$$

with $R^\mathcal{A}_{<a} = \sum_{x<a} R^\mathcal{A}_x = \{(x,y) : x < a \,\&\, y \in R^\mathcal{A}_x\}$.

The language $\mathcal{L}_1$ is obtained from $\mathcal{L}_0$ by adding the predicate constants $R^\mathcal{A}$ and $R^\mathcal{A}_<$ for each bounded formula $\mathcal{A}(X, a, b)$ in $\mathcal{L}_0 \cup \{X\}$.

Let $F : Ord \to L$ denote (a variant of) Gödel's onto map from the class Ord of ordinals to the class L of constructible sets. The language $\mathcal{L}_1$ is chosen so that the set-theoretic membership relation $\in$ on L is interpretable by a Δ_0-formula in $\mathcal{L}_1$: There are Δ_0-formulae $\in (X, a, b), = (X, a, b)$, and a predicate

$E = R^{\mathcal{A}}$ for an $\mathcal{A}$ such that, *cf.* [12],

$$F(a) \in F(b) \Leftrightarrow \in (E, a, b) \,\&\, F(a) = F(b) \Leftrightarrow = (E, a, b)$$

Thus instead of developing an ordinal analysis of set theory we can equally develop a proof theory for theories of ordinals.

Every multiplicative principal number α is closed under each function constant in $\mathcal{L}_0$, *cf.* [12]. Let $\alpha = \langle \alpha; 0, 1, +, \cdot, \ldots, R^{\mathcal{A}} \mid \alpha, \ldots \rangle$ denote the $\mathcal{L}_1$-model with the universe α.

Axioms of a base theory T_0 are logical ones and mathematical axioms for function and predicate constants in $\mathcal{L}_0$.

Inference rules in T_0. There are twelve kinds of inference rules: $(\wedge)$, $(\vee)$, $(\forall)$, $(\exists)$,

$$\frac{\Gamma, x \not< t, A(x)}{\Gamma, \forall x < t A(x)}\ (b\forall) \qquad \frac{\Gamma, s < t \quad \Gamma, A(s)}{\Gamma, \exists x < t A(x)}\ (b\exists)$$

with the usual *eigenvariable condition* for the *eigenvariable* x of $(b\forall)$. The term s in $(b\exists)$ is said to be the *instance term* of the rule.

$$\frac{\Gamma, \neg\mathcal{A}(R^{\mathcal{A}}_{<t}, t, s)}{\Gamma, s \notin R^{\mathcal{A}}_t}\ (\neg R) \qquad \frac{\Gamma, \mathcal{A}(R^{\mathcal{A}}_{<t}, t, s)}{\Gamma, s \in R^{\mathcal{A}}_t}\ (R)$$

$$\frac{\Gamma, s \not< t, u \notin R^{\mathcal{A}}_s}{\Gamma, (s, u) \notin R^{\mathcal{A}}_{<t}}\ (\neg R_<) \qquad \frac{\Gamma, s < t \quad \Gamma, u \in R^{\mathcal{A}}_s}{\Gamma, (s, u) \in R^{\mathcal{A}}_{<t}}\ (R_<)$$

$$\frac{\Gamma, \neg\forall x < a A(x), A(a) \quad \Gamma, s < t \quad \Gamma, \neg A(s)}{\Gamma}\ (ind)$$

The variable a is the *eigenvariable* of the rule.

$$\frac{\Gamma, \neg A \quad A, \Lambda}{\Gamma, \Lambda}\ (cut)$$

5 Π^{Ω}_2-ordinal of a theory

First recall the definition of Π_2-reflecting ordinals in Richter and Aczel [40]. We say that an ordinal $\alpha \in Ord$ is Π_2*-reflecting* if

$$\forall A \in \Pi_2 \text{ with parameters from } \alpha[\alpha \models A \Rightarrow \exists \beta < \alpha(\beta \models A)]$$

If a parameter $\gamma < \alpha$ occurs in A, then it should be understood that $\gamma < \beta$.

Following G. Jäger [27] we define the Π^{Ω}_2-ordinal of a theory of ordinals.

Definition 5.1 (Π_2^Ω-*ordinal of a theory*) Let T be a sound and recursive theory of ordinals. For a sentence A let A^α denote the result of replacing unbounded quantifiers $Qx\,(Q \in \{\forall, \exists\})$ in A by $Qx < \alpha$. Let Ω denote the (individual constant corresponding to the) ordinal ω_1^{CK}. Define the Π_2^Ω-*ordinal* $|T|_{\Pi_2^\Omega}$ *of* T by

$$|T|_{\Pi_2^\Omega} =_{df} \inf\{\alpha \leq \omega_1^{CK} : \forall A \in \Pi_2(T \vdash A^\Omega \Rightarrow \alpha \models A)\} < \omega_1^{CK}$$

For the case $T \not\vdash \exists\omega_1^{CK}$, e.g., the theory T_2 for Π_2-reflecting ordinals defined below, we set $A^\Omega =_{df} A$ and the individual constant Ω is absent.

Here note that $|T|_{\Pi_2^\Omega} < \omega_1^{CK}$ since we have assumed $\forall A \in \Pi_2(T \vdash A^\Omega \Rightarrow \Omega \models A)$ and $\Omega = \omega_1^{CK}$ is recursively regular, i.e., Π_2-reflecting, *cf.* [12].

In order to get an upper bound for the Π_2^Ω-ordinal $|T|_{\Pi_2^\Omega}$ of a theory T we attach a *term* $o(\Gamma; P)$ to each sequent Γ occurring in a proof P in the theory T, which ends with a Π_2^Ω-sentence. The term $o(\Gamma; P)$ is built up from atomic diagrams and *variables* by applying constructors in a system $(O(T), <)$ of o.d.s for T. Variables occurring in the term $o(\Gamma; P)$ are eigenvariables occurring below Γ. Thus the term $o(\Gamma_{end}; P)$ attached to the endsequent of P is a closed term, i.e., denotes an o.d. Also each redex in our transformation is on the main branch, i.e., the rightmost branch of a proof tree, and is the lowermost one. Therefore when we resolve an inference rule J no free variable occurs below J.

Applied constructors in building the term $o(\Gamma; P)$ correspond to the inference rules occurring above Γ. For example at an inference rule $(b\exists)$

$$\frac{\Gamma, s < t \quad \Gamma, A(s)}{\Gamma, \exists x < tA(x)}\ (b\exists)$$

if $gr(A)$ is the complexity measure of a formula A we set

$$o(\Gamma, \exists x < tA(x)) = o(\Gamma, s < t)\#o(\Gamma, A(s))\#s\#gr(A(s)) \tag{3}$$

Note that the instance term s may contain variables, e.g., $s \equiv y \cdot z$. Also at an inference rule $(b\forall)$

$$\frac{\Gamma, x \not< t, A(x)}{\Gamma, \forall x < tA(x)}\ (b\forall)$$

we substitute the term t for the eigenvariable x in the term $o(\Gamma, \forall x < tA(x))$;

$$o(\Gamma, \forall x < tA(x)) = o(\Gamma, x \not< t, A(x))[x := t] \tag{4}$$

Finally set

$$o(P) = o(\Gamma_{end}; P) \in O(T)\,|\,\Omega (= \{\alpha \in O(T) : \alpha < \Omega\})$$

Now our theorem for an upper bound is stated as follows.

Theorem 5.1 *If P is a proof of a Π_2^Ω-sentence A^Ω in T, then A^α is true with $\alpha = o(P)$.*

6 Π_2-reflection

In this section we explain our approach to ordinal analysis by taking the theories of Π_2-reflecting ordinals as an example.

The fact that Ω is Π_2-reflecting is expressed by the following inference rule:

$$\frac{\Gamma, A^{\Omega} \quad \neg\exists z(t < z < \Omega \wedge A^{z}), \Gamma}{\Gamma} \; (\Pi_2 - rfl)$$

for any Π_2-formula $A^{\Omega} \equiv A \equiv \forall x \exists y B(x, y, t)$ with a *parameter term t*. T_2 denotes the theory obtained from T_0 by adding the inference rule $(\Pi_2 - rfl)$.

Let $\mathcal{L}_c$ denote the extended language of $\mathcal{L}_1$ obtained by adding an individual constant β for each o.d. $\beta < \Omega$.

$$\mathcal{L}_c = \mathcal{L}_1 \cup \{\beta \in O(\Omega) : \beta < \Omega\}$$

Let $KP\omega$ denote Kripke-Platek set theory with the axiom of infinity.

Theorem 6.1 *(Jäger [27])*

$$|KP\omega|_{\Pi_2^{\Omega}} = |T_2|_{\Pi_2^{\Omega}} = \text{the Howard ordinal } d_{\Omega}\varepsilon_{\Omega+1}$$

We show the half for the upper bound.

Theorem 6.2

$$\forall A \in \Pi_2(T_2 \vdash A^{\Omega} \Rightarrow \exists \alpha \in O(\Omega) \,|\, d_{\Omega}\varepsilon_{\Omega+1}\, A^{\alpha}).$$

6.1 Local predicativity

First our explanation follows Jäger [27], i.e., Pohler's local predicativity method.

For $\alpha \in O(\Omega)$, $\rho \in \{0\} \cup \{\Omega + n : n < \omega\}$ and a finite set Γ of sentences in the language $\mathcal{L}_c$, we introduce a derivability relation $\vdash^{\alpha}_{\rho} \Gamma$ which says:

1. ρ counts unbounded cut degrees $Gr(A)$ of cut formulae A in the infinitary derivation of Γ, where $Gr(A) = 0$ if A is bounded, and

$$Gr(QxB(x)) = \begin{cases} \Omega & \text{if } B \text{ is bounded} \\ Gr(B) + 1 & \text{otherwise} \end{cases}$$

2. $(\exists)$ $\vdash^{\alpha_0}_{\rho} \Gamma, A(\beta) \;\&\; \alpha_0, \beta \ll \alpha \Rightarrow \vdash^{\alpha}_{\rho} \Gamma, \exists x A(x)$

3. $(\forall)$ $f \in \mathcal{F} \;\&\; \forall \beta < \Omega \vdash^{f(\beta)}_{\rho} \Gamma, A(\beta) \;\&\; f(\Omega) \ll \alpha \Rightarrow \vdash^{\alpha}_{\rho} \Gamma, \forall x A(x)$ where $\mathcal{F}$ denotes a set of functions on $O(\Omega)$ defined below.

4. $(\Pi_2 - rfl)$ $f, g \in \mathcal{F} \,\&\, \forall\alpha < \Omega[\vdash^{f(\alpha)}_{\rho} \Gamma, \exists y B(\alpha, y) \,\&\, \vdash^{g(\alpha)}_{\rho} \neg A^{\alpha}, \Gamma] \,\&$ $\Omega \leq \rho \,\&\, f(\Omega), g(\Omega) \ll \alpha_0$ $\Rightarrow \vdash^{\alpha_0}_{\rho} \Gamma$

Inspection of the proofs of the Elimination and Embedding Lemmata 6.1, 6.3 below shows that the class $\mathcal{F}$ of functions on $O(\Omega)$ can consist of simple functions:

1. $O(\Omega) \subset \mathcal{F}$, i.e., each constant function is in $\mathcal{F}$.

2. each variable is in $\mathcal{F}$, i.e., each projection function $f(a_1, \ldots, a_n) = a_i$ is in $\mathcal{F}$.

3. $f, g \in \mathcal{F} \Rightarrow f \# g \in \mathcal{F}$ and $\omega^f \in \mathcal{F}$.

This simplicity is derived from the uniformity of infinitary derivations as images of finite proof figures.

Then it is easy to see with $\omega_0(\alpha) = \alpha, \omega_{n+1}(\alpha) = \omega^{\omega_n(\alpha)}$ that:

Lemma 6.1 *Elimination Lemma:* $\vdash^{\alpha}_{\Omega+n} \Gamma \Rightarrow \vdash^{\omega_n(\alpha)}_{\Omega} \Gamma$

The following crucial lemma is proved by induction on α:

Lemma 6.2 *Collapsing-Bounding Lemma:* $\vdash^{\alpha}_{\Omega} \Gamma \,\&\, \Gamma \subset \Sigma_1 \Rightarrow d_{\Omega}\alpha \models \Gamma$

where $\beta \models \Gamma \Leftrightarrow_{df} \bigvee \Gamma^{\beta}$ is true in the model $\langle O(\Omega) \,|\, \beta; +, \cdot, j, \ldots, R^A \mid \beta, \ldots \rangle$, and $\Gamma^{\beta} = \{\exists x_1 < \beta B_1, \ldots, \exists x_n < \beta B_n\}$ if $\Gamma = \{\exists x_1 B_1, \ldots, \exists x_n B_n\}$ with bounded $B_1, \ldots, B_n$.

Thus the length α of the derivation in $\vdash^{\alpha}_{\rho} \Gamma$ counts also $\beta < \Omega$ such that a rule

$$\frac{\Gamma_0, A(\beta)}{\Gamma, \exists x A(x)} \; (\exists)$$

occurs in the derivation. *cf.* the item 2.($\exists$) in the (not fully presented) definition of the derivability relation.

Next we consider an embedding of proofs in T_2 into infinitary derivations. Assume $T_2 \vdash \Gamma(a)$ where a denotes a free variable in the sequent Γ. In the embedding one substitutes each o.d. $\alpha \in O(\Omega) \mid \Omega$ for a variable x as usual. [Of course this is the case when x is the eigenvariable of an unbounded quantifier $\forall x$. We substitute $\alpha < \beta$ when x is the eigenvariable of a bounded quantifier $\forall x < \beta$.] The resulting infinitary derivation has length depending on α, i.e., $\vdash^{f(\alpha)} \Gamma(\alpha)$ for some $f : O(\Omega) \to O(\Omega)$. This is because the above ($\exists$), the axiom schema (*ind*) of transfinite induction on $<$ and the constant R^A. Thus we get:

Lemma 6.3 *Embedding Lemma:*

$$T_2 \vdash \Gamma(\bar{a}) \Rightarrow \exists n \in \omega \exists f \in \mathcal{F} \forall \bar{\alpha} < \Omega \vdash^{f(\bar{\alpha})}_{\Omega+n} \Gamma(\bar{\alpha})$$

where $\bar{a}$ denotes the variables occurring in Γ.

We have finished defining the relation $\vdash^{\alpha}_{\rho} \Gamma$, and we verify that the properties (<1)-(<4) on o.d.s ensure the Collapsing-Bounding Lemma 6.2.

If the last inference of the infinitary derivation is an $(\exists)$, then (<4) and the condition $\beta \ll \alpha$ in $(\exists)$ take care of this case. Assume that it ends with a $(\Pi_2 - rfl)$

$$f, g \in \mathcal{F} \,\&\, \Gamma \subset \Sigma_1 \,\&\, \forall \alpha < \Omega[\vdash^{f(\alpha)}_{\Omega} \Gamma, \exists y B(\alpha, y) \,\&\, \vdash^{g(\alpha)}_{\Omega} \neg A^{\alpha}, \Gamma] \,\&$$
$$f(\Omega), g(\Omega) \ll \alpha_0 \Rightarrow \vdash^{\alpha_0}_{\Omega} \Gamma]$$

We have to show: $d_{\Omega}\alpha_0 \models \Gamma$. By IH we have

$$\forall \alpha < \Omega[d_{\Omega} f(\alpha) \models \Gamma, \exists y B(\alpha, y) \,\&\, d_{\Omega} g(\alpha) \models \neg A^{\alpha}, \Gamma] \tag{5}$$

We have by (preservation) $d_{\Omega} f(\Omega) < d_{\Omega}\alpha_0$. Assume $d_{\Omega} f(\Omega) \not\models \Gamma$ otherwise we would have $d_{\Omega}\alpha_0 \models \Gamma$ by Σ-persistency. First we show

$$\forall \alpha < d_{\Omega} f(\Omega)(d_{\Omega} f(\alpha) \leq d_{\Omega} f(\Omega)), \tag{6}$$

i.e., $d_{\Omega} f(\Omega)$ is closed under $\alpha \mapsto d_{\Omega} f(\alpha)$.

Assume $\alpha < d_{\Omega} f(\Omega)$. Since no argument a of the function $f(a)$ occurs in the scope of d_{Ω}, we certainly have $f(\alpha) \leq f(\Omega)$ by $\alpha < \Omega$. By the same token we have $K_{\Omega} f(\alpha) \subseteq K_{\Omega} f(\Omega) \cup K_{\Omega}\alpha$. (<2) and (<3) yield $K_{\Omega} f(\alpha) < d_{\Omega} f(\Omega)$. From this $d_{\Omega} f(\alpha) \leq d_{\Omega} f(\Omega)$ by the definition. Thus we have shown (6).

By (5), (6) and $d_{\Omega} f(\Omega) \not\models \Gamma$ we get

$$d_{\Omega} f(\Omega) \models A (\equiv \forall x \exists y B(x, y))$$

From this and (5) for $\alpha = d_{\Omega} f(\Omega) < \Omega$ we have

$$d_{\Omega} g(d_{\Omega} f(\Omega)) \models \Gamma$$

It remains to show

$$d_{\Omega} g(d_{\Omega} f(\Omega)) < d_{\Omega}\alpha_0$$

First we have $g(d_{\Omega} f(\Omega)) \leq g(\Omega) < \alpha_0$ and $K_{\Omega} g(d_{\Omega} f(\Omega)) \subseteq K_{\Omega} g(\Omega) \cup \{d_{\Omega} f(\Omega)\}$. Then $K_{\Omega} g(\Omega) < d_{\Omega} g(\Omega) < d_{\Omega}\alpha_0$ by (<2) and (preservation), and $d_{\Omega} f(\Omega) < d_{\Omega}\alpha_0$. Thus by the definition we conclude $d_{\Omega} g(d_{\Omega} f(\Omega)) < d_{\Omega}\alpha_0$.

Ad Elimination Lemma : As usual it suffices to show

$$\vdash^{\alpha_0}_{\Omega+n} \Gamma, A \,\&\, \vdash^{\alpha_1}_{\Omega+n} \Delta, \neg A \,\&\, Gr(A) = \Omega + n \;\Rightarrow \vdash^{\alpha_0 \# \alpha_1}_{\Omega+n} \Gamma, \Delta$$

The crucial case is when $A \equiv \forall x B(x)$, and
$\vdash^{\alpha_1}_{\Omega+n} \Delta, \neg A$ is derived by $(\exists)$ from $\vdash^{\alpha_2}_{\Omega+n} \Delta, \neg B(\beta), \neg A$ with $\alpha_2, \beta \ll \alpha_1$, and $\vdash^{\alpha_0}_{\Omega+n} \Gamma, A$ is derived by $(\forall)$ from $\forall \alpha < \Omega \vdash^{f(\alpha)}_{\Omega+n} \Gamma, B(\alpha)$ with $f \in \mathcal{F} \,\&\, f(\Omega) \ll \alpha_0$.

It suffices to show $f(\beta)\#\alpha_1 \ll \alpha_0\#\alpha_1$. The point is that we cannot expect to have $f(\beta) \ll f(\Omega)$ since, so to speak, the o.d. β is an alien or an excess to $f(\Omega)$, i.e., to the infinitary derivation of Γ, A. For example β may be $d_\Omega f(\Omega)$. We do have $f(\beta) \leq f(\Omega) < \alpha_0\#\alpha_1$. Also $K_\Omega(f(\beta\#\alpha_1)) \subseteq K_\Omega\{f(\Omega), \beta, \alpha_1\}$. By $(\ll \#)$ we have $K_\Omega f(\Omega) < d_\Omega\alpha_0 \leq d_\Omega(\alpha_0\#\alpha_1)$ and $K_\Omega\beta < d_\Omega\alpha_1 \leq d_\Omega(\alpha_0\#\alpha_1)$. Also by (< 2) and $(\ll \#)$ we have $K_\Omega\alpha_1 < d_\Omega\alpha_1 \leq d_\Omega(\alpha_0\#\alpha_1)$. Thus $f(\beta)\#\alpha_1 \ll \alpha_0\#\alpha_1$. The excess β is compensated only when we consider together the part $\Delta, \neg B(\beta), \neg A$ in which β occurs originally.

Thus we have proved the Theorem 6.2 by a local predicativity method.

6.2 Finitary analysis

In this subsection we expound a finitary analysis of the theory T_2. Consider the finite proof figures as the preimages of infinite proofs under the embedding. This may help us to understand the finitary analysis.

Let P be a proof ending with a Π^Ω_2-sentence A^Ω. To each sequent Γ in P, we assign a term $o(\Gamma; P) \in \mathcal{F}$ so that A^α is true with $\alpha = o(P) = o(\Gamma_{end}; P)$. This is proved by induction on α along the line **(G1)-(G4)** in section 1.

To analyse the rule $(\Pi_2 - rfl)$ we introduce a new rule:

$$\frac{\Gamma, A^\Omega}{\Gamma, A^{d_\Omega \alpha}} \,(c)^\Omega_{d_\Omega \alpha}$$

where $\Gamma \subset \Sigma^\Omega_1$ sentences, $A^\Omega \equiv \forall x \exists y B$ is a Π^Ω_2-sentence and the following condition has to be enjoyed:

$$o(\Gamma, A^\Omega) \ll \alpha \tag{7}$$

This rule is plausible in view of the Collapsing-Bounding Lemma 6.2. When a $(\Pi_2 - rfl)$ is to be analyzed,

$$\frac{\Lambda, A^\Omega \quad \neg\exists z(t < z < \Omega \wedge A^z), \Lambda}{\Lambda} \,(\Pi_2 - rfl)\, J_0$$

roughly speaking, we set $\alpha = o(\Gamma, A^\Omega)$, where Γ is a Σ_1 sequent below J_0 and substitute $d_\Omega\alpha$ for the variable z [originally z is replaced by Ω, *cf.* (4) in section 5], and replace the $(\Pi_2 - rfl)$ by a (cut).

The line J_0 of the inference rule $(\Pi_2 - rfl)$ receives the o.d.

$$o(J_0) = o(\Lambda, A^\Omega)\#o(\neg\exists z(t < z < \Omega \wedge A^z), \Lambda)\#\Omega\#t$$

The inference rule $(\Pi_2 - rfl)$ is resolved as follows:

$$\dfrac{\dfrac{\begin{matrix}\Lambda, A^\Omega \\ \vdots \\ \Gamma, A^\Omega\end{matrix}}{\Gamma, A^{d_\Omega \alpha}}\,(c)^\Omega_{d_\Omega \alpha} \qquad \begin{matrix}\delta \\ \vdots \\ \neg A^{d_\Omega \alpha}, \Lambda \\ \vdots \\ \neg A^{d_\Omega \alpha}, \Gamma\end{matrix}}{\Gamma}\, J$$

where

1. $\alpha = o(\Gamma, A^\Omega)\#t$.

2. $(c)^\Omega_{d_\Omega \alpha}$ is the new inference rule, which says, if Π^Ω_2-sentence A^Ω is derivable with Σ^Ω_1 side formulae Γ and an o.d. α, then we have $\Gamma, A^{d_\Omega \alpha}$, viz. after substituting any $\delta < d_\Omega \alpha$ coming from the right upper part of the (*cut*) J for the universal quantifier $\forall x < \Omega$ in A^Ω, we should have $\beta < d_\Omega \alpha$ for any instance term $\beta < \Omega$ of the existential quantifier $\exists y < \Omega$ in A^Ω.

3. The right upper part of J is obtained by inversion, i.e., substituting the individual constant $d_\Omega \alpha$ for the variable z. $t < d_\Omega \alpha < \Omega$ follows from $t < \Omega$ and the fact that t is contained in α, *cf.* (< 4).

Then the points are that we have to retain the condition (7) $o(\Gamma, A^\Omega) \ll \alpha$ in the inference $(c)^\Omega_{d_\Omega \alpha}$ and if we have, in the left upper part of J,

$$\dfrac{\dfrac{\Gamma, B(\beta_0)}{\Gamma, \exists y B(y)}\,(\exists)}{\Gamma, \exists y < d_\Omega \alpha B}\,(c)$$

then it should be the case that $\beta_0 < d_\Omega \alpha$, i.e., $d_\Omega \beta \in K_\Omega \beta_0 \Rightarrow d_\Omega \beta < d_\Omega \alpha$, *cf.* (< 4). First of all, $d_\Omega \beta$ occurs in a proof only because $d_\Omega \beta$ was generated at a (c) and then substituted at the right upper part of a $(\Pi_2 - rfl)$. The latter condition $d_\Omega \beta < d_\Omega \alpha$ is ensured by the former (7) since, by (3) in section 5, $d_\Omega \beta \ll o(\Gamma, A) \ll \alpha$. The former condition (7) is retained since the only unbounded universal quantifier in Γ, A is the outermost one $\forall x$ in A and any o.d. $\geq d_\Omega \alpha$ is forbidden to be substituted for x by the restriction $\exists x < d_\Omega \alpha$ in $\neg A^{d_\Omega \alpha}$.

Let us examine how to fit the definition of $d_\Omega \beta < d_\Omega \alpha$ in the present setting.

1. If $d_\Omega\beta$ has already occurred when a $(c)^\Omega_{d_\Omega\alpha}$ is introduced, then $d_\Omega\beta < d_\Omega\alpha$ by (< 2), i.e., $d_\Omega\beta \in K_\Omega\alpha < d_\Omega\alpha$, *cf.* (3).

2. Assume another $(\Pi_2 - rfl)$ is to be analyzed for which the $(\Pi_2 - rfl)$ is above a $(c)^\Omega_{d_\Omega\alpha}$ (for simplicity side formulae are omitted in the following figures):

$$\cfrac{\cfrac{A_1^\Omega \quad \neg\exists z_1(t_1 < z_1 < \Omega \wedge A_1^{z_1})}{\vdots}\,(\Pi_2 - rfl)}{\cfrac{\exists y < \Omega B}{\exists y < d_\Omega\alpha B}\,(c)^\Omega_{d_\Omega\alpha}\, I}$$

Let β denote the o.d. attached to the uppersequent $\exists y < \Omega B$ of the $(c)^\Omega_{d_\Omega\alpha} I$. Introduce a new $(c)^\Omega_{d_\Omega\beta} J_1$ immediately above the old one I:

$$\cfrac{\cfrac{\cfrac{\begin{matrix}A_1^\Omega\\ \vdots\\ \exists y < \Omega B, A_1^\Omega\end{matrix}}{\exists y < \Omega B, A_1^{d_\Omega\beta}}\,(c)^\Omega_{d_\Omega\beta} J_1}{\exists y < d_\Omega\alpha B, A_1^{d_\Omega\beta}}\,(c)^\Omega_{d_\Omega\alpha} \qquad \cfrac{\begin{matrix}\neg A_1^{d_\Omega\beta}\\ \vdots\\ \neg A_1^{d_\Omega\beta}, \exists y < \Omega B\end{matrix}}{\neg A_1^{d_\Omega\beta}, \exists y < d_\Omega\alpha B}\,(c)^\Omega_{d_\Omega\alpha}}{\exists y < d_\Omega\alpha B}$$

The o.d. $d_\Omega\beta$ is substituted for the variable z_1 and hence it may be an instance term for the existential quantifier $\exists y < \Omega$ in $\exists y < \Omega B$. Therefore we have to have $d_\Omega\beta < d_\Omega\alpha$. We have $\beta < \alpha \,\&\, K_\Omega\beta < d_\Omega\alpha$ by (7) $\beta \ll \alpha$. Hence this follows from (< 4).

Along these lines Theorem 6.2 can be shown by a finitary analysis, *cf.* [12].

References

[1] T. Arai, On Takeuti's fundamental conjecture (Japanese), Sûgaku 40 (1988), 322-337, MR 90f:03091 and Zbl. Math. 689 03f:03026.

[2] T. Arai, Cut elimination for *SBL*, manuscript, 1988.

[3] T. Arai, Proof theory for theories of ordinals I. Π_2-reflecting ordinals, manuscript, Feb.1989.

[4] T. Arai, Systems of ordinal diagrams, manuscript, Aug. 1996.

[5] T. Arai, Proof theory for theories of ordinals I: reflecting ordinals, manuscript, Nov. 1996.

[6] T. Arai, Proof theory for theories of ordinals II: Σ_1 stability, manuscript, Feb. 1997.

[7] T. Arai, Proof theory for theories of ordinals III: Π_1 collection, manuscript, Mar. 1997.

[8] T. Arai, Consistency proof via pointwise induction, Arch Math Logic 37 (1998), 149-165.

[9] T. Arai, Variations on a theme by Weiermann, J. Symb. Logic 63 (1998), 897-925.

[10] T. Arai, Ordinal diagrams for recursively Mahlo universes, submited.

[11] T. Arai, Ordinal diagrams for Π_3-reflection, submitted.

[12] T. Arai, Proof theory for theories of ordinals I: recursively Mahlo ordinals, submitted.

[13] T. Arai, Proof theory for theories of ordinals II: Π_3-reflection, submitted.

[14] T. Arai, Ordinal diagrams for Π_N-reflection, submitted.

[15] T. Arai, Proof theory for theories of ordinals III: Π_N-reflection, submitted.

[16] W. Buchholz, Normalfunktionen und konstruktive Systeme von Ordinalzahlen. In: Proof theory symposium, Kiel 1974, J. Diller and G. H. Müller (eds.). Lecture Notes in Mathematics 500 (1975), Springer, 4-25.

[17] W. Buchholz, Über Teilsysteme von $\bar{\theta}\{g\}$), Arch math Logik Grundlagenforsch 18 (1976), 85-98.

[18] W. Buchholz, Eine Erweiterung der Schnitteliminationmethode, Habilitationsschrift München, 1977.

[19] W. Buchholz, An independence result for $(\Pi^1_1 - CA) + BI$. Ann. Pure Appl. Logic 33(1987), 131-155.

[20] W. Buchholz, Proof-theoretic analysis of termination proofs, Ann. Pure Appl. Logic 75(1995), 57-65.

[21] W. Buchholz, Explaining Gentzen's consistency proof within infinitary proof theory. In: Computational Logic and Proof Theory, G. Gottlob, A. Leitsch and D. Mundici (eds.). Lecture Notes in Computer Science 1289 (1997), Springer, 4-17.

[22] W. Buchholz, Explaining the Gentzen-Takeuti reduction steps, preliminary version, May 1997.

[23] W. Buchholz and K. Schütte, Ein Ordinalzahlensystem für die beweistheoretische Abgrenzung der Π^1_2-Separation und Bar-Induktion, Sitzungsber. d. Bayer. Akad. d. Wiss., Math.-Nat. Kl., 1983, 99-132.

[24] G. Gentzen, Der erste Widerspruchsfreiheitsbeweis für die klassische Zahlentheorie, Arch. math. Logik und Grundlagenforsch, 16 (1974), 97-118.

[25] G. Gentzen, Die Widerspruchsfreiheit der reinen Zahlentheorie, Mathematische Annalen 112 (1936), 493-565.

[26] G. Gentzen, Neue Fassung des Widerspruchsfreiheitsbeweises für die reine Zahlentheorie, Forschungen zur Logik und zur Grundlegung der exakter Wissenschaften, Neue Folge 4 (1938), 19-44.

[27] G. Jäger, Zur Beweistheorie der Kripke-Platek Mengenlehre über den natürlichen Zahlen, Arch math Logik Grundlagenforsch, 22 (1982), 121-139.

[28] G. Kreisel, Review of the book 'The Collected Papers of Gerhard Gentzen', ed. and transl. by M. E. Szabo, Journal of Philosophy 68 (1971), 238-265.

[29] H. Levitz, On the relationsship between Takeuti's ordinal diagrams $O(n)$ and Schütte's system of ordinal notations $\Sigma(n)$. In: Intuitionism and Proof Theory, A. Kino, J. Myhill and R.E. Vesley (eds.), North-Holland (1970), 377-405.

[30] H. Levitz and K. Schütte, A characterization of Takeuti's ordinal diagrams of finite order. Arch math Logik Grundlagenforsch 14(1970), 75-97.

[31] P. Päppinghaus, Completeness properties of classical theories of finite type and the normal form theorem, Dissertationes Mathematicae 207, Warsaw (1987).

[32] H. Pfeiffer, Ausgezeichnete Folgen für gewisse Abschnitte der zweiten und weiterer Zahlenklassen. Dissertation Hannover 1964.

[33] H. Pfeiffer, Ein Bezeichnungssystem für Ordinalzahlen. Arch math Logik Grundlagenforsch 12(1969), 12-17.

[34] H. Pfeiffer, Ein Bezeichnungssystem für Ordinalzahlen. Arch math Logik Grundlagenforsch 13(1970), 74-90.

[35] H. Pfeiffer, Vergleich zweier Bezeichnungssysteme für Ordinalzahlen. Arch math Logik Grundlagenforsch 15(1972), 41-56.

[36] W. Pohlers, Pure proof theory, aims, methods and results, Bull. Symb. Logic 2 (1996), 159-188.

[37] M. Rathjen, Proof theory of reflection, Ann. Pure Appl Logic 68(1994), 181-224.

[38] M. Rathjen, Recent advances in ordinal analysis: $\Pi^1_2 - CA$ and related systems, Bull. Symb. Logic 1(1995), 468-485.

[39] M. Rathjen and A. Weiermann, Proof-theoretic investigations on Kruskal's theorem. Ann. Pure Appl. Logic 60 (1993), 49-88.

[40] W.H. Richter and P. Aczel, Inductive definitions and reflecting properties of admissible ordinals. In: Generalized Recursion Theory, J.E. Fenstad and P.G. Hinman, (eds.), North-Holland, Amsterdam, 1974, 301-381.

[41] K. Schütte, Ein konstruktives System von Ordinalzahlen I. Arch math Logik Grundlagenforsch 11 (1968), 126-137.

[42] K. Schütte, Eine beweistheoretische Abgrenzung des Teilsystems der Analysis mit Π^1_2-Separation und Bar-Induktion, Sitzungsber. d. Bayer. Akad. d. Wiss., Math.-Nat. Kl. 1987, 11-41.

[43] G. Takeuti, On the fundamental conjecture of GLC, I, J. Math. Soc. Japan 7(1955), 249-275.

[44] G. Takeuti, Ordinal diagrams, J. Math. Soc. Japan 9 (1957), 386-394.

[45] G. Takeuti, Consistency proofs of subsystems of classical analysis, Ann. Math. 86 (1967), 299-348.

[46] G. Takeuti, Proof Theory, second edition, North-Holland, Amsterdam, 1987.

[47] A. Weiermann, Ein neuer Zugang zu Kollabierungsfunktionen. Dissertation. Westfälische Wilhelms-Universität Münster, 1990.

[48] A. Weiermann, How to characterize provably total functions by local predicativity, J. Symb. Logic 61 (1996), 52-69.

What Mathematical Truth Could Not Be - II[1]

or

Through a Glass Darkly[2]

Paul Benacerraf
Princeton, NJ

"For now we see through a glass darkly, but then [we shall see] face-to-face." Paul, *Corinthians I*.13.12[3]

"When we look at a thing, we must examine its essence and treat its appearance merely as an usher at the threshold, we must grasp the essence of the thing; this is the only reliable and scientific method of analysis." Mao TseTung, "A Single Spark Can Start a Prairie Fire" January 5, 1930, Selected Works, Vol. I, p. 119

"...By analysis we mean analyzing the contradiction in things. And sound analysis is impossible without a real understanding of the pertinent contradictions." Mao TseTung, Speech at the Chinese Communist Party's National Conference on Propaganda Work (March 12, 1957, 1st. pocket ed., p. 20)

That is our task today then -- to look at the thing, to examine its essence by treating its appearance as an usher at the threshold -- by analyzing the pertinent contradictions. The glass through (in) which I would like us to look today is then (as in the quotation from Paul) a mirror offered by George Boolos's "Must We Believe in Set Theory?" -- a wonderful, but, as I shall argue, unusually (even-for-George) enigmatic paper.[4]

I will set the stage by reminding us very briefly of three accounts or conceptions of sets that figure in what I shall be discussing today -- not so much for their specific details, but rather for the sake of broad-brush contrast. Nothing I say will depend on the exact detail.

All three (in direct affront to Russell's theory of descriptions) keeps being called "*the*(?) iterative conception". They find an early expression in the following passage:

"... a set is something obtainable from the integers (or some other well-defined objects) by iterated application of the operation 'set of', not by dividing the totality of all existing things into two categories."[5]

Roughly, given some things (already in hand, honestly come by – that is the force of "well-defined objects"), there is a set of them satisfying any condition you like. Moreover, this process of obtaining sets of "well-defined objects" invariably yields *more* well-defined objects (the sets themselves) and can, by the same principle, be iterated indefinitely. The three versions of this conception described below differ from this only by being more detailed and explicit, laying out their assumptions for all to see – and, of course, by having different assumptions.

The first is Gödel's own conception, as expressed in his "Gibbs Lecture," his 1951 address to the American Mathematical Society. It came four years after the publication of "What is Cantor's Continuum Problem?", the celebrated paper from which the brief snippet quoted above was taken. I present Gödel's words as they are quoted in George's paper. As George points out,

> Gödel describes both a process for arriving at axioms for set theory and a picture of the set-theoretic universe. (It is of course a version of the iterative conception.) ...
>
> > '...evidently this procedure can be iterated beyond ω, in fact up to any transfinite ordinal. So it may be required as the next axiom that the iteration is possible for *any* ordinal, that is, for any order-type belonging to some well-ordered set. But are we at an end now? By no means. For we have now a new operation of forming sets, namely forming sets out of some initial set A and some well-ordered set B by applying the operation "set of" to A as many times as the well-ordered set B indicates. And setting B equal to some well-ordering of A, we can now iterate this new operation, and again iterate it into the transfinite. This will give rise to a new operation again, which we can treat in the same way, and so on. So the next step will be to require that *any* operation producing sets out of sets can be iterated up to any ordinal number (that is, order-type of a well-ordered set). But are we at an end now? No, ..." [131-2]

That is Gödel's. A second, possibly less-generous rendition, is drawn from Joseph Shoenfield's enticing and pellucid account of ZFC in the *Handbook of Mathematical Logic*.[6] The third, surely the sparest in its yield, is George's own account of the iterative conception as presented most recently in "Iteration Again."[7] I will describe Shoenfield's in a little more detail, since, as the intermediate case, it serves as a foil for both of the others.

For Shoenfield, sets (all but one, I imagine) have members. They are "formed" at "stages" that are partially ordered, and the members of a set x (those that are sets, at least) are all formed at stages earlier than any stage at which x is formed. Given a collection Γ of sets and a collection S of stages such that each member of Γ is formed at a stage that is a member of S, if there is a stage T after all the stages in S, then Γ can be formed at T.

As Shoenfield points out, this naturally prompts the question: Given a collection S of stages, just when *is* there a stage, T, later than each member of S? His answer: After all the stages in S *provided we can imagine a situation in which all of the stages in S have been completed.*

Vague? Perhaps. Shoenfield distinguishes three cases that satisfy this criterion:

(a) S consists of a single stage (it is then followed by a successor stage.)
(b) S consists of an infinite sequence of stages (it is followed by a limit stage).
(c) There is a set x and for each y in x, a *stage* S_y, (e.g., the earliest stage at which y was formed). In that case T consists of all the S_y's.

Finally, there is George's. All we need to say about it is that it is basically similar in structure, although considerably weaker in that it won't sustain Shoenfield's third case (which looks suspiciously like the Axiom of Replacement). On this scale, it seems clear, at least from our Olympian standpoint, that the universes generated (or "described", choose your own poison) are roughly as follows: Boolos $\subset$ Shoenfield $\subset$ Gödel.[8]

There are *other* conceptions, of course, some of which are orthogonal to these. Frege's is a justly famous example. Other examples are systems produced by W. V. Quine ("New Foundations for Mathematical Logic" and "Mathematical Logic"); yet another, favored by Solomon Feferman and others, retains almost all of ZFC's usual axioms, including Replacement, but abandons the view that every set has a power set. Take it as a given of mathematical anthropology that all these conceptions are abroad in the land. The array is sufficiently bewildering to start us off in immediate sympathy with George's question, now heard as a lament: "*Must* we believe in set theory?"

I first encountered "Must We Believe in Set Theory?" when George delivered it at the Chapel Hill Colloquium in Philosophy in October 1993, a presentation that David Auerbach followed with an excellent set of comments. What George seemed to be saying in that paper struck me at the time as preposterous on its face, and I distinctly remember thinking: "He can't *really* believe this." -- I was convinced that the whole thing had to be a very subtle and elaborate put-on. Since George was surely not timid about such matters this wasn't an outlandish thought. Still, I never followed it up with George. The tragedy that occasioned this paper intervened, so to this day I am not quite sure whether George had his tongue firmly planted in his cheek (where it spent much of the time anyway) or whether he really meant what he said.

Because the view he seemed to be adopting and that struck me as preposterous involved elements that I recognized as ones to which I was committed by some of the central arguments of MT (this is the mirror effect "through a glass darkly"), George's presentation gave me pause -- that I should at once be finding a view preposterous and recognizing it as my own naturally caused me some embarrassment. Did *George* actually believe the things he was saying – and that I was finding preposterous? Since that didn't seem likely either, all I was left with was a suspicion that George was selling out to the hated anti-realists (who are all around us), and covertly to boot: Why attack a view head on when you can discredit it much more effectively by an oblique remark, or better, by explaining it in detail? Some views are their own worst enemies -- philosophers know from bitter experience that often the most effective way to undermine a view is to advocate it.

I will explain why I am committed to taking the question seriously, why I thought the view preposterous and, by the above tortuous reasoning, why I found George oblique -- and at the end, why I am no longer sure about either.

But first the view.

Recall his title: "Must We Believe in Set Theory?" For the purposes of that paper (and this), set theory was represented by ZFC -- Zermelo-Fraenkel Set Theory with the Axiom of Choice -- and the claim of George's that I found so startling was that the cardinals whose existence were implied by this relatively tame theory, which I thought everybody except a handful of philosophers (Nelson Goodman, Bas VanFraassen, Hartry Field, *et al.*) believed,[9] were so large that George simply wasn't prepared to go along with the gag. Belief ran out *long* before theory. The particular example that he cited (of a cardinal in whose existence he couldn't believe) was the cardinal κ ($=\aleph_\kappa$), the limit of the sequence $<\aleph_0, \aleph_{\aleph_0}, \aleph_{\aleph_{\aleph_0}}, \ldots>$ (if it has a limit, we must add if we are suspending belief).

That is, κ, if it exists, is the least ordinal greater than all of the f(i), where $f(0)=\aleph_0$ and $f(i+1)=\aleph_{f(i)}$. "...if κ exists, there are at least as many as κ sets. *Are* there so many sets?" [120] George asked.

George was a superb pedagogue, as anyone who knew him even slightly will attest -- but he was also an incurable ham. I can still see him, delighted, gleeful and pixie-like, pointing out that, although huge-for-George, κ is "teensy," at least by the standards of those who study what are called in the trade "large cardinals." He had no quarrel with the proof-theory for ZFC; so he didn't doubt that *if* ZFC's axioms are *true*, κ exists. κ, after all, is one of the smaller cardinals whose existence can be "proved" in ZFC. He just couldn't believe there was such a large cardinal.

I was numb: Since he readily acknowledged the existence of ω (or $\aleph_0$) George clearly hadn't turned finitist, though in retrospect that might have been better -- less of a shock; finitists, after all, are a principled lot. But what *principle* would single κ out for this kind of discriminatory treatment? Yet, asking him to believe that κ exists was pushing his credulity too far. κ, he claimed, is just too *big*.

Too big *for what?* Was he claiming that it was "too big to exist"?[10] Surely not (though he might have thought it), because he was well aware that to advance that view, he would have needed some argument to the effect that things ran out *before* κ, and he didn't really have much of an argument and knew it. If the claim was that to have κ members was to be so bloated as to defy the laws of existence, something had to be said about those laws. Since ZFC is unlikely to oblige by being inconsistent, contradictions would not rush in to fill the gap, as they had in support of the "limitation of size" principle (any collection that didn't exist – that wasn't a set -- was just too big to be a set, on pain of contradiction).

No, although belief in κ was almost certainly consistent, consistency isn't all there is to the matter: κ was just too big for George *to believe in*. Autobiography is fine, but it has little probative weight. So George's point, despite his characteristic modesty, had to be that κ is too big to believe in, not just for George, but also *for any right-thinking person*, a point to which we shall return presently. Here's George again:

> "...it's a *pretty big* number, by the lights of those with no previous exposure to set theory, so big, it seems to me, that it calls into question the truth of *any* theory one of whose assertions is the claim that there are at least κ objects."[120, emphasis on 'any' added]

That's pretty direct. But why the qualifier "with no previous exposure to set theory"? Why should seeming so big **to the uninitiated** call *anything* into question? Why should those lights be thought to shed any light on how things really are with sets? Moreover, that qualifier surely didn't describe *George*. Was he suggesting that what is relevant in these judgments is one's uneducated, pre-set-theoretical, intuitions about what's big and what is not? Is that the appropriate standard of "bigness"? That's not much of a criterion, as having no previous exposure to set theory is an innocence easily lost: one need but enter "the paradise that Cantor has created for us."[11] Why give *any* weight to the fact that a number strikes those innocent eyes as big if to the truly experienced it is really *teensy*? Shouldn't experience with such large objects -- or not to prejudge the issue, with *theory committed to the existence of* such large objects --count for *something*? Yet those were his words. It is as if experience -- doing what set-theorists do naturally (at the office) -- was a **corrupting** influence. As if taking a course in set theory was more like being brainwashed than like being informed. That's a hard line.

Of course, the connection between *seeming big to the inexperienced* and being of dubious lineage might not have been intended by George to carry the full weight of the doubt. Maybe his word-processor swallowed some words; perhaps, before final processing, the sentence had read

> "...it's a *pretty big* number, by the lights of those with no previous exposure to set theory **to be sure, but also by the lights of any right-thinking person**, so big, it seems to me, that it calls into question the truth of any theory one of whose assertions is the claim that there are at least κ objects."

Words to that effect *might* have gotten deleted, if not by his word-processor, then by George himself, upon rereading them, aghast at the immodesty they would evince. But if it was his own doing, he did it without lavishing his characteristically meticulous attention on what would remain after the cut. [I should add that there is no suggestion in his paper that κ might be *the smallest* cardinal too big (believably) to exist. Or even that there might *be* a smallest, or that some *other* cardinal, $<\kappa$, naturally, is *the largest* cardinal one could or should think exists. George is aware of these traps and hedges against all of that by pointing to the vagueness of some of these intensional "mental-state" notions of folk psychology.]

George was very clear that the problem to which he was calling our attention couldn't be resolved by regarding the sentence of ZFC (call it 'σ') that carried the assertion of κ's existence as just some "formal" object, itself not worthy of "belief" because it was merely a part of some formalism that we were free to interpret as we wished, or not at all. Or, as Skolemites would urge on us, one that we were free to *re*interpret as "really" being committed only to some countable domain, since on that view, none of ZFC's sets could be tarred as "too big" without dragging all the other infinite ones under with it.

On the contrary; George took elaborate pains to insist that ***ZFC*** is committed to the existence of κ, and hence to the existence of at least κ objects, because one of its *theorems* (σ) *says* just *that*. On his (correct) view, ZFC is an *interpreted* first-order theory whose sole non-logical vocabulary is the epsilon of class membership and whose quantifiers range over all the sets that there are. So σ, for whose proof one must invoke the axioms of Infinity, Power Set, Union, and Replacement, is ZFC's witness to this heresy. If there simply *aren't* all that (κ-) many things, then σ is false (because it implies that there are); and ZFC is false too (because it implies/contains σ).[12]

Diplomat that he was, George didn't state outright that no right thinking person would believe in the existence of κ; after all, many of his best friends and companions (at least appear to) claim they do. He provided us instead with an interesting assortment of indirect arguments, many of which I will survey below.

The first is a discussion of *the ethics of assertion*, the main point of which is that you don't (normally, and honestly) assert what you don't (believe you) know to be the case: If you are aware that something is *merely* your opinion, or less, you don't simply blurt it out with the normal assertoric intonation contour; it is customary, and expected, to mark that epistemic distance explicitly with some standard phrase like "I think that ...", "I'm not sure, but...", etc.. . The point of launching into what may seem like a digression is to invite each of us to reflect on just how *we* feel about κ. What would *we* say about the matter were we consciously to limit our claims to what we (thought we) *knew* to be so? He concludes that probably no one -- *not even the most dedicated of set-theorists* -- would go around blabbing things so extravagant as σ.

This is not presented as an argument for *dis*believing σ -- for believing it *false*, for believing that κ *doesn't* exist. George knows he doesn't really have much of one. It is presented instead as "good reasons *for failing to assent to* the existence of κ." [121, emphasis added] So we find him asking "But perhaps company can make up for the absence of reasons."[122]

Think about it. Would *you* assert categorically that κ exists? Or would you hold back somewhat and limit your claim to something like "According to ZFC, σ ....." or "In ZFC you can prove σ, and hence, according to ZFC at least, there are at least κ many things in the universe." Rare indeed is the set theorist who would not take refuge in some such hedge.[13] The thrust of this argument is to nudge us to the realization that *we* won't feel much better about κ than George did, so there must be something to the view he is advancing. Such a realization, if true, would be telling. Although irrelevant to whether κ exists, it would seriously affect both our epistemic stance and our dialectical position toward that proposition -- if we are already in agreement, why argue?

Notice that σ's truth conditions remain intact throughout all of the hedges -- σ is always understood as being about sets, and to be making a claim that might be true or false. Because we may not *know* which it is, we hedge. Otherwise, why hedge? Semantically, the same holds with $\aleph_0$, but according to George at least, there is no need to hedge *there*. When it comes to $\aleph_0$, George is right up there among its most vocal boosters, on the front lines proclaiming its existence. And according to him, so do *we* -- or at least, so do most of us. We should -- and would -- exhibit no parallel hesitation with $\aleph_0$.

This argument derives its dialectical force from focusing our attention on our hesitation, when challenged, to give κ our fullest, unequivocal, backing. But it can have that force in a way helpful to George's point only if we would be differently disposed toward less bloated objets. If, when pushed, we would be equally reluctant to give our unequivocal backing to *any* existence claim in set theory, our hesitation to give κ our fullest support loses much of its sting. Here I must disagree with George, because I fear that is precisely how most people would respond: Once the context is shifted from that of a normal set-theoretic discussion to another, in which the legitimacy of the entire enterprise is clearly being called into question (as it is when, in plain view of the fact that it is a *theorem*, you are nonetheless challenged to give your unequivocal backing to some proposition), our natural philosophic caution takes over. We tend to pull back and hedge our bets, using words much as those I have suggested above. I submit that most would do the same if similarly pushed on the existence of the null set, or of ω, or of $\aleph_0$. You can hear the argument …

- *Skeptic* -- ω .. yes, I know you can 'prove' its existence by a few applications of some of the axioms. … But does it *really* exist?

- *Us* -- Well, I don't know what you mean here, but surely, according to ZFC it does.

etc..

But, put that aside and suppose (with George) that (like George) we would exhibit no parallel hesitation with some elementary core of ZFC. Why then should we balk at κ? Why treat κ and ω differently? George gives us precious few reasons either for subscribing to the pittance he allows us -- believing what he takes to be the believable part of ZFC -- or for refusing to believe the "rest". His paper contains a number of truly elegant rejections of a host of skeptical arguments that he imagines might be addressed to him. Some of these concern whether it makes sense to wonder about the size of the universe at all, or *specifically*, whether it is at most κ. He reviews holist arguments, as well as arguments stemming from Carnap's famed distinction between "internal" and "external" questions, according to which the real choice is between accepting or rejecting "setspeak", or what he calls the "framework" of set theory, on pragmatic grounds: if we reject it, the question of whether κ is a believable measure of at least a portion of the universe doesn't arise; whereas if we accept that framework, we should not pick over its theorems one by one with a suspicious eye, for there is no standpoint from which we could (justifiably) do the picking. On that view, having made our bed, we should lie in it and accept whatever it may give us.

George rejects them all -- rightly, I think -- and argues that it makes perfect sense to ask whether σ is true, just as it makes perfect sense to ask it of many others of the other sentences of ZFC, theorems or not. He doesn't go so far as to suggest that they *all* have truth-values, so we may speculate on whether he believes that. But σ, and any *less* problematic sentences are surely ok.[14] We understand the question; nothing about it *and nothing about* σ or about the axioms used in its proof is *unclear*.[15] But, although, as we have seen, the axioms imply *an* answer to the question of whether there are at least κ sets in the universe, according to George, they don't *settle* the question, because the answer they imply is not credible. Despite that, whether σ is true remains a *mathematical* question, that we do and should treat as such: σ is, if true, a *mathematical truth*, not to be settled by philosophical fiat. And if it's not a mathematical *truth*, it's a *mathematical* falsehood -- it is false because the sets give out before there get to be κ of them. Why *should* they? Or, how *could* they? Perhaps not every set has a union; or not every set has a power set, or some of the relevant instances of Replacement are false. What, after all, tells us these must all be *true*? It can't be the simple fact that they are Axioms of ZFC? If they were chosen as axioms because they were *thought to be* true, the question of their truth is an antecedent one, about which we might have been mistaken. If they were chosen to characterize an idea, the question of their truth is entirely inappropriate, except perhaps in some derivative sense, as a measure of their fidelity to that idea. Clearly, it is in the former spirit that George was taking them, or else, why be concerned about their believability? What would there be to believe or disbelieve?

Given the drift of these arguments, and since his skepticism (or his agnosticism, at least) about κ (and its cousins) is firmly coupled to a declaration of faith that $\aleph_0$ exists, and hence also that, at least insofar as it underwrites the existence of $\aleph_0$ (and *its* acceptable cousins) ZFC is just fine, one should not suspect George of more general and thoroughgoing philosophical skepticism -- unless, of course, this is all the put-on I thought it might be. He remains, at least on the surface, firmly an enemy of the nominalist. Though a friend of the concrete, he also embraces the abstract -- to the point of intimacy, it seems -- as he claims that we evidently *know* a lot about a lot of abstract things -- languages, pieces of software, radio programs, symphonies, bank balances, theorems, and the like; so a philosopher who wishes to cast doubt on specifically mathematical knowledge simply because of the (admitted) abstractness of its objects has an enormous amount to "explain away" if she is to save the appearances:

> "...no sense of 'sensible' or 'experience' has been shown to exist under which it is not correct to say that we can have sensible experience of such objects, such things as the zither melody in Tales from the Vienna Woods, the front page of the sports section of this morning's *Globe*, a broad grin, or a proof in set theory of the existence of κ.
>
> "It is thus no surprise that we should be able to reason mathematically about many of the things we experience, for they are already abstract." [129]

The point is well taken (I, for one, have never disputed it). Along the way, George instructively contrasts his attitude and position with that of my wimpy colleague David Lewis. Why wimpy (David has been called many things, but seldom that)? Simply because although he believes, on what he takes to be excellent grounds, that singletons -- and classes more generally -- are an abomination, a metaphysical monstrosity, he nevertheless *refuses* to confront the set theorist to urge him to abandon these false, even incoherent, theories -- *formally* consistent though they may be. In fact, pushed by an unseemly deference to the working-mathematician-in-the-street, Lewis appears to have brought himself actually to *believe* in these monstrosities, despite all the good reasons he offers for a contrary stance. So his wimpiness lies less in a reluctance to emerge from the closet of unbelievers than in an excessively modest *philosophic* principle that settles all conflict between philosophy and science (may we call set theory science?) in favor of science: This is the philosopher-interpreter, not the philosopher-critic. George, I suspect, feels much the same about their respective roles -- at least he quotes Lewis with some approbation. But this is all irrelevant to the status of κ, since for George, the issue is whether we have adequate *scientific* grounds for believing in κ: exactly the sort of question that often arises *within* Science, and hence not a Science vs. Philosophy issue, despite the fact that it is presented, perhaps for rhetorical purposes, as one concerning that philosophic obsession -- *knowledge*:

> What we are contemplating here, however, is nothing so radical as the rejection of singletons, but only the claim to be a body of *knowledge* on the part of a portion of set theory that treats objects far removed from ordinary experience, the rest of physical science, the rest of mathematics, and the rest of a certain more "concrete" part of set theory. [130]

I said before that George did offer some positive arguments for *dis*believing in the existence of κ. The first troubles me because, although hinted at in the above paragraph, as it first strikes one, it is so unlike George -- not the sort of argument to which he was normally given. Although different in some important respects, it appears to be a version of the "indispensability argument" so popular in recent decades, especially in Cambridge (Massachusetts) and in New York (with a touch of Occamism thrown in as well, I fear). Here, in full, is George's allusion to it.

> Furthermore, to the best of my knowledge, nothing in the rest of mathematics or science requires the existence of such high orders of infinity. The burden of proof should be, I think, on someone who would adopt a theory so removed from experience and the requirements of the rest of science (including the rest of mathematics) as to claim that there are κ objects."[122]

We have, on the one hand, the clear *semantic*, or, if you, prefer, *hermeneutic* claim that ZFC is a theory about the domain of sets. It is our theory, ***but not the*** *only one we have.* Sentences of ZFC are true iff the universe (the cumulative hierarchy?[16]) is as ZFC depicts it. If replacement, or Union, or Power Set, or even -- yes -- Infinity makes an extravagant (in the sense of 'false') claim, it is so only because the postulated sets just aren't *there.*[17] In particular, on this view, whether or not some ordinal exists doesn't hinge on whether we would make good use of it if only it were available (and, presumably, if we could also bring ourselves to believe it existed). The claim that it exists is either correct or it is not. That is the first half of the problem, surely.

On the other hand, we have this "burden of proof" remark, with its blatant Occamite coloration. How do they marry? *What could the existence of* κ *conceivably have to do with the "requirements of the rest of science (including the rest of mathematics) .." ?*[18] "[R]emoved from experience," maybe. But is experience the appropriate guru to consult when choosing axioms? Much ink has been spilled on that question by Penelope Maddy, valiantly investigating and deepening our insight into how one might substantiate some remarks of Gödel's that others have, less charitably, simply ignored or even ridiculed. But how can our "non-mathematical" experience be a guide in the present context, where we are dealing with "pure" sets? To put it differently, to what is our theory of (pure) sets answerable? How *might* there be fewer sets than ZFC says there are, and if that is the sad state of affairs, how might we find out?

I see two possibilities, neither of which is very favorable to the overall coherence of the view under consideration. The first is an old story. Set theory is not an "autonomous" discipline, but rather the hand-servant of the "foundations" of mathematics -- tailored to (*invented* for?) making coherent sense of the entire mathematical universe, which, for the mathematical realist, consists of all sorts of structures (reals, rationals, complex numbers, etc.). We use set theory to model it (and alternatively, for the reductivist, to construct it). The second regards the evidence in favor of what axioms we each accept as deriving from some "picture" of sets such as is provided by some version of *the (?)* "iterative conception," of which, as we have seen, there are many.

On the first of these, she who postulates more sets than are "needed" for this modeling task is guilty of ontic profligacy and rightly to be chastised for her extravagance. This is a view with many problems of its own; but even if they could be solved, *can it be George's view?* Even if we ignored the fact that he explicitly *rejects* it (see the next quotation), how would it square with his brave declarations of the semantic (and metaphysical) autonomy of the theory? -- with the thought that, whatever we may *postulate*, the sets themselves may run out and fail to satisfy the axioms? Or with the following remarks about (his own version of) the iterative conception

"The interest of the iterative conception is that it shows that the axioms of Zermelo(-Fraenkel) are not just a collection of principles chosen for

> their apparent consistency and ability to deliver desired theorems concerning arithmetic, analysis, and Cantorian transfinite numbers, but not otherwise distinguished from other equally powerful consistent theories. The conception is *natural* in the simple sense that people can and do easily understand, and readily regard as at least plausible, the view of sets it embodies."[127, his emphasis]

These are not the words of someone for whom set theory's *raison d'être* is to be a servant discipline.

But -- and here's the rub -- neither is it easily squared with the stark "realist" stance we have seen him take. For there is a tension between offering *these* reasons for believing that sets satisfy a given conception and arguing (against someone who finds Replacement "plausible") that we could be mistaken in accepting it nonetheless, because sets just might not satisfy Replacement:

> "...Even if the iterative conception is supplemented so that replacement follows, what reason have we to think that any such story is *correct*?" [127, his emphasis]

Can't the same be said of the (weak) iterative conception that *George* finds plausible: What if he is *mistaken* in accepting it as a limitation? What does the "plausibility-to-us" adverted to here have to do with truth? That is something that we need to be told, if this view is to stay aloft. Leaving us in the dark about this makes George no worse than anyone else; yet without some account of the link between plausibility and truth, this is just a position very hard to occupy for a "hard-line" realist about the meaning of set-theoretic language. This is, and continues to be, the agonizing pull of "Mathematical Truth", at least for me.

To be sure, George is careful to say only that he finds himself "inclined to believe" the pronouncements of the weak iterative conception (and inclined to *dis*believe that there could be more sets than it furnishes). More autobiography. But a reasonable person apportions her belief -- and disbelief -- *to the evidence*. Why does this (finding the weak iterative "picture" plausible) count as evidence? Alternatively, on what evidence does he find it plausible? Maybe the real truth is that we are (genetically -- as the human genome project is sure eventually to discover) anti-set-ic people, programmed with set-blinders that only a very few (like George's J.?) are ever able to shed. Perhaps the preference for "desert landscapes" that some profess is merely a symptom of this deficiency. If so, and if it is *truth* that we seek, finding crowded universes *im*plausible should not have a high standing in the space of reasons.

The skeptic will object that, because there is no "external" standpoint from which to judge whatever theory of sets is currently the "best" that we accept, sets are whatever our best theory of the moment says they are -- anyone who thinks otherwise is afflicted with "metaphysical realism", a virulent philosophical disease uncovered and ably catalogued by another past President of this Association, but still alarmingly common despite his valiant efforts to stamp it out. Faced with this objection, George insists (correctly, I think) that we always have a perfect right to question whether the axioms of our currently best or most widely accepted theory of sets are true:

> "Whatever their strength or source may be, the plausibility considerations about how many things there are that conflict with various theorems of ZFC have as much right to be considered a part of "our conceptual scheme" as does ZFC. ZFC conflicts with certain intuitions about cardinality that we happen to have; those intuitions form part of a fragmentary, inchoate, rival theory."[124]

I believe this is just the right response. But how does it connect with the grounds for refusing to believe in κ that George offered in the passages previously cited? Does the other construal I mentioned above -- that evidence in favor of what axioms we each accept stems from (or expresses) some "picture" of sets such as is provided by some version of "the iterative conception" work any better? George himself confesses to being *unable to believe* that there are more sets than would be required by *his* account of the iterative conception:

> "...I incline -- *for whatever accident of psychology* -- to find principles of set theory acceptable if they can be "read off" *that* presentation."[127, first emphasis added]

Autobiography again.

As I have been suggesting, I find it very difficult to believe that George was not sensitive to the juxtaposition of the arguments that he uses, on the one hand to combat anti-realist construals of σ and insist on his own "common-sense" picture of what we *mean* by σ -- what would make σ true if it *were* true, and how it might well be false, regardless of what axioms we may accept; and on the other, in defense of his acceptance of the axioms that he does accept and of his reasons for balking when he sees that they imply the existence of κ.

So, here is my problem: I share George's view of the meaning of σ, as well as of the other sentences of the theory of which it forms a part. I can also put myself in the position of someone who believes she understands the sentences of this theory (perhaps up to some degree of complexity) and who has intuitions, at least to a limited extent, regarding which are true and which must be false. Part of this involves thinking it is possible to make sense of plausibility considerations of the sort George offers when he explains which *he* believes as he criticizes Gödel's reasons for urging hyperprofligate hyper-iteration on us. It is my inclination to accept both of these things -- and, because I cannot make them mesh, to be appalled that I do so -- that in turn makes me wonder if George might not have seen that and, consequently, chosen to string us along by exhibiting these difficulties so that we may see them too, but in a novel way, by taking them on and seeing them as *our* difficulties.

Personally, I am a rank amateur and have no special license that entitles me to wonder if ZFC might not be too generous after all -- other than that George seemed to think so, and I respect George's instincts about such matters far more than I do my own. Moreover, there are (reputedly) even greater skeptics, even among the card-carrying set-theorists: As Geoffrey Hellman reports in a letter "Silver for years has claimed that it might be possible to derive a formal contradiction in ZF!" If it is indeed possible to derive a formal contradiction in ZF, then ZFC is certainly too generous. But I take George *not* to have had any such doubts of his own, and hence not to be exhibiting a concern fueled by them. Indeed, there is a perfectly plausible reading of his paper according to which finding ZFC was inconsistent after all *wouldn't even settle the question of κ's existence* -- κ might exist anyhow, although the assumptions we had chosen to represent the theory it inhabited were ill-chosen (I don't really mean that if it exists, κ *inhabits a theory* -- that's just a figure of speech). In any case, I respect George's -- and Silver's -- instincts far more than I do my own.

But I also respect Gödel's and they are in direct conflict with George's. Listen to Gödel again, and to George commenting on Gödel. First Gödel (I skip a phrase here and there):

> ...evidently this procedure [of applying the "set of" operation to get the power set of the argument set] can be iterated beyond ω. So it may be required as the next axiom that the iteration is possible for *any* ordinal... But are we at an end now? By no means. For we have a new operation of forming sets, namely forming a set out of some initial set A and some well-ordered set B by applying the operation "set of" to A as many times as ... B indicates. And, setting B equal to some well ordering of A we can now iterate this new operation, and again iterate it into the transfinite. This will give rise to a new operation again, which we can treat in the same way, and so on. So the next step will be to require that *any* operation producing sets out of sets can be iterated up to any ordinal number ... But are we at an end now? No, ..." [131-2]

Maybe not; George tired long before that. But do not be fooled; the target of his next remarks is less Gödel's picture of the set-theoretic universe than it is his *epistemological* pronouncement that new axioms "force themselves upon us as being true." George asks

"Does this view of how matters are with respect to sets really *force itself upon us as being true*? Do we find that, on reflection, we are unable to deny in our heart of hearts that matters must be as Gödel described them? Or do we suspect that, however it may have been at the beginning of the story, by the time we have come this far the wheels are spinning and we are no longer listening to a description of anything that is the case?"[132, his emphasis]

It is clear where George stands; but is it obvious what (other than freedom of opinion) entitles him to stand there? Moreover, how are we to adjudicate this dispute? Clearly, Gödel believes that iteration is a *good thing*, and that as long as it *can* be carried out, as long as some process of composing sets from "previously" obtained material can be consistently described, it "generates" (*i.e.* leads to) only sets that there must be -- and hence that must be there. And even this may fail to get all of them. But the ones it "reaches" are "there" to be reached. Since George finds even Replacement repugnant -- or unbelievable -- he gets off this trolley rather early. Compare them both with Shoenfield, whose critical condition -- that, given a set x and for each element y of x, a stage S_y, then there is a stage S which is later than all of the S_y and to which they all belong – evidences his concern to expound and motivate ZFC, not necessarily for what may lie beyond the outer banks, which, in the face of the (then only suspected) independence results, was surely an important part of *Gödel's* concern.

Recall Skolem's account in 1922[19] of the initial rationale for adding Replacement to Z (Zermelo's initial theory). He remarked that, although the axioms of Zermelo's theory were sufficient to establish the existence of each of the following sets, ω, P(ω), P(P(ω), P(P(P(ω))), etc... -- i.e., for each i, the result of the i-th iteration of the power set operation applied to ω -- they did not imply the existence of a set containing them all. So, in some sense, Z could yield up *each* of a bunch of things, all of them honestly come by -- "generated" in some sense in some uniform way from something we had started with (ω) -- but no (consistent) principle yet laid down implied that there was a set to which they *all* belonged. Still, wasn't it obvious that there *was* such a set? Suppose we tried to explain what it is to be a set, as follows: Whenever you have some things that "hang together" in some way,[20] there is a set of exactly those things, provided, of course, that one avoided illicit closure. Does being able (i.e. having the proof-theoretic means) to show of each member in the above sequence of power sets that it exists constitute "having" *that* thing in some relevant sense? Although, in the present context it is evidently contentious to suppose that proving some appropriate theorem of ZFC *per se* constitutes a proof of the existence of anything implied by that theorem to exist, let us put aside that concern and agree for the sake of argument that it does.

Does it follow from the fact that we came (or could come) by each of them honestly that they hang together in the appropriate way -- which is what would be needed to justify application of the plural principle above? Surely not. Sticking to the temporal metaphor, even if there is for each, theoretically, a time or point at which you have *it*, there may be no time (or point) at which you have them all, *without appeal to some principle of collection just like the one we are trying to "justify."* Shoenfield's third principle is what gets him over this hump. It states simply that if, for some set x there is, for each member y of x, a stage S_y then there must be some stage T that is later than any stage in the collection of stages that are images of members of x, and that contains all of them. Fair enough; but it is also a "principle of collection" so close in its formulation to Replacement itself that it is hard to see how one could use it to substantiate or argue for Replacement. To be sure, it makes no explicit mention of a *function* whose domain is x and range T, but only because it manages to say the same thing without using the words 'function', 'domain' and 'range'. This makes it perfectly fine, as an account of the ideas underlying ZFC, but less fine if offered as an argument for the reasonableness or *evidence* of the axioms of ZFC. Of course, carelessly employed,[21] these principles skirt disaster: if (a) being possession of methods that can be used to "prove" the existence of each is enough for "having them all", in the relevant sense, and if (b) "having them all" in this sense is claimed to be sufficient warrant for proclaiming the existence of a set that contains all of the things thus "had," we are on a road that leads straight to the dreaded universal set.

Although it is plausible to regard my formulation ("...all of them "generated" in some uniform way...") as at least one possible rendition of Cantor's "totality of elements that can be bound up together by means of a law", Replacement is genuinely new and doesn't spring full-blown from *this* informal account; not, at least, from this version of it, though it does from Shoenfield's, and, of course, Gödel's. But, if Replacement isn't implied, by the same token, neither is Infinity, for it too presupposes what I have called a 'principle of collection.' It is plausible to suppose that if such a formulation "justified" Infinity, it would also yield, if not full Replacement, then at least portions of it.

But suppose even that we persuaded ourselves that some principle we had formulated *did* imply Replacement, what then? Recall the remark regarding stages that Shoenfield uses to argue for both the limit principle and his Third Principle, which he uses in turn to justify Replacement: namely, that there is a stage S' following all of the stages in some collection S of stages *provided we can imagine a situation in which all the stages in S have been completed.* Waiving any doubts about whether the situation in which all the stages in S have been completed is indeed one we can *imagine* (and hence one to which his informal account *applies*), are we then constrained to accept the axiom? I argued above that finding a principle on which it would be evident that Replacement held --- one that implied Replacement -- would surely *not* suffice, as it would leave unanswered the question of whether the sets that would exist if Replacement were true *did indeed exist* – for what in the story tells us that sets satisfy the principle? What does Shoenfield tell us, keeping in mind that this appears in a handbook of mathematical logic and not in a tract on the philosophy of mathematics?

> Ideally, an axiom system is formed as follows. First we select the basic concepts and explain their nature as fully as possible. Then we write down axioms for the concepts. *If all goes well, our explanation will make it clear that the axioms are true.* [J. Shoenfield, *op. cit.*, p. 322, emphasis added]

I find that at once compelling and deeply mysterious -- compelling because it seems just right as an account of what one does, mysterious because it merely pushes the burden back one level, onto the explanation of the basic concepts. For that account, if it is to serve to make it clear that the axioms are *true,* has to be not only *clear*, but also in some sense *correct,* especially if the axioms' being true is understood as we have been understanding it (with George). If, on the other hand, their being true amounted simply to "being true *to*" some given conception or other, the one set out in the explanation, say, regardless of whether or not there existed in "paradise" anything answering to that conception, there would be no mystery, or very little. But now the question naturally arises: Why be true to *that* conception, as opposed to any other, since the argument can be run in exact parallel for all? If skepticism regarding this latter sort of justification is not immediately persuasive, recall the centuries of squabbles over the ontological argument -- it is the very concept of God that demands that He exist.

If their being "true" is what it is thought by some to be, namely some complex modal fact, like being *semantically* consistent, then perhaps it is intelligible how the explanation might make it clear that the axioms are true (*tout court*).[22] Although that might well be accurate, we are still too far from a sufficiently revealing theory of logical possibility and validity, especially as it would apply in the present case, to make that a very satisfying answer.

So, must we believe in set theory? And what of κ? *Is* it unreasonably large? And what of these conceptions of set: Z, ZFC, G, or ones orthogonal to them that reject, e.g. Power Set? Which, if any is (are) "correct"? And how are we to choose among them? On one view, it is the (set-theoretic) universe that determines which is correct, and hence whether or not there are at least κ sets; however, that view tells us precious little about how *we* fit into the picture, about how we would ever know which it is, if we ever would. Perhaps we follow our instincts and hope. On the other, where we fit in is not a problem. Each conception is true to itself (assuming they are all at least consistent) and only a fool wonders about whether κ *really* exists. It exists-for-Gödel and exists-for-Shoenfield; it doesn't exist-for-George.[23] Each of these pictures of how it is with κ has its own attractions. Those who abhor mystery will be pulled in the latter direction. But the price they must pay for this "clarity" is an obligation to explain why we all use what appear to be the same logical and semantic structures to express our respective views about sets, especially if, despite appearances, we are not *really* disagreeing. The cost could be too steep.

On the other hand, those who are willing to tolerate the mystery of access to maintain (what they take to be) communication and the possibility of genuine disagreement will be pulled in the other direction; the price for them is to explain why we should believe what we find plausible, and reject what we do not. Those who (like myself) are pulled both ways, can be sure to have the worst of both worlds.

The reason why listening to George made the author of "Mathematical Truth" squirm is that [24] in that article, I brand holistic (and conventionalist, and related accounts) of truth in mathematics as defective – precisely because they don't help us understand why the feature they identify with truth is correctly so identified. They are motivated, I argue, by the [understandable] desire to replace truth in these domains with a property, such as Theoremhood-in-ZFC, whose presence we can detect. Except by bare assertion, they make no connection between truth and, e.g. theoremhood. But claiming won't make it so, or explain why it *is* so, if indeed it is.

I also argue, as I do here in passing, that the semantics for the languages in which these theories are normally couched should be taken to be the "standard" semantics for predicate calculi of order *n*, for whatever *n* is appropriate, at least until shown by cogent argument to be different. But a commitment of such an account is that individual sentences of formalized languages, e.g. for set theory, are but translations into the formal idiom of their Mathematese counterparts (cf. fn. 12 above), which is where mathematics is done for real. If the formal sentences (such as our σ) can be said to have truth values at all, it is only because they represent Mathematese sentences such as "The sequence $\langle \aleph_0, \aleph_{\aleph_0}, \aleph_{\aleph_{\aleph_0}}, \ldots \rangle$ has a limit; call it κ." Anyone who wonders or doubts whether there are as many as κ sets is, on this account, (1) treating it as a question about which we *can* wonder and, because it has a proof and we know that proof, (2) also committed to wondering or doubting whether the principles employed in that proof are true.

At this point, one is strongly tempted to respond that (*pace* George, Feferman, *et al.*) doubting the existence of κ is simply *bizarre*. What principled reason could one have for so doing – an intuitive, immediate, grasp of the set-theoretic hierarchy that shows it to be smaller than κ? That is even more bizarre. No; when in this mood, whether κ exists feels less like a real question, more like something calling for a *decision* than a judgment of truth and falsity. But that is just the sort of response I was at pains in MT to argue was unsatisfactory, because it turns truth into convention. For on such an account, ZFC becomes a theory that we could adopt or not according to whether or not it answered to our conception of set, not one that is true or false in the domain of sets. George, by deftly and dramatically handling the fine structure of the issues involved, leads one into the position where one is inclined to respond in this [conventionalist] way. In so arguing, he does two things: (1) he *presupposes* an account of the semantics of the language of set theory that I have argued is correct, and (2) exhibits, by making us feel them, the problems to which we are led, either on its supposition if we go along with his doubts, or in response, by making a number of "anti-realist" responses appear attractive in comparison. Hence my discomfiture.

Finally, what to think about George? Was he putting us on? I think it was only in a moment of weakness on my part that I thought he had sold out to the anti-realists but, unwilling to be up front about it, chose a devious means to advance their cause: to present "our" view at its worst, with its worst problems exposed for all to see. I think it was, instead, another example of his unfailing honesty and willingness to follow reason where it took him, however uncomfortable the journey may be from time to time. By elaborating a case in which it makes sense to doubt the consequences of our axioms, he reminds us that doubting the axioms makes perfect sense. This in turn shows both that they mean what they obviously mean, and that what grip we have on the theory cannot be *simply* through them – and does it more dramatically than one could through skepticism born either of independence results or philosophical "scruple".

That still leaves it mysterious just *what* that grip is; but that is nothing new.

I miss George, a lot, and always will.

What Mathematical Truth Could Not Be --II

Abstract

As the title suggests, the work from which this talk is drawn is a review of issues arising out of the author's What Numbers Could Not Be [*Philosophical Review* 1965] and Mathematical Truth [*Journal of Philosophy* 1973], but principally out of the latter. (a version of WMTCNB-*I* appears in *Benacerraf and his critics*, A. Morton and S. Stich, eds., Blackwells, 1996).

In a recent paper, "Must we believe in set theory?" (recently published in a collection of his papers, *Logic, logic and logic*, R. C. Jeffrey, Ed., with Introductions and Afterword by John Burgess, Harvard, 1998), George Boolos declares that according to ZFC there exists a cardinal κ, which is equal to $\aleph_\kappa$, and is the limit of the sequence $<\aleph_0, \aleph_{\aleph_0}, \aleph_{\aleph_{\aleph_0}}, \ldots>$. That is (if it exists) κ is the least ordinal greater than all of the f(i), where f(0)= $\aleph_0$ and f(i+1)=$\aleph_{f(i)}$. To go directly to the source: "...if κ exists, there are at least as many as κ sets. *Are* there so many sets?" Boolos asks. It is, of course, uncontroversial that ZFC is committed to the existence of κ, just as it is committed to the existence of ω. Boolos applauds the latter but seems to regard the former as an *over*commitment -- a commitment with which he, personally, cannot go along. Any theory committed to the existence of *as many as* κ things is simply not to be believed, on that ground alone. I take it that such a view is not uncontroversial -- indeed that it is not even pellucid what the view *comes to*. I try in this talk to examine the reasons Boolos offers for it and, through them, how we should interpret the view and how defensible it might be to hold it. My focus throughout is, of course, not the belief specifically in κ itself, but how this case illustrates certain tensions in our views of mathematical truth, meaning, and belief.

NOTES

[1] About the title: The present paper is part of a larger work that began when I was invited, for presentation at a conference in Munich in 1994 to revisit issues I had first discussed many years ago in "What Numbers Could Not Be" and "Mathematical Truth". My Munich talk barely made it through WNCNB, forcing a change in title from the projected "WMTCNB", *tout court*, to "WMTCNB-I"; the present paper is its sequel, or the start of its sequel, anyhow. "WMTCNB-I" began with a review of questions I had first explored in WNCNB -- mainly what, if anything, arithmetic might be *about*, and thus what, if anything made the truths of arithmetic *true*, and contained some additional remarks on the much narrower issue of the import and relevance of *meta*mathematical results in philosophical argument. Both bear on mathematical truth and what it could or could not be, but neither engages directly the matters of knowledge and belief in mathematics that had been one important focus of MT. I will not recap any of that here, but move on instead to revisit more directly some of the epistemic issues that are joined in MT. ("WMTCNB-I" was first published in *Benacerraf and His Critics*, A. Morton and S. Stich, eds., Blackwell's, Oxford, 1996, and has appeared, in a slightly different version, in the proceedings of the Munich conference, *The Philosophy of Mathematics Today*, Oxford University Press, Oxford 1998, edited by the conference's organizer, Matthias Schirn.

Although there is much that I should address in the vast literature that MT has spawned in the 25 or so years since its appearance, for today's talk, I can think of no more fitting way to revisit those issues than through the prism of a recent paper by George Boolos that makes no mention of it, but that disturbed me more than somewhat when I first heard it.

[2] I am grateful to the ASL for the opportunity to talk today, and to the British Logic Colloquium for its sponsorship of this lecture. An ancestor of this paper was written for a Symposium at MIT, Friday, March 21, 1997, in Memory of our late President, George Boolos, the other symposiasts being Charles Parsons and Sergei Artemov. I had known, loved, and admired George for close to forty years (he was an undergraduate in the first Philosophy of Mathematics course I ever taught). I thank the organizers for having included me, many at the session for lively discussion from the floor, and especially Vann McGee and Geoffrey Hellman for their subsequent very thoughtful written comments. Later versions were read at the UCLA Philosophy of Mathematics Seminar and at Pomona College, where I also received valuable comments. Naturally, I remain responsible for the errors that remain.

[3] Thanks to Gideon Rosen for identifying the source, and to Lisa Eckstrom for the information that the Greek is 'enigma'.

[4] Written as a contribution to a splendid volume in honor of Charles Parsons, this has already appeared in a recent posthumous collection of George's writings entitled "Logic, Logic, and Logic", R.C. Jeffrey, ed., with Introductions and Afterword by John Burgess, Harvard, 1998. References will be to this printing.

[5] Gödel, of course. "What is Cantor's Continuum Problem?", reprinted in *Philosophy of Mathematics: Selected Readings*, P. Benacerraf and H. Putnam, eds., 2nd edition, Cambridge University Press, 1983, pp. 474-5; future references to this as "B&P".

[6] "Axioms of Set Theory", by J. R. Shoenfield, in *Handbook of Mathematical Logic*, J. Barwise, ed., North Holland Publishing Company, Amsterdam and New York, 1977, pp. 321-70.

[7] "Iteration Again," *Philosophical Topics* 17 (2), Fall 1989, pp. 5-21.

[8] It takes a bit of psychic vision to see that Gödel outruns Shoenfield. It is surely meant to be more generous than ZFC, since he is describing a procedure for coming up with "new" axioms in a context in which he has reason to believe that ZFC necessarily will fall short of the mark. But it is not obvious to me from its description that it actually does outstrip Shoenfield, i.e. that what he says cannot be accommodated by Shoenfield's third principle. But nor is it clear that it *can*. Unless it is thought that there is some inherent reason why a more generous conception than Shoenfield's cannot be *described*, let us assume that this passage from Gödel's does just that. As I said above, the details won't matter much today.

[9] As I indicated above, others must also reject it, in particular Feferman, since he rejects Power Set. And presumably Quine, although his systems are very hard to compare with any more "standard" systems.

[10] After all, there is precedent, in the "limitation of size" principle, according to which the lesson to be learned from the paradoxes was that there was a single forbidden *size* that all forbidden collections (the universal"set", the "set" of all ordinals, the "set" of all cardinals", etc...) could be shown to share.

[11] Hilbert, of course. "On the infinite", reprinted in B&P, p. 191.

[12] I fully agree, of course. But I would not have *put* the matter quite that way myself. I take the claim that there are (at least) κ objects in the universe to be a statement in *Mathematese* -- that extension of English, French, Russian, German,

---

Japanese, Chinese, Hebrew, etc., in which mathematicians communicate with one another. George's -- ZFC's, actually --formal sentence σ is *at best* a sentence we constructed to bear the interpretation "there is a cardinal that is the smallest cardinal larger than every cardinal in the sequence f(0), f(1), ..., f(17), ...etc.," as defined in the text above. Putting it George's way unnecessarily invites a confusion regarding what σ (the *formal* sentence) could possibly *mean*, and in particular, by a now all too familiar argument, whether it could *possibly* be saying that there are more than $\aleph_0$ things. This is a confusion to which George is evidently not subject, or else he wouldn't be concerned that σ, and with it, ZFC, grossly overestimated the size of the universe. Whatever the merits of that argument [and here I have in mind the collection of Skolemite arguments that deny, at least to sentences of ZFC and hence to σ, sufficient expressive power to enable them to *over*-commit themselves], it is one that may be at least deferred, if not avoided altogether, by putting the question from the start as one that arises in Mathematese, and hence in the *in*formal theory that formalized ZFC is meant to represent. For a longer and more tedious treatment of these matters, please see WMTCNB-I.

[13] Never one to sweep contrary evidence under the rug, George tells of "J.", a set-theorist who, frustrated him by steadfastly refusing to back down or hedge when challenged on whether he really believed σ -- Would he, for example, *say to an inquiring child*, matter-of-fact-ly, that yes, *there really is* such a number as κ. "I have no problem with that, George," J. is quoted as saying. But George was undaunted, even by J., as he goes on

> I did have the impression that J. was perhaps not entirely speaking *in propria persona*, but rather was making an announcement, as from the standpoint of a set theorist. I also had the sense that nothing I could think of to say could dislodge J. from that standpoint. (But maybe J. really does believe that there are κ and many more sets in existence.) [130]

[14] And I agree. Because I thought we would do the same if challenged with regard to the null set or ω, I was reluctant some paragraphs back to go along with George on what conclusions to draw from the fact that, if challenged, we would hedge our bets with regard to the existence of κ. But that was an *empirical* claim, based on what I think most would do under those circumstances, not on what thought it would be right to do. Here, I side with George -- with regard to the existence of the null set or ω: Though, as I suggested, we *might* balk at the null set or ω, we *should* not. But that we *might* also balk at the null set or ω

surely undercuts the import of our reluctance to stand up for κ.

[15] Here I must enter a caveat of my own; clarity is not itself a very clear notion, and we must use it sparingly. If our inability to resolve questions about the height or width of the cumulative hiearchy is already a symptom of some unclarity in the notion of set, then there is reason to believe that it is, to that extent, unclear. Although we may not perceive any localized muddiness right now it could well be that something will emerge from future research that it would be appropriate to label as resolving a heretofore unsuspected unclarity in the notion of subset (and hence of set, used in several of the axioms), or of function (used in Replacement).

[16] That the universe of sets just *is* the cumulative hierarchy -- that all the sets there are are so arranged -- is presumably a substantive question, whose answer should not be *assumed* simply to be "Yes" in the context of the evaluation of axioms. Although it might be "analytic" that if there are any sets at all, some sub-portion of them are suitably arranged as in the hierarchy, it is by no means so that sets on every permissible conception satisfy it; so if there is a question of mathematical "fact" about which, if any, of the permissible conceptions the sets *actually* satisfy, it is not to be settled simply by comparing the sets that exist *on a given conception* with "the hierarchy," or "the hierarchy of sets less than some given rank" (however such a "comparison" might be carried out).

[17] Gertrude Stein is alleged to have said [about Oakland, CA, although it is often said that she said it of Los Angeles]: "When you get there, there is no there there." It is difficult to resist wondering whether the remark wouldn't be even more apt here.

[18] Forget that it is consistent to suppose that the cardinality of the reals is $\aleph_{(\kappa+1)}$, and hence that if, as it is commonly supposed, we "need" the reals, we may "need" κ as well. If something like that were really so, then κ wouldn't be so "remote from the requirements of science and the rest of mathematics" after all. (But if there are really *so* many reals, *more than κ* of them, perhaps we wouldn't "need" *all* of them after all, and we are back to square one.)

[19] In "Axiomatized Set Theory", in *From Frege to Gödel*, J. van Heijenoort, ed., Harvard, 1967, pp. 290-301, esp. pp. 297-8.

[20] Compare Cantor: "By a manifold or aggregate I understand generally any multiplicity which can be thought of as one (jedes Viele, welches sich als Eines denken lasst), that is to say, any totality of definite elements that can be bound up together by means of a law." quoted in Jourdain's Introduction to Cantor's *Contributions to the Founding of the Theory of Transfinite Numbers*, P. E. B.

Jourdain, trans., Dover, 1915, p. 54.

[21] What I have in mind by "carelessly employed" is treating it as a principle that reflexively closes the universe with self-applications – just the thing, of course, that "iterative" conceptions paradigmatically avoid. So, nothing to fear there from an iterativist. There is no harm in supposing that any "antecedently given" objects you like can be collected into a set, including that the domain of any set of axioms for set theory (on a given interpretation) forms a set, provided that is supposed from a respectful distance, e.g. from the standpoint of some appropriate other theory.

[22] Something Geoffrey Hellman urged in a letter to me (but that is set out in detail in *Mathematics Without Numbers: towards a modal-structural interpretation*, Oxford, Clarendon Press; NewYork, Oxford University Press, 1989).

[23] That, of course, is an oversimplification, since on this view there is no reason to think that 'κ' has a constant denotation across all of the contexts in which it is being used. So there is no contextually determinate "it" to serve as the anaphoric referent of all the uses of 'it' in that sentence.

[24] I thank an anonymous referee for numerous corrections, and for the thoughtful suggestion that I make clearer and more explicit the connection between what I say in this paper and MT. These next few paragraphs are written in response to that suggestion.

Proof Search in Constructive Logics *

Roy Dyckhoff
Computer Science Division, University of St Andrews, Scotland
email: rd@dcs.st-and.ac.uk

Luís Pinto
Departamento de Matemática, Universidade do Minho, Portugal
email: luis@math.uminho.pt

Abstract

We present an overview of some sequent calculi organised not for "theorem-proving" but for proof search, where the proofs themselves (and the avoidance of known proofs on backtracking) are objects of interest. The main calculus discussed is that of Herbelin [1994] for intuitionistic logic, which extends methods used in hereditary Harrop logic programming; we give a brief discussion of some similar calculi for other logics. We also point to some related work on permutations in intuitionistic Gentzen sequent calculi that clarifies the relationship between such calculi and natural deduction.

1 Introduction

It is widely held that ordinary logic programming is based on classical logic, with a Tarski-style semantics (answering questions "What judgments are provable?") rather than a Heyting-style semantics (answering questions like "What are the proofs, if any, of each judgment?"). If one adopts the latter style (equivalently, the BHK interpretation: see [35] for details) by regarding proofs as answers to questions, or as solutions to problems, then proof-enumeration rather than theorem-proving is the issue. See [12] for discussion of differences between the two styles of semantics.

*Both authors were supported by the European Commission via the ESPRIT BRA 7232 GENTZEN and by the Centro de Matemática da Universidade do Minho, Portugal.

Some authors (e.g. [25]) have shown that as an adequate basis for pure Prolog one can, instead of classical logic, take the Horn fragment of minimal logic, and that this can be extended up to the hereditary Harrop fragment of minimal (or, equivalently, intuitionistic) logic, thus providing [24] a good logical basis for software engineering features such as scoping and modularity. In such work, the emphasis is on provability: the semantics is Tarski-style rather than Heyting-style. The uniform proof system of [25], a Gentzen-style sequent calculus with side-conditions restricting the applicability of certain rules, is presented merely as an efficient mechanism for answering questions about provability rather than as a mechanism for enumerating proofs.

Our purpose in this paper is to argue that one should go further: that the semantics of such languages should be Heyting-style, and that the appropriate proof-search calculi are those that have not only the syntax-directed features of Gentzen-style sequent calculi but also a natural 1-1 correspondence between the derivations and the real objects of interest, normal natural deductions.

Such calculi are to be found in the work of Howard [18] and of Herbelin [15], [16]. Howard's idea (attributed to Curry) is close to the terminology of logic programming, but does not generalise beyond the $(\wedge, \supset, \forall)$-fragment of minimal logic. Herbelin's work was motivated by the application to functional programming, replacing the ordinary typed λ-calculus by a new calculus $\overline{\lambda}$ of terms representing derivations in a sequent calculus, where the cut-elimination rules form a clean and simple calculus of explicit substitutions. The same calculus (without the *Cut* rule) may, we observe, be used effectively as a proof-search calculus.

In this paper we look at such calculi for some constructive logics, where a logic is (loosely) regarded as *constructive* if the disjunction and existence properties hold for it. As is well-known, classical logic is not constructive in this sense: disjunctive formulae $p \vee \neg p$ are provable without either of the literals p and $\neg p$ being provable, and some existential formulae $\exists x.U(x)$ (such as $\exists x.\forall y.(px \supset py)$) are provable without $U(t)$ being provable for any particular t. Another aspect of constructivity that we consider is the emphasis on proofs (as "constructions") rather than just on provability. Extension of the ideas to (for example) classical logic is an interesting challenge, explored in [15]: one of the problems is that of agreeing on a suitable proof system to provide the "semantics". We choose not to discuss whether classical logic is "constructive" as argued in [13].

We consider therefore (first-order) minimal logic, intuitionistic logic, an intuitionistic modal logic and intuitionistic linear logic. Elsewhere [29] we report on work extending these ideas to a dependent type theory based on that of [32].

The calculi that we consider to be good candidates for proof search have derivations in 1-1 correspondence with the normal natural deductions. The usual calculi of Gentzen and Kleene do have derivations interpretable as normal natural deductions, but the interpretation is many-one. We show elsewhere [5], in the case of one calculus (roughly that of Kleene) for intuitionistic logic, how the fibres of this interpretation are generated by the Kleene-style permutations. In other words, the permutabilities in (cut-free) LJ are an obstacle to the use of LJ as a proof-search calculus. For this reason, we have in [4, 5, 6] described the new proof-search calculi as "permutation-free": any "permutation" in such calculi would be dismissed as semantically unsound. Note that in contrast it is the non-permutabilities of a calculus that are obstacles ([36], p. 50) to its use for efficient automated theorem proving.

Further, in the case of disjunction, there are (see [5]) Kleene-style permutations in LJ that are not semantically sound, according to the usual equational theory of the typed lambda calculus. This is a second feature of LJ that makes it a poor proof search calculus.

Proofs of all results mentioned here appear elsewhere, mainly in [4, 5, 6, 19, 20, 29].

2 Background

Notations are as in [34], except that we use $A \supset B$ for implicational formulae. $\bot$ is a logical constant and not an atomic formula. $\neg A$ abbreviates $A \supset \bot$.

A formula is *Horn* when it is the closure of a formula of the form $B \supset H$, where the *head* H is atomic and the *body* B is a (possibly trivial) conjunction of atoms; when B is trivial we replace it by *true*, and $true \supset H$ we replace by H. A formula is *hereditary Harrop* (resp., an *hH goal*) when every occurrence therein of either a disjunction or an existential subformula is negative (resp., positive): cf. [24, 25] for details, but note that the definition therein differs inessentially from ours by prohibiting formulae such as $p_1 \supset (p_2 \wedge p_3)$ and $\bot$ (and in being less memorable).

There are several ways [27] to extend a pure logical calculus with a theory given by axioms. One way [12, 14], assuming that the axioms are Horn formulae, is to interpret them as new inference rules, as reflected in the traditional logic programming term "rules". We adopt instead the approach of Gentzen, where a *theory* is a list Δ of closed formulae: the *achievement* of a *goal* G (where G is any closed formula) w.r.t. such a theory is just a proof of the sequent $\Delta \Rightarrow G$. We shall also call a theory a *logic program*. The (Heyting-style) semantics of a logic program Δ is then just the association to each goal G of the set of proofs of $\Delta \Rightarrow G$.

As the proofs, we consider as primary the normal natural deductions, for the simple reasons that (i) (under the Curry-Howard correspondence) such proofs correspond to values in the sense of functional programming, (ii) natural deductions are well understood and (iii) they lack the redundancy of traditional Gentzen-style sequent calculi (arising from the permutations [21] therein). We consider also the expanded normal deductions of [30].

"Normal" is defined as in [34]. Note that this restricts application of $\bot E$ to cases where the conclusion is atomic. $\bot$ in fact causes many problems. Our guiding principle here is that there should be exactly one normal natural deduction proof of $\top =_{def} \bot\supset\bot$; there are two possible candidates:

$$\frac{[\bot]}{\bot\supset\bot}\supset I \qquad\qquad \frac{\dfrac{[\bot]}{\bot}\,\bot E}{\bot\supset\bot}\supset I$$

of which we choose the first, both for simplicity and because $\bot$ is not an atom and so we may reject the second. A consequence of this choice is that if we try to restrict sequent calculus axioms $\Delta, A \Rightarrow A$ in LJ to atomic A, then we have also to allow their use when $A = \bot$.

The cut-free calculus LJ of Gentzen [11] (hereafter "G1i", as in [34]) is a starting point for automated proof search: proof search in G1i is "syntax-directed", in the sense that active formulae are always the same as, or immediate subformulae of, the principal formula, and G1i derivations can be interpreted as normal NJ proofs. The variations G2i and G3i, incorporating the structural rules into the logical rules, are even better [34] for this purpose. However, the permutations mentioned above impose an undesired redundancy on backtracking: the same NJ proof may be rediscovered several, possibly infinitely many, times. See [5] for a discussion of these permutations, with permutation reductions and a normalisation argument (extended to strong normalisation for a related system in [33]). The present paper therefore looks at calculi without these permutability properties.

3 Herbelin's calculus

Motivated by interest in obtaining a calculus of explicit substitutions based on sequent calculus, Herbelin has introduced a new λ-calculus of terms and a type system for it that we can regard as a deduction system with proof terms. Rather than use his name "LJT" (already used for a system introduced in our [3]), we use another name: we call our modification of his system "MJ", since it is intermediate between LJ and NJ.

The *formulae* $A, B, ..., G, ...$ of MJ are as usual. The two categories M and Ms of *terms* are described by the two grammars

$$M ::= (V; Ms) \mid \lambda V.M \mid in_i(M) \mid pair(M, M) \mid \lambda W.M \mid pairq(T, M)$$
$$Ms ::= ax \mid ae \mid (M :: Ms) \mid when(V.M, V.M) \mid p_i(Ms) \mid apq(T, Ms) \mid spl(W.V.M$$

in which $i = 1, 2$; V is the category of (proof) variables $x, y, \ldots$, W is the category of (individual) variables $u, v, w, \ldots$ and T is the usual category of terms $t, \ldots$ built from W by means of some function symbols. Variable binding occurs at occurrences of $V.M$, $W.M$ and $W.V.M$, with the usual conventions.

Contexts Δ are partial functions from V to formulae, written (in arbitrary order) as a sequence of *declarations* $x : A$. There are two forms of sequent: the forms $\Delta \Rightarrow M : G$ and $\Delta \xrightarrow{A} Ms : G$, respectively called "ordinary sequents" and "stoup sequents". The position above the arrow in the second form of sequent is called the "stoup". The terms of the calculus admit a natural 1-1 onto translation to the terms of the typed lambda calculus (with types corresponding to the formulae of first-order logic). Details of this translation can be found in [6], including checks that it works not just at the level of terms but also of inference rules.

The *axioms* for this calculus are of the form $\Delta \xrightarrow{A} ax : A$. The inference rules, in which R is for "Right", S is for "Stoup" and *Sel* is for "Selection" (from the context into the stoup), are as follows:

$$\frac{\Delta, x : A \xrightarrow{A} Ms : B}{\Delta, x : A \Rightarrow (x; Ms) : B}\ Sel \qquad \frac{}{\Delta \xrightarrow{\perp} ae : C}\ S\perp$$

$$\frac{\Delta, x : A \Rightarrow M : B}{\Delta \Rightarrow \lambda x.M : A \supset B}\ R\supset \qquad \frac{\Delta \Rightarrow M : A \quad \Delta \xrightarrow{B} Ms : C}{\Delta \xrightarrow{A \supset B} (M :: Ms) : C}\ S\supset$$

$$\frac{\Delta \Rightarrow M : A_i}{\Delta \Rightarrow in_i(M) : A_1 \vee A_2}\ R\vee_i \qquad \frac{\Delta, x : A \Rightarrow M : C \quad \Delta, y : B \Rightarrow M' : C}{\Delta \xrightarrow{A \vee B} when(x.M, y.M') : C}\ S\vee$$

$$\frac{\Delta \Rightarrow M : A \quad \Delta \Rightarrow M' : B}{\Delta \Rightarrow pair(M, M') : A \wedge B}\ R\wedge \qquad \frac{\Delta \xrightarrow{A_i} Ms : C}{\Delta \xrightarrow{A_1 \wedge A_2} p_i(Ms) : C}\ S\wedge_i$$

$$\frac{\Delta \Rightarrow M : U(w)}{\Delta \Rightarrow \lambda w.M : \forall U}\ R\forall \qquad \frac{\Delta \xrightarrow{U(t)} Ms : C}{\Delta \xrightarrow{\forall U} apq(t, Ms) : C}\ S\forall$$

$$\frac{\Delta \Rightarrow M : U(t)}{\Delta \Rightarrow pairq(t, M) : \exists U}\ R\exists \qquad \frac{\Delta, x : U(w) \Rightarrow M : C}{\Delta \xrightarrow{\exists U} spl(w.x.M) : C}\ S\exists$$

in which the usual side conditions are imposed by the notation (e.g. that w is new in $R\forall$) and U stands for an abstraction of a formula w.r.t. an individual variable, so $U(t)$ is a formula for any term t. (U is for "unsaturated", as used by Frege.) Note that the *Sel* rule corresponds to the contraction rule of LJ.

In root-first proof search we shall say that a rule is *applicable* when the conclusion of one of its instances matches the current sequent.

4 Non-determinism in Herbelin's calculus

MJ has clear sources of non-determinism. Consider root-first search for a proof of a sequent $\Delta \Rightarrow M : G$ (where M is yet unknown). At most two rules are applicable: the *Sel* rule for selecting a member $x : A$ of the context (and copying it into the stoup), and the rule that introduces the principal connective of G, if any. Similarly, in searching for a proof of a sequent $\Delta \xrightarrow{A} Ms : B$, where Ms is unknown, we only consider the formula A. If this is an implication or disjunction or an atom or absurdity ($\perp$) or existential, we have no choice; if a conjunction, then we must choose a conjunct; if a universal formula, then we must choose an instantiating term t. (This choice of t can be delayed, using the usual technique of unification proposed by Herbrand and developed by Prawitz [31].) Thus the only non-determinism is associated with the *Sel* rule, the $S\wedge$ rules and (less seriously) the $S\forall$ rule.

The use of stoup sequents is thus a form of *focusing*, as introduced by Andreoli [1]. However, note that some rules have an ordinary sequent as premise and a stoup-sequent as conclusion, thus allowing a transition back (during root-first proof search) to ordinary sequents and abandonment of the focus on the stoup formula. These rules are those dealing with (in the stoup) an implication, a disjunction or an existential formula.

The first (implication) is not too problematic, since the $S\supset$ rule involves (as we move from conclusion to left premise) a change in the goal. This is exactly what happens in logic programming when a program clause $B\supset H$ is selected, with H matching the current goal and the new goal being (an instantiation of) B. B is often in practice organised so that the search terminates. The matching of H with the goal corresponds to the right premise of $S\supset$ being an axiom.

The possibility of the stoup formula being a disjunction, however, is more inconvenient: there is no corresponding change of goal, and the effect is to add a new declaration $x : A$ to the context, followed by the selection of a declaration from the context, perhaps even that which led to the disjunctive stoup formula. Similar remarks apply to existential formulae.

5 Herbelin's calculus and logic programming

The main form of non-determinism that we can avoid is the alternation between stoup-sequents and ordinary sequents arising from the presence of disjunctions or existential formulae in the stoup. Such formula occurrences can be excluded by restrictions to hereditary Harrop formulae in the program and to hH goals as goals. Such restrictions are presented in [25] as a means to allow the search to be *goal-directed*, in the sense that the only rules that may be applied when the goal is compound are right rules: but even without the goal-directedness there are good reasons, just discussed, for these restrictions.

One further way of reducing the non-determinism (for an ordinary sequent) is to (try to) require that *Sel* only applies when the succedent is atomic. We could consider what restrictions are required in order not to affect the derivability of sequents $\Delta \Rightarrow M : A$ (but allowing M to change): but the essence of our approach is to consider first what class of normal deductions we are interested in and then what sequent calculus like MJ corresponds to it in a bijective fashion.

To this end, we may further require that the minimal formulae (of normal natural deductions) are atomic (or $\perp$); in this case we have the expanded normal form deductions of Prawitz [30] (equivalently, proof terms in $\beta\eta$-long normal form).

If we restrict use of axioms in MJ to the matching of atomic formulae (or of $\perp$), and the use of $S\perp$ to cases where the goal formula is atomic, and syntactically restrict the programs Δ to hereditary Harrop formulae and goals G to hH goals, then we may restrict uses of *Sel* to cases with atomic succedent. We now impose these restrictions. Let P range over atomic formulae. The proof system MJ^r, with axioms $\Delta \xrightarrow{P} ax : P$ and $\Delta \xrightarrow{\perp} ax : \perp$, is then:

$$\frac{\Delta, x : A \xrightarrow{A} Ms : P}{\Delta, x : A \Rightarrow (x; Ms) : P}\ Sel \qquad \frac{}{\Delta \xrightarrow{\perp} ae : P}\ S\perp$$

$$\frac{\Delta, x : A \Rightarrow M : B}{\Delta \Rightarrow \lambda x.M : A \supset B}\ R\supset \qquad \frac{\Delta \Rightarrow M : A \quad \Delta \xrightarrow{B} Ms : P}{\Delta \xrightarrow{A \supset B} (M :: Ms) : P}\ S\supset$$

$$\frac{\Delta \Rightarrow M : A_i}{\Delta \Rightarrow in_i(M) : A_1 \vee A_2}\ R\vee_i$$

$$\frac{\Delta \Rightarrow M : A \quad \Delta \Rightarrow M' : B}{\Delta \Rightarrow pair(M, M') : A \wedge B}\ R\wedge \qquad \frac{\Delta \xrightarrow{A_i} Ms : P}{\Delta \xrightarrow{A_1 \wedge A_2} p_i(Ms) : P}\ S\wedge_i$$

$$\frac{\Delta \Rightarrow M : U(w)}{\Delta \Rightarrow \lambda w.M : \forall U}\ R\forall \qquad \frac{\Delta \xrightarrow{U(t)} Ms : P}{\Delta \xrightarrow{\forall U} apq(t, Ms) : P}\ S\forall$$

$$\frac{\Delta \Rightarrow M : U(t)}{\Delta \Rightarrow pairq(t, M) : \exists U}\ R\exists$$

We thus achieve (in another notation) a slight generalisation of the uniform proof system of [25]. Proofs of hH goal formulae in MJ^r naturally correspond, in a 1-1 fashion, to the expanded normal form deductions. As a corollary we get (for hereditary Harrop logic) a 1-1 correspondence [7] between uniform proofs with back-chaining [24] and expanded normal natural deductions.

Search for a proof of $\Delta \Rightarrow M : G$ in this calculus (M being unknown) is goal-directed: it decomposes the goal G until it is atomic (or $\perp$), then selects a declaration $x : A$ from the program Δ and copies A into the stoup. A may be compound: but following now always the rightmost branch (in the case that the stoup formula is an implication) will either reduce it to $\perp$ (in which case we apply the rule $S\perp$) or reduce it to an atom, whose non-matching with the atomic (or $\perp$) goal would force backtracking to the last choice point. Such a point will either be where we chose $S\wedge_1$ rather than $S\wedge_2$ (or *vice versa*) or where we selected wrongly from the program into the stoup. Some pre-processing can of course speed this up even more, e.g. by arranging that formulae $A\supset(B\supset C)$ are replaced by $(A \wedge B)\supset C$ and that quantifiers appear inside rather than outside conjunctions; our point however is that the essence of the uniform proof system is hidden by such optimisations and lies in the (restricted) version of Herbelin's calculus just given.

This description of the proof search assumes that we search depth-first, but with minor modifications it also applies to breadth-first search.

Note that where a goal is existential (as goals implicitly are in, say, Prolog) any proof of the goal, say $\exists x.U(x)$, will end with an $R\exists$ step with the proof term $pairq(t, M)$ from which the "answer substitution" of t for x is trivially extracted. A similar extraction may be done where there are several existential quantifiers. It is an implementational rather than a logical issue how much of the rest of the proof term is made apparent to the user.

6 Semantics

Implicit in the above explanation is the assumption that normal natural deductions give the semantics, in contrast to the usual views that the semantics is Tarski-style (using minimal Herbrand models) and that the automatic method of "resolution" is the best way to answer questions about provability. One may see the development of resolution as one way to automate certain kinds of reasoning efficiently: this has led both to a view that automated reasoning *was* resolution (with various strategies and parameter adjustments to obtain efficiency for hard problems) and to a view that resolution and logic programming were connected. Resolution being classically based (with its conversions to CNF and negation of the goal), the constructive nature of logic programming was hidden and thus ignored by many (but not [25]). However, the operational semantics incorporated in Prolog interpreters actually gives answers with multiplicities that reflect the above proof system rather than the Tarski-style semantics. (Note that our system doesn't capture the order of solutions: nor does it capture the more subtle aspects of depth-first search: see [8] for solutions to this kind of problem.)

For example, the restricted version MJ^r of MJ provides answers not just about what but how many times and why, just as natural deduction does. Consider the classical problem "Is there a member of the list $[1,2,1,2]$?" in the context of the usual definition Δ (below) of *member* as a predicate relating items and lists. Classically, the answer is "Yes": a more detailed answer, in the classical version of constructivity, would be "Yes: 1 and 2." Even more detailed would be the answer "Yes, the members are 1 (twice) and 2 (twice)." Better still (but we don't achieve this in MJ^r) would be "Yes: the members are 1, 2, 1 (again) and 2 (again), in that order." MJ^r actually has the terms

$$\begin{array}{l} pairq(1,(m_1; apq(1, apq([2,1,2], ax)))). \\ pairq(2,(m_2; apq(2, apq([2,1,2], apq(1,(m_1; apq(2, apq([1,2], ax)) :: ax)))))). \\ pairq(1, ...). \\ pairq(2, ...). \end{array}$$

as proof-terms M for which $\Delta \Rightarrow M : \exists x.member(x,[1,2,1,2])$ is derivable, where Δ is the program

$$\begin{array}{l} m_1 : \forall y.\forall z.member(y, cons(y,z)). \\ m_2 : \forall y.\forall z.\forall w.(member(y,z) \supset member(y, cons(w,z))). \end{array}$$

and $[1,2,1,2]$ abbreviates $cons(1, cons(2, cons(1, cons(2, nil))))$ as usual. (The reader is invited to work out the missing terms and to apply the translation [6] into the standard natural deduction or lambda calculus terminology.)

The point here is not that it is useful to have the four proof terms, the traces of the computations, in full detail, but just that they are different. (However, extension of these ideas [29] to a more complex logic based on dependent type theory takes more account of the actual values.) Nor is it our point that the notation of MJ is the best way of presenting the terms: it is not as familiar as ordinary lambda notation, for example. We consider however that it is proof-theoretically attractive, to incorporate restrictions into the rules rather than into strategies for using the rules, and thus to have the calculus as an explicit search calculus.

7 Proof search in lax logic

Curry introduced in 1952 an intuitionistic modal logic—lax logic (LL)—recently rediscovered by several authors, with applications in hardware design [9, 23], constraint logic programming [10]; it is also [2] the type system CL of Moggi's computational lambda calculus [26]. It is essentially intuitionistic logic with a modal operator $\circ$ (read "somehow"), axiomatised by the axioms $A \supset (\circ A)$, $(\circ \circ A) \supset (\circ A)$ and $(A \supset B) \supset ((\circ A) \supset (\circ B))$. It may easily be formalised as a Hilbert-style system, as a natural deduction system and as a Gentzen-style sequent calculus.

Howe [19] has extended the "permutation-free" approach described above to zero-order lax logic, allowing effective search for normal natural deductions. The new rules are just

$$\frac{\Delta \Rightarrow M : A}{\Delta \Rightarrow smhr(M) : \circ A}\, R\circ \qquad \frac{\Delta, x : A \Rightarrow M : \circ B}{\Delta \xrightarrow{\circ A} smhl(x.M) : \circ B}\, S\circ$$

Extension [20] to the first-order case is routine. There are alternative approaches to the problem of effective problem solving in LL; one open research problem is the correct application of the "contraction-free" techniques from [3] at least to decide solvability of zero-order problems in LL. We consider however that the "permutation-free" approach, finding "all" proofs and not just deciding solvability, is the right way forward, being closer both to the logic programming motivation and to the computational lambda-calculus concern with proofs as terms.

8 Proof search in intuitionistic linear logic

Intuitionistic linear logic (ILL) is another constructive logic, with applications in logic programming [17]. Howe [20] has developed a (cut-free) sequent calculus for all of ILL with its derivations in 1-1 correspondence with the normal natural deductions of ILL, equivalent, on the fragment of ILL considered in [17], to the calculus therein.

9 Proof search in dependent type theory

Our interest in these proof search problems arises not only from ordinary logic programming but also from the desire to automate proof search in dependent type theory. It began with work [28] on the integration of functional and logic programming using a type-theoretic perspective, where both proof search and function evaluation are seen as two aspects of the same issue: the search for normal proofs (or normal lambda terms). It is our opinion that a proper integration of the two paradigms is best done by restriction to logically pure fragments of each, seen in each case as a fragment of type theory [22].

[29] presents a sequent calculus for typing the normal terms of the $\lambda\Pi$-calculus of [32], with an extension for the $\lambda\Pi\Sigma$-calculus, i.e. the extension of the $\lambda\Pi$-calculus with Σ-types. These sequent calculi have the same property as that explored above for MJ, admitting no permutations of the order in which rules are used. No clausal forms are required, in contrast to the resolution calculi of [32].

10 Acknowledgments

We thank Andrew Adams, Jacob Howe, Dale Miller, Sara Negri, Christian Urban and Jan von Plato for useful discussions and collaboration, and the authors of [8, 10, 15, 27, 33] for making them available before publication.

References

[1] J.-M. Andreoli. *Logic programming with focusing proofs in linear logic*, Journal of Logic and Computation, vol. 2, pp 297–347, 1992.

[2] N. Benton, G. M. Bierman, V. de Paiva. *Computational types from a logical perspective*, Technical Report TR-365, Computer Laboratory, University of Cambridge, 1995.

[3] R. Dyckhoff. *Contraction-free sequent calculi for intuitionistic logic*, Journal of Symbolic Logic, vol. 57, no. 3, pp 795–807, 1992.

[4] R. Dyckhoff, L. Pinto. *Cut-elimination and a permutation-free sequent calculus for intuitionistic logic*, Studia Logica, vol. 60, pp 107–118, 1998.

[5] R. Dyckhoff, L. Pinto. *Permutability of proofs in intuitionistic sequent calculi*, Theoretical Computer Science, to appear.

[6] R. Dyckhoff, L. Pinto. *A permutation-free sequent calculus for intuitionistic logic*, Research Report CS/96/9, Computer Science Division, St Andrews University, 1996, available from "`http://www-theory.dcs.st-and.ac.uk/~rd/`".

[7] R. Dyckhoff, L. Pinto. Uniform proofs and natural deductions, in: D. Galmiche and L. Wallen, eds., *Proceedings of CADE-12 workshop on "Proof search in type theoretic languages"* (Nancy, 1994) 17–23.

[8] B. Elbl. *A declarative semantics for depth-first logic programs*, submitted (1997), Journal of Logic Programming, to appear.

[9] M. Fairtlough, M. Mendler. *An intuitionistic modal logic with applications to the formal verification of hardware*, Proceedings of the 1994 Conference on Computer Science Logic, Springer LNCS 933, pp 354–368, 1995.

[10] M. Fairtlough, M. Mendler, M. Walton. *First-order lax logic as a framework for constraint logic programming*, Technical Report MIPS-9714, Department of Mathematics and Computer Science, University of Passau, 1997.

[11] G. Gentzen. *The collected papers of Gerhard Gentzen* (ed. M. Szabo, North-Holland, Amsterdam, 1969).

[12] J.-Y. Girard, Y. Lafont, P. Taylor. *Proofs and Types*, Cambridge University Press, 1987.

[13] J.-Y. Girard. *A new constructive logic: classical logic*, Math. Structures Comput. Sci., vol. 1, pp 255–296, 1991.

[14] L. Hallnäs, P. Schroeder-Heister. *A proof-theoretic approach to logic programming. I. Clauses as rules*, Journal of Logic & Computation, vol. 1(2), pp 261–283, 1990.

[15] H. Herbelin. *A λ-calculus structure isomorphic to sequent calculus structure*, preprint (October 1994); available from "`http://capella.ibp.fr/~herbelin/LAMBDA-BAR-FULL.dvi.gz`".

[16] H. Herbelin. *A λ-calculus structure isomorphic to Gentzen-style sequent calculus structure*, Proceedings of the 1994 Conference on Computer Science Logic, Springer LNCS 933, pp 61–75, 1995.

[17] J. Hodas, D. Miller. *Logic programming in a fragment of intuitionistc linear logic*, Information and Computation, vol. 110, pp 327–365, 1994.

[18] W. A. Howard. *The formulae-as-types notion of construction*, in: "To H. B. Curry, Essays on Combinatory Logic, Lambda Calculus and Formalism", (edited by J. R. Hindley and J. P. Seldin), Academic Press (1980).

[19] J. Howe. *A permutation-free calculus for lax logic*, Research Report CS/98/1, University of St Andrews, 1998.

[20] J. Howe. *Proof enumeration issues for intuitionist linear logic*, PhD thesis (in preparation), University of St Andrews, 1998.

[21] S. C. Kleene. *Permutability of inferences in Gentzen's calculi LK and LJ*, Mem. Amer. Math. Soc., pp 1–26, 1952.

[22] P. Martin-Löf. "Intuitionistic type theory", Bibliopolis, Naples, 1984.

[23] M. Mendler. *A timing refinement of intuitionistic proofs and its application to the timing analysis of combinational circuits*, Theorem Proving with Analytic Tableaux and Related Methods, Springer LNCS 1071, pp 261–277, 1996.

[24] D. Miller. *Abstractions in logic programs*, in: P. Odifreddi, ed., *Logic and computer science*, Academic Press, pp 329–359, 1990.

[25] D. Miller, G. Nadathur, F. Pfenning, A. Scedrov. *Uniform proofs as a foundation for logic programming*, Annals of Pure and Applied Logic, vol. 51, pp 125–157, 1991.

[26] E. Moggi. *Notions of computation and monads*, Information and Computation, vol. 93, pp 55–92, 1991.

[27] S. Negri, J. von Plato. *Cut elimination in the presence of axioms*, submitted, 1998.

[28] L. Pinto. *Proof-theoretic investigations into integrated logical and functional programming*, PhD thesis, University of St Andrews, 1996.

[29] L. Pinto, R. Dyckhoff. *Sequent calculi for the normal terms of the $\lambda\Pi$- and $\lambda\Pi\Sigma$-calculus*, Proceedings of CADE-15 workshop on Proof Search in Type Theoretic Languages, ed. D. Galmiche et al, CNRS, University of Nancy I, 1998.

[30] D. Prawitz. *Ideas and results in proof theory*, in: J. E. Fenstad, Proceedings of the Second Scandinavian logic symposium, North-Holland, pp 235–308, 1971.

[31] D. Prawitz. *An improved proof procedure*, Theoria, vol. 26, pp 102–139, 1960.

[32] D. Pym and L. Wallen. *Proof search in the $\lambda\Pi$-calculus*, in: G. Huet and G. Plotkin, eds., *Logical frameworks*, Cambridge University Press, pp 309–340, 1991

[33] H. Schwichtenberg. *Termination of permutative conversions in intuitionistic Gentzen calculi*, Theoretical Computer Science, to appear.

[34] A. S. Troelstra, H. Schwichtenberg. "Basic proof theory", Cambridge University Press, 1996.

[35] A. S. Troelstra, D. van Dalen. "Constructivism in mathematics, vol. 1", North-Holland, 1988.

[36] L. Wallen. "Automated deduction in non-classical logics", MIT Press, 1989.

David's Trick

Sy D. Friedman*

M.I.T.

In David [82] a method is introduced for creating reals R which not only code classes in the sense of Jensen coding but in addition have the property that in $L[R]$, R is the unique solution to a Π^1_2 formula. In this article we cast David's "trick" in a general form and describe some of its uses.

Theorem. *Suppose* $A \subseteq \mathrm{ORD}$, $\langle L[A], A\rangle \models ZFC + 0^\#$ *does not exist,* φ *a sentence and suppose that for every infinite cardinal* κ *of* $L[A]$, $H_\kappa^{L[A]} = L_\kappa[A]$ *and* $\langle L_\kappa[A], A \cap \kappa\rangle \models \varphi$. *Then there exists a* Π^1_2 *formula* ψ *such that:*

(a) *If* R *is a real satisfying* ψ *then there is* $A \subseteq \mathrm{ORD}$ *as above, definable over* $L[R]$ *in the parameter* R.

(b) *For some* ZFC*-preserving,* $\langle L[A], A\rangle$*-definable, cofinality-preserving forcing* P, $P \Vdash \exists R\psi(R)$.

Moreover if A *preserves indiscernibles then* ψ *has a solution in* $L[A, 0^\#]$, *preserving indiscernibles.*

Remark

(1) We require that $H_\kappa^{L[A]}$ equal $L_\kappa[A]$ for infinite $L[A]$-cardinals solely to permit cofinality-preservation for P; if cofinality-preservation is dropped then such a requirement is unnecessary, by coding A into A^* with this requirement and then applying our result to A^*.

(2) A class A *preserves indiscernibles* if the Silver indiscernibles are indiscernible for $\langle L[A], A\rangle$. It follows from the technique of Theorem 0.2 of Beller-Jensen-Welch [82] (see Friedman [98]) that if A preserves indiscernibles then A is definable from a real $R \in L[A, 0^\#]$, preserving indiscernibles.

*Research supported by NSF Contract #9625997-DMS

Proof. Our plan is to create an $\langle L[A], A\rangle$-definable, tame, cofinality-preserving forcing P for adding a real R such that whenever $L_\alpha[R] \models ZF^-$ there is $A_\alpha \subseteq \alpha$, definable over $L_\alpha[R]$ (via a definition independent of α) such that $L_\alpha[R] \models$ for every infinite cardinal κ, $H_\kappa = L_\kappa[A_\kappa]$ and φ is true in $\langle L_\kappa[A_\alpha], A_\alpha \cap \kappa\rangle$. This property ψ of R is Π^1_2 and gives us (a), (b) of the Theorem. The last statement of the Theorem will follow using Remark (2) above.

P is obtained as a modification of the forcing from Friedman [97], used to prove Jensen's Coding Theorem (in the case where $0^\#$ does not exist in the ground model). The following definitions take place inside $L[A]$.

Definition (Strings). *Let α belong to* Card = *the class of all infinite cardinals. S_α consists of all $s : [\alpha, |s|) \to 2$, $\alpha \leq |s| < \alpha^+$ such that $|s|$ is a multiple of α and:*

(a) $\eta \leq |s| \to L_\delta[A \cap \alpha, s \restriction \eta] \models \mathrm{Card}\, \eta \leq \alpha$ for some $\delta < (\eta^+)^L \cup \omega_2$.

(b) If $\mathcal{A} = \langle L_\beta[A \cap \alpha, s \restriction \eta], s \restriction \eta\rangle \models (ZF^-$ and $\eta = \alpha^+)$ then over $\mathcal{A}$, $s \restriction \eta$ codes a predicate $A(s \restriction \eta, \beta) = A^* \subseteq \beta$ such that $A^* \cap \alpha = A \cap \alpha$ and for every cardinal κ of $L_\beta[A^*]$, $H_\kappa^{L_\beta[A^*]} = L_\kappa[A^*]$ and $\langle L_\kappa[A^*], A^* \cap \kappa\rangle \models \varphi$.

Remark When in (b) above we say that $s \restriction \eta$ *codes* A^* we are referring to the canonical coding from the proof of Theorem 4 of Friedman [97] of a subset of β by a subset of $(\alpha^+)^{\mathcal{A}} = \eta$ (relative to $A \cap \alpha$).

The remainder of the definitions from the proof of Theorem 4 of Friedman [97] remain the same in the present context. We now verify that the proofs of the lemmas from Friedman [97] can successfully accommodate the new restriction (clause (b)) on elements of S_α.

Lemma 1 (Distributivity for R^s). *Suppose $\alpha \in$ Card, $s \in S_{\alpha^+}$. Then R^s is α^+-distributive in $\mathcal{A}^s$.*

Proof. Proceed as in the proof of Lemma 5 of Friedman [97]. The only new point is to verify that in the proof of the Claim, t_λ satisfies clause (b) (of the new definition of S_α). The fact that s belongs to S_{α^+} and that t_λ codes $\bar{H}_\lambda$ imply that clause (b) holds for t_λ whenever β is at most $\bar{\mu}_\lambda$ = the height of $\bar{H}_\lambda$. But as $|t_\lambda|$ is definably singular over $L_{\bar{\mu}_\lambda}[t_\lambda]$ these are the only β's that concern us. □

Lemma 2 (Extendibility of P^s). *Suppose $p \in P^s$, $s \in S_\alpha$, $X \subseteq \alpha$, $X \in \mathcal{A}^s$. Then there exists $q \leq p$ such that $X \cap \beta \in \mathcal{A}^{q_\beta}$ for each $\beta \in \mathrm{Card} \cap \alpha$.*

Proof. Proceed as in the proof of Lemma 6 of Friedman [97]. In the definition of q, the only instances of clause (b) to check are for s_β when Even $(Y \cap \beta)$ codes s_β, s_β satisfying clause (a) of the definition of membership in S_β. But the embedding $\bar{\mathcal{A}}_\beta \to \mathcal{A}$ is Σ_1-elementary and instances of clause (b) refer to ordinals less than the height of $\mathcal{A}$; so the fact that s belongs to S_α implies that s_β belongs to S_β. □

Lemma 3 (Distributivity for P^s). *Suppose $s \in S_{\beta+}$, $\beta \in$ Card.*

(a) *If $\langle D_i \mid i < \beta \rangle \in \mathcal{A}^s$, D_i i^+ dense on P^s for each $i < \beta$ and $p \in P^s$ then there is $q \leq p$, q meets each D_i.*

(b) *If $p \in P^s$, f small in $\mathcal{A}^s$ then there exists $q \leq p$, $q \in \Sigma^p_f$.*

Proof. Proceed as in the proof of Lemma 7 of Friedman [97]. In the Claim we must verify that p^λ_γ satisfies clause (b). But once again this is clear by the Σ_1-elementary of $\bar{H}_\lambda(\gamma)$ and the fact that $L_{\bar{\mu}}[A \cap \gamma, p^\lambda_\gamma] \models |p^\lambda_\gamma|$ is Σ_1-singular, where $\bar{\mu} =$ height of $\bar{H}_\lambda(\gamma)$. □

The argument of the proof of Lemma 3 can also be applied to prove the distributivity of P, observing that when building sequences of conditions $\langle p^i \mid i < \lambda \rangle$, λ limit to meet an $\langle L[A], A \rangle$-definable sequence of dense classes, one has that p^λ_γ codes $\bar{H}^\lambda(\gamma)$ of height $\bar{\mu}$, where $L_{\bar{\mu}+1}[A \cap \gamma, p^\lambda_\gamma] \models |p^\lambda_\gamma|$ is not a cardinal. Thus there is no additional instance of clause (b) to verify beyond those considered in the proof of Lemma 3.

Thus P is tame and cofinality-preserving. The final statement of the Theorem also follows, using Remark (2) immediately after the statement of the Theorem. □

Applications

(1) Local Π^1_2-Singletons. David [82] proves the following: There is an L-definable forcing P for adding a real R such that R is a Π^1_2-singleton in every set-generic extension of $L[R]$ (via a Π^1_2 formula independent of the set-generic extension). This is accomplished as follows: One can produce an L-definable sequence $\langle T(\kappa) \mid \kappa$ an infinite L-cardinal$\rangle$ such that $T(\kappa)$ is a κ^{++}-Suslin tree in L for each κ and the forcing $\prod T(\kappa)$ for adding a branch $b(\kappa)$ through each $T(\kappa)$ (via product forcing, with Easton support) is tame and cofinality-preserving. Now for each n let $X_n \subseteq \omega_1^L$ be class-generic over L, X_n codes a branch through $T(\kappa)$ iff κ is of the form $(\aleph^L_{\lambda+n})$, λ limit. The forcing $\prod P_n$, where P_n adds X_n, can be shown to be tame and cofinality-preserving. Finally over

$L[\langle X_n \mid n \in \omega \rangle]$ add a real R such that $n \in R$ iff R codes X_n. Then one has that in $L[R]$, $n \in R$ iff $T(\aleph^L_{\lambda+n})$ is not $\aleph^L_{\lambda+n}$-Suslin for sufficiently large λ. Clearly this characterization will still hold in any set-generic extension of $L[R]$. David's trick is used to strengthen this to a Π^1_2 property of R.

(2) A Global Π^1_2-Singleton. Friedman [90] produces a Π^1_2-singleton R, $0 <_L R <_L 0^\#$. This is accomplished as follows: assume that one has an index for a $\Sigma_1(L)$ class function $(\alpha_1 \cdots \alpha_n) \mapsto r(\alpha_1 \cdots \alpha_n)$, that produces $r(\alpha_1 \cdots \alpha_n) \in 2^{<\omega}$ for each $\alpha_1 < \cdots < \alpha_n$ in ORD, such that $R = \bigcup\{r(i_1 \cdots i_n) \mid i_1 < \cdots < i_n \text{ in } I = \text{Silver indiscernibles}\}$. For each $r \in 2^{<\omega}$ there is a forcing $\mathbb{Q}(r)$ for "killing" all $(\alpha_1 \cdots \alpha_n)$ such that $r(\alpha_1 \cdots \alpha_n)$ is incompatible with r. (To "kill" $(\alpha_1 \cdots \alpha_n)$, one adds a CUB subset of α_1 disjoint from $\{\beta < \alpha_1 \mid \beta, \alpha_1$ satisfy the same formulas in L_{α_n} with parameters from $\alpha_2 \cdots \alpha_{n-1}\}$.) No $(i_1 \cdots i_n)$ from I^n can be killed (when i_1 has uncountable L-cofinality). Now build R such that $r \subseteq R$ iff R codes a $\mathbb{Q}(r)$-generic. Then R is the unique real with this property. David's trick is used to strengthen this to a Π^1_2 property.

(3) New Σ^1_3 facts. Friedman [98] shows that if M is an inner model of ZFC, $0^\# \notin M$, then there is a Σ^1_3 sentence false in M yet true in a forcing extension of M. This is accomplished as follows: let $\langle C_\alpha | \alpha\ L\text{-singular}\rangle$ be a $\Box$-sequence in L; i.e., C_α is CUB in α, $otC_\alpha < \alpha$, $\bar\alpha \in \lim C_\alpha \to C_{\bar\alpha} = C_\alpha \cap \bar\alpha$. Define $n(\alpha) = 0$ if otC_α is L-regular and otherwise $n(\alpha) = n(otC_\alpha) + 1$. Then for some n, $\{\alpha \mid n(\alpha) = n\}$ is stationary in M. And for each n, there is a ZFC-preserving forcing extension of M in which $\{\alpha \mid n(\alpha) \le n\}$ is non-stationary, and is in fact disjoint from the class of limit cardinals. David's trick is used to strengthen the latter into a Σ^1_3 property, namely: There exists a real R such that for every ordinal α, if $L_\alpha[R] \models ZF^-$ then $L_\alpha[R] \models \beta$ a limit cardinal $\to \beta$ L-regular or $n(\beta) \ge n$).

References

[82] R. David, A Very Absolute Π^1_2-Singleton, *Annals of Pure and Applied Logic* **23** pp. 101-120.

[82] A. Beller, R. Jensen, P. Welch, *Coding the Universe*, book, *Cambridge University Press*

[90] S. Friedman, The Π^1_2 Singleton Conjecture, *Journal of the American Mathematical Society*, Vol.3, No.4, pp. 771-791.

[97] S. Friedman, Coding without Fine Structure, *Journal of Symbolic Logic*, Vol.62, No.3, pp.808-815.

[98] S. Friedman, New Σ^1_3 Facts, to appear.

[99] S. Friedman, *Fine Structure and Class Forcing*, book, in preparation.

A SEMANTICAL CALCULUS FOR INTUITIONISTIC PROPOSITIONAL LOGIC

JÖRG HUDELMAIER

WSI, University of Tübingen
Sand 13, D-72076 Tübingen, Tel. +49 7071 2977361
joerg@informatik.uni-tuebingen.de

1. Introduction

Classical propositional logic was founded by George Boole as the theory of logical validity of formulae: a formula is *valid*, if and only if it is verified (i.e. mapped to 1) by all propositional valuations (i.e. homomorphisms from the algebra of formulae into suitable so called Boolean algebras), if and only if it is not falsified by any such valuation. The well known so called *classical tableau calculus* is then based on a syntactic analysis of the conditions a propositional valuation has to satisfy in order to falsify a given formula.

If in the course of this analysis conditions arise which no valuation satisfies, vic. both truth values associated to one and the same propositional variable, then the given formula is shown valid. Otherwise the formula is falsified by all valuations which satisfy the conditions.

By this approach validity of sets of formulae (being a property defined using the external concept of a valuation) may be reduced to an internal recursive description of such sets of formulae. (The observation that instead of single formulae one has to consider sets of formulae, so called *sequents,* is due to Gerhard Gentzen, the most convenient form of such a tableaux calculus is due to Raymond Smullyan.)

Thus the classical tableaux calculus TK establishes a correspondence between the model theory and the proof theory of classical propositional logic; this correspondence is expressed by the following two basic properties:

A) From failing *TK-deductions* Boolean counterexamples may be read off.

B) From the *TK-rules* the definition of Boolean semantics may be read off.

Typeset by $\mathcal{A}_{\mathcal{M}}\mathcal{S}$-TEX

Thus by property A) any branch of a failing deduction which does not end in an axiom of TK yields a counter model for the top most sequent if the T-signed formulae on this branch take the value *true* and the F-signed formulae take the value *false*. By property B), however, truth value assignments for complex formulae are reduced to truth value assignments of subformulae according to the form of the corresponding TK-rule. For this purpose the rules of TK may be read backwards saying that any assignment which gives all the formulae of the conclusion their indicated truth values also gives their indicated values to all formulae of one of the premisses. Thus in particular the truth value of the principal formula of a rule is reduced to the truth value of one or both of its immediate subformulae. Obviously property B) is much stronger than property A). It provides a reduction of model theory to proof theory which e.g. trivializes soundness and completeness proofs for this calculus.

Now there are various forms of tableaux calculi for intuitionistic propositional logic described in the literature. These calculi, however, usually only have the first of the above two properties. Thus there is no such embedding of semantics into syntax known for intuitionistic logic as is known for classical logic. This means that semantical methods – successfully applied in classical logic – are much less straightforward in intuitionistic logic. Therefore – as witnessed by the well known textbooks such as [7] – completeness proofs are much harder for intuitionistic logic then they are for classical logic.

Moreover the classical tableaux calculus allows at the same time searching for deductions as well as searching for counter models of a given sequent. This cannot be simulated by conventional sequent or tableaux calculi for intuitionistic propositional logic.

Both problems are solved, however, by a new tableaux calculus for intuitionistic propositional logic to be introduced in the present article.

2. A classical tableaux calculus

Consider now a formulation of the calculus TK for classical propositional logic consisting of axioms of the form M, Ta, Fa, where a is a propositional variable and rules

$$\text{E}\wedge \quad \frac{M, Tb, Ta}{M, Ta \wedge b} \qquad \frac{M, Fa \qquad M, Fb}{M, Fa \wedge b} \quad \text{I}\wedge$$

$$\text{E}\vee \quad \frac{M, Ta \qquad M, Tb}{M, Ta \vee b} \qquad \frac{M, Fa, Fb}{M, Fa \vee b} \quad \text{I}\vee$$

$$\text{E}\rightarrow \quad \frac{M, Fa \qquad M, Tb}{M, Ta \rightarrow b} \qquad \frac{M, Ta, Fb}{M, Fa \rightarrow b} \quad \text{I}\rightarrow$$

$$\text{E}\neg \quad \frac{M, Fa}{M, T\neg a} \qquad\qquad \frac{M, Ta}{M, F\neg a} \quad \text{I}\neg$$

To confirm the above property B) for this calculus we use an obvious analogy between sequents and rules of sequent calculi on the one hand and goals and rules of logic programming languages on the other hand: premiss sequents in TK are formed by the same context free formalism by which subsequent goals are obtained from previous goals in logic programming. Thus e.g. the passage from a sequent $\langle Ta \to b, Fa_1, \ldots, Fa_m, Tb_1, \ldots, Tb_n\rangle$ to its left premiss $\langle Fa, Fa_1, \ldots, Fa_m, Tb_1, \ldots, Tb_n\rangle$ could be accomplished by applying to the goal $true(a \to b), false(a_1), \ldots, false(a_m), true(b_1), \ldots, true(b_n)$ the rule $true(a \to b) \;:-\; false(a)$. (Note that context freeness of the tableaux rules does not extend beyond propositional logic. In fact completeness of predicate logic cannot be shown by constructive means. (cf. [3]))

Using a set of similar rules any goal corresponding to a nonderivable sequent of TK may be reduced to a goal corresponding to a sequent without complex formulae which is not an axiom of TK, i.e. does not contain two complementary propositional variables. Now the only way to recognize such basic goals g by a logic program is to have some sequent v consisting of pairwise noncomplementary propositional variables given in advance, such that the sequent corresponding to g is a subsequent of s. Therefore we have to add a second argument v, i.e. a valuation to the above predicates $true(\_)$ and $false(\_)$, and we obtain a program of the following form:

```
t(et(U,V),B)     :-  t(U,B),t(V,B).
t(vel(U,V),B)    :-  t(U,B).
t(vel(U,V),B)    :-  t(V,B).
t(imp(U,V),B)    :-  f(U,B).
t(imp(U,V),B)    :-  t(V,B).
t(non(U),B)      :-  f(U,B).
f(et(U,V),B)     :-  f(U,B).
f(et(U,V),B)     :-  f(V,B).
f(vel(U,V),B)    :-  f(U,B),f(V,B).
f(imp(U,V),B)    :-  t(U,B),f(V,B).
f(non(U),B)      :-  t(U,B).

t(at(U),B)       :-  member(B,t(U)).
f(at(U),B)       :-  member(B,f(U)).
```

For this program evaluation of any goal according to any search rule terminates and thus we obtain a well founded definition of the relevant concepts:

A formula $u \wedge v$ is *true* with respect to a valuation b,
if and only if both u and v are true with respect to b.

A formula $u \vee v$ is *true* with respect to a valuation b,
if and only if u is true with respect to b or v is true with respect to b.

A formula $u \rightarrow v$ is *true* with respect to a valuation b,
if and only if u is false with respect to b or v is true with respect to b.

A formula $\neg u$ is *true* with respect to a valuation b,
if and only if u is false with respect to b.

A formula $u \wedge v$ is *false* with respect to a valuation b,
if and only if u is false with respect to b or v is false with respect to b.

A formula $u \vee v$ is *false* with respect to a valuation b,
if and only if both u and v are false with respect to b.

A formula $u \rightarrow v$ is *false* with respect to a valuation b,
if and only if u is true with respect to b and v is false with respect to b.

A formula $\neg u$ is *false* with respect to a valuation b,
if and only if u is true with respect to b.

A propositional variable is *true* or *false* with respect to a valuation b
depending on the truth value given it by b.

In usual formulations of classical semantics only the concept of truth of a formula is defined, while falsity is declared as nontruth. Here, however, we cannot presuppose that false sequents and nontrue sequents coincide – this requires insight into the validity of the cut rule and the provability of all sequents $\langle Ta, Fa \rangle$ for arbitrary formulae a, and both of these facts cannot simply be read off from the program. Apart from this peculiarity the present definition of classical semantics is the same as the ordinary definition.

The possibility of converting our calculus into the above logic program obviously rests on three structural properties of TK:

a) Any rule of TK has a single principal formula.
b) All rules of TK are invertible.
c) Any sequence of applications of TK-rules terminates.

3. Intuitionistic calculi

Now for intuitionistic propositional logic no calculus is known which has all these properties – actually properties b) and c) seem to exclude each other. Moreover the well known so called contraction free calculi (cf. [1,4]), while having property c), do not have property a) (this shows in the usual rule

$$\frac{M, Tp, Tb}{M, Tp, Tp \to b}$$

of such calculi.)

In the sequel we will see that in order to obtain a corresponding semantics from a given calculus K properties b) and c) may be relaxed to

b′) All rules of K are quasi invertible.
c′) All non derivable sequents of K are refutable.

These notions are given by

a) A rule R of a sequent calculus K is quasi invertible, if from the derivability of a sequent $\langle s, a \rangle$, where a is the principal formula of an application of R, follows the derivability of all sequents $\langle s, b_0, \ldots, b_n \rangle$, where $\langle b_0, \ldots, b_n \rangle$ are the side formulae of some premiss of this application of R.

b) For a sequent s, a K-inference I by some K-rule R, and a premiss p of R the corresponding premiss sequent is denoted by $d(I, s, p)$.

c) A sequent s is refutable in 0 steps by a sequent calculus K, if s is neither an axiom of K nor the conclusion of an inference of K.

The sequent s is refutable in $n+1$ steps by K, if s is refutable in 0 steps or for any K-inference I with conclusion s there is some premiss p such that the premiss sequent $d(I, s, p)$ is refutable in n steps.

Using the above definition of $d(I, s, p)$ we may formulate derivability of a sequent s as follows: s is derivable by a calculus K in 0 steps, if s is an axiom of K; s is derivable in $n+1$ steps, if it is derivable in 0 steps or there is a K-inference I such that all sequents $d(I, s, p)$ are derivable in n steps.

Now completeness proofs for intuitionistic logic are commonly based on a multisuccedent calculus MJ (cf. [2,5]) with axioms M, Ta, Fa and $M, T\bot$ and rules

E∧ $$\frac{M, Ta \wedge b, Tb, Ta}{M, Ta \wedge b} \qquad \frac{M, Fa \wedge b, Fa \qquad M, Fa \wedge b, Fb}{M, Fa \wedge b}$$ I∧

E∨ $$\frac{M, Ta \vee b, Ta \qquad M, Ta \vee b, Tb}{M, Ta \vee b} \qquad \frac{M, Fa \vee b, Fa, Fb}{M, Fa \vee b}$$ I∨

E→ $$\frac{M, Ta \to b, Fa \qquad M, Ta \to b, Tb}{M, Ta \to b} \qquad \frac{M^T, Ta, Fb}{M, Fa \to b}$$ I→

(where M^T results from M by removing all F-signed formulae.)
In this calculus as in all subsequent calculi the connective $\neg$ is not employed; instead as usual $\neg a$ is implemented as $a \to \bot$.

4. The calculi TJ and TJ^{m}

MJ obviously has the above properties a) and b′) but many nonderivable sequents of this calculus are not refutable either, e.g. all sequents containing a signed formula $Ta \to b$. Therefore we consider a slight variant TJ of MJ with the usual axioms and the rules

$$\text{E}\wedge \quad \frac{M, Tb, Ta}{M, Ta \wedge b} \qquad\qquad \frac{M, Fa \qquad M, Fb}{M, Fa \wedge b} \quad \text{I}\wedge$$

$$\text{E}\vee \quad \frac{M, Ta \qquad M, Tb}{M, Ta \vee b} \qquad\qquad \frac{M, Fa, Fb}{M, Fa \vee b} \quad \text{I}\vee$$

$$\text{E}\to \quad \frac{M, Ta \to b, Fa \qquad M, Tb}{M, Ta \to b} \quad \frac{M^T, Ta, Fb \qquad M, Ta, Fb}{M, Fa \to b} \quad \text{I}\to$$

(where M^T results from M by removing all F-signed formulae.)
Thus –apart from dropping various principal formulae from premisses– TJ differs from MJ only in the rule I$\to$ which now has two premisses instead of one; the additional premiss of this rule is the same as the corresponding premiss of TK. The calculi MJ and TJ are obviously equivalent, but unfortunately the new calculus still does not have property c′). Instead it has a weaker property

c$'_{\mathrm{w}}$) All non derivable sequents of TJ are weakly refutable.

Here the notion of weak refutability is given by

a) An inference I of a sequent calculus K is redundant for a sequent s, if s contains all side formulas of some premiss of I.

b) A sequent s is weakly refutable in 0 steps by a sequent calculus K, if s is not an axiom of K and all K-inferences with conclusion s are redundant for s.

The sequent s is weakly refutable in $n+1$ steps by K, if s is weakly refutable in 0 steps or for any irredundant K-inference I with conclusion s there is some premiss p such that $d(I, s, p)$ is weakly refutable in n steps.

Obviously any sequent derivable by a calculus K which admits the so called contraction rule is already derivable by using only irredundant inferences. Moreover all refutable sequents are weakly refutable, and no weakly refutable sequent is derivable. It is, however, not true that all weakly refutable sequents of an arbitrary calculus K are refutable by the same calculus (E.g. the sequent $\langle Ta \to b, Fb\rangle$ is weakly refutable in 1 step by MJ but it is not refutable.) Nor is it true that all nonderivable sequents of some calculus K are weakly refutable by K. (e.g. the sequent $\langle T((d \to c) \to a) \to a, T(a \to b) \to a, Fa\rangle$, while not derivable by MJ is still not weakly refutable.)

Although the calculus TJ is well suited for constructing Kripkean countermodels for nonderivable sequents, the notion of weak refutability of a sequent cannot be directly implemented into a logic program in the above manner. (At least not for a reasonably perspicuous notion of valuation.) Thus TJ has the above property A), but it still lacks property B). There is, however, a straightforward way to turn TJ into a calculus for which all nonderivable sequents are refutable. For this purpose we introduce so called marked arrows $\overset{*}{\to}$ by which we replace the ordinary arrow connectives in certain situations. Thus we obtain the calculus $\mathrm{TJ^m}$, consisting of the usual axioms and the rules

$$\text{E}\wedge \quad \frac{M, Tb, Ta}{M, Ta \wedge b} \qquad\qquad \frac{M, Fa \qquad M, Fb}{M, Fa \wedge b} \quad \text{I}\wedge$$

$$\text{E}\vee \quad \frac{M, Ta \qquad M, Tb}{M, Ta \vee b} \qquad\qquad \frac{M, Fa, Fb}{M, Fa \vee b} \quad \text{I}\vee$$

$$\text{E}\to \quad \frac{M, Ta \overset{*}{\to} b, Fa \qquad M, Tb}{M, Ta \to b} \qquad \frac{M^T, Ta, Fb \qquad M, Ta, Fb}{M, Fa \to b} \quad \text{I}\to$$

(where M^T results from M by removing all marks and all F-signed formulae).

Obviously the mark on the arrow in a formula $Ta \overset{*}{\to} b$ of a sequent s indicates that s contains both Fa and Tb. Therefore an implicative formula need not be analysed again as long as its arrow is marked. Only after the mark has been removed by an I→-inference does it have to be considered again.

First we note

Lemma 1. *If a sequent s is derivable by TJ^m and the sequent s' results from s by replacing all marked connectives by the corresponding unmarked connectives, then s' is derivable by TJ.*

Proof: If s is an axiom of $\mathrm{TJ^m}$, then s' is an axiom of TJ. If s is derivable in $n+1$ steps by $\mathrm{TJ^m}$, but is not an axiom, and $d(I, s, p)$ is derivable in n

steps, then by the induction hypothesis the sequent $d(I, s, p)'$ obtained from $d(I, s, p)$ by removing all marks is derivable by TJ. But this sequent is the same as $d(I, s', p)$; therefore the latter sequent is derivable by TJ, too. Since this holds for arbitrary premisses p the sequent s' is derivable by TJ. □

This shows in particular that all sequents without marked connectives which are derivable by TJ^{m}, are also derivable by TJ. The proof of the converse inclusion requires another

Lemma 2. *If a sequent* $\langle s, a, d(I_k, d(I_{k-1}, \dots, d(I_2, d(I_1, \langle a\rangle, p_1), p_2), \dots, p_{k-1}), p_k)\rangle$ *is derivable by* TJ^m, *where no* p_i *is a left premiss of* $I{\rightarrow}$, *then the sequent* $\langle s, d(I_k, d(I_{k-1}, \dots, d(I_2, d(I_1, \langle a\rangle, p_1), p_2), \dots, p_{k-1}), p_k)\rangle$ *is also derivable by* TJ^m.

Proof: Let D be the sequent $d(I_k, d(I_{k-1}, \dots, d(I_2, d(I_1, \langle a\rangle, p_1), p_2), \dots, p_{k-1}), p_k)$. If a is of degree g and $\langle s, D, a\rangle$ is derivable in n steps. then we use an induction along the lexicographic ordering of the pairs $\langle g, n\rangle$:

If $\langle s, D, a\rangle$ is an axiom of TJ^{m} and a is not a propositional variable, then $\langle s, D\rangle$ is also an axiom. If, however, a is a propositional variable, then D is the sequent $\langle a\rangle$, and therefore $\langle s, D\rangle$ is still an axiom.

If $\langle s, D, a\rangle$ is not an axiom but is derivable in $n+1$ steps, then there is an inference I with principal formula in s such that all premisses $d(I, \langle s, D, a\rangle, p)$ are derivable in n steps. If a is not contained in $\langle s, D, a\rangle$, then this sequent coincides with $\langle s, D\rangle$. If both a and all formulae of D are contained in $d(I, \langle s, D, a\rangle, p)$, then by the induction hypothesis a may be dropped, resulting in the sequent $d(I, \langle s, D\rangle, p)$. If, finally, a is contained in $d(I, \langle s, D, a\rangle, p)$, but not all formulae of D are contained in this sequent, then there is a subsequent D' of D such that $D' = d(J_{k'}, d(J_{k'-1}, \dots, d(J_2, d(J_1, \langle a\rangle, q_1), q_2), \dots, q_{k'-1}), q_{k'})$ where the J_i and q_i form a subset of the I_i and p_i and such that $d(I, \langle s, D, a\rangle, p) = \langle s', D', a\rangle$. Then by the induction hypothesis the sequent $\langle s', D'\rangle$ is derivable, and this sequent coincides with the sequent $d(I, \langle s, D\rangle, p)$. Thus all premisses of an inference I with conclusion $\langle s, D\rangle$ are derivable, and therefore this sequent is also derivable.

If the principal formula of I is in D and is of the form $Tb \wedge c$ or $Fb \wedge c$ or $Tb \vee c$ or $Fb \vee c$ or $Tb \rightarrow c$, then the premisses $d(I, \langle s, D, a\rangle, p)$ are of the form $\langle s, d(I, D, p), a\rangle$, and thus the induction hypothesis applies, and we obtain the sequents $\langle s, d(I, D, p)\rangle = d(I, \langle s, D\rangle, p)$, and from these sequents we obtain the required sequent $\langle s, D\rangle$.

If the principal formula of I is in D and is of the form $Fb \rightarrow c$, then again for the right premiss p^r the induction hypothesis applies to $d(I, \langle s, D, a\rangle, p^r) = \langle s, d(I, D, p^r), a\rangle$. Thus we obtain as before the sequent $d(I, \langle s, D\rangle, p^r)$. For the left premiss p^l either the sequent $d(I, \langle s, D, a\rangle, p^l)$ contains both the formula a and all formulae of D, and in this case the induction hypothesis applies to this sequent and yields the sequent $d(I, \langle s, D\rangle, p^l)$, or $d(I, \langle s, D, a\rangle, p^l)$ does not contain a, and then the sequents $d(I, \langle s, D, a\rangle, p^l)$ and $d(I, \langle s, D\rangle, p^l)$ coincide, or finally $d(I, \langle s, D, a\rangle, p^l)$ contains a but not all formulae of D: In this case there is again a subsequent D' of D of the form $d(J_{k'}, d(J_{k'-1}, \dots, d(J_2, d(J_1, \langle a\rangle, q_1), q_2), \dots, q_{k'-1}), q_{k'})$ where the J_i and q_i form subsets of the I_i and p_i with $d(I, \langle s, D, a\rangle, p^l) = \langle s', D', a\rangle$. Then by the induction hypothesis the sequent $\langle s', D'\rangle$ is derivable, and this sequent is the same as the sequent $d(I, \langle s, D\rangle, p^l)$. Thus in all cases this sequent is derivable, and so is $\langle s, D\rangle$.

If a is the principal formula of I and $D = \langle a\rangle$ and a is not of the form $Fb \to c$, then we exchange the rôles of D and a in the premisses $d(I, \langle s, D, a\rangle, p)$, and thus we may delete D from these sequents and obtain the premisses $d(I, \langle s, a\rangle, p)$ and therefore $\langle s, a\rangle$ which is the required sequent $\langle s, D\rangle$. If, however, a is of the form $Fb \to c$, then in the right premiss $d(I, \langle s, D, a\rangle, p^r)$ we again exchange the rôles of a and D and thus by the induction hypothesis we may again eliminate D and obtain $d(I, \langle s, a\rangle, p^r)$. The left premiss $d(I, \langle s, D, a\rangle, p^l)$, however, is the same as the premiss $d(I, \langle s, a\rangle, p^l)$, and therefore the sequent $\langle s, a\rangle = \langle s, D\rangle$ is derivable.

If a is the principal formula of I and D is different from $\langle a\rangle$ and a is of the form $Tb{\wedge}c$, then D is of the form $d(J^b_{k'}, d(J^b_{k'-1}, \dots, d(J^b_2, d(J^b_1, \langle Tb\rangle, q^b_1), q^b_2), \dots, q^b_{k'-1}), q^b_{k'}), d(J^c_{l'}, d(J^c_{l'-1}, \dots, d(J^c_2, d(J^c_1, \langle Tc\rangle, q^c_1), q^c_2), \dots, q^c_{l'-1}), q^c_{l'})$, thus by double application of the induction hypothesis we obtain the required sequent $\langle s, D\rangle$.

If a is of the form $Fb{\wedge}c$, then D is either of the form $d(I_k, d(I_{k-1}, \dots, d(I_3, d(I_2, \langle Fb\rangle, p_2), p_3), \dots, p_{k-1}), p_k)$ or of the form $d(I_k, d(I_{k-1}, \dots, d(I_3, d(I_2, \langle Fc\rangle, p_3), p_3), \dots, p_{k-1}), p_k)$. In both cases the induction hypothesis applies to the corresponding premiss of I, and thus we obtain the sequent $\langle s, D\rangle$.

If a is of the form $Tb \vee c$, then D is either of the form $d(I_k, d(I_{k-1}, \dots, d(I_3, d(I_2, \langle Tb\rangle, p_2), p_3), \dots, p_{k-1}), p_k)$ or of the form $d(I_k, d(I_{k-1}, \dots, d(I_3, d(I_2, \langle Tc\rangle, p_2), p_3), \dots, p_{k-1}), p_k)$, and as before we may apply the induction hypothesis to the corresponding premiss of I and obtain $\langle s, D\rangle$.

If a is of the form $Fb \vee c$, then D is of the form $d(J^b_{k'}, d(J^b_{k'-1}, \dots, d(J^b_2, d(J^b_1, \langle Fb\rangle, q^b_1), q^b_2), \dots, q^b_{k'-1}), q^b_{k'})$, $d(J^c_{l'}, d(J^c_{l'-1}, \dots, d(J^c_2, d(J^c_1, \langle Fc\rangle, q^c_1), q^c_2), \dots, q^c_{l'-1}), q^c_{l'})$, and as above we apply the induction hypothesis twice and obtain the required sequent $\langle s, D\rangle$.

If a is of the form $Tb \to c$, then D is either of the form $d(I_k, d(I_{k-1}, \dots, d(I_3, d(I_s, \langle Tb\rangle, p_3), p_3), \dots, p_{k-1}), p_k)$, or of the form $d(J^c_{k'}, d(J^c_{k'-1}, \dots, d(J^c_2, d(J^c_1, \langle Fc\rangle, q^c_1), q^c_2), \dots, q^c_{k'-1}), q^c_{k'})$, $d(J^{b\to c}_{l'}, d(J^{b\to c}_{l'-1}, \dots, d(J^{b\to c}_2, d(J^{b\to c}_1, \langle Tb \overset{*}{\to} c\rangle, q^{b\to c}_1), q^{b\to c}_2), \dots, q^{b\to c}_{l'-1}), q^{b\to c}_{l'})$. In the first case we may apply the induction hypothesis and remove the formula Tb from the right premiss of I. In the second case we may first apply the induction hypothesis with respect to the formula $Tb \overset{*}{\to} c$ in the left premiss of I, then, as c is of lower degree than a, we may apply the induction hypothesis a second time with respect to the formula Fc and thus obtain the required sequent $\langle s, D\rangle$.

If, finally, a is of the form $Fb \to c$, then D is of the form $d(J^b_{k'}, d(J^b_{k'-1}, \ldots, d(J^b_2, d(J^b_1, \langle Tb\rangle, q^b_1), q^b_2), \ldots, q^b_{k'-1}), q^b_{k'}), \quad d(J^c_{l'}, d(J^c_{l'-1}, \ldots, d(J^c_2, d(J^c_1, \langle Fc\rangle, q^c_1), q^c_2), \ldots, q^c_{l'-1}), q^c_{l'})$, and therefore we may apply the induction hypothesis twice to the right premiss of I and obtain the required sequent $\langle s, D\rangle$. □

Now we can show:

Lemma 3. *If a sequent is derivable by TJ, then it is derivable by TJ^m.*

Proof: All axioms and all rules of TJ with the exception of E→ are also axioms and rules of TJ^{m}. Moreover the right premiss of E→ coincides for both calculi. Therefore it suffices to show that derivability of the left E→-premiss of TJ by TJ^{m} implies derivability of the left E→-premiss of TJ^{m} by TJ^{m}:

Thus if a sequent $\langle s, Ta \to b, Fa\rangle$ is derivable by TJ^{m}, then $\langle s, Ta \overset{*}{\to} b, Ta \to b, Fa\rangle$ is derivable, too, and by the preceding lemma the sequent $\langle s, Ta \overset{*}{\to} b, Fa\rangle$ is derivable. □

Consequently we have:

Theorem 1. *The calculi TJ and TJ^m are equivalent for sequents without marked arrows.*

5. The refutable sequents of TJ^{m}

Now that we have established equivalence of TJ^{m} and TJ with respect to derivability of sequents, we are going to characterize the refutable sequents of TJ^{m} so as to confirm the above property c$'$) for this calculus. We start by showing

Lemma 4. *If a sequent s is refutable by TJ^m and for all formulae $Ta \overset{*}{\to} b$ of s the formula Fa is contained in s, then the sequent s' obtained from s by replacing all marked arrows by unmarked arrows is weakly refutable by TJ.*

Proof: If s is refutable in 0 steps by TJ^{m}, then s is not an axiom and the only nonatomic formulae of s are of the form $Ta \overset{*}{\to} b$. Thus by the above condition the formulae Fa are contained in s, and therefore all TJ-inferences with conclusion s' are redundant.

If s is refutable in $n+1$ steps, but not in 0 steps by TJ^{m}, and some sequent $d(I, s, p)$ is refutable in n steps by TJ^{m}, then the induction hypothesis applies to this sequent, and therefore $d(I, s, p)'$ is weakly refutable by TJ, and the latter sequent coincides with $d(I, s', p)$. Now the only TJ-inferences with conclusion s' which do not occur among these TJ^{m}-inferences I with conclusion s are those inferences whose principal formulae $Ta \to b$ are obtained from marked formulae $Ta \overset{*}{\to} b$ of s. But again these inferences are redundant for s'. Therefore s' is weakly refutable by TJ. □

Again this shows that $\mathrm{TJ^m}$-refutable sequents without marked arrows are weakly TJ-refutable. Now we show that all nonderivable sequents of $\mathrm{TJ^m}$ are in fact refutable by this calculus. We start with a straightforward

Lemma 5. *If a sequent $\langle a, s\rangle$ is refutable by TJ^m, then the sequent s is refutable.*

Proof: If $\langle a, s\rangle$ is refutable in 0 steps, then so is s. If it is refutable in $n+1$ steps and not in 0 steps, then for any inference with conclusion $\langle a, s\rangle$, in particular for any inference I with principal formula different from a, there is a corresponding premiss p such that $d(I, \langle a, s\rangle, p)$ is refutable in n steps. Now those sequents $d(I, \langle a, s\rangle, p)$ which do not contain a are the same as the sequents $d(I, s, p)$, while from those $d(I, \langle a, s\rangle, p)$ which do contain a this formula may be deleted by the induction hypothesis, resulting again in the required sequent $d(I, s, p)$. Now all inferences with conclusion s are among these inferences I, and therefore s itself is refutable. □

Moreover we can show a kind of invertibility for some premisses of $\mathrm{TJ^m}$:

Lemma 6. *If a sequent $d(I, s, p)$ is refutable by TJ^m, where p is not the left premiss of the I→-rule, then the sequent s is refutable, too.*

Proof: If $d(I, s, p)$ is refutable in 0 steps, then the only non atomic formulae of $d(I, s, p)$ are of the form $Ta \overset{*}{\to} b$. Thus s has exactly one complex formula which is not of the form $Ta \overset{*}{\to} b$, and this formula is the principal formula of I. Therefore s is $\mathrm{TJ^m}$-refutable in 1 step.

If $d(I, s, p)$ is refutable in $n + 1$ steps but not in 0 steps, then for any $\mathrm{TJ^m}$-inference J with conclusion $d(I, s, p)$ there is a premiss q of the corresponding $\mathrm{TJ^m}$-rule such that $d(J, d(I, s, p), q)$ is refutable in n steps. Among all such inferences J we consider those whose principal formula is already contained in s, i.e. those whose principal formula is not a side formula of p: If for such a J the premiss q is not the left premiss of the I→-rule, then this is the same sequent as $d(I, d(J, s, q), p)$, and by the induction hypothesis the sequent $d(J, s, q)$ is refutable. If, however, q *is* the left premiss of the I→-rule and p is a premiss of an E∧- or E∨-rule or the right premiss of an E→-rule, then $d(J, d(I, s, p), q)$ is again the same sequent as $d(I, d(J, s, q), p)$, thus by the induction hypothesis the sequent $d(J, s, q)$ is again refutable. If p is the left premiss of the E→-rule or a premiss of the I∧- or I∨-rule, then $d(J, d(I, s, p), q)$ is the same sequent as $d(J, s, q)$, and therefore this sequent is refutable, too. If p is the right premiss of the I→-rule, then the sequent $d(J, s, q)$ results from $d(J, d(I, s, p), q)$ by deleting one formula, and therefore by lemma 5 the sequent $d(J, s, q)$ is refutable. Now all inferences with conclusion s different from I are among these inferences J. But the sequent $d(I, s, p)$ has been assumed refutable; thus for all inferences with conclusion s there exists a refutable premiss sequent, and consequently the sequent s is also refutable. □

Now we can proof a kind of converse contraction:

Lemma 7. *If a sequent $\langle a, s\rangle$ is refutable by TJ^m, then the sequent $\langle a, a, s\rangle$ is also refutable by TJ^m.*

Proof: If $\langle a, s\rangle$ is refutable in n steps and a has degree g, then we use an induction along the lexicographic ordering of the pairs $\langle g, n\rangle$:

If $n = 0$, then the formula a is either a propositional variable or a formula of the form $b \overset{*}{\rightarrow} c$. Thus the required sequent $\langle a, a, s\rangle$ is obviously refutable. If a is a signed propositional variable and the sequent $\langle a, s\rangle$ is refutable in $n + 1$ steps but not in 0 steps, then we consider all sequents $d(I, \langle a, s\rangle, p)$ refutable in n steps which contain the formula a and all sequents $d(J, \langle a, s\rangle, q)$ refutable in n steps which do not contain this formula. To the former sequents the induction hypothesis applies, and we obtain the sequents $d(I, \langle a, a, s\rangle, p)$; the latter sequents instead coincide with the sequents $d(J, \langle a, a, s\rangle, q)$. Thus all sequents required to establish refutability of $\langle a, a, s\rangle$ are refutable and therefore this sequent is also refutable.

If a is a formula of the form $Tb \overset{*}{\rightarrow} c$ and $\langle a, s\rangle$ is refutable in $n+1$ steps but not in 0 steps, then we have sequents $d(I, \langle a, s\rangle, p)$ refutable in n steps which contain the formula a and sequents $d(J, \langle a, s\rangle, q)$ refutable in n steps which instead contain the formula $Tb \rightarrow c$. To both kinds of sequents the induction hypothesis applies, resulting in the sequents $d(I, \langle a, a, s\rangle, p)$ and $d(J, \langle a, a, s\rangle, q)$. Thus all sequents required to show $\langle a, a, s\rangle$ refutable are refutable, and therefore $\langle a, a, s\rangle$ is refutable, too.

If a is of the form $Tb \wedge c$ or $Fb \wedge c$ or $Tb \vee c$ or $Fb \vee c$, then the sequents $\langle Tb, Tc, s\rangle$, $\langle Fb, s\rangle$ or $\langle Fc, s\rangle$, $\langle Tb, s\rangle$ or $\langle Tc, s\rangle$, $\langle Fb, Fc, s\rangle$, respectively, are refutable. Now the degrees of the formulae b and c are lower than the degree of a. Therefore the induction hypothesis applies to these sequents and we obtain sequents $\langle Tb, Tc, Tb, Tc, s\rangle$, $\langle Fb, Fb, s\rangle$ or $\langle Fc, Fc, s\rangle$, $\langle Tb, Tb, s\rangle$ or $\langle Tc, Tc, s\rangle$, $\langle Fb, Fc, Fb, Fc, s\rangle$ respectively. From these sequents by lemma 6 we obtain the required sequents $\langle Tb \wedge c, Tb \wedge c, s\rangle$, $\langle Fb \wedge c, Fb \wedge c, s\rangle$, $\langle Tb \vee c, Tb \vee c, s\rangle$ and $\langle Fb \vee c, Fb \vee c, s\rangle$.

If a is of the form $Tb \rightarrow c$ and $\langle a, s\rangle$ is refutable in $n + 1$ steps and the sequent $\langle s, Tc\rangle$ is refutable in n steps, then by the induction hypothesis $\langle s, Tc, Tc\rangle$ is refutable, and thus by lemma 6 $\langle s, Tb \rightarrow c, Tb \rightarrow c\rangle$ is refutable. If instead the sequent $\langle s, Tb \overset{*}{\rightarrow} c, Fb\rangle$ is refutable in n steps, then by the induction hypothesis the sequent $\langle s, Tb \overset{*}{\rightarrow} c, Tb \overset{*}{\rightarrow} c, Fb\rangle$ is refutable, and moreover the formula b is of lower degree than a, whence the induction hypothesis applies again to this latter sequent, resulting in the refutable sequent $\langle s, Tb \overset{*}{\to} c, Tb \overset{*}{\to} c, Fb, Fb\rangle$. From this sequent by lemma 6 we obtain the required sequent $\langle s, Tb \to c, Tb \to c\rangle$.

If a is of the form $Fb \to c$ and the sequent $\langle s, Tb, Fc\rangle$ is refutable, then by using the induction hypothesis twice we obtain the sequent $\langle s, Tb, Tb, Fc, Fc\rangle$, and from this sequent by lemma 6 we obtain the sequent $\langle s, Fb \to c, Fb \to c\rangle$.

If a is of the form $Fb \to c$ and $\langle a, s\rangle$ is refutable in $n+1$ steps and the sequent $\langle s, Tb, Fc\rangle$ is not refutable in n steps, then we consider all inferences I with conclusion $\langle a, s\rangle$ and principal formula different from a for which $d(I, \langle a, s\rangle, p)$ is refutable in n steps and contains a. For these inferences we may by induction hypothesis add a second copy of a to $d(I, \langle a, s\rangle, p)$ resulting in the sequent $d(I, \langle a, a, s\rangle, p)$. Moreover we consider all inferences J with conclusion $\langle a, s\rangle$ and principal formula different from a for which $d(J, \langle a, s\rangle, q)$ is refutable in n steps and does not contain a. For these inferences the sequents $d(J, \langle a, s\rangle, q)$ coincide with the sequents $d(J, \langle a, a, s\rangle, q)$. Finally we consider the inference with principal formula a and its premiss sequent $\langle s^T, Tb, Fc\rangle$. This is the same sequent as $\langle\langle s, Fb \to c\rangle^T, Tb, Fc\rangle$. Thus for all inferences with conclusion $\langle a, a, s\rangle$ there is a refutable premiss and therefore $\langle a, a, s\rangle$ is refutable. □

Using this lemma we can establish some more inversion principles:

Lemma 8. *a) If a sequent* $\langle s, Tb \to c\rangle$, $\langle s, Tb \overset{*}{\to} c\rangle$ *is refutable by* TJ^m*, then the sequent* $\langle s, T(a \wedge b) \to c\rangle$*, respectively* $\langle s, T(a \wedge b) \overset{*}{\to} c\rangle$ *is refutable by* TJ^m.

b) If a sequent $\langle s, Ta \to (b \to c)\rangle$, $\langle s, Ta \overset{*}{\to} (b \to c)\rangle$ *is refutable by* TJ^m*, then the sequent* $\langle s, T(a \wedge b) \to c\rangle$*, respectively* $\langle s, T(a \wedge b) \overset{*}{\to} c\rangle$ *is refutable by* TJ^m.

c) If a sequent $\langle s, Ta \to c, Tb \to c\rangle$, $\langle s, Ta \overset{*}{\to} c, Tb \overset{*}{\to} c\rangle$ *is refutable by* TJ^m*, then the sequent* $\langle s, T(a \vee b) \to c\rangle$*, respectively* $\langle s, T(a \vee b) \overset{*}{\to} c\rangle$ *is refutable by* TJ^m.

d) If a sequent $\langle s, Ta, Tb \to c\rangle$, $\langle s, Ta, Tb \overset{*}{\to} c\rangle$ *is refutable by* TJ^m*, then the sequent* $\langle s, Ta, T(a \to b) \to c\rangle$*, respectively* $\langle s, Ta, T(a \to b) \overset{*}{\to} c\rangle$ *is refutable by* TJ^m.

Proof: a) $\langle s, Tb \to c\rangle$ is not refutable in 0 steps. If $\langle s, Tb \overset{*}{\to} c\rangle$ is refutable in 0 steps, then so is $\langle s, Ta, T(a \wedge b) \overset{*}{\to} c\rangle$.

If a sequent $\langle s, Tb \to c\rangle$ is refutable in $n+1$ steps, then either the sequent $\langle s, Tb \overset{*}{\to} c, Fb\rangle$ or the sequent $\langle s, Tc\rangle$ is refutable in n steps. In the first case by the induction hypothesis the sequent $\langle s, T(a \wedge b) \overset{*}{\to} c, Fb\rangle$ is refutable, and therefore by lemma 6 the sequent $\langle s, T(a \wedge b) \overset{*}{\to} c, Fa \wedge b\rangle$ is refutable, and again by lemma 6 the required sequent $\langle s, T(a \wedge b) \to c\rangle$ is refutable by $\mathrm{TJ^m}$. In the second case the claim follows directly from the same lemma 6.

If a sequent $\langle s, Tb \overset{*}{\to} c\rangle$ is refutable in $n+1$ steps but not in 0 steps, then for all inferences I with conclusion $\langle s, Tb \overset{*}{\to} c\rangle$ some sequent $d(I, \langle s, Tb \overset{*}{\to} c\rangle, p)$ is refutable in n steps. Now some of these sequents contain the formula $Tb \overset{*}{\to} c$, while some contain the formula $Tb \to c$ instead. By the induction hypothesis these formulae may be replaced by $T(a \wedge b) \overset{*}{\to} c$, $T(a \wedge b) \to c$, respectively, and thus we obtain all sequents $d(I, \langle s, T(a \wedge b) \overset{*}{\to} c\rangle, p)$ for all inferences I with conclusion $\langle s, T(a \wedge b) \overset{*}{\to} c$ and therefore this sequent is refutable.

b) $\langle s, Ta \to (b \to c)\rangle$ is not refutable in 0 steps; if $\langle s, Ta \overset{*}{\to} (b \to c)\rangle$ is refutable in 0 steps, than the same holds for $\langle s, T(a \wedge b) \overset{*}{\to} c\rangle$. If $\langle s, Ta \to (b \to c)\rangle$ is refutable in $n+1$ steps, then either $\langle s, Ta \overset{*}{\to} (b \to c), Fa\rangle$ or $\langle s, Tb \to c\rangle$ is refutable in n steps. In the first case the induction hypothesis shows that $\langle s, T(a \wedge b) \overset{*}{\to} c, Fa\rangle$ is refutable, and therefore by lemma 6 $\langle s, T(a \wedge b) \overset{*}{\to} c, Fa \wedge b\rangle$ is refutable and again by lemma 6 $\langle s, T(a \wedge b) \to c\rangle$ is refutable. In the second case part a) of this lemma immediately shows that the required sequent is refutable.

If $\langle s, Ta \overset{*}{\to} (b \to c)\rangle$ is refutable in $n+1$ steps, we may again replace all formulae $Ta \overset{*}{\to} (b \to c)$ resp. $Ta \to (b \to c)$ in the sequents $d(I, \langle s, Ta \overset{*}{\to} (b \to c)\rangle, p)$ by $T(a \wedge b) \overset{*}{\to} c$ resp. $T(a \wedge b) \overset{*}{\to} c$ and thereby obtain the necessary sequents $d(I, \langle s, T(a \wedge b) \overset{*}{\to} c\rangle, p)$ establishing refutability of $\langle s, T(a \wedge b) \overset{*}{\to} c\rangle$.

c) The sequent $\langle s, Ta \to c, Tb \to c\rangle$ is not refutable in 0 steps; if the sequent $\langle s, Ta \overset{*}{\to} c, Tb \overset{*}{\to} c)\rangle$ is refutable in 0 steps, then this holds also for $\langle s, T(a \vee b) \overset{*}{\to} c)\rangle$.

If $\langle s, Ta \to c, Tb \to c\rangle$ is refutable in $n+1$ steps, then either $\langle s, Ta \to c, Tc\rangle$ is refutable in n steps and then by lemma 5 $\langle s, Tc\rangle$ is refutable and by lemma 6 $\langle s, T(a \vee b) \to c\rangle$ is refutable, or $\langle s, Ta \to c, Tb \overset{*}{\to} c, Fb\rangle$ is refutable in n steps. In this case either $\langle s, Tc, Tb \overset{*}{\to} c, Fb\rangle$ is refutable and then as before $\langle s, Tc\rangle$ and $\langle s, T(a \vee b) \to c\rangle$ are refutable, or $\langle s, Ta \overset{*}{\to} c, Tb \overset{*}{\to} c, Fa, Fb\rangle$ is refutable in $n-1$ steps. Thus by the induction hypothesis $\langle s, T(a \vee b) \overset{*}{\to} c, Fa, Fb\rangle$ is refutable and then by lemma 6 $\langle s, T(a \vee b) \overset{*}{\to} c, Fa \vee b\rangle$ is refutable and again by the same lemma the required sequent $\langle s, T(a \vee b) \to c\rangle$ is refutable.

If $\langle s, Ta \overset{*}{\to} c, Tb \overset{*}{\to} c\rangle$ is refutable in $n+1$ steps, then in all sequents $d(I, \langle s, Ta \overset{*}{\to} c, Tb \overset{*}{\to} c\rangle, p)$ either both formulae $Ta \overset{*}{\to} c$ and $Tb \overset{*}{\to} c$ occur with marked connectives or both occur with unmarked connectives. Thus by the induction hypothesis we may replace these formulae by $T(a \vee b) \overset{*}{\to} c\rangle$, $T(a \vee b) \to c\rangle$, respectively, and obtain the required sequents $d(I, \langle s, T(a \vee b) \overset{*}{\to} c\rangle, p)$ establishing refutability of $\langle s, T(a \vee b) \overset{*}{\to} c\rangle$.

d) We show the following generalization of this proposition: *If D is the sequent $d(I_k, d(I_{k-1}, \dots, d(I_2, d(I_1, \langle a\rangle, p_1), p_2), \dots, p_{k-1}), p_k)$ where no p_i is a left premiss of I$\rightarrow$, and the sequent $\langle s, D, b \rightarrow c\rangle$ is refutable by TJ^m, then the sequent $\langle s, D, (a \rightarrow b) \rightarrow c\rangle$ is refutable by TJ^m.*

If the sequent $\langle s, D, b \xrightarrow{*} c\rangle$ is refutable by TJ^m, then the sequent $\langle s, D, (a \rightarrow b) \xrightarrow{*} c\rangle$ is refutable by TJ^m.

Now $\langle s, D, b \rightarrow c\rangle$ is not refutable in 0 steps, and if $\langle s, D, b \xrightarrow{*} c\rangle$ is refutable in 0 steps, then so is $\langle s, D, (a \rightarrow b) \xrightarrow{*} c\rangle$.

If $\langle s, D, b \rightarrow c\rangle$ is refutable in $n+1$ steps, then either $\langle s, D, c\rangle$ or $\langle s, D, b \xrightarrow{*} c, Fb\rangle$ is refutable in n steps. If the former, then by lemma 6 the required sequent $\langle s, D, (a \rightarrow b) \rightarrow c\rangle$ is refutable. If, however, $\langle s, D, b \xrightarrow{*} c, Fb\rangle$ is refutable in n steps, then by the induction hypothesis the sequent $\langle s, D, (a \rightarrow b) \xrightarrow{*} c, Fb\rangle$ is refutable, and therefore by lemma 7 $\langle s, D, D, (a \rightarrow b) \xrightarrow{*} c, Fb\rangle$ is refutable and by lemma 6 $\langle s, D, a, (a \rightarrow b) \xrightarrow{*} c, Fb\rangle$ is refutable, and by the same lemma $\langle s, D, (a \rightarrow b) \xrightarrow{*} c, Fa \rightarrow b\rangle$ and $\langle s, D, (a \rightarrow b) \rightarrow c\rangle$ are refutable.

If $\langle s, D, b \xrightarrow{*} c\rangle$ is refutable in $n+1$ steps, then we consider all inferences I with this conclusion: if the principal formula of I is in D, but the corresponding premiss is not a left premiss of I$\rightarrow$, then $d(I, \langle s, D, b \xrightarrow{*} c\rangle, p) = \langle s, d(I, D, p), b \xrightarrow{*} c\rangle$ and thus by the induction hypothesis the sequent $\langle s, d(I, D, p), (a \rightarrow b) \xrightarrow{*} c\rangle$ is refutable, and this is again the sequent $d(I, \langle s, D, (a \rightarrow b) \xrightarrow{*} c\rangle, p)$. If the principal formula is in s and the premiss p is not a left premiss of I$\rightarrow$, then $d(I, \langle s, D, b \xrightarrow{*} c\rangle, p) = \langle d(I, s, p), D, b \xrightarrow{*} c\rangle$ and therefore by the induction hypothesis the sequent $\langle d(I, s, p), D, (a \rightarrow b) \xrightarrow{*} c\rangle = d(I, \langle s, D, (a \rightarrow b) \xrightarrow{*} c\rangle, p)$ is refutable. If the premiss p finally is a left premiss of I$\rightarrow$, then there is a subsequent D' of D such that $D' = d(J_{k'}, d(J_{k'-1}, \dots, d(J_2, d(J_1, \langle a\rangle, q_1), q_2), \dots, q_{k'-1}), q_{k'})$ where the J_i and q_i form a subset of the I_i and p_i and such that $d(I, \langle s, D, b \xrightarrow{*} c\rangle, p) = \langle s', D', b \rightarrow c\rangle$. Then by the induction hypothesis the sequent $\langle s', D', (a \rightarrow b) \rightarrow c\rangle$ is refutable, and this is the sequent $d(I, \langle s, D, (a \rightarrow b) \xrightarrow{*} c\rangle, p)$. Since the formula $(a \rightarrow b) \xrightarrow{*} c$ is not a principal formula, these inferences I comprise all inferences with conclusion $\langle s, D, (a \rightarrow b) \xrightarrow{*} c\rangle$. Therefore this sequent is refutable. □

Now we are ready to proof our

Theorem 2. *If a sequent is not derivable by TJ, then it is refutable by TJ^m.*

Proof: We use a characterization of the nonderivable sequents of TJ due to Dyckhoff and Pinto (cf. [6]) as the derivable sequents of the calculus CRIP consisting of axioms of the form $\langle p_1 \rightarrow b_1, \dots, p_n \rightarrow b_n, M \Rightarrow N\rangle$, where all

p_i are propositional variables, M and N are disjoint sets of propositional variables and none of the p_i occurs in M and rules

$$\frac{M, a, b \Rightarrow N}{M, a \wedge b \Rightarrow N} (1) \quad \frac{M \Rightarrow a, N}{M \Rightarrow a \wedge b, N} (2) \quad \frac{M \Rightarrow b, N}{M \Rightarrow a \wedge b, N} (3) \quad \frac{M, a \Rightarrow N}{M, a \vee b \Rightarrow N} (4)$$

$$\frac{M, b \Rightarrow N}{M, a \vee b \Rightarrow N} (5) \qquad \frac{M \Rightarrow a, b, N}{M \Rightarrow a \vee b, N} (6) \qquad \frac{M, p, b \Rightarrow N}{M, p, p \rightarrow b \Rightarrow N} (7)$$

$$\frac{M, b \Rightarrow N}{M, (c \rightarrow d) \rightarrow b \Rightarrow N} (8) \frac{M, c \rightarrow d, d \rightarrow b \Rightarrow N}{M, (c \vee d) \rightarrow b \Rightarrow N} (9) \frac{M, c \rightarrow (d \rightarrow b) \Rightarrow N}{M, (c \wedge d) \rightarrow b \Rightarrow N} (10)$$

$$\frac{M_1, c_1, d_1 \rightarrow b_1 \Rightarrow d_1 \quad \ldots \quad M_m, c_m, d_m \rightarrow b_m \Rightarrow d_m \quad M, e_1 \Rightarrow f_1 \quad \ldots \quad M, e_n \Rightarrow f_n}{M \Rightarrow e_1 \rightarrow f_1, \ldots, e_n \rightarrow f_n, N}$$

where p in rule (7) is a propositional variable and where in the last rule, i.e. rule (11) M is of the form $\langle (c_1 \rightarrow d_1) \rightarrow b_1, \ldots, (c_m \rightarrow d_m) \rightarrow b_m, M' \rangle$, M_i is M with $(c_i \rightarrow d_i) \rightarrow b_i$ deleted, all formulas of M' are either propositional variables or formulas of the form $p \rightarrow b$, where p is a propositional variable not occurring in M', and N is a set of propositional variables disjoint from M'.

Now the axioms $p_1 \rightarrow b_1, \ldots, p_n \rightarrow b_n, M \Rightarrow N$ of CRIP are obviously refutable by $\mathrm{TJ^m}$ in n steps, and if a premiss of one of the rules (1–8) is refutable, then by lemma 6 so is its conclusion. If a premiss of one of the rules (9,10) is refutable, then by lemma 8 the same holds for its conclusion. If, finally, the premisses of rule (11) of the form $\langle M_i, c_i, d_i \rightarrow b_i \Rightarrow d_i \rangle$ are refutable, then by the preceding lemma the sequents $\langle M_i, c_i, (c_i \rightarrow d_i) \rightarrow b_i \Rightarrow d_i \rangle = \langle M, c_i \Rightarrow d_i \rangle$ are refutable. If, furthermore, all premisses of the form $\langle M, e_j \Rightarrow f_j \rangle$ are refutable, then we have a refutable premiss for every inference with conclusion $\langle M^* \Rightarrow c_1 \rightarrow d_1, \ldots, c_m \rightarrow d_m, e_1 \rightarrow f_1, \ldots, e_n \rightarrow f_n, N \rangle$, where M^* results from M by marking all second arrows in the formulae $(c_i \rightarrow d_i) \rightarrow b_i$, and therefore this latter sequent is refutable. But the sequent $\langle M \Rightarrow e_1 \rightarrow f_1, \ldots, e_n \rightarrow f_n, N \rangle$ is obtained from this sequent by lemma 6. Therefore it is also refutable. □

By theorem 1 the nonderivable sequents of TJ and $\mathrm{TJ^m}$ coincide, therefore theorem 2 yields:

Corollary. *The refutable sequents of TJ^m are the same as the nonderivable sequents of the same calculus.*

6. The *refutable sequents of $\mathrm{TJ^m}$

Now we want to establish a correspondence between refutation trees of

TJ^{m} and Kripke models. But unfortunately this correspondence does not hold in both directions; while any refutation tree corresponds to a Kripke model, the converse correspondence cannot be obtained. E.g. a sequent consisting of two formulas $Fa \to b$ and $Fc \to d$ has two obvious countermodels M_1 and M_2, where M_1 has a root k labelled with the empty set and two successor nodes of k labelled with $\langle Ta, Fb \rangle$ and $\langle Tc, Fd \rangle$, while M_2 has the same root as M_1, but only one successor node labelled with $\langle Ta, Tc, Fb, Fd \rangle$. Then M_1 corresponds to the obvious TJ^{m}-refutation of this sequent, while M_2 does not correspond to any such refutation. Therefore we have to adjust either the notion of a Kripke model or the notion of refutability. The second approach seems to be much simpler. For this purpose we use the concept of quasi invertibility introduced above. Thus we obviously have the

Lemma 9. *All rules of TJ^m are quasi invertible.*

Now we define a notion of generalized premiss sequents:

a) For a set of inferences $\langle I_1, \ldots, I_k \rangle$ of TJ^{m} and corresponding premisses $\langle p_1, \ldots, p_k \rangle$, all different from the left premiss of I→, and a sequent s the sequent $d^*(\langle I_1, \ldots, I_k \rangle, s, \langle p_1, \ldots, p_k \rangle)$ is obtained from s by replacing all principal formulae of the I_i by the side formulae of the corresponding premisses p_i.

b) For a set of I→-inferences $\langle I_1, \ldots, I_k \rangle$ of TJ^{m} and corresponding left premisses $\langle p_1, \ldots, p_k \rangle$ and for a sequent s, the sequent $d^*(\langle I_1, \ldots, I_k \rangle, s, \langle p_1, \ldots, p_k \rangle)$ is obtained from s by deleting all F-signed formulae which are not principal formulae of one of the I_i and replacing afterwards all these principal formulae by the corresponding side formulae of the p_i.

Making use of this definition we arrive at the following notion of generalized refutability:

A sequent s is *refutable in 0 steps, if s is not an axiom of TJ^{m} and is not the conclusion of any inference of this calculus.

The sequent s is *refutable in $n+1$ steps, if it is *refutable in 0 steps or it is not an axiom and there exists a partition of the inferences with conclusion s into sets $\langle I_1^h, \ldots, I_{k_h}^h \rangle$ $(1 \le h \le l)$ such that for any inference I_i^h there is a premiss p_i^h such that all sequents $d^*(\langle I_1^h, \ldots, I_{k_h}^h \rangle, s, \langle p_1^h, \ldots, p_{k_h}^h \rangle)$ are *refutable in n steps.

Using the quasi invertibility of all rules of TJ^{m} we can show:

Lemma 10. *If a sequent $d^*(\langle I_1, \ldots, I_k \rangle, s, \langle p_1, \ldots, p_k \rangle)$ is refutable by TJ^m, then all sequents $d(I_i, s, p_i)$ are refutable.*

Proof: If one sequent $d(I_i, s, p_i)$, where p_i is not the left premiss of I→, were not refutable, then according to the above corollary of theorem 2 this sequent would be derivable by TJ^m. But the sequent $d^*(\langle I_1, \ldots, I_k\rangle, s, \langle p_1, \ldots, p_k\rangle)$ results from $d(I_i, s, p_i)$ by replacing the principal formulae of the I_j with $j \neq i$ by the corresponding side formulae. Since all rules of TJ^m are quasi invertible this would imply that $d^*(\langle I_1, \ldots, I_k\rangle, s, \langle p_1, \ldots, p_k\rangle)$ is derivable and therefore not refutable.

If p_i is the left premiss of I→, then all p_j with $j \neq i$ are also left premisses of I→, and thus $d^*(\langle I_1, \ldots, I_k\rangle, s, \langle p_1, \ldots, p_k\rangle)$ results from $d(I_i, s, p_i)$ by weakening and would be derivable. □

Now we can prove that refutability and *refutability for TJ^m coincide:

Theorem 3. *A sequent is refutable by TJ^m if and only if it is *refutable.*

Proof: If a sequent s is refutable, then it is obviously *refutable. If s is *refutable in 0 steps, then it is refutable. If, finally, s is refutable in $n+1$ steps, because all sequents $d^*(\langle I_1^h, \ldots, I_{k_h}^h\rangle, s, \langle p_1^h, \ldots, p_{k_h}^h\rangle)$ are *refutable in n steps, then by the preceding lemma all sequents $d(I_i^h, s, p_i^h)$ are refutable, and thus s is refutable. □

7. *Refutations and the semantics of TJ^m

*Refutation trees of TJ^m may be constructed by successively adding single formulae instead of entire sequents. For this purpose we have to give some additional structure to these *refutation trees: therefore we label the edge between a node l and its predecessor k by 0, if the sequent at l is obtained as the left premiss of an I→-inference from the sequent at k, and we label the same edge by 1 in all other cases.

Now all formulae at a node l of such a *refutation tree T different from the root of T are either side formulae of some inference with principal formula at the predecessor k of l or parameter formulae already present at k.

Conversely any T-signed formula a at a nonminimal node k of T which is a principal formula of some inference of TJ^m gives rise to some side formulae at some successor l of k and to the formula a itself at all successor nodes $l' \neq l$, and any T-signed formulae a at a nonminimal node k of T which is a principal formula of some inference gives rise to these side formulae at l and to the formula a at all successor nodes $l' \neq l$ connected to k by an edge labelled with 1.

Moreover all T-signed formulae at a nonminimal node k which are not principal formulae of an inference of TJ^m give rise to the same formulae at all successor nodes l of k, and all F-signed formulae which are not principal formulae of such an inference give rise to the same formulae at successor nodes l of k for which the edge $\langle k, l\rangle$ is labelled with 1.

Now we are ready to implement construction of such *refutation trees into a logic program. Our valuations will now be trees whose nodes are labelled with sets of pairwise noncomplementary signed propositional variables and whose edges are labelled with 0 or 1 and which fulfil an obvious monotonicity condition:

a) If k is the predecessor of a node l and the edge $\langle k, l\rangle$ is labelled with 1, then the set of signed propositional variables at k is a subset of the set of signed propositional variables at l.

b) If the edge $\langle k, l\rangle$ is labelled with 0, then the set of T-signed propositional variables at k is a subset of the set of T-signed propositional variables at l.

Thus our program has the following form

```
t(et(U,V),[B,M,T])       :-  remove(T,[B1,1,T1],TR]),t(U,[B1,1,T1]),
                             t(V,[B1,1,T1]), t01(et(U,V),TR).
t(vel(U,V),[B,M,T])      :-  remove(T,[B1,1,T1],TR]),t(U,[B1,1,T1]),
                             t01(vel(U,V),TR).
t(vel(U,V),[B,M,T])      :-  remove(T,[B1,1,T1],TR]),t(V,[B1,1,T1]),
                             t01(vel(U,V),TR).
t(imp(U,V),[B,M,T])      :-  remove(T,[B1,1,T1],TR]),f(U,[B1,1,T1]),
                             t(mimp(U,V),[B1,1,T1]),t01(imp(U,V),TR).
t(imp(U,V),[B,M,T])      :-  remove(T,[B1,1,T1],TR]),t(V,[B1,1,T1]),
                             t01(imp(U,V),TR).
t(non(U),[B,M,T])        :-  remove(T,[B1,1,T1],TR]),f(U,[B1,1,T1]),
                             t(mnon(U,[B1,1,T1]), t01(non(U),TR).

f(et(U,V),[B,M,T])       :-  remove(T,[B1,1,T1],TR]),f(U,[B1,1,T1]),
                             f1(et(U,V),TR).
f(et(U,V),[B,M,T])       :-  remove(T,[B1,1,T1],TR]),f(V,[B1,1,T1]),
                             f1(et(U,V),TR).
f(vel(U,V),[B,M,T])      :-  remove(T,[B1,1,T1],TR]),f(U,[B1,1,T1]),
                             f(V,[B1,1,T1]),f1(vel(U,V),TR).
f(imp(U,V),[B,M,T])      :-  remove(T,[B1,1,T1],TR]),t(U,[B1,1,T1]),
                             f(V,[B1,1,T1]),f1(imp(U,V),TR).
f(imp(U,V),[B,M,T])      :-  remove(T,[B1,0,T1],TR]),t(U,[B1,0,T1]),
                             f(V,[B1,0,T1]),f1(imp(U,V),TR).
f(non(U),[B,M,T])        :-  remove(T,[B1,1,T1],TR]),t(U,[B1,1,T1]),
                             f1(non(U),TR).
f(non(U,V),[B,M,T])      :-  remove(T,[B1,0,T1],TR]),t(U,[B1,0,T1]),
                             f1(non(U),TR).

t(mimp(U,V),[B,M,T])     :-  t0(imp(U,V),T),t1(mimp(U,V),T).
t(at(U),[B,M,T])         :-  member(B,t(U)),t01(at(U),T).
```

```
f(at(U),[B,M,T])              :-   member(B,f(U)),f1(at(U),T).
t0(F,[])                      :-   .
t0(F,[[B1,0,T1])|(TR]         :-   t(F,[B1,0,T1]),t0(F,TR).
t0(F,[[B1,1,T1]|TR]           :-   t0(F,TR).
t1(F,[])                      :-   .
t1(F,[[B1,0,T1]|TR]           :-   t1(F,TR).
t1(F,[[B1,1,T1]|TR]           :-   t(F,[B1,1,T1]),t1(F,TR).
t01(F,[])                     :-   .
t01(F,[[B1,M1,T1]|TR]         :-   t(F,[B1,M1,T1]),t01(F,TR).
f1(F,[])                      :-   .
f1(F,[[B1,0,T1]|TR]           :-   f1(F,TR).
f1(F,[[B1,1,T1]|TR]           :-   f(F,[B1,1,T1]),f1(F,TR).
```

Here the predicates t(_, _) and f(_, _) have the same meaning as before in the program for TK. The new predicates t0(_, _), t1(_, _), and t01(_, _), however, take in their first arguments formulae F and in their second arguments sequences T of valuations, and they hold for those F which are verified by all valuations from T whose root is connected to the present node by an edge labelled with 0, 1, 0 or 1, respectively. The predicate f1(_, _) holds for those formulae F which are falsified by all valuations from T whose root is connected to the present node by an edge labelled with 1.

Now for a given intuitionistic valuation B with root node k we call a *successor* of B any subvaluation B' of B with root k' for which there is an edge between k and k'. We call B' a *1-successor*, if this edge is labelled with 1. A *0-successor* of B, instead, is any subvaluation B' of B with root k' which is a successor of some subvaluation B'' of B with root k'', where the edge between k' and k'' is labelled with 0 and where there is a path between k and k'' whose edges are all labelled with 1.

Thus we immediately obtain the following definition of intuitionistic semantics from this program:

A formula $u \wedge v$ is *true* with respect to a valuation B,
if and only if both u and v are true with respect to some 1-successor of B and $u \wedge v$ is true with respect to all other successors of B.

A formula $u \vee v$ is *true* with respect to a valuation B,
if and only if u or v is true with respect to some 1-successor of B and $u \vee v$ is true with respect to all other successors of B.

A formula $u \rightarrow v$ is *true* with respect to a valuation B,
if and only if u is false or v is true with respect to some 1-successor of B

and $u \to v$ is true with respect to all 0-successors of B.

A formula $\neg u$ is *true* with respect to a valuation B,
if and only if u is false with respect to some 1-successor of B
and $\neg u$ is true with respect to all 0-successors of B.

A formula $u \wedge v$ is *false* with respect to a valuation B,
if and only if u or v is false with respect to some 1-successor of B
and $u \wedge v$ is false with respect to all other 1-successors of B.

A formula $u \vee v$ is *false* with respect to a valuation B,
if and only if both u and v are false with respect to some 1-successor of B
and $u \vee v$ is false with respect to all other 1-successors of B.

A formula $u \to v$ is *false* with respect to a valuation B,
if and only if u is true and v is false with respect to some successor of B
and $u \to v$ false with respect to all other 1-successors of B.

A formula $\neg u$ is *false* with respect to a valuation B,
if and only if u is true with respect to some successor of B
and $u \to v$ is false with respect to all other 1-successors of B.

A propositional variable is *true* with respect to a valuation B,
if and only if it is given the value *true* by
the classical valuation at the root of B.

A propositional variable is *false* with respect to a valuation B,
if and only if it is given the value *false* by
the classical valuation at the root of B.

8. Collapsing 1-successors

Our intuitionistic valuations differ from the usual Kripke models by the labels attached to the edges. Thus our valuations have two kinds of edges whereas in Kripke models all edges are of the same type. Now any two (Kripkean) worlds connected by one of our 1-edges carry the same classical valuation. Moreover they share all outgoing edges. Thus, in the Kripkean sense, they are not different worlds but two copies of one and the same world. We may, therefore, eliminate edges labelled with 1 by a simple collapsing operation, thus obtaining the usual Kripkean semantics for intuitionistic propositional logic: We first define the notion of *addition* of a signed propositional variable a to a valuation B: a T-signed propositional variable a is added to B by adding it to the root of B and to all successors of B. An F-signed propositional variable a is added to B by adding it to the root of B and to all 1-successors of B.

Obviously, if during this process no axioms are created at the nodes of the resulting tree B', then B' is again a valuation.

Collapsing of a minimal 1-successor B_1 with root node k of B now works as follows: Let k' be the predecessor of k in B and let B_1' be the subvaluation of B with root k'. Then first the signed propositional variables occurring at k which do not occur at k' are added to B_1'. Since the edge between k and k' is labelled 1, this implies that k and k' now contain the same objects. Then the successors of B_1 are copied to successors of B_1', resulting in a valuation B_1''. Now B_1'' is equivalent to B_1, i.e. any formula b which is verified or falsified by B_1 is already verified, falsified, respectively, by B_1''. Thus we may delete the node k from B_1''. Moreover since k was chosen minimal, no other 1-successors have been introduced. Therefore by successive such operations we may eliminate all 1-successors from our given valuation B.

Then in the definition of our semantics we just have to replace all references to 1-successors of some valuation B by references to B itself, and we obtain the well known Kripkean definition of this semantics.

References

1. Dyckhoff, R.: Contraction free sequent calculi for intuitionistic logic. In: *Journal of Symbolic Logic* **57**(1992), pp. 795–807.

2. Fitting, M.: *Intuitionistic logic, model theory and forcing*, North-Holland 1969.

3. Friedman,H.: Some systems of second order arithmetic and their use. In: *Proceedings of the International Congress of Mathematicians, Vancouver 1974*, Canadian Mathematical Congress 1975, Vol. I, pp. 235–242.

4. Hudelmaier, J.: An $n \log n$-SPACE decision procedure for intuitionistic propositional logic. In:*Journal of Logic and Computatation* **3** (1993) 63–75.

5. Kripke, S.: Semantical analysis of intuitionistic logic. I. In: Crossley, J.N & M.A.E. Dummett (eds.): *Formal systems and recursive functions*, North Holland 1965, pp.92–130.

6. Pinto, L. & Dyckhoff, R.: Loop-free construction of counter-models for intuitionistic propositional logic. In: Behara & Fritsch & Lintz (eds.): *Symposia Gaussiana, Conf. A*, de Gruyter 1995, pp. 225–232.

7. Schütte, K.: *Vollständige Systeme modaler und intuitionistischer Logik*, Springer 1968.

An Iteration Model Violating the Singular Cardinals Hypothesis

Peter Koepke
University of Bonn

1 Introduction

Models of Set Theory showing exotic behaviour at singular cardinals are usually constructed via forcing. The archetypical method is Prikry-Forcing [Pr1970], which has been generalized in various ways, as for example by Gitik and Magidor [GiMa1992]. It was observed early that Prikry generic sequences can be obtained as successive critical points in an iteration of the universe V by a normal ultrafilter ([Ka1994], see [De1978] for an exhausting analysis). In this paper iterations by stronger extenders are studied similarly and yield the following theorem:

Main Theorem:

Assume that there is an elementary embedding $\pi\colon V \to M$, $V \models GCH$, M transitive, $\pi \restriction \kappa = id$, $\pi(\kappa) \geq \kappa^{++}$, ${}^{\kappa}M \subseteq M$. Then there is an inner model N of

$$\mathbf{ZF} \wedge \neg\mathbf{AC} \wedge \forall \nu < \lambda\ 2^{\nu} = \nu^{+} \wedge \neg 2^{\lambda} = \lambda^{+} \wedge \lambda \text{ has cofinality } \omega.$$

This says that N violates, in a choiceless way, the **S**ingular **C**ardinals **H**ypothesis (**SCH**), since **SCH** implies that the generalized continuum hypothesis is true at singular strong limit cardinals. The model N will roughly be defined as the intersection of all models obtained by finitely iterating the embedding π.

The proof of the Main Theorem stretches over the rest of this paper. In section 2 we investigate iterations of elementary embeddings. In section 3 the intersection model N is defined and shown to be a model of **ZF**. Sections 4 and 5 are used to establish the cardinality properties and the negative result about choice in N, respectively.

From now on let us assume that $\pi\colon V \to M$ is as above. We may also assume that π is $\in$-definable from some parameters.

2   Iterations

To analyze the intersection model it is advantageous to have efficient representations of the elements of M and further iterates. Therefore we may have to modify π a bit:

Lemma 1 *There is an elementary map* $\pi' \colon V \to M'$, M' *transitive*, $\pi' \upharpoonright \kappa = id$, $\pi'(\kappa) = \pi(\kappa) \geq \kappa^{++}$, $^{\kappa}M' \subseteq M'$ *with the added property:*

$$M' = \{\pi'(f)(x) | f \colon (\kappa)^{<\omega} \to V, \, x \in (\pi'(\kappa))^{<\omega}\}$$

Proof: Let $X = \{\pi(f)(x) | f \colon (\kappa)^{<\omega} \to V, \, x \in (\pi(\kappa))^{<\omega}\}$. Since the family of functions $f \colon (\kappa)^{<\omega} \to V$ contains sufficiently long initial segments of Skolem functions for V, X is an elementary submodel of M: $X \prec M$. Let $\sigma \colon X \simeq M'$, M' transitive, $\pi' = \sigma \circ \pi$:

$$
\begin{array}{ccc}
V & \xrightarrow{\ \pi\ } & X \prec M \\
& \underset{\pi'}{\searrow} & \big\Vert \sigma \\
& & M'
\end{array}
$$

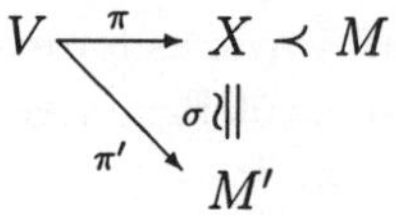

We show that M' is κ-closed, the other properties in the lemma are easily verified for π', M'. It suffices to show:

Claim: $^{\kappa}X \subseteq X$

For $i < \kappa$ consider $\pi(f_i)(x_i) \in X$ as above. Let $g \colon \kappa \leftrightarrow H_\kappa$ be a bijection, then $(x_i)_{i<\kappa} \in H^{M}_{\pi(\kappa)} = \pi(g)''\pi(\kappa)$. Let $\xi_0 < \pi(\kappa)$, such that $(x_i)_{i<\kappa} = \pi(g)(\xi_0)$ and define a function $h \colon \kappa \to V$ by cases: if $g(\xi) \colon \kappa_\xi \to V$ for some $\kappa_\xi < \kappa$, then

$$h(\xi) \colon \kappa_\xi \to V, \, h(\xi)(i) = f_i(g(\xi)(i)).$$

Otherwise set $h(\xi) = \emptyset$. Then we have $\pi(f_i)(x_i) = (\pi(f_i))((\pi(g)(\xi_0))(i)) = (\pi(h)(\xi_0))(i)$ and so $(\pi(f_i)(x_i))_{i<\kappa} = \pi(h)(\xi_0) \in X$ as required. $\square$

By the Lemma we may assume that π already satisfies the

Assumption: $M = \{\pi(f)(x) | f \colon (\kappa)^{<\omega} \to V, \, x \in (\pi(\kappa))^{<\omega}\}$.

A definable elementary embedding of V may be applied to its own definition and thus be iterated. This process can be repeated transfinitely along the ordinals. All iterates will be transitive inner models. For A a definable class define its image under π as $\pi(A) = \bigcup\{\pi(A \cap V_\alpha) | \alpha \in \mathrm{On}\}$. Then $\pi(A)$ is definable in M just like A is definable in V, with all parameters mapped by π.

Definition 1 *The* iteration $(M_i, \pi_{i,j})_{i \leq j < \theta}$, $\theta \leq \infty$ *of* V *by* π *is defined recursively until breakdown:* $M_0 = V$, $\pi_{0,0} = id$;
$M_{i+1} = \pi_{0,i}(M)$, $\pi_{i,i+1} = \pi_{0,i}(\pi)$, $\pi_{j,i+1} = \pi_{i,i+1} \circ \pi_{j,i}$ *for* $j < i$, $\pi_{i+1,i+1} = id \restriction M_{i+1}$;
if j *is a limit ordinal then* $(M_j, \pi_{i,j})_{i<j}$ *is a direct limit of* $(M_i, \pi_{i,i'})_{i \leq i' < j}$, *and* $\pi_{j,j} = id \restriction M_j$.
If any of these M_i *is wellfounded we also require it to be transitive. If there exists a minimal* i *with* M_i *non-wellfounded set* $\theta = i+1$ *and stop the construction; otherwise let* $\theta = \infty$. *The* M_i *for* $i < \theta$ *are the* iterates *of* V *by* π.

Indeed this construction does not break down:

Theorem 1 *The embedding* π *is* iterable, *i.e., every iterate of* V *by* π *is transitive, and* $\theta = \infty$.

Proof: Assume not. Then there is a unique last iterate $M_j = (M_j, \in')$ of V that is illfounded. By the construction j cannot be a successor ordinal. The image range($\pi_{0,j}$) lies $\in'$-cofinally in M_j. Let $\alpha \in \mathrm{On}$ be minimal such that $\pi_{0,j}(\alpha)$ is in the illfounded part of $(M_j, \in')$. There is $\eta \in' \pi_{0,j}(\alpha)$ such that η is still in the illfounded part of $(M_j, \in')$. Let $i < j$ and $\beta \in \mathrm{On}$ so that $\pi_{i,j}(\beta) = \eta \in' \pi_{0,j}(\alpha)$. In M_i, M_j is the unique illfounded iterate of M_i by $\pi_{i,i+1}$. By absoluteness properties of iterations as defined above, β witnesses the existential statement:

$$M_i \models \exists \gamma < \pi_{0,i}(\alpha)\colon \text{"}\pi_{i,j}(\gamma) \text{ is in the illfounded part of the unique non-wellfounded iterate of } M_i \text{ by } \pi_{i,i+1}\text{"}$$

Since $\pi_{0,i}\colon V \mapsto M_i$ is elementary, the corresponding statement holds in V: $\exists \gamma < \alpha$: "$\pi_{0,j}(\gamma)$ is in the illfounded part of the unique non-wellfounded iterate of V by π". This contradicts the minimality of α. □

The critical points of the maps $\pi_{i,i+1}$ are given by $\kappa_i = \pi_{0,i}(\kappa)$. The following facts are proved by a straightforward induction along the iteration (see also [Je1978]):

Lemma 2 *For* $i, j \in \mathrm{On}, i < j$:

(a) $\kappa_i < \kappa_j$;

(b) $\pi_{i,j} \restriction \kappa_i = id$, $\pi_{i,j}(\kappa_i) = \kappa_j$;

(c) $V_{\kappa_i} \cap M_i = V_{\kappa_i} \cap M_j$;

(d) $\mathcal{P}(\kappa_i) \cap M_i = \mathcal{P}(\kappa_i) \cap M_j$;

(e) *$M_i \supseteq M_j$, $M_i \neq M_j$;*

(f) *κ_i is a cardinal in M_j;*

(g) *If j is a limit ordinal, then $\kappa_j = \lim_{i<j} \kappa_i$;*

(h) *If $i < \omega$, then M_i is κ-closed: ${}^{\kappa}M_i \subseteq M_i$.*

The representation property of Lemma 1 can be generalized to all iterates.

Lemma 3 *For all $i < \infty$: $M_i = \{\pi_{0,i}(f)(x) | f\colon (\kappa)^{<\omega} \to V,\ x \in (\kappa_i)^{<\omega}\}$.*

Proof: By induction. The initial cases $i = 0, 1$ are trivial by our Assumption. The limit case is easy because M_i is a direct limit of earlier iterates.
For the successor step assume the claim for i and let $z \in M_{i+1} = \pi_{0,i}(M)$. By the Assumption and the elementarity of $\pi_{0,i}$ we may assume $z = \pi_{i,i+1}(g)(y)$ for some $g \in M_i$, $g\colon (\kappa_i)^{<\omega} \to M_i$, $y \in (\kappa_{i+1})^{<\omega}$. By the induction hypothesis $g = \pi_{0,i}(h)(z)$ for some $h\colon (\kappa)^{<\omega} \to V$, $z \in (\kappa_i)^{<\omega}$. Hence

$$z = \pi_{i,i+1}(g)(y) = \pi_{i,i+1}(\pi_{0,i}(h)(z))(y) = (\pi_{0,i+1}(h)(z))(y) = \pi_{0,i+1}(f)(x)$$

with $x = z^\frown y$ and $f\colon (\kappa)^{<\omega} \to V$ defined by $f(v^\frown u) := (h(v))(u)$ if this is welldefined and $\text{length}(v) = \text{length}(z)$, $\text{length}(u) = \text{length}(y)$; $f(v^\frown u) = \emptyset$ otherwise. □

We shall need the following "algebraic" facts about the system of iteration maps (see [De1978] for more general statements of this kind):

Lemma 4 **(a)** *If $i < \omega$ then $\pi_{i,\omega}(\pi_{i,\omega}) = \pi_{\omega,\omega+\omega}$.*

(b) *If $x \in M_\omega$ then $\pi_{i,\omega}(x) = \pi_{\omega,\omega+\omega}(x)$ for almost all[1] $i < \omega$.*

Proof: (a) $M_i \models$ "$\pi_{i,\omega}$ is the iteration map from V [which is M_i here] to its ω-th iterate." By $\pi_{i,\omega}$ this is mapped elementarily to $M_\omega \models$ "$\pi_{i,\omega}(\pi_{i,\omega})$ is the iteration map from V [which is M_ω now] to its ω-th iterate." By the absoluteness properties of iterations this map is just $\pi_{\omega,\omega+\omega}\colon M_\omega \to M_{\omega,\omega+\omega}$.
(b) Let $x \in M_\omega, x = \pi_{j,\omega}(y)$ for some $j < \omega$, $y \in M_j$. For $j < i < \omega$ we see:

$$\begin{aligned}
\pi_{i,\omega}(x) &= \pi_{i,\omega}(\pi_{j,\omega}(y)) \\
&= \pi_{i,\omega}(\pi_{i,\omega}(\pi_{j,i}(y))) \\
&= (\pi_{i,\omega}(\pi_{i,\omega}))(\pi_{i,\omega}(\pi_{j,i}(y))) \qquad (*) \\
&= \pi_{\omega,\omega+\omega}(\pi_{j,\omega}(y)) \\
&= \pi_{\omega,\omega+\omega}(x);
\end{aligned}$$

in $(*)$, $\pi_{i,\omega}$ is applied to the term "$\pi_{i,\omega}$ evaluated at the argument $\pi_{j,i}(y)$" □

[1]= all but finitely many

3 The Intersection Model

From the iteration of V by π we can define the *intersection model* $N := \bigcap_{i<\omega} M_i$.

Lemma 5 *For $i < \omega$: $M_\omega \subseteq N \subseteq M_i$ and N is uniformly definable in M_i from π_i as the intersection of the finite iterates of M_i.*

Theorem 2 *N is an inner model of Zermelo-Fraenkel set theory* **ZF**.

Proof: N is transitive and contains the class On, which implies extensionality and foundation in N. For the other axioms the existence of certain abstraction terms $t = \{x \in N | \varphi^N(x, \bar{a})\}$ for $\bar{a} \in N$ has to be shown in N. For all $i < \omega$, N is definable in M_i. Hence t exists in all M_i and $t \in N = \bigcap_{i<\omega} M_i$. □

The status of the Axiom of Choice (**AC**) will be discussed later. N and its inner model M_ω are in close relationship reminiscent of Prikry- or Gitik-Magidor generic extensions. Set $\lambda = \kappa_\omega$. Then

Lemma 6 **(a)** $N \cap V_\lambda = M_\omega \cap V_\lambda$.

(b) *$N \models \lambda$ is a strong limit cardinal, $N \models \forall \nu < \lambda\ 2^\nu = \nu^+$.*

Proof: (a) $\supseteq$ is clear by Lemma 5. Let $x \in N \cap V_\lambda$. For some $i < \omega$: $x \in N \cap V_{\kappa_i}$. Then by Lemma 2(c), $x \in M_i \cap V_{\kappa_i} = M_\omega \cap V_{\kappa_i}$.
(b) is true since the corresponding statements hold in M_ω and are absolute between M_ω and N by (a). □

Every $z \in M_\omega$ is the limit of a *thread* $\pi_{0,\omega}^{-1}(z), \pi_{1,\omega}^{-1}(z), \pi_{2,\omega}^{-1}(z), \ldots$. These threads provide us with natural Prikry sequences for M_ω; N can see the system of these sequences modulo finite changes.

Definition 2 *Let $k < \omega$. For $\alpha < \kappa_{\omega+1}$ set $c_\alpha^k := \{\pi_{i,\omega}^{-1}(\alpha) | k \leq i < \omega$ and $\alpha \in$ range$(\pi_{i,\omega})\}$. Define $C^k := (c_\alpha^k | \alpha < \kappa_{\omega+1})$. This definition can be carried out inside M_k, hence $C^k \in M_k$. $\pi_{i,\omega}^{-1}(\alpha) < \kappa_{i+1} < \lambda$ and so $c_\alpha^k \subseteq \lambda$. For any $x \subseteq \lambda$ define $\tilde{x} = \{y \subseteq \lambda | x \Delta y$ is finite$\}$. We call $\tilde{C} := (\tilde{c}_\alpha^0 | \alpha < \kappa_{\omega+1})$ the* Prikry-System *derived from iterating V by π.*

Obviously $\tilde{c}_\alpha^0 = \tilde{c}_\alpha^k$ and $\tilde{C} := (\tilde{c}_\alpha^k | \alpha < \kappa_{\omega+1})$, so $\tilde{C}$ can be defined from $C^k \in M_k$ for all k and we obtain $\tilde{C} \in N = \bigcap_{k<\omega} M_k$.

Lemma 7 **(a)** *If $\alpha < \beta < \kappa_{\omega+1}$, then $\tilde{c}_\alpha^0 \neq \tilde{c}_\beta^0$.*

(b) *There is a surjective map $s\colon \mathcal{P}(\lambda) \to \kappa_{\omega+1}$, $s \in N$.*

(c) *$N \models \lambda$ is singular of cofinality ω.*

Proof: (a) The threads c^k_α and c^k_β differ on an endsegment.
(b) For $x \subseteq \lambda$ let $s(x)$ be the unique α such that $x \in \vec{c}^0_\alpha$, if this exists, and 0 otherwise.
(c) The sequence $(\kappa_i | i < \omega)$ is cofinal in κ_ω (by Lemma 2(g)) and $(\kappa_i | i < \omega) \in \vec{c}^0_\lambda \in N$. □

4 Cardinal Preservation

Our assumptions on π imply that $\kappa_{\omega+1} \geq (\lambda^{++})^{M_\omega}$. If $(\lambda^{++})^{M_\omega} = (\lambda^{++})^N$ then Lemma 7(b) provides us with the desired negation of **SCH**. Therefore we show cardinal preservation between M_ω and N. The proof of the following "covering theorem" is based on "naming" elements of N by the normal form given in Lemma 3 and counting "names".

Lemma 8 *Let $f: \eta \to \theta$, $\eta, \theta \in \mathrm{On}$, $f \in N$. Then there is a function $F: \eta \to M_\omega$, $F \in M_\omega$ such that*

(a) *$\forall \xi < \eta$: $f(\xi) \in F(\xi)$;*

(b) *$\forall \xi < \eta$: $card^{M_\omega}(F(\xi)) \leq \lambda$.*

Proof: By Lemma 3, f can be represented in the various M_i, $i < \omega$, as: $f = \pi_{0,i}(f_i)(x_i)$ with $f_i: [\kappa]^{<\omega} \to V$, $x_i \in [\kappa_i]^{<\omega}$. For $\xi < \eta$, $\zeta < \theta$ we have

$$\begin{aligned}(\xi, \zeta) \in f \quad &\leftrightarrow \quad \pi_{i,\omega}(\xi, \zeta) \in \pi_{i,\omega}(f) = \pi_{0,\omega}(f_i)(x_i) \\ &\leftrightarrow \quad \pi_{\omega,\omega+\omega}(\xi, \zeta) \in \pi_{0,\omega}(f_i)(x_i), \text{ for almost all } i < \omega.\end{aligned}$$

Define $F: \eta \to V$ by
$F(\xi) = \{\zeta < \theta | \exists i < \omega \exists x \in [\lambda]^{<\omega}: \pi_{0,\omega}(f_i)(x) \text{ is a function and } \pi_{\omega,\omega+\omega}(\xi, \zeta) \in \pi_{0,\omega}(f_i)(x)\}$. Then $F \in M_\omega$ since it is definable in M_ω using the parameters η, θ, λ and $(\pi_{0,\omega}(f_i))_{i<\omega} = \pi_{0,\omega}((f_i)_{i<\omega})$. Property (a) holds by the preceding equivalences, (b) is immediate from the definition of F. □

Theorem 3 *$card^{M_\omega} = card^N$.*

Proof: The inclusion $\supseteq$ is clear since $M_\omega \subseteq N$.
If $\theta \leq \lambda$ is a cardinal in M_ω, then it is a cardinal in N by Lemma 6(a). If $\theta > \lambda$ is not a cardinal in N, there is $f: \eta \to \theta$ onto, $\eta < \theta$, $f \in N$. Take $F: \eta \to M_\omega$, $F \in M_\omega$ as in Lemma 8. By Lemma 8(a) $\theta \subseteq \bigcup_{\xi<\eta} F(\xi)$, and by (b) M_ω satisfies $\mathrm{card}(\bigcup_{\xi<\eta} F(\xi)) \leq \eta \cdot \lambda < \theta \cdot \theta = \theta$. So θ is not a cardinal in M_ω. □

Concerning the proof of our main theorem this yields

Theorem 4 $N \models \neg 2^\lambda = \lambda^+$

Proof: Assume $N \models 2^\lambda = \lambda^+$ instead. In N, there is a surjective map $\lambda^+ \overset{\text{onto}}{\longrightarrow} \mathcal{P}(\lambda)$ and by Lemma 7(b) there is a surjective map $\mathcal{P}(\lambda) \overset{\text{onto}}{\longrightarrow} \kappa_{\omega+1}$. So $N \models \text{card}(\kappa_{\omega+1}) \leq \lambda^+$ and by the preceding cardinal absoluteness $M_\omega \models \text{card}(\kappa_{\omega+1}) \leq \lambda^+$. This can be pulled back by $\pi_{0,\omega}$ to V where we get $\text{card}(\kappa_1) = \text{card}(\pi(\kappa)) \leq \kappa^+$ contradicting our assumptions on π. □

5 No Choice

Theorem 5 *In N, $\tilde{C} \upharpoonright \lambda^{++}$ has no choice function, hence in N the Axiom of Choice fails for sequences of length λ^{++}.*

Proof: Assume for a contradiction, that $h\colon \lambda^{++} \to \mathcal{P}(\lambda)$, $h \in N$ is a choice function for $\tilde{C} \upharpoonright \lambda^{++}$. Then $\forall \xi < \lambda^{++}\colon h(\xi) \in \tilde{C}(\xi) = \tilde{c}^0_\xi$. We use λ^{++} to denote $(\lambda^{++})^N = (\lambda^{++})^{M_\omega}$. Since $h \in M_i$ for all i we get by our normal form result of section 2: $h = \pi_{0,i}(f_i)(x_i)$ for $i < \omega$, with $f_i\colon (\kappa)^{<\omega} \to V$, $x_i \in (\kappa_i)^{<\omega}$. Consider $\xi < \xi' < \lambda^{++}$. By Lemma 7(a) $h(\xi) \neq h(\xi')$. There is $i < \omega$ such that $\kappa_i \cap h(\xi) \neq \kappa_i \cap h(\xi')$ and by Lemma 4(b) we can assume (increasing i if necessary) that $\pi_{i,\omega}(\xi) = \pi_{\omega,\omega+\omega}(\xi)$ and $\pi_{i,\omega}(\xi') = \pi_{\omega,\omega+\omega}(\xi')$. Then $\kappa_i \cap (\pi_{0,i}(f_i)(x_i))(\xi) \neq \kappa_i \cap (\pi_{0,i}(f_i)(x_i))(\xi')$. Applying $\pi_{i,\omega}$ we get

$$(*) \qquad \kappa_i \cap (\pi_{0,\omega}(f_i)(x_i))(\pi_{\omega,\omega+\omega}(\xi)) \neq \kappa_i \cap (\pi_{0,\omega}(f_i)(x_i))(\pi_{\omega,\omega+\omega}(\xi')).$$

In M_ω, we define the function $H\colon \lambda^{++} \to M_\omega$, $H(\xi) = (h^\xi_i | i < \omega)$, where $h^\xi_i\colon (\lambda)^{<\omega} \to \mathcal{P}(\lambda)$ is defined by $h^\xi_i(x) = \kappa_i \cap (\pi_{0,\omega}(f_i)(x))(\pi_{\omega,\omega+\omega}(\xi))$. By $(*)$ H is injective and its domain is λ^{++}. But this contradicts $M_\omega \models \text{card}(\text{range}(H)) \leq ((2^\lambda)^\lambda)^\omega = 2^\lambda = \lambda^+$, using **GCH** inside M_ω □

6 Further Aspects

1. More detailed studies of the choice situation show $N \models \lambda^+ - \mathbf{AC}$, the axiom of choice for λ^+-sequences and $N \models < \lambda^+ - \mathbf{DC}$, the axiom of dependent choice for sequences shorter than λ^+. This is not true for sequences of length λ^+, i.e. $N \models \neg\lambda^+ - \mathbf{DC}$. Indeed it is not possible to force over N with partial choice functions for $\tilde{C}$ of size $\leq \lambda$ without collapsing λ^+ and thus destroying the $\neg$**SCH**-situation.
2. $N = M_\omega[\tilde{C}]$, i.e., N is the smallest transitive model of **ZF** containing M_ω and $\tilde{C}$.
3. Gitik-Magidor forcing over M_ω with the canonical extender at λ derived from $\pi_{\omega,\omega+1}$ yields a model $N^* = M_\omega[(c_\alpha | \alpha < \kappa_{\omega+1})]$, where each c_α is an

ω-sequence cofinal in κ. It is possible in the context of countable ground models to find N^* such that $\forall\alpha < \kappa_{\omega+1}\colon c_\alpha \in \tilde{C}(\alpha)$. Then N is a natural submodel of N^* and the generic object for Gitik-Magidor forcing is basically a choice function for the Prikry system $\tilde{C}$. We shall discuss this in a subsequent article.

4. Ideas from this paper can be applied to other "Prikry-like" forcings as e.g. Magidor forcing [Ma1975] and Radin forcing.

References

[De1978] P. Dehornoy, *Iterated Ultrapowers and Prikry forcing* Annals of Math. Logik **15** (1978) pp. 109-160

[GiMa1992] M. Gitik, M Magidor, *The singular cardinal hypothesis revisited* in Set Theory of the Continuum, Ed.:Judah, Just, Woodin Springer, Berlin 1992

[Je1978] T. Jech: Set Theory, Academic Press, New York 1978

[Ka1994] A. Kanamori, The higher infinite, Springer, Berlin 1992

[Ma1975] M. Magidor, *Changing cofinality of cardinals* Fundamenta Mathematicae **99** (1978) pp. 61-71

[Pr1970] K. Prikry, *Changing measurable cardinals into accessible cardinals* Dissertationes Mathematicae **68** (1970) pp. 5-52

AN INTRODUCTION TO CORE MODEL THEORY

BENEDIKT LÖWE AND JOHN R. STEEL

ABSTRACT. In this paper we give an informal introduction to core model theory at the level of Woodin cardinals.

1. INTRODUCTION

Zermelo–Fraenkel set theory with choice, or ZFC, is the commonly accepted system of axioms for set theory, and hence for all of mathematics. Most of the axioms of ZFC express closure properties of the universe of sets. (The exceptions are Extensionality and Foundation, which in effect limit the objects under consideration.) Although all mathematical assertions can be expressed in the language of ZFC, and "most" of them can be decided using only the axioms of ZFC, there are nevertheless interesting mathematical assertions which cannot be decided using ZFC alone. The most famous of these is the Continuum Hypothesis.

Gödel's response to the incompleteness of ZFC with respect to assertions like the Continuum Hypothesis was that one should seek well–justified extensions of ZFC which decide these assertions (*cf.* [Gö47]). This is known as "Gödel's Program" and is still one of the most important tasks of higher set theory. Gödel suggested *strong axioms of infinity*, now more commonly known as *large cardinal axioms*, as candidates for basic principles to be added to the foundation provided by ZFC. In the years since [Gö47], large cardinal axioms have been extensively investigated, and have proved very fruitful in deciding in natural ways propositions about the real numbers left undecided by ZFC. They do not decide the Continuum Hypothesis, however.

There are many other natural extensions of ZFC which have been studied; for example, there are the forcing axioms MA, PFA, and MM. We do not think any of these extensions are currently well–justified in the way the large cardinal axioms are, but they are certainly interesting and useful. Indeed, the development of any consistent theory can be justified in a Hilbertian vein: one can regard such a theory as a tool for proving true Π^0_1 statements.[1] In this

---

1991 *Mathematics Subject Classification.* **03E15 03E45** 01A65 03–03 03E55 03E60.

The first author was partially supported by DAAD–Grant Ref.316–D/96/20969 in the program HSPII/AUFE and the Studienstiftung des Deutschen Volkes.

The authors would like to thank Martin Zeman (Berlin) for his comments and Robert Schmidt (Potsdam) for his TEXnical support.

[1]Of course, not all consistent theories are worthy of development; the point here is just that a theory need not be true to be worthy of development.

connection, the following remarkable phenomenon has emerged over the past thirty or forty years: the family of all remotely natural extensions of ZFC seems to be prewellordered by consistency strength[2] moreover, it seems that each theory in this family is equivalent in consistency strength to a theory whose axioms are large cardinal axioms. One of our goals in this paper is to exhibit some of the mathematical structure underlying this remarkable phenomenon.[3]

One can think of the large cardinal axioms as extrapolating from, and strengthening, the closure principles on the universe of sets which are inherent in ZFC. One way to understand such principles is to study minimal universes satisfying them. It seems to be the case that for each large cardinal axiom A there is a canonical minimal universe satisfying A, and that the structure of this universe can be analyzed in detail. The consistency strength order on large cardinal axioms corresponds to the inclusion order on the canonical minimal universes satisfying them. We shall call these canonical minimal universes, and their "iterates", *core models.*[4]

In general, one "computes" the consistency strength of a theory T extending ZFC by proving $T \equiv_{\mathsf{Cons}} \mathsf{A}$ for some large cardinal axiom A. Thinking of A as providing the standard measure, one says that $T \leq_{\mathsf{Cons}} \mathsf{A}$ yields an upper bound, and $\mathsf{A} \leq_{\mathsf{Cons}} T$ a lower bound, on the consistency strength of T. Upper bounds are generally proved by forcing over a model of A. Lower bounds are nearly always proved by constructing a core model satisfying A inside an arbitrary model of T. In this paper we shall give an introduction to the core model techniques.[5]

The simplest core model is Gödel's universe $\mathbf{L}$ of constructible sets (*cf.* [Gö40]). Many of the important features of core model theory are already present in the theory of $\mathbf{L}$: the existence of a fine–grained stratification of the model with strong condensation properties, the existence of a definable wellorder of the reals in the model, the absoluteness of the construction of the

---

[2]Perhaps the most natural way to define the consistency strength order is by

$$S \leq_{\mathsf{Cons}} T \iff P_S \subseteq P_T,$$

where P_S (resp. P_T) is the set of Π^0_1 consequences of S (resp.T).

[3]In fact, for "remotely natural" extensions S and T of ZFC, it seems to be the case that $S \leq_{\mathsf{Cons}} T$ if and only if every Σ^1_3 consequence of S is a consequence of T. One can replace Σ^1_3 by Σ^1_{n+3} in this last statement if one restricts attention to theories of consistency strength greater than the large cardinal hypothesis "there are n Woodin cardinals". Thus as we climb the consistency strength hierarchy of "natural" theories, our theories *converge*, not just in their Π^0_1 consequences, but in their consequences of ever greater complexity. The mathematical structure underlying this aspect of the remarkable phenomenon has to do with the correctness properties of core models.

[4]This usage of "core model" differs somewhat from the original one of Dodd and Jensen ([Do82], [DoJen81], [Jen95]).

[5]For a less technical introduction directed at a broader audience, see the survey in [Jen95]. The introduction to [MaSt94] also contains a general description of core model theory, and some of its history. Further history can be found in the introduction to [MiSt94].

model (between universes resembling each other sufficiently, which in the case of **L** means having the same ordinals), and the existence of covering theorems. Since the work of Gödel and Jensen (*cf.* [Gö40], [Jen72] and [DevJen75]) on **L**, many larger core models have been constructed and studied. Although significant new ideas have appeared, these basic themes remain. There is a great deal of unity of method in core model theory, and a great deal of resemblance among core models.

At the foundation of core model theory is the existence of a constructive *comparison process*, a method for *iterating* two approximations (called *mice*) $\mathfrak{M}$ and $\mathfrak{N}$ to the model in question so as to produce elementary embeddings $i\colon \mathfrak{M} \to \mathfrak{P}$ and $j\colon \mathfrak{N} \to \mathfrak{Q}$ such that the hierarchy on $\mathfrak{P}$ is an initial segment of that on $\mathfrak{Q}$ or *vice versa*.[6] The problem of extending core model theory to stronger large cardinal axioms comes down to the problem of extending the comparison process to more complicated mice. There are at least two ways to set goals and measure progress here. One can aim at mice satisfying certain large cardinal axioms, or one can aim at mice containing reals of a certain complexity. This latter measure is closely related to the logical complexity of the wellorder of the reals satisfied to exist by the core model in question. It also corresponds closely to the constructivity of the comparison process, a fact which lends it special technical importance.

In the period between 1966 and 1980 core models larger than **L** were constructed by Silver, Kunen, Dodd, Jensen, Mitchell and Baldwin[7]. These models can satisfy large cardinal hypotheses as strong as "there are many strong cardinals", and certain aspects of their form seem suitable for models satisfying much stronger large cardinal hypotheses.[8] However, the models themselves cannot satisfy hypotheses much stronger than the existence of many strong cardinals, and in the descriptive–set–theoretic measure of progress they do not go very far at all: each of these core models satisfies "there is a Δ^1_3 wellorder of the reals", and under reasonable large cardinal assumptions, there is a single Δ^1_3 real enumerating all the reals occurring in any of them. Behind these limitations is the fact that the mice approximating these core models can all be compared using only linear iteration, which is a particularly simple comparison process.[9] In [MaSt94], [MiSt94], [St93] and

---

[6]Gödel's constructible universe **L** is something of a degenerate case here: the mice are simply transitive models of $\mathbf{V} = \mathbf{L}$, and we can take $\mathfrak{M} = \mathfrak{P}$, $\mathfrak{N} = \mathfrak{Q}$, and $i = j = \mathrm{id}$.

[7]*Cf.* [Si71a], [Si71b], [Ku70], [Mi74], [Mi78], [Do81], [Bal83], [Bal86], [DoJen81], [Mi84], [Jen88] and [Jen90]. A very thorough introduction to the theory of these smaller core models will appear in [Ze9?].

[8]The models are constructed from *coherent sequences of extenders* (*cf.* Section 2.2 below). This framework, due mainly to [Mi74] and [Mi78], seems adequate for models satisfying "there is a superstrong cardinal".

[9]More precisely, certain artificial limitations on the "overlapping" of extenders on the coherent sequence from which the model is constructed are imposed. These limitations make comparison via linear iteration possible. They make the appearance of extenders

[St96] the authors make use of a more complicated comparison process involving iterations with a nonlinear tree structure. This enables them to develop a good theory of core models satisfying large cardinal hypotheses as strong as "there are many Woodin cardinals" (*cf.* Definition 2.27), and containing all reals which are Δ^1_n for some natural number n. These models still cannot satisfy "there is a superstrong cardinal", and there is still an upper bound on the complexity of the reals which occur in them.[10] This work represents more or less the frontier of current core model theory.

The purpose of this introductory article is to give the interested reader an informal introduction to the concepts behind the construction of these models. We will be very sketchy at times, and this paper is not to be understood as a means to learn core model theory, but merely as a means to acquire a rough idea of what it is, and what its central open problems are. The word "Proof" is to be understood as an abbreviation for "Proof Sketch" throughout this paper.

In particular, we hope to convey the important features of these ideas without sinking into the morass of finestructural detail in which they are embedded. The level–by–level finestructural analysis of the models one constructs is the locus of some of the most important ideas of core model theory, and so we shall not to able to avoid it entirely. We shall, however, suppress as much detail as possible. We hope that the resulting outline will be useful to anyone acquainted with the theory of iterated ultrapowers contained in Kunen's paper [Ku70] on $\mathbf{L}[U]$, and the finestructure theory for $\mathbf{L}$ contained in Jensen's paper [Jen72]. The knowledge about $\mathbf{L}[U]$ can be easily obtained from [Kan94, §§12, 19, 20].

One hypothesis whose consistency strength can be bounded from below using the core model theory of this paper is the hypothesis of $\mathbf{L}(\mathbb{R})$*-generic absoluteness.* The fact that this instance of generic absoluteness can be proved from large cardinal hypotheses[11] is strong evidence that the theory of $\mathbf{L}(\mathbb{R})$ which follows from these hypotheses is complete for natural statements, and hence has a privileged position. As an application of the core model theory we outline here, we shall sketch a proof of the following theorem of Hugh Woodin and the second author:

---

witnessing large cardinal hypotheses stronger than the existence of many strong cardinals impossible.

[10] This is again due to an artificial limitation on the degree of overlapping in the coherent sequence from which the model is constructed; *cf.* Definition 2.31.

The precise upper bound on the logical complexity of the reals in these models involves the game quantifier for certain clopen games of length ω_1; it goes well beyond notions of definability familiar from descriptive set theory, like definability over $\mathbf{L}(\mathbb{R})$. *Cf.* [Nee98].

[11] The existence of arbitrarily large Woodin cardinals is sufficient.

Theorem 1.1. If for every set–sized forcing notion $\mathbb{P}$ and every G which is $\mathbb{P}$–generic over $\mathbf{V}$ we have

$$(\mathbf{L}_{\omega_1}(\mathbb{R}))^{\mathbf{V}} \equiv (\mathbf{L}_{\omega_1}(\mathbb{R}))^{\mathbf{V}[G]},$$

then there is an inner model with a Woodin cardinal.

In other words, the consistency strength of generic absoluteness is at least the consistency strength of "There is a Woodin cardinal".[12]

Of course, we hope that this paper sparks an interest in core model theory in some readers. The constructions described in this paper can be found in much greater detail in [MiSt94] and [St96], and will be described in somewhat more detail in the survey article [St9?a].

2. Preliminaries

We will introduce the main notions and some notation. Readers familiar with extenders and premice can skip the first two subsections.

2.1. **Extenders.** Many large cardinal axioms can be written in some form of an elementary embedding axiom. In fact, researchers in the field of large cardinals tend to use elementary embedding properties to define the cardinals they study. If j is an elementary embedding from the universe $\mathbf{V}$ to a transitive class M which is nontrivial (*i.e.* not the identity), then we denote the least ordinal κ such that $j(\kappa) \neq \kappa$ by $\mathrm{crit}(j)$ and call it *the critical point of* j.

To see the reflection inherent in the existence of such an embedding, notice that if $\mathbf{V} \models \varphi[\kappa]$, and M resembles $\mathbf{V}$ enough that $M \models \varphi[\kappa]$, then $M \models \exists\alpha < j(\kappa)(\varphi(\alpha))$, so there is an $\alpha < \kappa$ such that $\mathbf{V} \models \varphi[\alpha]$. Thus the more M resembles $\mathbf{V}$, the stronger the reflection in the cumulative hierarchy present at κ, and, it turns out, the greater the consistency strength of the axiom asserting that such an embedding exists. The "ultimate" resemblance $M = \mathbf{V}$ is impossible by a result of Kunen (*cf.* [Ku71]).

The large cardinal corresponding to an elementary embedding is its critical point. So we have:

- κ is a *measurable* cardinal iff there is an elementary embedding $j : \mathbf{V} \to M$ with $\mathrm{crit}(j) = \kappa$.
- κ is an *α–strong* cardinal iff there is an elementary embedding $j : \mathbf{V} \to M$ with $\mathrm{crit}(j) = \kappa$ and $\mathbf{V}_\alpha \subseteq M$ (*cf.* Figure 1).
- κ is a *strong* cardinal iff κ is α–strong for all $\alpha \in \mathrm{Ord}$.

[12]This instance of generic absoluteness actually implies that for each $n < \omega$, there is an inner model with n Woodin cardinals. Hugh Woodin has shown that generic absoluteness for arbitrary statements about $\mathbf{L}(\mathbb{R})$ implies that there is an inner model with ω Woodin cardinals.

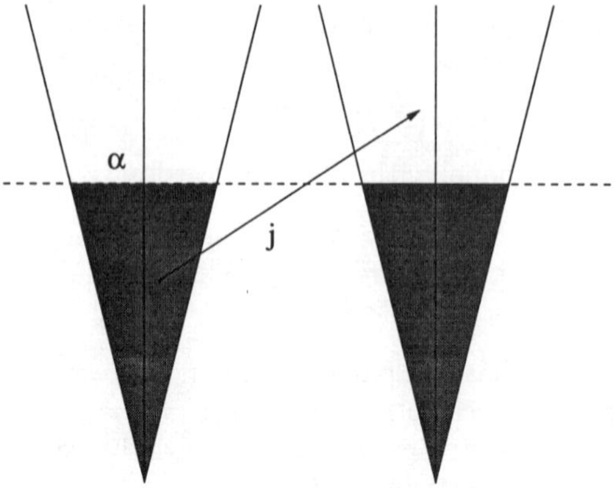

FIGURE 1. An α–strong embedding

- κ is a *superstrong* cardinal iff there is an elementary embedding $j : \mathbf{V} \to M$ with $\text{crit}(j) = \kappa$ and $\mathbf{V}_{j(\kappa)} \subseteq M$.
- κ is an *α–supercompact* cardinal iff there is an elementary embedding $j : \mathbf{V} \to M$ with $\text{crit}(j) = \kappa$ and $M^\alpha \subseteq M$.
- κ is a *supercompact* cardinal iff κ is α–supercompact for all $\alpha \in \text{Ord}$.
- κ is a *huge* cardinal iff there is an elementary embedding $j : \mathbf{V} \to M$ with $\text{crit}(j) = \kappa$ and $M^{j(\kappa)} \subseteq M$.

The axioms corresponding to the embedding properties above, namely, that the cardinals (or embeddings) in question exist, are listed in order of increasing consistency strength (for any nontrivial choice of α), and in most cases in order of least cardinal having the property in question.[13] The cardinals with which we shall be mainly concerned in this paper, *Woodin* cardinals, lie between strong and superstrong cardinals in consistency strength. We shall define them in Section 2.7 (*cf.* Definition 2.27).

Asserting the existence of elementary embeddings with certain additional properties is an elegant and powerful way to obtain systems transcending ZFC. But, as elementary embeddings from a proper class model into another proper class model are proper classes, we need some coding device to talk about them properly in our formal language. Readers familiar with measurable cardinals know that ultrafilters do this job to a certain extent: A nonprincipal κ–complete ultrafilter U on κ induces the ultrapower embedding $j_U : \mathbf{V} \to M :=$

[13]The resemblance of M to $\mathbf{V}$ suffices in many cases to reflect all the weaker properties to smaller cardinals. This is not obvious in some cases, however; the argument requires the use of something like extenders representing the weaker embeddings.

The notable exceptions are: The least superstrong is smaller than the least strong, and the least huge is smaller than the least supercompact. In these cases the fact that an extender from the weaker cardinal overlaps the stronger cardinal allows downward reflection.

Ult($\mathbf{V}, U$) with critical point κ, and conversely if we have any embedding j with crit(j) $= \kappa$ then we can define an ultrafilter on κ by setting

$$X \in U \iff \kappa \in j(X).$$

So ultrafilters are one way to witness the existence of an embedding. However, one can easily show that $U \notin \mathrm{Ult}(\mathbf{V}, U)$, and thus such an ultrapower embedding can never have the property $\mathbf{V}_{\kappa+2} \subseteq M$, because the ultrafilter (an element of $\mathfrak{P}(\mathfrak{P}(\kappa))$) is not in $\mathfrak{M}$. So to represent embeddings whose target models have the stronger closure properties described above, we need a more powerful coding device.

Given $j : M \to N$ with $\kappa = \mathrm{crit}(j)$ and $\lambda < j(\kappa)$, put

$$X \in E_a :\iff a \in j(X)$$

for $a \in [\lambda]^{<\omega}$ and $X \subseteq [\kappa]^{|a|}$, $X \in M$. We call the system

$$E := \langle E_a : a \in [\lambda]^{<\omega} \rangle$$

the $\langle \kappa, \lambda \rangle$*-extender over* M derived from j. We call κ the *critical point of* E and λ the *length of* E, in symbols: $\kappa = \mathrm{crit}(E)$, $\lambda = \mathrm{lh}(E)$.

If $a \subseteq b$, then there is a natural embedding from Ult(M, E_a) into Ult(M, E_b) (send $[f]$ to $[\tilde{f}]$, where $\tilde{f}$ comes from f by adding dummy variables at spots corresponding to ordinals in $b \setminus a$). Thus we can set

$$\mathrm{Ult}(M, E) := \mathrm{dirlim}_{a \in [\lambda]^{<\omega}} \mathrm{Ult}(M, E_a).$$

We write $[a, f]$ for the image of the element $[f]$ of Ult(M, E_a) under the natural map from Ult(M, E_a) into Ult(M, E). Thus the universe of Ult(M, E) consists of all $[a, f]$ such that $a \in [\mathrm{lh}(E)]^{<\omega}$ and $f \in M$. We have a natural embedding

$$i_E : M \to \mathrm{Ult}(M, E)$$

given by $i_E(x) = [\{\kappa\}, c_x]$, where c_x is the constant function $\alpha \mapsto x$. Since there is no danger of confusion, we will call the embedding i_E the *ultrapower embedding* and the model Ult(M, E) the *ultrapower of* M *by* E.
We have

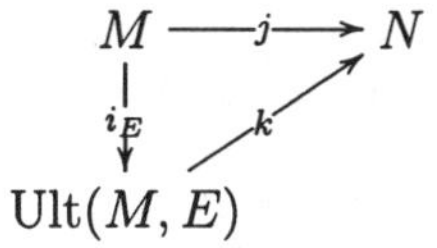

with $k([a, f]) = j(f)(a)$ and $k \restriction \lambda = \mathrm{id}$. So

$$X \in E_a \iff a \in i_E(X),$$

and hence E is the extender derived from i_E.

Because of the diagram above, any embedding can be fully represented as an extender ultrapower.[14] In particular, if $j\colon M \to N$ with $\mathbf{V}_\lambda \subseteq N$, $|\mathbf{V}_\lambda| = \lambda$, and E is the $\langle\kappa, \lambda\rangle$-extender derived from j, then $\mathbf{V}_\lambda \subseteq \mathrm{Ult}(M, E)$.

2.2. **Coherent Sequences and Premice.** The models of set theory we consider will be of the form $\mathbf{L}[\vec{E}]$ where $\vec{E} = \langle E_\alpha : \alpha \in \mathrm{Ord}\rangle$ is a coherent sequence of extenders. Roughly speaking, this means that the αth extender on the sequence is an extender for the αth level of the model, and that the extenders on the sequence are listed in order of increasing strength and without leaving gaps. This setup is due mainly to Mitchell (*cf.* [Mi74], [Mi78]), and it seems adequate to provide models satisfying any of the known large cardinal axioms. The limitations on present–day core model theory probably do not lie in the coherent–sequence–of–extenders framework.

As usual we put[15]:

$$\mathbf{J}_0[\vec{E}] := \emptyset,$$

$$\mathbf{J}_\lambda[\vec{E}] := \bigcup_{\beta<\lambda} \mathbf{J}_\beta[\vec{E}]$$

for λ a limit ordinal, and

$$\mathbf{J}_{\alpha+1}[\vec{E}] := \mathrm{rud}(\mathbf{J}_\alpha[\vec{E}]).$$

Here is an elliptical definition of the sort of extender sequences from which our core models are constructed:

Definition 2.1. A *good extender sequence* is a sequence $\vec{E}$ such that for all α we have either that $E_\alpha = \emptyset$ or that E_α is an extender over $\mathbf{J}_\alpha[\vec{E}]$ such that $(\mathrm{crit}(E_\alpha)^+)^{\mathbf{J}_\alpha[\vec{E}]} \leq \mathrm{lh}(E_\alpha)$, and

1. $i_{E_\alpha}(\vec{E}) \restriction \alpha = \vec{E} \restriction \alpha$,
2. $\left(i_{E_\alpha}(\vec{E})\right)_\alpha = \emptyset$,
3. $\alpha = (\mathrm{lh}(E_\alpha)^+)^{\mathrm{Ult}(\mathbf{J}_\alpha[\vec{E}], E_\alpha)}$, and
4. $\mathbf{J}_\alpha[\vec{E}]$ satisfies the first order conditions enumerated in [MiSt94, p.7].

[14]We ignore here the limitation $\mathrm{lh}(E) < i_E(\mathrm{crit}(E))$, which we imposed only to simplify matters. To represent superstrong, supercompact, or huge embeddings by extenders one would need to drop this limitation.

However, core model theory does not reach these cardinals anyway for other reasons, so for our purposes we may as well assume $\mathrm{lh}(E) < i_E(\mathrm{crit}(E))$. We also ignore the possibility that one might need a proper class extender to capture j.

[15]The reader who is not familiar with Jensen's **J**–hierarchy can replace every **J** by an **L** and the rudimentary closure with the definable closure without great loss. Anyway we have that

$$\mathfrak{P}(\mathbf{J}_\alpha[\vec{E}]) \cap \mathbf{J}_{\alpha+1}[\vec{E}] = \mathrm{def}(\mathbf{J}_\alpha[\vec{E}])$$

where $\mathrm{def}(X)$ denotes the first order definable subsets of X. For this *cf.* [Jen72, Section 2.4].

The most important of the first order conditions from [MiSt94] which we have not listed is the *initial segment condition.* If E is an extender and $\text{crit}(E) < \nu \leq \text{lh}(E)$, then we set $E \restriction \nu = \{\langle a, x\rangle \in E\colon a \in [\nu]^{<\omega}\}$, and call $E \restriction \nu$ an *initial segment* of E. The initial segment condition of [MiSt94] has as a consequence that every proper initial segment of E_α is a member of $\mathbf{J}_\alpha[\vec{E}]$.[16]

Conditions (1.) and (2.) are the coherence conditions. Condition (3.) guarantees that the ordinal α at which $E = E_\alpha$ is indexed is determined completely by E, and therefore must not be used to code random information into the model.[17] We are aiming to construct canonical models, so coding in random information cannot be allowed. There are variant indexing schemes possible; one sometimes convenient scheme involves indexing E at $i_E(\text{crit}(E))$, or at its successor in the E-ultrapower.[18] Such variant schemes lead to hierarchies which are level–by–level intertranslatable with ours.[19]

One important feature of Definition 2.1 is that we have only required E_α to be an extender over $\mathbf{J}_\alpha[\vec{E}]$. E_α will not measure any subsets of its critical point which are constructed in the model we are building after stage α. When we form an ultrapower using E_α, we shall generally form the ultrapower of the largest fragment of the model we can, the largest fragment containing only subsets of $\text{crit}(E_\alpha)$ constructed before α. We need finestructure theory to do this properly, and so adding such "partial" extenders to our sequence may seem to complicate matters. In fact, it leads to dramatic simplifications. The idea is due to Stewart Baldwin and William Mitchell.

Definition 2.2. A structure $\mathfrak{M} = \langle M, \in, \vec{F}, G\rangle$ is called a *premouse*, if it is of the form $\langle \mathbf{J}_\alpha[\vec{E}], \in, E \restriction \alpha, E_\alpha\rangle$, where $\vec{E}$ is a good extender sequence.

If $\mathfrak{M} = \langle M, \in, \vec{F}, G\rangle$ is a premouse then we call $\vec{F}$ the *$\mathfrak{M}$–sequence.* We shall often refer to "extenders on the $\mathfrak{M}$–sequence".

Definition 2.3. If $\mathfrak{M} = \langle \mathbf{J}_\alpha[\vec{E}], \in, E \restriction \alpha, E_\alpha\rangle$ is a premouse and $\beta \leq \alpha$ then set

$$\mathcal{J}_\beta^{\mathfrak{M}} := \langle \mathbf{J}_\beta[\vec{E}], \in, E \restriction \beta, E_\beta\rangle$$

Definition 2.4. $\mathfrak{M}$ is an *initial segment* of $\mathfrak{N}$ (in symbols $\mathfrak{M} \trianglelefteq \mathfrak{N}$) if there is a $\beta \leq \text{Ord}^{\mathfrak{N}}$ such that $\mathfrak{M} = \mathcal{J}_\beta^{\mathfrak{N}}$. We say that $\mathfrak{M}$ and $\mathfrak{N}$ *agree up to α* if $\mathcal{J}_\alpha^{\mathfrak{M}} = \mathcal{J}_\alpha^{\mathfrak{N}}$.

---

[16]More precisely, it states that every proper initial segment of E_α is either on $\vec{E}$ with index $< \alpha$, or "one ultrapower away" from being so.

[17]In [MiSt94], the authors take $\alpha = (\nu(E_\alpha)^+)^{\text{Ult}(\mathbf{J}_\alpha[\vec{E}], E_\alpha)}$, where $\nu(E_\alpha)$ is the supremum of the "generators" of E_α. This difference affects only details we shall ignore here.

[18]This idea is due to Sy Friedman, and has been developed by Jensen. *Cf.* [Jen88].

[19]The comparability of inner model operators proved in [St82] is good evidence this *must* be the case.

Definition 2.5. A structure $\mathfrak{M}$ is *k-sound* if for all $n \leq k$

$$\mathfrak{M} = \mathrm{Hull}_n^{\mathfrak{M}}(\varrho_n \cup \{p_n\})$$

where ϱ_n is the Σ_n–projectum of $\mathfrak{M}$ and p_n is the nth standard parameter of $\mathfrak{M}$. We say $\mathfrak{M}$ is *sound* just in case $\mathfrak{M}$ is k-sound for all $k < \omega$.[20]

All levels of the core models we shall construct will be sound premice. Nonetheless, we must study unsound premice as well, because iterates of sound premice may fail to be sound. This occurs when the critical point of one of the ultrapower embeddings lies above one of the projecta of the premouse whose ultrapower is being taken:

Proposition 2.6. Let $\mathfrak{M}$ be a premouse and E an extender on the $\mathfrak{M}$–sequence whose critical point is above the Σ_1–projectum ϱ of $\mathfrak{M}$. Then $\mathrm{Ult}(\mathfrak{M}, E)$ is not 1–sound.

Proof: A little finestructural analysis shows that in this case, the Σ_1–projectum of $\mathrm{Ult}(\mathfrak{M}, E)$ is still ϱ, and the 1st standard parameter is the image $i_E(p)$ of the 1st standard parameter p of $\mathfrak{M}$. It follows immediately that $\mathrm{Ult}(\mathfrak{M}, E)$ is not 1–sound; in fact, $\mathrm{Hull}_1^{\mathrm{Ult}(\mathfrak{M},E)}(\varrho \cup \{i_E(p)\})$ is just $\mathrm{ran}(i_E)$, and $\mathrm{crit}(E)$ is not a member of it. □

But our finestructural ultrapowers satisfy enough of Łos's theorem that the property that all proper initial segments are sound is passed from the structure to its ultrapower. Since we shall make heavy use of this property[21], we make it part of the definition of "$\mathfrak{M}$ is a premouse" that all proper initial segments of $\mathfrak{M}$ are sound.[22]

Given an arbitrary premouse $\mathfrak{M}$, there is a natural way to try to extract from $\mathfrak{M}$ a sound premouse $\mathfrak{C}(\mathfrak{M})$ which is in some sense equivalent to $\mathfrak{M}$. The correct procedure involves some finestructural subtleties, but we can easily sketch the basic idea. Set $\mathfrak{C}_0 = \mathfrak{M}$, and

$$\mathfrak{C}_{n+1} = \mathrm{Hull}_{n+1}^{\mathfrak{C}_n}(\varrho_{n+1} \cup \{p_{n+1}\}),$$

[20]For readers unfamiliar with these standard notions of finestructure theory of core models, [Dev84] or [Jen72] are good references. Roughly speaking, ϱ_n is the first ordinal α such that there is a new Σ_n subset of α which is not in $\mathfrak{M}$, and p_n is the least parameter in the order of construction from which you can define such a subset. These are not literally the concepts and definitions used in [MiSt94] for technical reasons which are beyond the resolution of the microscope we are using here.

[21]Or rather, we would do so if we were giving detailed proofs.

[22]This can be regarded as another clause in the elliptical Definition 2.1.

where ϱ_{n+1} and p_{n+1} are the Σ_{n+1}–projectum and standard parameter of $\mathfrak{C}_n$.[23] If $\mathfrak{M}$ is well–behaved, then the projecta $\varrho_{n+1}^{\mathfrak{C}_n}$ are decreasing, and hence they and the "cores" $\mathfrak{C}_n$ are eventually constant. We then set

$$\mathfrak{C}(\mathfrak{M}) = \text{ eventual value of } \mathfrak{C}_n,$$

and

$$\varrho_\omega^{\mathfrak{M}} = \text{ eventual value of } \varrho_{n+1}^{\mathfrak{C}_n},$$

as n approaches ω, and call $\mathfrak{C}(\mathfrak{M})$ and $\varrho_\omega^{\mathfrak{M}}$ the (ωth) *core* and *projectum* of $\mathfrak{M}$. We call a premouse $\mathfrak{M}$ *solid* if it is well–behaved in the way that guarantees that $\mathfrak{C}(\mathfrak{M})$ and $\varrho_\omega^{\mathfrak{M}}$ exist.[24]

One important consequence of the soundness of their proper initial segments is that premice satisfy a strong local form of GCH:

If there is a subset of κ in $\mathbf{J}_{\tau+1}^{\mathfrak{M}} \setminus \mathbf{J}_{\tau}^{\mathfrak{M}}$, then $\mathcal{J}_{\tau+1}^{\mathfrak{M}} \models |\tau| \leq \kappa$.

So for the models $\mathbf{L}[\vec{E}]$ we build, we have

$$\mathfrak{P}(\kappa) \cap \mathbf{L}[\vec{E}] \subseteq \mathbf{J}_{(\kappa^+)^{\mathbf{L}[\vec{E}]}}[\vec{E}],$$

and an immediate consequence of this is that our models will be models of GCH.

The levels of the usual hierarchy for $\mathbf{L}$ are all sound, but this is not true for the levels of the most obvious hierarchy on Silver's core model $\mathbf{L}[U]$, the universe of sets constructible from a normal ultrafilter U on a measurable cardinal.[25] Silver's model *is* one of the core models we study here, but we do so via a different hierarchy. The soundness of the levels of this new hierarchy is a direct consequence of the Baldwin–Mitchell idea of putting *partial* extenders on the coherent sequence from which we generate it. This new hierarchy has a substantially simpler finestructure, one which generalizes to arbitrary core models in a way the old finestructure theory of $\mathbf{L}[U]$ did not.[26]

2.3. **Iteration Trees.** We now move on to one of the main tools of core model theory, the Comparison Process.

The key to Kunen's theory of $\mathbf{L}[U]$ is the method of *iterated ultrapowers*. Given a structure $\mathfrak{M}_0 = \langle \mathbf{L}_\varsigma[U], \in, U \rangle$ with appropriate ultrafilter U, one

[23] It would be wrong to replace $\mathfrak{C}_n$ by $\mathfrak{M}$ in this definition. For example, $\mathfrak{M}$ might come from a sound mouse by iterating it above its Σ_1–projectum in such a way to encode some random information in a $\Sigma_2^{\mathfrak{M}}$ subset of ω. We want $\mathfrak{C}_2$ to be *canonical*, so we cannot take $\mathfrak{C}_2 = \mathrm{Hull}_2^{\mathfrak{M}}(\varrho_2 \cup \{p_2\})$. Instead, we first pass to $\mathfrak{C}_1 = \mathrm{Hull}_1^{\mathfrak{M}}(\varrho_1 \cup \{p_1\})$, which in effect undoes our iteration, then set $\mathfrak{C}_2 = \mathrm{Hull}_2^{\mathfrak{C}_1}(\varrho_2 \cup \{p_2\})$.

[24] We have only sketched an oversimplification of the correct definitions of the core and projectum of $\mathfrak{M}$, and the reader should see [MiSt94, Definition 2.8.1] for the true definitions. Solidity is defined in [MiSt94, Definition 2.8.2].

[25] The usual $\mathbf{L}[U]$ hierarchy is not 1–sound at $\kappa+1$, where κ is the measurable cardinal. For $0^\#$ is Σ_1–definable over $\langle \mathbf{J}_{\kappa+1}[U], \in, U \rangle$ but not a member of $\mathbf{J}_{\kappa+1}[U]$, so that the Σ_1–projectum of this structure is ω. Since $\langle \mathbf{J}_{\kappa+1}[U], \in, U \rangle$ is an uncountable structure, it cannot be the Σ_1–Skolem closure of any countable set.

[26] This finestructure is due to Solovay (unpublished) and [DoJen82].

can form ultrapowers by U and its images under the canonical embeddings repeatedly, taking direct limits at limit ordinals. One obtains thereby structures $\mathfrak{M}_\alpha$ and embeddings $i_{\alpha,\beta}\colon \mathfrak{M}_\alpha \to \mathfrak{M}_\beta$ for $\alpha < \beta$. We call the structures $\mathfrak{M}_\alpha$ *iterates* of $\mathfrak{M}_0$, and say that $\mathfrak{M}_0$ is *iterable* just in case all its iterates are wellfounded. Kunen's key *comparison lemma* states that if $\mathfrak{M}_0$ and $\mathfrak{N}_0$ are two iterable structures of this form, then there are iterates $\mathfrak{M}_\alpha$ and $\mathfrak{N}_\alpha$ such that one of the two is an initial segment of the other.[27]

One can form iterated ultrapowers of an arbitrary premouse $\mathfrak{M}_0$ similarly. In this case, the $\mathfrak{M}_\alpha$-sequence may have more than one extender, and we are allowed to choose any one of them to continue. If E_α is the extender chosen, then we take $\mathfrak{M}_{\alpha+1}$ to be $\mathrm{Ult}(\mathfrak{M}_\alpha, E_\alpha)$.[28] At limit stages we form direct limits and continue. We call any such sequence $\langle\langle\mathfrak{M}_\alpha, E_\alpha\rangle\colon \alpha < \beta\rangle$ a *linear iteration* of $\mathfrak{M}_0$, and the structures $\mathfrak{M}_\alpha$ in it *linear iterates* of $\mathfrak{M}_0$. We say $\mathfrak{M}_0$ is *linearly iterable* just in case all its linear iterates are wellfounded.[29]

Given linearly iterable premice $\mathfrak{M}_0$ and $\mathfrak{N}_0$, there is a natural way to try to compare the two via linear iteration. Having reached $\mathfrak{M}_\alpha$ and $\mathfrak{N}_\alpha$, and supposing neither is an initial segment of the other (as otherwise our work is finished), we pick extenders E and F representing the least disagreement between $\mathfrak{M}_\alpha$ and $\mathfrak{N}_\alpha$, and use these to form $\mathfrak{M}_{\alpha+1}$ and $\mathfrak{N}_{\alpha+1}$.

If the extenders of the coherent sequence of $\mathfrak{M}_0$ do not overlap one another too much, and similarly for $\mathfrak{N}_0$, then this process must terminate with all disagreements between some $\mathfrak{M}_\alpha$ and $\mathfrak{N}_\alpha$ eliminated, so that one is an initial segment of the other. This is the key to core model theory at the level of strong cardinals.

At bottom, the reason this comparison process must terminate is the following: if E and F are the extenders used at a typical stage α, then there will be some a and sets $\overline{X}$ and $\tilde{X}$ such that $X = i_{\eta,\alpha}(\overline{X}) = j_{\xi,\alpha}(\tilde{X})$, and X is measured differently by E_a and F_a.[30]

But then $a \in i_{\alpha,\alpha+1}(X) \Leftrightarrow a \notin j_{\alpha,\alpha+1}(X)$, so $i_{\eta,\alpha+1}(\overline{X}) \neq j_{\xi,\alpha+1}(\tilde{X})$, and the images of $\overline{X}$ and $\tilde{X}$ do not participate in a disagreement at stage $\alpha+1$ the

---

[27]This means that there is a filter F such that $\mathfrak{M}_\alpha$ and $\mathfrak{N}_\alpha$ are of the form

$$\mathfrak{M}_\alpha = \langle \mathbf{L}_\xi[F], \in, F\rangle$$

and

$$\mathfrak{N}_\alpha = \langle \mathbf{L}_\eta[F], \in, F\rangle$$

for some ξ and η. (Here and elsewhere we identify wellfounded, extensional structures with their transitive isomorphs.) In fact, in this simple case we can take α to be $\sup\{|\mathfrak{M}_0|, |\mathfrak{N}_0|\}^+$ and F to be the club filter on α.

[28]This must be qualified, since if E_α does not measure all subsets of its critical point in $\mathfrak{M}_\alpha$, then $\mathrm{Ult}(\mathfrak{M}_\alpha, E_\alpha)$ makes no sense. In this case we take the "largest" E_α-ultrapower of an initial segment of $\mathfrak{M}_\alpha$ we can in order to form $\mathfrak{M}_{\alpha+1}$. (See below.)

[29]In which case we identify these iterates with the premice to which they are isomorphic. Linear iterability should be taken to include the condition that no linear iteration of $\mathfrak{M}_0$ drops to proper initial segments infinitely often.

[30]We use i for the embeddings in the $\mathfrak{M}$-iteration, and j in the $\mathfrak{N}$-iteration.

way they did at stage α. *If* all future extenders used in either iteration have critical point above sup(a), then $i_{\eta,\beta}(\overline{X}) \neq j_{\xi,\beta}(\tilde{X})$ for all β, so the images of $\overline{X}$ and $\tilde{X}$ never again participate in a disagreement, and we have made real progress at stage α. A simple Fodor argument shows that if we never "move generators" in one of our iterations,[31] then eventually all disagreements are removed.[32] The lack of overlaps in the sequences of mice below a strong cardinal means that this process of iterating away the least disagreement does not move generators, and hence terminates in a successful comparison.

However, beyond a strong cardinal this linear comparison process definitely will lead to moving generators. There are tricks for making do with linear iterations a bit beyond strong cardinals, but the right solution is to give up linearity. If the extender E_α from the $\mathfrak{M}_\alpha$–sequence we want to use has critical point less than $\mathrm{lh}(E_\beta)$ for some $\beta < \alpha$, then we apply E_α not to $\mathfrak{M}_\alpha$, but to $\mathfrak{M}_\beta$, for the least such β: *i.e.*, we set $\mathfrak{M}_{\alpha+1} = \mathrm{Ult}(\mathfrak{M}_\beta, E_\alpha)$, where β is least such that $\mathrm{crit}(E_\alpha) < \mathrm{lh}(E_\beta)$.[33] We have an embedding $i_{\beta,\alpha+1}\colon \mathfrak{M}_\beta \to \mathfrak{M}_{\alpha+1}$. Thus this new iteration process gives rise to a *tree* of models, with embeddings along each branch of the tree. Along each branch the generators of the extenders used are not moved by later embeddings, and this is good enough to show that if a comparison process involving the formation of such "iteration trees" goes on long enough, it must eventually succeed.

What one needs to keep the construction of an iteration tree going past some limit ordinal λ is a branch of the tree which has been visited cofinally often before λ and is such that the direct limit of the premice along the branch is wellfounded. Thus the iterability we need for comparison amounts to the existence of some method for choosing such branches. We can formalize this as the existence of a winning strategy in a certain game.

We shall now define this "iteration game" and introduce the important notions of the theory of iteration trees in terms of this game.

Let $\mathfrak{M}$ be a premouse and ϑ an ordinal. We shall define the iteration game on $\mathfrak{M}$ of length ϑ. In this game, players I and II cooperate to produce an iteration tree $\mathfrak{T}$. This system consists of a tree order[34] T on ϑ together with, for $\alpha, \beta < \vartheta$:

---

[31]*I.e.*, if whenever E is used before E' in the $\mathfrak{M}$–iteration, then $\mathrm{lh}(E) \leq \mathrm{crit}(E')$, and similarly on the $\mathfrak{N}$–side.

[32]More precisely, there must be a stage $\alpha < \sup\{|\mathfrak{M}_0|, |\mathfrak{N}_0|\}^+$ at which $\mathfrak{M}_\alpha$ is an initial segment of $\mathfrak{N}_\alpha$, or *vice versa*.

[33]Again, if E_α fails to measure all sets in $\mathfrak{M}_\beta$, we take the ultrapower of the longest possible initial segment of $\mathfrak{M}_\beta$.

[34]An order T on an ordinal ϑ is called a *tree order* if

- For all $\alpha \neq 0$, we have $0T\alpha$,
- $\{\alpha : \alpha T\beta\}$ is wellordered by T for all $\beta < \vartheta$,
- successor ordinals are successors in the order T,
- if $\alpha T\beta$, then $\alpha < \beta$, and
- for any limit $\lambda < \vartheta$, $\{\alpha\colon \alpha T\lambda\}$ must be cofinal in λ.

- a premouse $\mathfrak{M}_\alpha^{\mathfrak{T}}$,
- if $\alpha + 1 < \vartheta$, an extender $F_\alpha^{\mathfrak{T}}$ on the $\mathfrak{M}_\alpha^{\mathfrak{T}}$–sequence,
- maps $i_{\alpha\beta}^{\mathfrak{T}} : \mathfrak{M}_\alpha^{\mathfrak{T}} \to \mathfrak{M}_\beta^{\mathfrak{T}}$ for $\alpha T \beta$ such that the branch of T leading from α to β has not dropped.[35]

The game is played as follows: To start, we set $\mathfrak{M}_0^{\mathfrak{T}} := \mathfrak{M}$. Then in move $\alpha + 1$, we suppose that we have already defined

- $F_\gamma^{\mathfrak{T}}$ for $\gamma < \alpha$,
- $\mathfrak{M}_\gamma^{\mathfrak{T}}$ for $\gamma \leq \alpha$,
- and the tree structure T up to α.

Now, player I picks some $F_\alpha^{\mathfrak{T}}$ from the $\mathfrak{M}_\alpha^{\mathfrak{T}}$–sequence such that $\mathrm{lh}(F_\alpha^{\mathfrak{T}}) > \mathrm{lh}(F_\gamma^{\mathfrak{T}})$ for all $\gamma < \alpha$. If he can't, he loses, and the game is over.

Let β be least such that $\mathrm{crit}(F_\alpha^{\mathfrak{T}}) < \mathrm{lh}(F_\beta^{\mathfrak{T}})$. Then set β to be the T–predecessor of $\alpha + 1$. We wish to take the $F_\alpha^{\mathfrak{T}}$–ultrapower of $\mathfrak{M}_\beta^{\mathfrak{T}}$, but (as we pointed out before) there is a problem in that it might be the case that $(\mathfrak{P}(\kappa))^{\mathfrak{M}_\beta^{\mathfrak{T}}} \not\subseteq \mathfrak{M}_\alpha^{\mathfrak{T}}$ for $\kappa = \mathrm{crit}(F_\alpha^{\mathfrak{T}})$ so that the ultrapower construction simply doesn't make sense.

So set $(\mathfrak{M}_{\alpha+1}^{\mathfrak{T}})^*$ to be the longest initial segment $\mathfrak{N}$ of $\mathfrak{M}_\beta^{\mathfrak{T}}$ for which $\mathfrak{P}(\kappa)^{\mathfrak{N}} \subseteq \mathfrak{M}_\alpha^{\mathfrak{T}}$. One can show that $\mathrm{lh}(F_\beta^{\mathfrak{T}}) \leq \mathrm{Ord}^{(\mathfrak{M}_{\alpha+1}^{\mathfrak{T}})^*}$ and $\mathfrak{P}(\kappa)^{\mathfrak{M}_\alpha^{\mathfrak{T}}} = \mathfrak{P}(\kappa)^{(\mathfrak{M}_{\alpha+1}^{\mathfrak{T}})^*}$. Now let

$$\mathfrak{M}_{\alpha+1}^{\mathfrak{T}} := \mathrm{Ult}((\mathfrak{M}_{\alpha+1}^{\mathfrak{T}})^*, F_\alpha^{\mathfrak{T}}).^{36}$$

We say the iteration tree $\mathfrak{T}$ *drops at* $\alpha + 1$ if $\mathfrak{M}_\beta^{\mathfrak{T}} \neq (\mathfrak{M}_{\alpha+1}^{\mathfrak{T}})^*$ and we say a branch b *drops at* $\alpha + 1$ if $\mathfrak{T}$ drops at $\alpha + 1$ and $\alpha + 1 \in b$.

Obviously, we do not necessarily have embeddings from $\mathfrak{M}_\beta^{\mathfrak{T}}$ to the models attached to the successors of β in T, because of possible dropping to an initial segment. But if $\mathfrak{T}$ does not drop at $\alpha + 1$ then we have the ultrapower

---

We use the standard interval notation in tree orders, *i.e.* $[\alpha, \beta] := \{\gamma \leq \beta : \alpha T \gamma T \beta\} \cup \{\alpha, \beta\}$ and likewise for open and half–open intervals.

[35] We will say later what we mean by this.

[36] To be more precise, let n be largest such that $\kappa < \varrho_n^{\mathfrak{M}_{\alpha+1}^*}$ and set

$$\mathfrak{M}_{\alpha+1}^{\mathfrak{T}} := \mathrm{Ult}_n((\mathfrak{M}_{\alpha+1}^{\mathfrak{T}})^*, F_\alpha^{\mathfrak{T}}),$$

where the subscript n in Ult_n indicates that the ultrapower is to be formed using functions which are Σ_n over $(\mathfrak{M}_{\alpha+1}^{\mathfrak{T}})^*$.

It is important in many contexts that the ultrapowers $\mathrm{Ult}_n((\mathfrak{M}_{\alpha+1}^{\mathfrak{T}})^*, F_\alpha^{\mathfrak{T}})$ taken at successor steps in the iteration game be "as large as possible", but one needs basic finestructural facts about the Σ_n–projectum $\varrho_n^{(\mathfrak{M}_{\alpha+1}^{\mathfrak{T}})^*}$ in order to keep track precisely of the degree of elementarity of the maps $i_{\gamma,\delta}^{\mathfrak{T}}$.

This is important for example for the proof of the comparison lemma 2.10. On the other hand, the proofs of the basic facts about the Σ_n–projectum require the comparison lemma. As a consequence the full development of the theory we outlined, requires an induction on the Levy hierarchy of mice.

embedding, and we can take it to define $i^{\mathfrak{T}}_{\beta,\alpha+1}$, and then define $i^{\mathfrak{T}}_{\gamma,\alpha+1} := i^{\mathfrak{T}}_{\beta,\alpha+1} \circ i^{\mathfrak{T}}_{\gamma,\beta}$ for $\gamma T \beta$. If $\mathfrak{T}$ drops at $\alpha+1$, we let $i^{\mathfrak{T}}_{\beta,\alpha+1}$ be undefined.

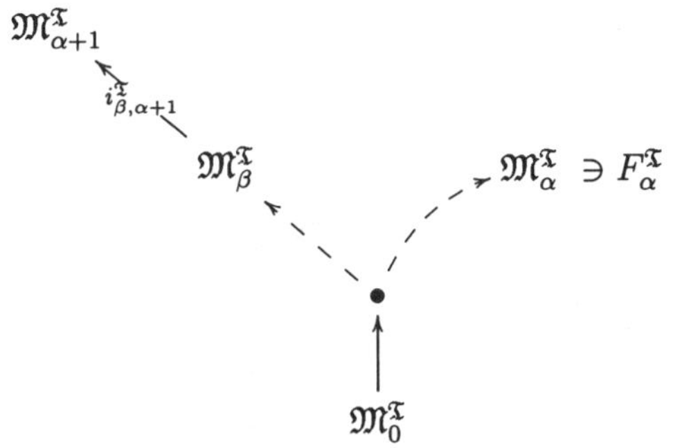

In a limit move λ, II picks a branch b of $\mathfrak{T}$ such that

- b is cofinal in λ,
- b drops at finitely many β, and
- the direct limit of the $\mathfrak{M}^{\mathfrak{T}}_{\alpha}$ under the $i^{\mathfrak{T}}_{\alpha,\beta}$ for sufficiently large $\alpha \in b$ is wellfounded.

If II fails to do so, then the game is over and he loses. We call such a branch b a *cofinal wellfounded branch* of $\mathfrak{T} \upharpoonright \lambda$.

Set $\mathfrak{M}^{\mathfrak{T}}_{\lambda} := \text{dirlim}_{\alpha \in b} \mathfrak{M}^{\mathfrak{T}}_{\alpha}$, set $\alpha T \lambda$ for all $\alpha \in b$ and take as $i^{\mathfrak{T}}_{\alpha,\lambda}$ the canonical embeddings for $\alpha T \lambda$ into the direct limit, if α is sufficiently large for the embedding to exist.

This completes the rules of the iteration game. If no one has lost after ϑ moves, player II wins. A tree $\mathfrak{T}$ constructed by playing this game where no player violates the rules is called an *iteration tree*.

Definition 2.7. $\mathfrak{M}$ is *ϑ–iterable* iff player II has a winning strategy in the iteration game of length ϑ. Such a strategy is called a *ϑ–iteration strategy* for $\mathfrak{M}$. $\mathfrak{M}$ is *iterable* iff $\mathfrak{M}$ is ϑ–iterable for all ϑ.

It is customary to call a premouse with useful iterability properties a *mouse*, and we shall adhere to this custom in informal discussion. We prefer, however, to give no formal definition of the term. There are many varieties of iterability beyond the one we have described formally above, and it is not clear which one should be enshrined in a formal definition of "mouse".

The possibility of dropping in an iteration tree adds some complexity, but it cannot be avoided. Indeed, even if we had demanded that all extenders on the sequence of a premouse be total on the premouse itself, the fact that in our nonlinear iterations we want to apply an extender E from the $\mathfrak{M}^{\mathfrak{T}}_{\alpha}$–sequence to some $\mathfrak{M}^{\mathfrak{T}}_{\beta}$ for $\beta < \alpha$ commits us to dropping, for there is no reasonable way to insure that E will measure all sets in $\mathfrak{M}^{\mathfrak{T}}_{\beta}$. But once we are

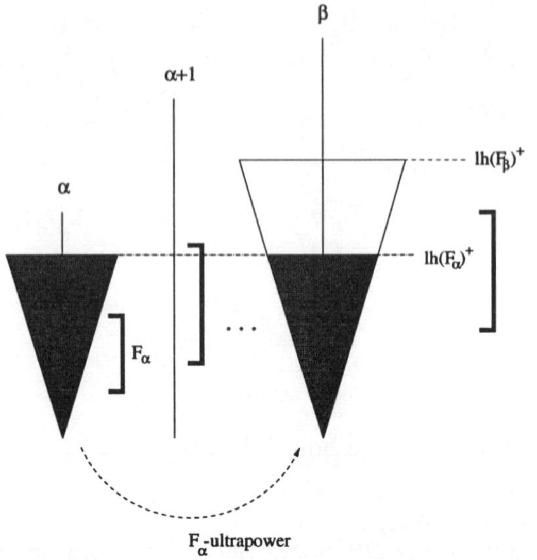

FIGURE 2. The agreement in iteration trees

committed to dropping to an initial segment of $\mathfrak{M}^{\mathfrak{T}}_{\beta}$, we need finestructure to make sure we do not drop too far, that we keep every last quantifier's worth of elementarity we can. This is why the Baldwin–Mitchell approach, which gives up the totality of the extenders on our coherent sequences in order to simplify finestructure, is an unadulterated gain once we enter the realm of nonlinear iterations.

One basic fact about iteration trees is the following agreement lemma:

Theorem 2.8. Let $\mathfrak{M}$ be a premouse, $\mathfrak{T}$ an iteration tree on $\mathfrak{M}$ with models

$$\langle \mathfrak{M}^{\mathfrak{T}}_{\gamma} : \gamma < \mathrm{lh}(\mathfrak{T}) \rangle,$$

let $\alpha < \beta < \mathrm{lh}(\mathfrak{T})$, and μ the index of $F^{\mathfrak{T}}_{\alpha}$ on the $\mathfrak{M}^{\mathfrak{T}}_{\alpha}$–sequence (so that μ is the cardinal successor of $\mathrm{lh}(F^{\mathfrak{T}}_{\alpha})$ in $\mathrm{Ult}(\mathcal{J}^{\mathfrak{M}^{\mathfrak{T}}_{\alpha}}_{\mu}, F^{\mathfrak{T}}_{\alpha})$); then

1. $\mathfrak{M}^{\mathfrak{T}}_{\alpha}$ and $\mathfrak{M}^{\mathfrak{T}}_{\beta}$ agree up to μ,
2. μ is a cardinal in $\mathfrak{M}^{\mathfrak{T}}_{\beta}$, and
3. $\mathfrak{M}^{\mathfrak{T}}_{\alpha}$ and $\mathfrak{M}^{\mathfrak{T}}_{\beta}$ do not agree below $\mu + 1$.

Proof : The proof is a routine induction using clause (1.) of Definition 2.1.
The agreement of $\mathfrak{M}^{\mathfrak{T}}_{\alpha}$ with later models passes through limit stages λ because if $\beta T \lambda$ and $\alpha < \beta$, then all extenders used in $[\beta, \lambda)_T$ have critical point above $\mathrm{lh}(F^{\mathfrak{T}}_{\alpha})$, so the agreement between $\mathfrak{M}^{\mathfrak{T}}_{\alpha}$ and $\mathfrak{M}^{\mathfrak{T}}_{\beta}$ gives the desired agreement between $\mathfrak{M}^{\mathfrak{T}}_{\alpha}$ and $\mathfrak{M}^{\mathfrak{T}}_{\lambda}$.

For the successor step, it is enough to see that $\mathfrak{M}_\beta^\mathfrak{T}$ agrees with $\mathfrak{M}_{\beta+1}^\mathfrak{T}$ up to ν, where ν is the index of $F_\beta^\mathfrak{T}$ on the $\mathfrak{M}_\beta^\mathfrak{T}$–sequence. Let $\gamma = \text{pred}_T(\beta+1)$, and let $\kappa = \text{crit}(F_\beta^\mathfrak{T})$. Since $\kappa < \text{lh}(F_\gamma^\mathfrak{T})$, our induction hypothesis implies that $\mathcal{J}_\nu^{\mathfrak{M}_\beta^\mathfrak{T}}$ agrees with $\mathfrak{M}_\gamma^\mathfrak{T}$ below their common value for κ^+, and hence agrees similarly with $(\mathfrak{M}_{\beta+1}^\mathfrak{T})^*$. It follows that $\text{Ult}(\mathcal{J}_\nu^{\mathfrak{M}_\beta^\mathfrak{T}}, F_\beta)$ agrees with $\text{Ult}((\mathfrak{M}_{\beta+1}^\mathfrak{T})^*, F_\beta) = \mathfrak{M}_{\beta+1}^\mathfrak{T}$ below the common image of κ^+ in the two ultrapowers. By the coherence condition on good extender sequences, $\mathfrak{M}_\beta^\mathfrak{T}$ agrees with $\text{Ult}(\mathcal{J}_\nu^{\mathfrak{M}_\beta^\mathfrak{T}}, F_\beta^\mathfrak{T})$ below ν, and since ν is less than the image of κ in this ultrapower, we have proved (1.).

The proof of (2.) is similar. For (3.), notice that by the second coherence condition in Definition 2.1, μ is the index of the empty "extender" on the sequence of $\text{Ult}(\mathcal{J}_\mu^{\mathfrak{M}_\alpha^\mathfrak{T}}, F_\alpha^\mathfrak{T})$, and hence μ indexes the empty extender in $\mathfrak{M}_\beta^\mathfrak{T}$. Since $F_\alpha^\mathfrak{T} \neq \emptyset$, $\mathfrak{M}_\alpha^\mathfrak{T}$ disagrees with $\mathfrak{M}_\beta^\mathfrak{T}$ at μ. □

The fact that the index μ of $F_\alpha^\mathfrak{T}$ is a cardinal in all later models of $\mathfrak{T}$ contrasts with the fact that it is not a cardinal in $\mathfrak{M}_\alpha^\mathfrak{T}$ itself. This is because one can use the representation of μ in the $F_\alpha^\mathfrak{T}$–ultrapower to define a map from $\text{lh}(F_\alpha^\mathfrak{T})$ onto μ.[37]

These observations yield the following useful corollary to Theorem 2.8. Let us say that two extenders are *compatible* if one is an initial segment of the other.

Corollary 2.9. Let $\mathfrak{T}$ be an iteration tree, and $F_\alpha^\mathfrak{T}$ and $F_\beta^\mathfrak{T}$ be extenders used in $\mathfrak{T}$ such that $\alpha \neq \beta$; then $F_\alpha^\mathfrak{T}$ is incompatible with $F_\beta^\mathfrak{T}$.

Proof: Say $\alpha < \beta$, so that $\text{lh}(F_\alpha^\mathfrak{T}) < \text{lh}(F_\beta^\mathfrak{T})$. If the two extenders are compatible, then $F_\alpha^\mathfrak{T}$ is a proper initial segment of $F_\beta^\mathfrak{T}$, and so by the initial segment condition of [MiSt94], $F_\alpha^\mathfrak{T} \in \mathfrak{M}_\beta^\mathfrak{T}$. By the observation above, this means the index of $F_\alpha^\mathfrak{T}$ on the $\mathfrak{M}_\alpha^\mathfrak{T}$–sequence is not a cardinal in $\mathfrak{M}_\beta^\mathfrak{T}$, contrary to Theorem 2.8. □

2.4. **The Comparison Process.** We come to the main tool of core model theory, the comparison process. We shall illustrate its use in the proof of the following "comparison lemma". There are many other uses for the comparison process.

Theorem 2.10 (Comparison Lemma). Let Ω be a regular cardinal, and let $\mathfrak{M}$ and $\mathfrak{N}$ be $\Omega+1$–iterable premice of ordinal height $\leq \Omega$; then there are iteration trees $\mathfrak{T}$ and $\mathfrak{U}$ of length $\leq \Omega+1$ on $\mathfrak{M}$ and $\mathfrak{N}$ respectively, with last models $\mathfrak{P}$ and $\mathfrak{Q}$, such that either

[37]It is possible that μ is the ordinal height of $\mathfrak{M}_\alpha^\mathfrak{T}$. In this case, the surjection of $\text{lh}(F_\alpha^\mathfrak{T})$ onto μ is not a member of $\mathfrak{M}_\alpha^\mathfrak{T}$, but is Σ_1–definable over $\mathfrak{M}_\alpha^\mathfrak{T}$.

1. $\mathfrak{P} \trianglelefteq \mathfrak{Q}$, $\mathfrak{P}$ has ordinal height $\leq \Omega$, and the branch of $\mathfrak{T}$ from $\mathfrak{M}$ to $\mathfrak{P}$ does not drop, or
2. $\mathfrak{Q} \trianglelefteq \mathfrak{P}$, $\mathfrak{Q}$ has ordinal height $\leq \Omega$, and the branch of $\mathfrak{U}$ from $\mathfrak{N}$ to $\mathfrak{Q}$ does not drop.

Moreover, if $\mathfrak{M}$ and $\mathfrak{N}$ both have ordinal height $< \Omega$, then $\mathfrak{T}$ and $\mathfrak{U}$ have length $< \Omega$.

Proof : Fix $\Omega+1$–iteration strategies Σ for $\mathfrak{M}$ and Γ for $\mathfrak{N}$. We construct $\mathfrak{T}$ and $\mathfrak{U}$ by induction, "iterating away the least disagreement" at successor steps and using Σ and Γ to pick branches at limit steps.

Let $\mathfrak{M}_0^{\mathfrak{T}} := \mathfrak{M}$ and $\mathfrak{M}_0^{\mathfrak{U}} := \mathfrak{N}$. At an arbitrary successor step our current approximations to $\mathfrak{T}$ and $\mathfrak{U}$ have last models $\mathfrak{M}_\alpha^{\mathfrak{T}}$ and $\mathfrak{M}_\beta^{\mathfrak{U}}$. If one is an initial segment of the other, we are done; otherwise let λ be least such that $\mathcal{J}_\lambda^{\mathfrak{M}_\alpha^{\mathfrak{T}}} \neq \mathcal{J}_\lambda^{\mathfrak{M}_\alpha^{\mathfrak{U}}}$. If $E_\lambda^{\mathfrak{M}_\alpha^{\mathfrak{T}}} = \emptyset$ then we do not extend $\mathfrak{T}$ at this step; otherwise, set $F_\alpha^{\mathfrak{T}} := E_\lambda^{\mathfrak{M}_\alpha^{\mathfrak{T}}}$. The rules of the iteration game now determine the T-predecessor of $\alpha+1$ (*i.e.* the model to which we apply $F_\alpha^{\mathfrak{T}}$), and $\mathfrak{M}_{\alpha+1}^{\mathfrak{T}}$, *etc.* Similarly, if $E_\lambda^{\mathfrak{M}_\beta^{\mathfrak{U}}} = \emptyset$ then we do not extend $\mathfrak{U}$ at this step; otherwise, set $F_\alpha^{\mathfrak{U}} := E_\lambda^{\mathfrak{M}_\alpha^{\mathfrak{U}}}$. Again, this choice of extender and the rules of the iteration game determine a one–model extension of $\mathfrak{U}$. Notice that our agreement lemma 2.8 implies that $F_\alpha^{\mathfrak{T}}$ and $F_\beta^{\mathfrak{U}}$ satisfy the increasing–length condition on the extenders in an iteration tree.

At an arbitrary limit step, one or both of our current approximations to $\mathfrak{T}$ and $\mathfrak{U}$ have limit length. If $\mathfrak{T}$ has limit length, we use Σ to produce a one–model extension of it; otherwise, we do not extend $\mathfrak{T}$. We proceed similarly with $\mathfrak{U}$.

Since Σ and Γ win the iteration game, we do not produce any illfounded ultrapowers at successor stages in the constructions of $\mathfrak{T}$ and $\mathfrak{U}$, and we always get cofinal wellfounded branches at limit stages.

If it terminates at some stage $< \Omega$, then one of the last models on $\mathfrak{T}$ and $\mathfrak{U}$ at that stage is an initial segment of the other, and by appealing to some finestructure theory we can verify the remaining clauses of the conclusion:

We must see that the side which comes out shorter does not drop, and here we use soundness in a crucial way. Suppose $\mathfrak{P}$ is the last model on $\mathfrak{T}$, and the branch from $\mathfrak{M}$ to $\mathfrak{P}$ drops. Consider the last drop along this branch. We dropped because the extender E we applied did not measure all sets in the model $\mathfrak{S}$ to which we wanted to apply it, and therefore we formed instead $\mathrm{Ult}_n(\mathfrak{R}, E)$ for the longest $\mathfrak{R} \trianglelefteq \mathfrak{S}$ and the largest n possible. The maximality of $\mathfrak{R}$ and n implies $\varrho_{n+1}^{\mathfrak{R}} \leq \mathrm{crit}(E)$, and therefore $\mathrm{Ult}_n(\mathfrak{R}, E)$ is not $n+1$–sound by the argument in the proof of Proposition 2.6. The extenders used on the branch from $\mathfrak{M}$ to $\mathfrak{P}$ after E have critical points $\geq \mathrm{lh}(E)$, and thus $\mathfrak{P}$ is not $n+1$–sound. It follows at once that $\mathfrak{P}$ is not a proper initial seqment of the last model $\mathfrak{Q}$ of $\mathfrak{U}$, for any proper initial segment of a premouse is fully sound. A somewhat more subtle finestructural argument shows that $\mathfrak{P} \neq \mathfrak{Q}$.

Suppose now that the process continues for Ω, and hence $\Omega+1$, many steps. Let $\mathfrak{P} = \mathfrak{M}_\Omega^\mathfrak{T}$ and $\mathfrak{Q} = \mathfrak{M}_\Omega^\mathfrak{T}$ be the respective last models. If one of these has ordinal height Ω, then by Theorem 2.8 it is an initial segment of the other, and we can finish the proof as before. The alternative is that each has ordinal height $> \Omega$. We shall show that this alternative leads to a contradiction, thereby completing the proof.

For this, the following general fact about the trees arising in a comparison process is useful: for any α and β, $F_\alpha^\mathfrak{T}$ is incompatible with $F_\beta^\mathfrak{U}$. The proof of this fact is very close to the proof of Corollary 2.9.

Now suppose $\Omega \in \mathrm{Ord}^\mathfrak{P} \cap \mathrm{Ord}^\mathfrak{Q}$, so that we have $\Omega \in \mathrm{ran}(i_{\xi,\Omega}^\mathfrak{T}) \cap \mathrm{ran}(i_{\tau,\Omega}^\mathfrak{U})$ for some $\xi \in [0,\Omega]_T$ and $\eta \in [0,\Omega]_U$. Now every branch of an iteration tree is closed below its supremum, and so $[0,\Omega]_T$ and $[0,\Omega]_U$ are closed unbounded in Ω. Using these facts it is easy to see that there is a closed unbounded set C in Ω such that:

- $C \subseteq [0,\Omega]_T \cap [0,\Omega]_U$,
- if $\kappa \in C$, then $\kappa = \mathrm{crit}(i_{\kappa,\Omega}^\mathfrak{T}) = \mathrm{crit}(i_{\kappa,\Omega}^\mathfrak{U})$, and $i_{\kappa,\Omega}^\mathfrak{T}(\kappa) = i_{\kappa,\Omega}^\mathfrak{U}(\kappa) = \Omega$.

Let $\kappa \in C$, and let E and F be the extenders with critical point κ used along $[0,\Omega]_T$ and $[0,\Omega]_U$ respectively. Because E and F are incompatible, we can find a set

$$A_\kappa \in \mathfrak{M}_\kappa^\mathfrak{T} \cap \mathfrak{M}_\kappa^\mathfrak{U}$$

such that for some $a \in \mathrm{lh}(E) \cap \mathrm{lh}(F)$,

$$A_\kappa \in E_a \Leftrightarrow A_\kappa \notin F_a,$$

so that if τ and σ are the next ordinals on $[0,\Omega]_T$ and $[0,\Omega]_U$ after κ respectively, then

$$a \in i_{\kappa,\tau}^\mathfrak{T}(A_\kappa) \Leftrightarrow a \in i_{\kappa,\sigma}^\mathfrak{U}(A_\kappa).$$

Because generators are not moved along the branches of an iteration tree, if $\nu > \kappa$ and $\nu \in C$, then

$$a \in i_{\kappa,\nu}^\mathfrak{T}(A_\kappa) \Leftrightarrow a \in i_{\kappa,\nu}^\mathfrak{U}(A_\kappa),$$

and therefore

$$i_{\kappa,\nu}^\mathfrak{T}(A_\kappa) \neq A_\nu \text{ or } i_{\kappa,\nu}^\mathfrak{U}(A_\kappa) \neq A_\nu.$$

On the other hand, each $\kappa \in C$ is a limit ordinal, so we can find $\xi_\kappa \in [0,\kappa)_T$ and $\eta_\kappa \in [0,\kappa]_U$ such that $A_\kappa \in \mathrm{ran}(i_{\xi_\kappa,\kappa}^\mathfrak{T}) \cap \mathrm{ran}(i_{\eta_\kappa,\kappa}^\mathfrak{U})$. By Fodor's lemma we can fix the value ξ of ξ_κ on a stationary set S_0, then we thin S_0 to a stationary S_1 on which the value η of η_κ is fixed, and then fix the pre-images of A_κ in $\mathfrak{M}_\xi^\mathfrak{T}$ and $\mathfrak{M}_\eta^\mathfrak{U}$ on a stationary $S_2 \subseteq S_1$. It follows that if $\kappa, \nu \in S_2$ and $\kappa < \nu$, then $i_{\kappa,\nu}^\mathfrak{T}(A_\kappa) = i_{\kappa,\nu}^\mathfrak{U}(A_\kappa) = A_\nu$, which is the desired contradiction. □

We shall call the pair of iteration trees $\langle \mathfrak{T}, \mathfrak{U} \rangle$ produced in the proof of Theorem 2.10 the $\langle \Sigma, \Gamma \rangle$*-coiteration* of $\mathfrak{M}$ with $\mathfrak{N}$. We say that $\mathfrak{N}$ $\langle \Sigma, \Gamma \rangle$*-iterates past* $\mathfrak{M}$ if the first alternative in the conclusion of Theorem 2.10 holds, and say that $\mathfrak{M}$ $\langle\Sigma,\Gamma\rangle$*–iterates past* $\mathfrak{N}$ if the second alternative holds. If both alternatives hold, then we say $\mathfrak{M}$ and $\mathfrak{N}$ have a *common* $\langle\Sigma,\Gamma\rangle$*–iterate*. Later on we shall be dealing mostly with premice having unique iteration strategies, and in this case we shall drop reference to the strategies in these locutions.

The comparison process is used in a crucial way in the proof that sufficiently iterable premice have cores.[38]

Theorem 2.11. If $\mathfrak{N}$ is a premouse all of whose countable elementary substructures are ω_1+1-iterable, then the core $\mathfrak{C}(\mathfrak{N})$ exists, and agrees with $\mathfrak{N}$ below $(\varrho^+)^{\mathfrak{N}}$, where $\varrho=\varrho_\omega^{\mathfrak{N}}$ is the projectum of $\mathfrak{N}$.[39]

Theorem 2.11 is one of the central results of basic finestructure theory, and its proof is far from trivial. Many authors have contributed to the evolution of the theorem.[40]

Since we have avoided even the subtleties involved in correct definitions is this area, we shall not attempt to sketch a proof of Theorem 2.11. However, the reader can gain some appreciation of the problem as follows. Let $\mathfrak{C}_1=\mathrm{Hull}_1^{\mathfrak{N}}(\varrho_1\cup\{p_1\})$, where ϱ_1 is the Σ_1–projectum of $\mathfrak{N}$ and p_1 is its standard parameter. One thing we need to see is that ϱ_1 is also the Σ_1–projectum of $\mathfrak{C}_1$, and for this we need to see that every subset of ϱ_1 in $\mathfrak{N}$ is in $\mathfrak{C}_1$. The natural proof of this involves *comparing* $\mathfrak{N}$ with $\mathfrak{C}_1$. If the iteration maps $i\colon\mathfrak{N}\to\mathfrak{P}$ and $j\colon\mathfrak{C}_1\to\mathfrak{Q}$ to the last models on the two sides have critical point $\geq\varrho_1$, then because both $\mathfrak{C}_1$ and $\mathfrak{N}$ define the same "new" Σ_1 subset of ϱ_1, we get that $\mathfrak{P}=\mathfrak{Q}$, and from this it is easy to see that $\mathfrak{N}$ and $\mathfrak{C}_1$ have the same subsets of ϱ_1.[41] Since $\mathfrak{N}$ and $\mathfrak{C}_1$ agree below ϱ_1, there is reason to hope that i and j have critical point $\geq\varrho_1$, but the existence of extenders E on the sequences of $\mathfrak{N}$ and $\mathfrak{C}_1$ overlapping ϱ_1 is a severe problem.[42] Why couldn't i or j use such an extender? The solution to this problem involves abandoning the simple coiteration above for a more complicated version, and making extensive use of the Dodd-Jensen Lemma 2.16. The idea of comparing $\mathfrak{N}$ and $\mathfrak{C}_1$ via iterations with critical point above ϱ_1 is still at the heart of it, however.

Mice which iterate past all other mice of no greater ordinal height will play an important rôle later on.

---

[38]*Cf.* the discussion following Definition 2.5 for the definition of $\mathfrak{C}(\mathfrak{N})$.

[39]*Cf.* Footnote 20 for a short discussion of finestructural terminology.

If $(\varrho^+)^{\mathfrak{N}}$ does not exist, that is if $\varrho=\mathrm{Ord}^{\mathfrak{N}}$ or ϱ is the largest cardinal of $\mathfrak{N}$, then this should be taken to mean that $\mathfrak{C}(\mathfrak{N})=\mathfrak{N}$.

[40]We shall name Tony Dodd, Ronald Jensen, Sy Friedman, Bill Mitchell, and Ernest Schimmerling (*cf.* [SchSt96]). The final version of Theorem 2.11 was proved in [NeeSt9?]; the argument in that paper rests heavily on [MiSt94, Theorem 8.1].

[41]For premice "below a strong cardinal", one can show that $\mathfrak{N}$ never moves in this coiteration, and so is an iterate of its core $\mathfrak{C}_1$ via an iteration with critical point above the projectum ϱ_1. More complicated mice, however, need not be iterates of their cores.

[42]*I.e.*, extenders such that $\mathrm{crit}(E)<\varrho_1\leq\mathrm{lh}(E)$.

Definition 2.12. Let $\mathfrak{M}$ be a premouse whose ordinal height is a regular cardinal Ω. We say that $\mathfrak{M}$ is *Ω-universal* just in case there is an $\Omega+1$–iteration strategy Σ for $\mathfrak{M}$ such that whenever $\mathfrak{N}$ is a premouse of ordinal height $\leq \Omega$ and Γ is an $\Omega+1$–iteration strategy for $\mathfrak{N}$, then $\mathfrak{M}$ $\langle \Sigma, \Gamma \rangle$–iterates past $\mathfrak{N}$.

Covering theorems can be used to prove that there are universal mice. In a covering theorem one proves that if there is some kind of bound on the complexity of the mice in $\mathbf{V}$, then there is a mouse which is "close" to $\mathbf{V}$ in some sense. The closeness to $\mathbf{V}$ required for universality is that the mouse should compute many successor cardinals correctly; this sort of covering is called *weak covering.*[43] The following very useful lemma shows that a form of weak covering implies universality.

Lemma 2.13. Let Ω be weakly Mahlo, and let $\mathfrak{M}$ be an $\Omega+1$–iterable premouse such that for stationarily many regular cardinals $\alpha < \Omega$, $(\alpha^+)^W = \alpha^+$; then W is Ω–universal.

Proof : Let Σ be an $\Omega+1$–iteration strategy for W, and let $\mathfrak{N}$ be a premouse of ordinal height $\leq \Omega$, and Γ an $\Omega+1$–iteration strategy for $\mathfrak{N}$. Let $\langle \mathfrak{T}, \mathfrak{U} \rangle$ be the $\langle \Sigma, \Gamma \rangle$–coiteration of W with $\mathfrak{N}$, and assume toward contradiction that W does not iterate past $\mathfrak{N}$. Thus the last model of $\mathfrak{T}$ is a proper initial segment of that of $\mathfrak{U}$, and the branch of $\mathfrak{T}$ leading to its last model does not drop. Since $\mathrm{Ord}^W = \Omega$, this means that the last model of $\mathfrak{U}$ has ordinal height $> \Omega$. Since $\mathrm{Ord}^{\mathfrak{N}} \leq \Omega$, we have $\mathrm{lh}(\mathfrak{U}) = \Omega+1$, and as in the proof of the comparison lemma 2.10 we have a closed unbounded set $C \subseteq \Omega$ such that for all $\kappa \in C$

$$\kappa = \mathrm{crit}(i^{\mathfrak{U}}_{\kappa,\Omega}) \text{ and } i^{\mathfrak{U}}_{\kappa,\Omega}(\kappa) = \Omega.$$

Now if $\kappa, \mu \in C$ and $\kappa < \mu$, then

$$i^{\mathfrak{U}}_{\kappa,\mu}((\kappa^+)^{\mathfrak{M}^{\mathfrak{U}}_\kappa}) = (\mu^+)^{\mathfrak{M}^{\mathfrak{U}}_\mu} = (\mu^+)^{\mathfrak{M}^{\mathfrak{U}}_\Omega},$$

where the second equality holds because $\mathrm{crit}(i^{\mathfrak{U}}_{\mu,\Omega}) = \mu$, so that $\mathfrak{M}^{\mathfrak{U}}_\mu$ and $\mathfrak{M}^{\mathfrak{U}}_\Omega$ have the same subsets of μ. Since all embeddings of an iteration tree are continuous at successor cardinals[44], this implies that

$$\sup(i^{\mathfrak{U}}_{\kappa,\mu}\text{''}(\kappa^+)^{\mathfrak{M}^{\mathfrak{U}}_\kappa}) = (\mu^+)^{\mathfrak{M}^{\mathfrak{U}}_\Omega} < \mu^+$$

[43] Covering theorems begin, of course, with Jensen's great leap forward ([DevJen75]). One cannot prove the existence of mice with the stronger covering properties studied in [DevJen75] and [DoJen81] without assuming the non–existence of mice satisfying that there is an inaccessible limit of measurable cardinals, or at least something close to that, so these stronger covering properties are not very useful at the Woodin cardinal level. Mitchell ([Mi84]) first realized that the covering proof still gives weak covering in the more general situation, and that this is very useful in basic core model theory. Lemma 2.13 is due to him.

[44] This is an easy induction; discontinuities come only from the ultrapower construction at points of cofinality $\mathrm{crit}(E)$, where E is the extender used, and $\mathrm{crit}(E)$ is never a successor cardinal of the model to which E is applied.

for all but the least $\mu \in C$.

Let's assume that $\mathrm{lh}(\mathfrak{T}) = \Omega+1$, the case that $\mathrm{lh}(\mathfrak{T}) < \Omega$ being similar but a bit simpler. Since $\mathrm{Ord}^{\mathfrak{M}^{\mathfrak{T}}_{\Omega}} = \Omega$ by the Comparison Lemma 2.10, $i^{\mathfrak{T}}_{0,\Omega}(\mu) < \Omega$ for all $\mu < \Omega$, and thus there is a closed unbounded set of $\mu \in [0, \Omega]_T$ such that

$$i^{\mathfrak{T}}_{0,\mu}{}''\mu \subseteq \mu.$$

Because Ω is Mahlo, stationarily many of these μ are regular, and for these

$$i^{\mathfrak{T}}_{0,\mu}(\mu) = \mu.$$

Stationarily many of these μ are such that $(\mu^+)^W = \mu^+$, and for these

$$(\mu^+)^{\mathfrak{M}^{\mathfrak{T}}_{\Omega}} = (\mu^+)^{\mathfrak{M}^{\mathfrak{T}}_{\mu}} = i^{\mathfrak{T}}_{0,\mu}(\mu^+) = \mu^+.$$

Since $\mathfrak{M}^{\mathfrak{T}}_{\Omega}$ is an initial segment of $\mathfrak{M}^{\mathfrak{U}}_{\Omega}$, we have that $(\mu^+)^{\mathfrak{M}^{\mathfrak{T}}_{\Omega}} \leq (\mu^+)^{\mathfrak{M}^{\mathfrak{U}}_{\Omega}}$ for all $\mu < \Omega$, so that $\mathfrak{M}^{\mathfrak{U}}_{\Omega}$ also computes stationarily many successor cardinals correctly, in contradiction to the previous paragraph. □

The reader may well be upset by the appearance of a Mahlo cardinal in the hypotheses of Lemma 2.13. There are variants of the lemma which do not make use of Mahlo cardinals, but for us later it will be Lemma 2.13 which is useful.

2.5. **The Mouse Order and the Dodd–Jensen Lemma.** The comparison lemma 2.10 can be stated in a different language: It says that the mouse order $\leq^*$ is linear on iterable premice. Although we won't use the mouse order in any applications in this paper, it is an important concept in core model theory and thus we shall devote this section to its introduction.

Definition 2.14. Let $\mathfrak{M}$ and $\mathfrak{N}$ be iterable premice with iteration strategies Σ and Γ such that for every $\alpha \in \mathrm{Ord}$ the restriction of Σ and Γ to games of length α is the unique α–iteration strategy.
Then set $\mathfrak{M} \leq^* \mathfrak{N}$ if and only if $\mathfrak{M}$ $\langle \Sigma, \Gamma \rangle$–iterates past $\mathfrak{N}$.
This relation is called the *mouse order*.

The term "order" indicates that the relation is transitive, though this is not at all obvious from the definition. In this section we will use the comparison lemma 2.10 and the Dodd–Jensen lemma 2.16 to show that $\leq^*$ is a linear prewellordering.[45]

Proposition 2.15. The relation $\leq^*$ is linear.

[45]The assumption that the iteration strategies be unique is made mainly to suppress complications. The proofs will work with assumptions that are a lot weaker than that, *cf.* [St93, Theorem 3.2] and [NeeSt9?]. But in our situation below one Woodin cardinal iteration strategies are always unique (*cf.* [MiSt94, §6]), so the restriction is of no harm here.

Proof : Immediate from Theorem 2.10. □

For transitivity we need the Dodd–Jensen lemma, a theorem about minimality of iteration maps that has uses far beyond this little application and has already been mentioned before:

Theorem 2.16 (Dodd–Jensen Lemma). Let $\mathfrak{M}$ be a premouse with a unique iteration strategy Σ and let $\mathfrak{T}$ be an iteration tree on $\mathfrak{M}$ according to Σ with last model $\mathfrak{M}^{\mathfrak{T}}_{\alpha}$.[46] Suppose there is an embedding $\pi : \mathfrak{M} \to \mathfrak{P}$ where $\mathfrak{P}$ is an initial segment of $\mathfrak{M}^{\mathfrak{T}}_{\alpha}$. Then:

1. $\mathfrak{P} = \mathfrak{M}^{\mathfrak{T}}_{\alpha}$,
2. $[0, \alpha]$ does not drop, and
3. for all $\eta \in \mathrm{Ord}^{\mathfrak{M}}$ we have $i^{\mathfrak{T}}_{0\alpha}(\eta) \leq \pi(\alpha)$.

Proof : We shall show the third claim as the proofs of the other claims are similar.
The basic technique used in the proof is the technique of copying iteration trees. Given an iterable premouse $\mathfrak{Q}$, an iteration tree $\mathfrak{U}$ on $\mathfrak{Q}$ and an embedding $\sigma : \mathfrak{Q} \to \mathfrak{R}$, we can form an iteration tree $\sigma\mathfrak{U}$ on $\mathfrak{R}$ with embeddings $\sigma_\beta : \mathfrak{M}^{\mathfrak{U}}_{\beta} \to \mathfrak{M}^{\sigma\mathfrak{U}}_{\beta}$ for all $\beta < \mathrm{lh}(\mathfrak{U})$ such that

- if $i^{\mathfrak{U}}_{\alpha\beta}$ is defined then so is $i^{\sigma\mathfrak{U}}_{\alpha\beta}$ and $\pi_\beta \circ i^{\mathfrak{U}}_{\alpha\beta} = i^{\sigma\mathfrak{U}}_{\alpha\beta} \circ \sigma_\alpha$,
- $\mathfrak{U}$ and $\sigma\mathfrak{U}$ have the same underlying trees and they drop at the same points.

To prove that such a tree $\sigma\mathfrak{U}$ exists, we construct the embeddings σ_α inductively.[47] For the successor step, we use the so–called "shift lemma":

Lemma 2.17. Let $\mathfrak{M}$ and $\mathfrak{N}$ be premice, F the last extender of $\mathfrak{N}$, κ its critical point, and let $\pi : \mathfrak{M} \to \mathfrak{M}^*$ and $\psi : \mathfrak{N} \to \mathfrak{N}^*$ be embeddings such that

- $\mathcal{J}^{\mathfrak{M}}_{(\kappa^+)^{\mathfrak{M}}} = \mathcal{J}^{\mathfrak{N}}_{(\kappa^+)^{\mathfrak{M}}}$
- $\mathcal{J}^{\mathfrak{M}^*}_{(\pi(\kappa)+)^{\mathfrak{M}^*}} = \mathcal{J}^{\mathfrak{N}^*}_{(\pi(\kappa)+)^{\mathfrak{M}^*}}$
- $\pi \restriction (\kappa^+)^{\mathfrak{M}} = \psi \restriction (\kappa^+)^{\mathfrak{M}}$, and
- $\mathrm{Ult}(\mathfrak{M}, F)$ and $\mathrm{Ult}(\mathfrak{M}^*, F)$ are defined and well–founded.

  Then there is an embedding σ such that the following diagram commutes:

$$\begin{array}{ccc} \mathfrak{M} & \xrightarrow{\pi} & \mathfrak{M}^* \\ \downarrow & & \downarrow \\ \mathrm{Ult}(\mathfrak{M}, F) & \xrightarrow{\sigma} & \mathrm{Ult}(\mathfrak{M}^*, F) \end{array}$$

[46] In fact, we need more iterability than we talked about up to now. In the proof we need that linear stacking of iteration trees gives iterations, which is not true in the context of our Section 2.3. The reader can find correct definitions for this generalized iteration game in [NeeSt9?] and [St9?a].

[47] In fact, it is not necessary that σ is fully elementary. It suffices that σ is a weak k–embedding, where k is the highest degree of elementarity occurring in the embeddings of non–dropping branches in $\mathfrak{U}$. For definitions, *cf.* [MiSt94, p. 52].

We shall say that the copying construction breaks down in the limit step if the copied branch is ill–founded. Otherwise we continue.

This process also gives us a method to pull back strategies from $\mathfrak{R}$ to $\mathfrak{Q}$:

If we have constructed $\sigma\mathfrak{U}$ up to λ and Σ is any strategy for player II on $\mathfrak{R}$, then Σ chooses a branch in $\sigma\mathfrak{U} \upharpoonright \lambda$. The appropriate strategy on $\mathfrak{Q}$ is called *the pullback of* Σ *via* σ and denoted by Σ^σ.

Because of the embeddings we are constructing, the property of being winning for player II also gets pulled back, so if Σ was an iteration strategy for $\mathfrak{R}$, Σ^σ will be an iteration strategy for $\mathfrak{Q}$.

Now we will use this technique to prove the Dodd–Jensen lemma:
We shall construct models $\mathfrak{M}_n$, "vertical" embeddings $\pi^n : \mathfrak{M}_n \to \mathfrak{M}_{n+1}$, "horizontal" embeddings $\sigma^n : \mathfrak{M}_n \to \mathfrak{M}_{n+1}$ and iteration trees $\mathfrak{T}_n$ of length α on $\mathfrak{M}_n$ with last models $\mathfrak{M}_{n+1}$ such that $\sigma^n = i^{\mathfrak{T}_n}_{0\alpha}$ inductively:

Set $\mathfrak{M}_0 := \mathfrak{M}$, $\pi^0 := \pi$ and $\sigma^0 := i^{\mathfrak{T}}_{0\alpha}$. Suppose the models have been constructed up to $n+1$ and the embeddings and iteration trees have been constructed up to n. Then we have

$$\begin{array}{ccc} \mathfrak{M}_{n+1} & & \\ \uparrow{\scriptstyle \pi^n} & & \\ \mathfrak{M}_n & \xrightarrow{\sigma^n} & \mathfrak{M}_{n+1} \end{array}$$

We now copy $\mathfrak{T}_n$ to $\mathfrak{M}_{n+1}$ to get the tree $\mathfrak{T}_{n+1} := \pi^n\mathfrak{T}_n$. This process doesn't break down because of our uniqueness assumption: Suppose that at a given limit step λ, the copied branch of $\pi^n\mathfrak{T}_n$ is ill–founded. By iterability there is a $\lambda+1$–iteration strategy Σ that picks a wellfounded branch of $\pi^n\mathfrak{T}_n \upharpoonright \lambda$, but then the pullback strategy Σ^{π^n} is also an iteration strategy. Uniqueness gives us that Σ must already have picked the copied branch.

Our construction process gives us an embedding $\pi^n_\alpha : \mathfrak{M}_{n+1} = \mathfrak{M}^{\mathfrak{T}_n}_\alpha \to \mathfrak{M}^{\pi^n\mathfrak{T}_n}_\alpha$ and an embedding $i^{\pi^n\mathfrak{T}_n}_{0\alpha} : \mathfrak{M}_n \to \mathfrak{M}^{\pi^n\mathfrak{T}_n}_\alpha$ to complete the diagram to

$$\begin{array}{ccc} \mathfrak{M}_{n+1} & \xrightarrow{i^{\pi^n\mathfrak{T}_n}_{0\alpha}} & \mathfrak{M}^{\pi^n\mathfrak{T}_n}_\alpha \\ \uparrow{\scriptstyle \pi^n} & & \uparrow{\scriptstyle \pi^n_\alpha} \\ \mathfrak{M}_n & \xrightarrow{\sigma^n = i^{\mathfrak{T}_n}_{0\alpha}} & \mathfrak{M}^{\mathfrak{T}_n}_\alpha \end{array}$$

We set $\mathfrak{M}_{n+2} := \mathfrak{M}^{\pi^n\mathfrak{T}_n}_\alpha$, $\pi^{n+1} := \pi^n_\alpha$ and $\sigma^{n+1} := i^{\pi^n\mathfrak{T}_n}_{0\alpha}$ and receive the following diagram:

$$\begin{array}{ccccccccccc}
\mathfrak{M}_0 & \xrightarrow{\sigma^0 = i^{\mathfrak{T}}_{0\alpha}} & \mathfrak{M}_1 & \xrightarrow{\sigma^1} & \mathfrak{M}_2 & \xrightarrow{\sigma^2} & \mathfrak{M}_3 & \xrightarrow{\sigma^3} & \mathfrak{M}_4 & \xrightarrow{\sigma^4} & \cdots \\
 & & \uparrow{\scriptstyle \pi^0 = \pi} & & \uparrow{\scriptstyle \pi^1} & & \uparrow{\scriptstyle \pi^2} & & \uparrow{\scriptstyle \pi^3} & & \\
 & & \mathfrak{M}_0 & \xrightarrow{\sigma^0} & \mathfrak{M}_1 & \xrightarrow{\sigma^1} & \mathfrak{M}_2 & \xrightarrow{\sigma^2} & \mathfrak{M}_3 & \xrightarrow{\sigma^3} & \cdots \\
 & & & & \uparrow{\scriptstyle \pi^0} & & \uparrow{\scriptstyle \pi^1} & & \uparrow{\scriptstyle \pi^2} & & \\
 & & & & \mathfrak{M}_0 & \xrightarrow{\sigma^0} & \mathfrak{M}_1 & \xrightarrow{\sigma^1} & \mathfrak{M}_2 & \xrightarrow{\sigma^2} & \cdots \\
 & & & & & & \uparrow{\scriptstyle \pi^0} & & \uparrow{\scriptstyle \pi^1} & & \\
 & & & & & & \mathfrak{M}_0 & \xrightarrow{\sigma^0} & \mathfrak{M}_1 & \xrightarrow{\sigma^1} & \cdots \\
 & & & & & & & & \uparrow{\scriptstyle \pi^0} & & \\
 & & & & & & & & \mathfrak{M}_0 & \xrightarrow{\sigma^0} & \cdots
\end{array}$$

We now suppose that there is an η such that $\pi(\eta) < i^{\mathfrak{T}}_{0\alpha}(\eta)$ and construct a descending sequence of ordinals in $\text{dirlim}_{n\in\omega}\mathfrak{M}_n$. This is a contradiction to the iterability of $\mathfrak{M}$.[48] To prove the claim, we construct for every $n \in \omega$ descending sequences $\langle \eta^n_0, \ldots, \eta^n_n \rangle$ in $\mathfrak{M}_n$ such that $\sigma^n(\eta^n_i) = \eta^{n+1}_i$ for $i \leq n$.

Set $\eta^1_0 := \pi(\eta)$ and $\eta^1_1 := i^{\mathfrak{T}}_{0\alpha}(\eta)$. If the sequences are constructed up to n, then set $\eta^{n+1}_i := \sigma^{n+1}(\eta^n_i)$ for $i \leq n$ and $\eta^{n+1}_{n+1} := \pi^{n+1} \circ \cdots \circ \pi^0(\eta)$. We have to show that the new sequence is descending, *i.e.* that $\eta^{n+1}_{n+1} < \eta^{n+1}_n$. But by commutativity and our assumption, we have

$$\begin{array}{rcl}
\eta^{n+1}_n = \sigma^{n+1}(\eta^n_n) & = & \sigma^{n+1} \circ \pi^n \circ \cdots \circ \pi^0(\eta) \\
 & = & \pi^{n+1} \circ \sigma^n \circ \pi^{n-1} \circ \cdots \circ \pi^0(\eta) \\
 & > & \pi^{n+1} \circ \pi^n \circ \pi^{n-1} \circ \cdots \circ \pi^0(\eta) \\
 & = & \eta^{n+1}_{n+1}
\end{array}$$

This completes the proof. □

Corollary 2.18. If $\mathfrak{T}$ and $\mathfrak{U}$ are two iteration trees on $\mathfrak{M}$ with last model $\mathfrak{N}$, then the branch leading to $\mathfrak{N}$ is either in both trees a dropping branch or in both trees a non–dropping branch.

Proof : If one branch is non–dropping, it gives rise to an embedding from $\mathfrak{M}$ into $\mathfrak{N}$. The Dodd–Jensen lemma 2.16 then tells us that the other branch can't drop either. □

Proposition 2.19. The mouse order $\leq^*$ is a linear prewellordering.

Proof : We only have to show well–foundedness, as reflexivity is trivial, linearity is Proposition 2.15 and transitivity follows abstractly from well–foundedness and linearity:

[48] This is the point where the notion of generalized iteration would come in if we had not suppressed this detail.

Take a counterexample $\mathfrak{M}_1 \leq^* \mathfrak{M}_2 \leq^* \mathfrak{M}_3$ to transitivity, *i.e.* $\mathfrak{M}_1 \not\leq^* \mathfrak{M}_3$. By linearity $\mathfrak{M}_3 \leq^* \mathfrak{M}_1$ and we can get a descending chain by concatenating ω copies of $\langle \mathfrak{M}_3, \mathfrak{M}_2, \mathfrak{M}_1 \rangle$, contradicting well–foundedness.

For the proof of well–foundedness, we take an infinite descending chain of models $\langle \mathfrak{M}_i : i \in \omega \rangle$. Let $\mathfrak{Q}_i^0$ be the witness that $\mathfrak{M}_i$ iterates past $\mathfrak{M}_{i+1}$, *i.e.* the coiteration of $\mathfrak{M}_i$ and $\mathfrak{M}_{i+1}$ results in a model $\mathfrak{N}$ on the $\mathfrak{M}_i$–side and the model $\mathfrak{Q}_i^0$ on the $\mathfrak{M}_{i+1}$–side where $\mathfrak{Q}_i^0 \trianglelefteq \mathfrak{N}$.

Now inductively suppose we had defined models $\langle \mathfrak{Q}_i^n : i \in \omega \rangle$. We compare $\mathfrak{Q}_i^n$ and $\mathfrak{Q}_{i+1}^n$ and call the common initial segment of the two iterates $\mathfrak{Q}_i^{n+1}$. This gives us the following diagram:

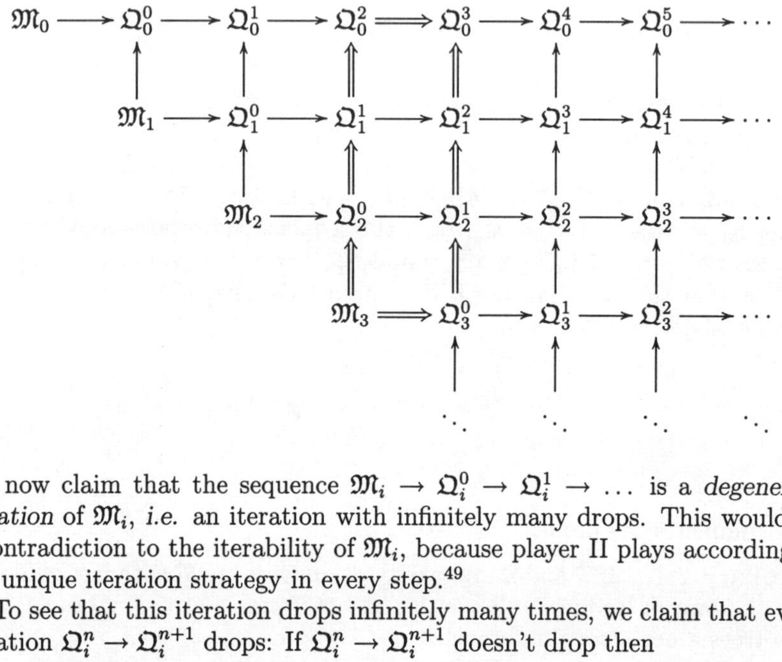

We now claim that the sequence $\mathfrak{M}_i \to \mathfrak{Q}_i^0 \to \mathfrak{Q}_i^1 \to \ldots$ is a *degenerate iteration* of $\mathfrak{M}_i$, *i.e.* an iteration with infinitely many drops. This would be a contradiction to the iterability of $\mathfrak{M}_i$, because player II plays according to the unique iteration strategy in every step.[49]

To see that this iteration drops infinitely many times, we claim that every iteration $\mathfrak{Q}_i^n \to \mathfrak{Q}_i^{n+1}$ drops: If $\mathfrak{Q}_i^n \to \mathfrak{Q}_i^{n+1}$ doesn't drop then

$$\mathfrak{M}_{n+i+1} \to \mathfrak{Q}_{n+i}^0 \to \cdots \to \mathfrak{Q}_i^n \to \mathfrak{Q}_i^{n+1}$$

is an iteration of $\mathfrak{M}_{n+i+1}$ without drops. But as $\mathfrak{M}_{n+i+1} \to \mathfrak{Q}_{n+i+1}^0$ drops by assumption, the iteration

$$\mathfrak{M}_{n+i+1} \to \mathfrak{Q}_{n+i+1}^0 \to \cdots \to \mathfrak{Q}_i^{n+1}$$

is a dropping iteration, contradicting Corollary 2.18.[50] □

[49]Note that this argument requires the generalized notion of iteration we mentioned in Footnote 46.

[50]An example ($n = 2$ and $i = 0$) of this argument is depicted in the above diagram by the double arrows.

2.6. **The construction of $\mathbf{K}^c$.** We have been studying mice in the abstract, but we have yet to produce any! In this section we shall describe the construction of a core model $\mathbf{K}^c$. This construction is sufficiently cautious about adding extenders to the model that one gets an iterable model in the end,[51], yet sufficiently daring that if one carries it out in a universe satisfying certain statements of consistency strength at least that of "there is a Woodin cardinal" one gets a model satisfying "there is a Woodin cardinal".[52]

The natural idea is to construct a good extender sequence $\vec{E}$ by induction. Given $\vec{E} \restriction \alpha$, we set $E_\alpha = \emptyset$ unless there is a certified[53] extender F such that $(\vec{E} \restriction \alpha)^\frown F$ is still a good extender sequence; if there is such an F we may set $E_\alpha = F$ for some such F. Here "certified" means roughly that F is the restriction to $\mathbf{J}_\alpha[\vec{E} \restriction \alpha]$ of a "background extender" F^* which measures a broader collection of subsets of its critical point than does F, and whose ultrapower agrees with $\mathbf{V}$ a bit past $\text{lh}(F)$. This background certificate demand is necessary in order to ensure that the premice we are constructing are iterable. Unfortunately, the background certificate demand conflicts with the demand that all levels of the model we are constructing be sound.[54] A $\mathbf{K}^c$–construction deals with this conflict by continually replacing the premouse $\mathfrak{N}_\alpha$ currently approximating the model being built by its core, the Skolem hull $\mathfrak{C}(\mathfrak{N}_\alpha)$.[55] Taking cores ensures soundness, while the background extenders one can resurrect by going back into the history of the construction ensure iterability.

This last claim must be qualified. We do not have a general proof of iterability for the premice $\mathfrak{N}_\alpha$ produced in a $\mathbf{K}^c$–construction. At the moment, in order to prove that such a premouse is appropriately iterable, we need to make an additional "smallness" assumption. One assumption that suffices, and which we shall spell out in more detail shortly, is that no initial segment of $\mathfrak{N}_\alpha$ satisfies "there is an extender E on my sequence such that $\text{lh}(E)$ is a Woodin cardinal". We shall call this property of $\mathfrak{N}_\alpha$ *tameness*. Iterability is essential from the very beginning, for our proof that $\mathfrak{C}(\mathfrak{N}_\alpha)$ exists involves comparison arguments, and hence relies on the iterability of $\mathfrak{N}_\alpha$. Thus, for all we know, a $\mathbf{K}^c$–construction might simply break down by reaching a non–tame premouse $\mathfrak{N}_\alpha$ such that $\mathfrak{C}(\mathfrak{N}_\alpha)$ does not exist.

The following definitions describe our background certificate condition. They come from [St96, §1].

---

[51]This is something between a conjecture and a theorem; see below.

[52]Again, there are qualifications to come. $\mathbf{K}^c$ is not actually the core model in which we shall be most interested in the end, but a stepping–stone to it.

[53]Whence the "c" in $\mathbf{K}^c$.

[54]Part of the requirement on F^* is that it be countably complete, and so $\text{crit}(F^*)$ must be uncountable; on the other hand, if α is least so that $E_\alpha \neq \emptyset$, then $\langle \mathbf{J}_\alpha[\vec{E} \restriction \alpha], \in, \vec{E} \restriction \alpha, E_\alpha \rangle$ has Σ_1–projectum ω, so that $\text{crit}(E_\alpha)$ must be countable if this structure is even 1–sound.

[55]$\mathfrak{C}(\mathfrak{M})$ is "defined" in the discussion following 2.5.

Definition 2.20. Let $\mathfrak{M} = \langle \mathbf{J}_\eta[\vec{E}], \in, \vec{E}, F\rangle$ be a premouse such that $F \neq \emptyset$, let $\kappa = \mathrm{crit}(F)$, and let $\nu = \mathrm{lh}(F)$. Let $\mathcal{A} \subseteq \bigcup_{n<\omega} \mathfrak{P}([\kappa]^n)^{\mathfrak{M}}$; then an $\mathcal{A}$–*certificate for* $\mathfrak{M}$ is a pair $\langle N, G\rangle$ such that

1. N is a transitive, power admissible set, $\mathbf{V}_\kappa \cup \mathcal{A} \subseteq N$, N is closed under ω–sequences, and G is an extender over N,
2. $F \cap ([\nu]^{<\omega} \times \mathcal{A}) = G \cap ([\nu]^{<\omega} \times \mathcal{A})$,
3. $\mathbf{V}_{\nu+1} \subseteq \mathrm{Ult}(N, G)$, and
4. for all $\gamma < \eta$ we have $\mathbf{J}_\gamma^{\mathfrak{M}} = \mathbf{J}_\gamma^{i(\mathbf{J}_\kappa^{\mathfrak{M}})}$, where i is the canonical embedding from N to $\mathrm{Ult}(N, G)$.

Definition 2.21. Let $\mathfrak{M}$ be a premouse whose last extender predicate is nonempty, and let κ be the critical point of this last extender. We say $\mathfrak{M}$ *is countably certified* iff for every countable $\mathcal{A} \subseteq \bigcup_{n<\omega} \mathfrak{P}([\kappa]^n)^{\mathfrak{M}}$, there is an $\mathcal{A}$–certificate for $\mathfrak{M}$. We say that $\mathfrak{M}$ *leaves gaps* iff κ is inaccessible, and either $(\kappa^+)^{\mathfrak{M}} < \kappa^+$ or $\{\mu < \kappa \colon (\mu^+)^{\mathfrak{M}} = \mu^+\}$ is stationary in κ.

In the situation described in Definition 2.20, we shall typically have $|N| = \kappa$, so that $\mathrm{Ord}^N < \mathrm{lh}(G)$. We are therefore not thinking of $\langle N, G\rangle$ as a structure to be iterated; N simply provides a reasonably large collection of sets to be measured by G. The conditions $\mathbf{V}_\kappa \subseteq N$ and $\mathbf{V}_{\nu+1} \subseteq \mathrm{Ult}(N, G)$ are crucial.

We are ready for one of the central definitions of this paper.

Definition 2.22. A $\mathbf{K}^c$–*construction* is a sequence $\langle \mathfrak{N}_\alpha \colon \alpha < \vartheta\rangle$ of premice such that

1. $\mathfrak{N}_0 = \langle \mathbf{V}_\omega, \in, \emptyset, \emptyset\rangle$;
2. if $\alpha + 1 < \vartheta$, then the core

$$\mathfrak{C}(\mathfrak{N}_\alpha) = \langle \mathbf{J}_\eta[\vec{E}], \in, \vec{E}, H\rangle,$$

exists, and either

(a) $H = \emptyset$, $\mathfrak{N}_{\alpha+1}$ is a countably certified premouse which leaves gaps, and $\mathfrak{N}_{\alpha+1}$ is of the form

$$\mathfrak{N}_{\alpha+1} = \langle \mathbf{J}_\eta[\vec{E}], \in, \vec{E}, F\rangle,$$

for some F such that $\mathrm{lh}(F)$ is as small as possible, or

(b) $H \neq \emptyset$, or there is no countably certified premouse which leaves gaps of the form $\langle \mathbf{J}_\eta[\vec{E}], \in, \vec{E}, F\rangle$ and

$$\mathfrak{N}_{\alpha+1} = \langle \mathbf{J}_{\eta+1}[\vec{E}^\frown H], \in, \vec{E}^\frown H, \emptyset\rangle;$$

3. if $\lambda < \vartheta$ is a limit ordinal, then $\mathfrak{N}_\lambda$ is the unique premouse $\mathfrak{P}$ such that

$$\mathrm{Ord}^{\mathfrak{P}} = \sup\{\omega\beta : \mathbf{J}_\beta^{\mathfrak{N}_\alpha} \text{ is defined and eventually constant as } \alpha \to \lambda\},$$

and for all β such that $\omega\beta < \mathrm{Ord}^{\mathfrak{P}}$,

$$\mathbf{J}_\beta^{\mathfrak{P}} = \text{ eventual value of } \mathbf{J}_\beta^{\mathfrak{N}_\alpha}, \text{ as } \alpha \to \lambda,$$

and the last extender predicate of $\mathfrak{P}$ is empty.

So at successor steps in a $\mathbf{K}^c$–construction one replaces the previous premouse with its core, and then either adds a countably certified extender to the resulting extender sequence or takes one step in its constructible closure. At limit steps one forms the natural limit of the stabilized initial segments of the previous premice.

We have required our $\mathbf{K}^c$–constructions to be maximal, in the sense that they must add to the coherent sequence a certified extender of minimal length whenever it is possible to do so. It would be reasonable to drop this condition, or perhaps relax it by requiring more of the background certificates.[56] However, we have no need for the more general kind of $\mathbf{K}^c$–construction.

To what extent are the $\mathbf{K}^c$–constructions canonical? Modulo a natural iterability conjecture, one can show that any two $\mathbf{K}^c$–constructions are compatible, in that one is an initial segment of the other (*cf.* [MiSt94, §9]). We shall have no use for this fact, however.

Because we replace $\mathfrak{N}_\alpha$ by its core at each step in a $\mathbf{K}^c$–construction, the models of the construction may not grow by end–extension, and we need a little argument to show, for example, that a construction of proper class length converges to a premouse of proper class size. Our Theorem 2.11 on the agreement of $\mathfrak{N}$ with $\mathfrak{C}(\mathfrak{N})$ is the key here.

Theorem 2.23. Let κ be an uncountable regular cardinal or $\kappa = \mathrm{Ord}$, and let $\langle \mathfrak{N}_\alpha \colon \alpha < \kappa \rangle$ be a $\mathbf{K}^c$–construction; then there is a unique premouse $\mathfrak{N}_\kappa$ of ordinal height κ such that $\langle \mathfrak{N}_\alpha \colon \alpha \leq \kappa \rangle$ is a $\mathbf{K}^c$–construction.

Proof : For any limit ordinal κ and $\mathbf{K}^c$–construction $\langle \mathfrak{N}_\alpha \colon \alpha < \kappa \rangle$, there is a unique premouse $\mathfrak{N}_\kappa$ satisfying the limit ordinal clause of Definition 2.22. We need only show that $\mathfrak{N}_\kappa$ has ordinal height κ in the case κ is an uncountable cardinal or $\kappa = \mathrm{Ord}$. It is clear that $|\mathfrak{N}_\alpha| < \kappa$ for all $\alpha < \kappa$, so $\mathfrak{N}_\kappa$ has ordinal height $\leq \kappa$.

For $\nu < \kappa$, let

$$\vartheta_\nu = \inf\{\varrho_\omega^{\mathfrak{N}_\alpha} \colon \nu \leq \alpha < \kappa\}.$$

So $\vartheta_0 = \omega$, and the ϑ's are nondecreasing. By Theorem 2.11, $\mathfrak{N}_\nu$ agrees with all later $\mathfrak{N}_\alpha$ below ϑ_ν, so if $\kappa = \sup\{\vartheta_\nu \colon \nu < \kappa\}$, we are done. Since κ is regular, the alternative is that the ϑ's are eventually constant; say $\vartheta_\nu = \varrho$ for all ν such that $\eta \leq \nu < \kappa$. Now notice that if $\eta \leq \nu < \kappa$ and $\varrho_\omega^{\mathfrak{N}_\nu} = \varrho$, then $\mathfrak{C}(\mathfrak{N}_\nu)$ is a proper initial segment of $\mathfrak{N}_{\nu+1}$.[57] Moreover, $\mathfrak{C}(\mathfrak{N}_\nu)$ has cardinality ϱ in $\mathfrak{N}_{\nu+1}$ by soundness. It follows from Theorem 2.11 that $\mathfrak{C}(\mathfrak{N}_\nu)$ is an initial segment of $\mathfrak{N}_\alpha$, for all $\alpha \geq \nu$. Since there are cofinally many $\nu < \kappa$ such that

[56] For example, one might require that the background extenders be total extenders over $\mathbf{V}$. This enables one to lift iteration trees on the $\mathfrak{N}_\alpha$'s to iteration trees on $\mathbf{V}$, and thereby simplifies the proof of iterability somewhat. This is the approach taken in [MiSt94]. Another sometimes useful device is to impose a lower bound on the critical points of the background extenders.

[57] Assume the last extender predicate of $\mathfrak{N}_\nu$ is empty here, as it obviously is for cofinally many such ν.

$\varrho = \varrho_\omega^{\mathfrak{N}_\nu}$, we again get that $\mathfrak{N}_\kappa$ has height κ. □

It is not hard to see that the ϑ_ν defined in the proof above are just the infinite cardinals of $\mathfrak{N}_\kappa$.

It would be natural at this point to fix some $\mathbf{K}^c$–construction of length Ord + 1, and define $\mathbf{K}^c$ itself to be the premouse in this construction indexed at Ord. We shall in effect eventually do this, but at certain points we shall need third order properties of Ord which go beyond third-order ZFC, and indeed come close to the assertion that Ord is a measurable cardinal.[58] So we shall eventually fix a measurable cardinal Ω, and let $\mathbf{K}^c$ be the Ωth model in some $\mathbf{K}^c$–construction.

2.7. **The iterability of $\mathbf{K}^c$.** It is clear by now that we have gotten nowhere unless we can prove that the premice we have constructed are sufficiently iterable. Here we encounter what is perhaps the central open problem of core model theory. We formulate it as a conjecture:

Conjecture 2.24. If $\mathfrak{N}$ occurs in a $\mathbf{K}^c$–construction, then every countable elementary substructure of $\mathfrak{N}$ is $\omega_1 + 1$-iterable.

A proof of this conjecture would yield the basics of core model theory at the level of superstrong cardinals, and it would no doubt extend to do the same for supercompact cardinals.[59] In particular, we could apply Theorem 2.11 and Theorem 2.23 to see that $\mathbf{K}^c$–constructions cannot break down, and, if carried on for Ord stages, converge to proper class premice.

In general, iterability proofs break up into an *existence* proof and a *uniqueness* proof for "sufficiently good" branches in iteration trees on the premice under consideration. The existence proof breaks itself breaks into two parts, a direct existence argument in the countable case and a reflection argument in the uncountable case.

The direct existence argument applies to countable iteration trees on countable elementary submodels of the premice under consideration, and proceeds by using something like the countable completeness of the extenders involved in the iteration to transform an ill–behaved iteration into an infinite descending $\in$–chain. When coupled with the uniqueness proof, this

[58]The need for these assumptions is a defect in core model theory at the level of Woodin cardinals which we hope will some day be removed.

[59]A new problem arises between supercompact and huge cardinals, but this is not the place to go into it. We should also note that one really needs a sharper version of the conjecture which keeps track of the degree of elementarity of the substructures; we have suppressed this finestructure here.

shows that any countable elementary submodel of a premouse under consideration has an ω_1–iteration strategy, namely, the strategy of choosing the unique cofinal "sufficiently good" branch.[60]

The reflection argument extends this method of iterating by choosing sufficiently good branches to the uncountable: given an iteration tree $\mathfrak{T}$ on $\mathfrak{M}$, we go to $\mathbf{V}[G]$ where G is $\mathrm{Col}(\kappa, \omega)$–generic over $\mathbf{V}$ and κ is large enough that $\mathfrak{M}$ and $\mathfrak{T}$ have become countable, and find a sufficiently good branch there. This branch is unique, and hence by the homogeneity of the collapse it is in $\mathbf{V}$. In order to execute this argument[61] one needs a certain level of absoluteness between $\mathbf{V}$ and $\mathbf{V}[G]$. Once one gets past mice with Woodin cardinals, "sufficiently good" can no longer be taken simply to mean "wellfounded", and in fact "sufficiently good" is no longer a Σ^1_2–notion at all. Because of this, the generic absoluteness required by our reflection argument needs large cardinal/mouse existence principles which go beyond ZFC.[62]

The conjecture above overlaps slightly with the uncountable case because it is ω_1+1–iterability, rather than ω_1–iterability, which is at stake. One needs ω_1+1–iterability to guarantee the comparability of countable mice; the Fodor argument that shows coiterations terminate requires a wellfounded branch of length ω_1. Nevertheless, we believe that the conjecture is provable in ZFC.[63]

At present, we can only prove the conjecture for premice of limited complexity. We shall call these special premice "tame". Our direct existence argument in the countable case seems perfectly general, but our uniqueness results are less definitive, and it is here that we resort to the tameness assumption. We begin by stating the existence theorem in the countable case.

We shall say that b is a *maximal branch* of an iteration tree $\mathfrak{T}$ if b has limit order type but is not continued in $\mathfrak{T}$ (so a cofinal branch is always maximal). If b is a branch of $\mathfrak{T}$ and $\sup(b) = \lambda < \mathrm{lh}(\mathfrak{T})$, then b is maximal iff b is different from $[0, \lambda)_T$, the branch chosen by $\mathfrak{T}$. We call a system a *putative iteration tree* if it has all the properties of an ordinary iteration tree except that it may have a last, ill–founded model.[64]

---

[60]Of course a sufficiently good branch must be wellfounded, but in general more is required, for we want to be able to find cofinal wellfounded branches later in the iteration game as well.

[61]See Theorem 2.35 below for a concrete example of such an argument.

[62]For example, if it is consistent that there is a Woodin cardinal, then it is consistent that there is a premouse $\mathfrak{N}$ occurring on a $\mathbf{K}^c$–construction which is not fully iterable. *Cf.* Section 4.

[63]We suspect that if κ is strictly less than the infimum of the critical points of the background extenders, then the κ–iterability of the size κ elementary submodels of premice in a $\mathbf{K}^c$–construction is provable in ZFC.

[64]Thus a play of the iteration game at which player II has just lost at a successor step is a putative iteration tree, but not an iteration tree.

Theorem 2.25 (Branch Existence Theorem). Let $\pi : \mathfrak{M} \to \mathfrak{N}_\alpha$ be an elementary embedding where $\mathfrak{M}$ is countable and $\mathfrak{N}_\alpha$ is a model of the $\mathbf{K}^c$–construction. Let $\mathfrak{T}$ be a countable putative iteration tree on $\mathfrak{M}$. Then either

1. there is a maximal branch b of $\mathfrak{T}$ such that
   (a) b does not drop and there is a $\sigma : \mathfrak{M}_b^{\mathfrak{T}} \to \mathfrak{N}_\alpha$ such that

$$\begin{array}{ccc} \mathfrak{M} & \xrightarrow{\pi} & \mathfrak{N}_\alpha \\ & \searrow^{i_{0b}^{\mathfrak{T}}} & \uparrow{\sigma} \\ & & \mathfrak{M}_b^{\mathfrak{T}} \end{array}$$

   commutes, or
   (b) b drops, and there is a $\beta < \alpha$ with $\sigma : \mathfrak{M}_b^{\mathfrak{T}} \to \mathfrak{N}_\beta$,
   or
2. $\mathfrak{T}$ has a last model $\mathfrak{M}_\vartheta^{\mathfrak{T}}$ and
   (a) the branch $[0, \vartheta]_T$ does not drop, and there is $\sigma : \mathfrak{M}_\vartheta^{\mathfrak{T}} \to \mathfrak{N}_\alpha$ such that

$$\begin{array}{ccc} \mathfrak{M} & \xrightarrow{\pi} & \mathfrak{N}_\alpha \\ & \searrow^{i_{0\vartheta}^{\mathfrak{T}}} & \uparrow{\sigma} \\ & & \mathfrak{M}_\vartheta^{\mathfrak{T}} \end{array}$$

   commutes, or
   (b) the branch $[0, \vartheta]_T$ does drop, and there is a $\beta < \alpha$ with $\sigma : \mathfrak{M}_\vartheta^{\mathfrak{T}} \to \mathfrak{N}_\beta$

The proof of Theorem 2.25 is a direct construction which transforms a counterexample to the theorem into an infinite descending $\in$–chain. The details of the construction are rather complicated, and seem to shed little light on the rest of the theory, so we shall not attempt to describe them. They can be found in [MaSt94, §4], where the theorem was first proved for "coarse" mice occurring in a $\mathbf{K}^c$–construction using only full background extenders over $\mathbf{V}$, and in [St96, §§2,4], where the full result Theorem 2.25 was first proved.

Because the Branch Existence Theorem 2.25 asserts only the existence of *maximal* wellfounded branches, not necessarily *cofinal* ones, it does not even give the ω_1–iterability of countable elementary submodels of premice in a $\mathbf{K}^c$–construction. For that we need an accompanying uniqueness theorem for branches, and here our results are less definitive. What we can show, roughly, is that a failure of uniqueness yields a premouse with a Woodin cardinal. We now make this more precise.

Definition 2.26. Let $\kappa < \delta$ and $A \subseteq \mathbf{V}_\delta$, then κ is called *A–reflecting in δ* iff for all $\nu < \delta$ there is a $j : \mathbf{V} \to \mathfrak{M}$ such that $\mathrm{crit}(j) = \kappa$ and $j(A) \cap \mathbf{V}_\nu = A \cap \mathbf{V}_\nu$,

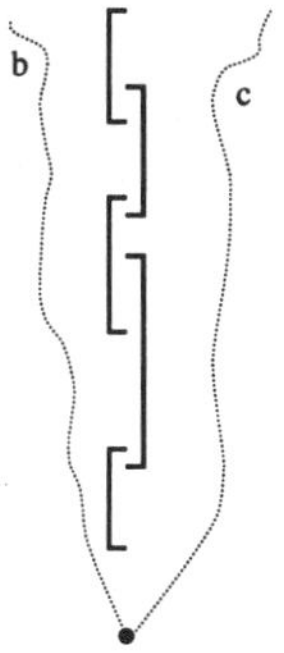

FIGURE 3. The overlapping pattern of two distinct well–founded branches

Definition 2.27. A cardinal δ is called a Woodin cardinal, iff for all $A \subseteq \mathbf{V}_\delta$ there is a $\kappa < \delta$ such that κ is A–reflecting in δ.

Woodin cardinals lie between strong and superstrong cardinals in the consistency strength hierarchy.[65] A Woodin cardinal need not itself be strong, or even measurable, but it is easy to see that if δ is Woodin, then δ is Mahlo, and δ is a limit of cardinals κ such that $\forall\alpha < \delta(\kappa$ is α–strong).[66] Measurable Woodin cardinals are well beyond the reach of current core model theory.

The main result connecting Woodin cardinals with the uniqueness of cofinal wellfounded branches in iteration trees is the following theorem of [MaSt94].

Theorem 2.28 (Branch Uniqueness Theorem). Let $\mathfrak{T}$ be an iteration tree of limit length λ, and let b and c be distinct cofinal wellfounded branches of $\mathfrak{T}$. Let

$$\delta := \sup\{\mathrm{lh}(F_\alpha^{\mathfrak{T}}) : \alpha < \lambda\},$$

and suppose $A \subseteq \mathbf{V}_\delta^{\mathfrak{M}_b^{\mathfrak{T}}} = \mathbf{V}_\delta^{\mathfrak{M}_c^{\mathfrak{T}}}$ and $A \in \mathfrak{M}_b^{\mathfrak{T}} \cap \mathfrak{M}_c^{\mathfrak{T}}$. Then

$$\mathfrak{M}_b^{\mathfrak{T}} \models \exists\kappa < \delta(\kappa \text{ is } A\text{–reflecting in } \delta)$$

Thus δ is Woodin in $\mathfrak{M}_b^{\mathfrak{T}} \cap \mathfrak{M}_c^{\mathfrak{T}}$.

Proof : The extenders used on b and c have an overlapping pattern pictured in Figure 3:

To see this, pick any successor ordinal

$$\alpha_0 + 1 \in b \setminus c,$$

and then let

$$\beta_n + 1 = \min\{\gamma \in c : \gamma > \alpha_n + 1\}$$

[65]For a diagram of the known large cardinals and their consistency strengths, *cf.* [Kan94, p. 471].

[66]Woodinness of δ is a Π_1–fact about $V_{\delta+1}$, so the least Woodin cardinal is not even weakly compact.

and

$$\alpha_{n+1} + 1 = \min\{\eta \in b : \eta > \beta_n + 1\},$$

for all $n < \omega$. Now for any n, the T-predecessor of β_n+1 is on c and $\leq \alpha_n+1$, hence $\leq \alpha_n$, so by the rules of the iteration game,

$$\text{crit}(F_{\beta_n}) < \text{lh}(F_{\alpha_n}).$$

Similarly, for any n,

$$\text{crit}(F_{\alpha_{n+1}}) < \text{lh}(F_{\beta_n}).$$

Now extenders used along the same branch of an iteration tree do not overlap (*i.e.*, if E is used before F, then $\text{lh}(E) < \text{crit}(F)$), so we have

$$\begin{aligned}\text{crit}(F_{\beta_n}) &< \text{lh}(F_{\alpha_n}) < \text{crit}(F_{\alpha_{n+1}}) < \text{lh}(F_{\beta_n}) \\ &< \text{crit}(F_{\beta_{n+1}}) < \text{lh}(F_{\alpha_{n+1}}) < \text{crit}(F_{\alpha_{n+2}})\end{aligned}$$

which is the overlapping pattern pictured.

Now $\sup\{\alpha_n : n < \omega\} = \sup\{\beta_n : n < \omega\}$, and since branches of iteration trees are closed below their suprema in the order topology on Ord, the common supremum of the α_n and β_n is λ. Let us assume α_0 was chosen large enough that we have

$$A = i_{\beta_0+1,c}(A^*) = i_{\alpha_1+1,b}(A^{**})$$

for some A^* and A^{**}. Let

$$\kappa = \text{crit}(F_{\beta_0}) = \text{crit}(i_{\beta_0+1,c});$$

we shall show that κ is A–reflecting in δ in the model $\mathfrak{M}_b$.

Let $E_0 = F_{\beta_0} \restriction \text{crit}(F_{\alpha_1})$. Because of the overlapping pattern, E_0 is a proper initial segment of F_{β_0}, and by initial segment condition on premice and the agreement of the models of an iteration tree, $E_0 \in \mathfrak{M}_b$. Moreover, if $j\colon \mathfrak{M}_b \to \text{Ult}(\mathfrak{M}_b, E_0)$ is the canonical embedding, then because A and A^* agree below κ, $j(A)$ and $i_{\beta_0+1,c}(A^*)$ agree below $\text{crit}(F_{\alpha_1})$. That is, $j(A)$ agrees with A below $\text{crit}(F_{\alpha_1})$, and hence E_0 witnesses that κ is A–reflecting up to $\text{crit}(F_{\alpha_1})$ in $\mathfrak{M}_b$.

To get A–reflection all the way up to δ, we set

$$E_{2n} = F_{\beta_n} \restriction \text{crit}(F_{\alpha_{n+1}}) \text{ and } E_{2n+1} = F_{\alpha_{n+1}} \restriction \text{crit}(F_{\beta_{n+1}}),$$

for all n. Each of the E_n is in $\mathfrak{M}_b$ for the same reason E_0 is in $\mathfrak{M}_b$. Therefore the extender E which represents the embedding coming from "composing" the ultrapowers by the E_i for $0 \leq i \leq 2n$, is in $\mathfrak{M}_b$. The argument above generalizes easily to show that E witnesses that κ is A–reflecting up to $\text{crit}(F_{\alpha_{n+1}})$. Since $\text{crit}(F_{\alpha_{n+1}})$ approaches δ as n approaches ω, κ is A-reflecting in δ in the model $\mathfrak{M}_b$. □

Definition 2.29. A premouse $\mathfrak{M}$ is 1–*small* iff whenever κ is the critical point of a (nonempty) extender on the $\mathfrak{M}$–sequence, then $\mathcal{J}_\kappa^{\mathfrak{M}} \models$ "There are no Woodin cardinals".

This means that a mouse is 1–small just in case it hasn't reached the sharp of a proper class model with a Woodin cardinal.

Definition 2.30. A premouse $\mathfrak{M}$ is *properly 1–small* iff $\mathfrak{M}$ is 1–small, and $\mathfrak{M} \models$ "There are no Woodin cardinals, and there is a largest cardinal".

The branch existence and uniqueness theorems rather easily yield $\omega_1 + 1$–iteration strategies for countable elementary submodels of the properly 1–small premice occurring in $\mathbf{K}^c$–constructions.[67] The strategy is just to choose at each limit stage $\lambda \leq \omega_1$ the unique cofinal branch of the iteration tree built so far. At countable stages λ, the Branch Existence Theorem 2.25 guarantees existence, modulo uniqueness at earlier stages, and the Branch Uniqueness Theorem 2.28 guarantees uniqueness.[68] At stage $\lambda = \omega_1$, we go to $\mathbf{V}[G]$, where G is $\mathrm{Col}(\omega_1, \omega)$–generic over $\mathbf{V}$. A strengthened form of Theorem 2.25 gives us a cofinal wellfounded branch in $\mathbf{V}[G]$ for the tree of length $\omega_1^{\mathbf{V}}$ (*cf.* [St93]). By Theorem 2.28 it is unique, hence definable in $\mathbf{V}[G]$ from the tree, and hence by the homogeneity of the collapse it is in $\mathbf{V}$.

One can push this sort of argument further. Notice that we have not used the full strength of the Branch Existence Theorem yet; the direct limit $\mathfrak{M}_b$ along the branch b it gives is not just wellfounded, but is itself embedded into a level of the same $\mathbf{K}^c$–construction. Thus if there is more than one cofinal branch of the tree built so far satisfying the conclusion of Theorem 2.25, we can simply pick one such branch b and start all over. As long as from this point on our opponent in the iteration game plays extenders which can be interpreted as forming a tree on $\mathfrak{M}_b$, this will work. We run into trouble, however, if at some later point our opponent plays an extender E which, according to the rules of the iteration game, should be applied to some model reached before we reached $\mathfrak{M}_b$. If we set things up correctly, however, this extender E will overlap the local Woodin cardinal $\delta := \sup\{\mathrm{lh}(F_\alpha) \colon \alpha \in b\}$.

Definition 2.31. A premouse $\mathfrak{M}$ is *tame* iff whenever E is an extender on the $\mathfrak{M}$–sequence, say the last extender of $\mathcal{J}_\lambda^{\mathfrak{M}}$, then

$$\mathcal{J}_\lambda^{\mathfrak{M}} \models \forall \delta \geq \mathrm{crit}(E)(\delta \text{ is not Woodin})$$

Otherwise $\mathfrak{M}$ is called *wild*.

Proposition 2.32. If there is a wild mouse, then there is a model with a proper class of Woodin cardinals.

Proof : Let $\mathfrak{M}$ be the wild mouse with a $\langle \kappa, \lambda \rangle$–extender E overlapping a Woodin cardinal δ. Let

$$\Psi_{\alpha,\beta} :\equiv \exists \gamma(\alpha < \gamma < \beta(\gamma \text{ is Woodin}))$$

[67] These constitute an initial segment of the construction.

[68] The uniqueness argument is a bit more subtle than might at first appear, because of the possibility that one of the cofinal branches might drop. Finestructure to the rescue!

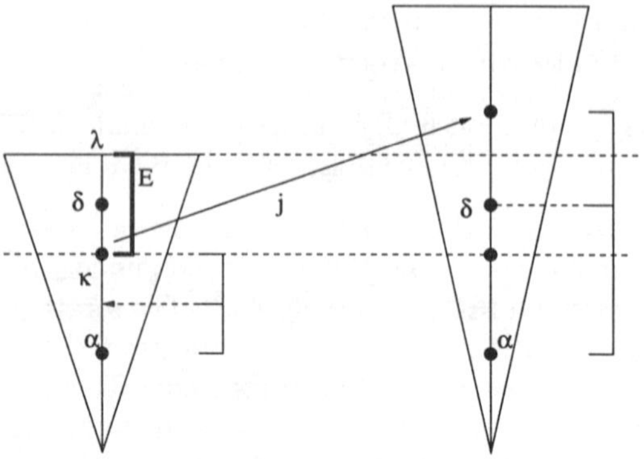

FIGURE 4. Reflection of Woodin cardinals in wild mice

Now we take the ultrapower by E, and we have for all $\alpha < \kappa$ that $\mathrm{Ult}(\mathfrak{M}, E) \models \Psi_{\alpha, j_E(\kappa)}$, as δ is still Woodin in the ultrapower.[69] But then we can transport these sentences back to $\mathfrak{M}$ giving $\mathfrak{M} \models \Psi_{\alpha,\kappa}$ for all $\alpha < \kappa$. This reflection is pictured in Figure 4.

Then $\mathbf{V}_\kappa^{\mathfrak{M}}$ is a model with a proper class of Woodin cardinals. □

Proposition 2.33. If $\mathfrak{M}$ is a tame premouse, then it cannot satisfy "there is a limit of Woodin cardinals which is itself a Woodin cardinal".

Proof : Working in $\mathfrak{M}$, suppose that there is a sequence $\langle \delta_\alpha : \alpha < \delta \rangle$ with limit δ which is itself a Woodin cardinal. (Note that δ is regular and hence the sequence has length δ.) As we noted in the short discussion of Woodin cardinals after Definition 2.27, δ is the limit of $< \delta$–strong cardinals $\langle \kappa_\alpha : \alpha < \delta \rangle$. Then there is some $\alpha < \delta$ such that $\kappa_0 < \delta_\alpha$ and κ_0 is $\delta_{\alpha+1}$–strong. Hence the extender witnessing this overlaps δ_α, and $\mathfrak{M}$ can't be tame. □

We mention without proof that a tame mouse can satisfy "there is a limit of Woodin cardinals which is a strong cardinal".

The argument we sketched very vaguely above can be turned into a proof of the following theorem:

Theorem 2.34. Every countable elementary substructure of a tame premouse occurring in a $\mathbf{K}^c$–construction is $\omega_1 + 1$–iterable.

This theorem is proved in [St93] and [SchSt96].

Although the $\omega_1 + 1$–iterability of countable elementary submodels of premice in a $\mathbf{K}^c$–construction suffices for many basic applications of iterability,

[69]This is the case because the strength of the extender is greater than δ; here we use our assumption that the extender overlaps the Woodin cardinal.

there are important contexts in which we need more.[70] The standard argument here once again involves collapsing. By restricting the complexity of the mice under consideration, one gets for each countable elementary submodel $\mathfrak{M}$ of $\mathfrak{N}$ an ω_1–iteration strategy $\Sigma(\mathfrak{M})$ which is uniformly definable from $\mathfrak{M}$ by some formula φ. We then define a putative ϑ–iteration strategy on $\mathfrak{N}$, for arbitrary ϑ, by going to $\mathbf{V}[G]$ where G is $\mathrm{Col}(\sup\{|\mathfrak{N}|, \vartheta\}, \omega)$–generic over $\mathbf{V}$, and using the prescription defined by φ from $\mathfrak{N}$ there. *If* we have enough generic absoluteness, this prescription will pick a unique branch in $\mathbf{V}[G]$, and by homogeneity of the collapse, this branch will be in $\mathbf{V}$.

The generic absoluteness required is closely related to the complexity of the formula φ defining the iteration strategies $\Sigma(\mathfrak{M})$, and thus to the complexity bound we imposed on $\mathfrak{M}$. If we want to make do with Shoenfield absoluteness, we must work with something like properly 1–small $\mathfrak{M}$. To handle mice beyond that, we need forms of generic absoluteness which are not provable in ZFC, but are themselves equivalent to the existence of nontrivial mice over arbitrary sets. We then need to construct these mice, and that leads the whole theory into an inductive proof that the universe is closed under building various mice over arbitrary sets. This process is called the *Core Model Induction*. Since it would take us too far afield to describe this further, we shall stick here to the iterability one can get using Shoenfield absoluteness.

Theorem 2.35. Suppose that there is no proper class premouse satisfying "There is a Woodin cardinal". Let $\mathfrak{N}$ be a premouse occurring in a $\mathbf{K}^c$–construction, and let $\vartheta \in \mathrm{Ord}$; then $\mathfrak{N}$ is ϑ–iterable.

Proof : If $\mathfrak{N}$ is not 1–small, then we can linearly iterate the last extender of its first non–1–small level Ord times, and thereby produce a proper class premouse with a Woodin cardinal. So we may assume $\mathfrak{N}$ is 1–small. Similarly, we can extend $\mathfrak{N}$ by adding ordinals on top (*i.e.* without adding new extenders) until we reach a premouse occurring in a $\mathbf{K}^c$–construction which satisfies "there are no Woodin cardinals". We may as well assume $\mathfrak{N}$ is this structure.[71]

Because of its smallness properties, Theorems 2.25 and 2.28 imply that every countable iteration tree of limit length on a countable elementary submodel of $\mathfrak{N}$ has a unique cofinal wellfounded branch. This gives ω_1–iteration strategies for the countable elementary submodels of $\mathfrak{N}$. We attempt to construct a similar strategy applying to arbitrary trees on $\mathfrak{N}$. For this it is enough to show

[70]We can handle trees and premice of size less than the critical points of the background extenders used in the construction by collapsing their size to be countable, as in the argument sketched above, but in important contexts one needs more than this.

[71]Any iteration strategy for this structure induces one for $\mathfrak{N}$. However, if $\mathfrak{N}$ itself satisfies "there is a Woodin cardinal", it may have iteration strategies which do not lift to this structure.

Claim : If $\mathfrak{T}$ is an iteration tree on $\mathfrak{N}$ of limit length, then $\mathfrak{T}$ has a unique cofinal wellfounded branch.

We prove the claim: Let $\delta := \sup\{\mathrm{lh}(F_\alpha^{\mathfrak{T}}): \alpha < \mathrm{lh}(\mathfrak{T})\}$, and let $\mathfrak{M}$ be the premouse of ordinal height δ to which the models of $\mathfrak{T}$ converge.[72] By adding ordinals on top of $\mathfrak{M}$ we reach a premouse $\mathfrak{Q}$ extending $\mathfrak{M}$ which satisfies "δ is not Woodin", as otherwise we have a proper class premouse with a Woodin cardinal.

Now let $\pi: R \to \mathbf{V}_\eta$ where η is sufficiently large, R is countable and transitive, and

$$\pi(\langle \overline{\mathfrak{N}}, \overline{\mathfrak{M}}, \overline{\mathfrak{Q}}, \overline{\mathfrak{T}} \rangle) = \langle \mathfrak{N}, \mathfrak{M}, \mathfrak{Q}, \mathfrak{T} \rangle.$$

We want to see that $\mathfrak{T}$ has a unique cofinal wellfounded branch, so it is enough to see that R satisfies "$\overline{\mathfrak{T}}$ has a unique cofinal wellfounded branch". By our branch existence and uniqueness theorems 2.25 and 2.28, $\overline{\mathfrak{T}}$ does indeed have a unique cofinal wellfounded branch b, so it is enough to show that $b \in R$.

Now $\mathfrak{M}_b^{\overline{\mathfrak{T}}} \models$ "$\overline{\delta}$ is not Woodin", by our smallness assumption on $\mathfrak{N}$ and the elementarity of $i_{0,b}$. It follows from this and Theorem 2.28 that b is the unique cofinal branch c of $\overline{\mathfrak{T}}$ such that $\overline{\mathfrak{Q}}$ is an initial segment of $\mathfrak{M}_c^{\overline{\mathfrak{T}}}$. Letting G be $\mathrm{Col}(|\overline{\mathfrak{Q}}|, \omega)$-generic over R, we get by Σ_1^1–absoluteness that $b \in R[G]$. This is true for all such G, so $b \in R$, and we are done with the claim, and thus with the proof.[73] □

The proof of Theorem 2.35 can be generalized to situations in which one does not have the very strong smallness assumption that there is no proper class model with a Woodin cardinal. What plays the rôle of the "branch oracle" structure $\mathfrak{Q}$ in the proof then involves extenders beyond those of the "lined up part" $\mathfrak{M}$ of the models of $\mathfrak{T}$. For example, if there is no proper class model with two Woodin cardinals, and over every premouse $\mathfrak{M}$ we can build an iterable–above–$\mathrm{Ord}^{\mathfrak{M}}$ proper class mouse with one Woodin cardinal, then the proof of Theorem 2.35 shows that every premouse on a $\mathbf{K}^c$–construction is ϑ–iterable for all $\vartheta \in \mathrm{Ord}$.

3. The core model K

3.1. **The Universality of $\mathbf{K}^c$.** For the rest of this paper, we shall assume that there is a measurable cardinal. This assumption plays no rôle in core model theory at the level of strong cardinals and below, but it is not known how to do without it at the level of Woodin cardinals. We therefore fix for

---

[72]So for $\alpha < \delta$, $\mathcal{J}_\alpha^{\mathfrak{M}}$ is the common value of $\mathcal{J}_\alpha^{\mathfrak{M}_\beta^{\mathfrak{T}}}$ for all β such that $\alpha < \mathrm{lh}(F_\beta^{\mathfrak{T}})$.

[73]Modulo the usual missing finestructure. The main point there is that δ may actually be Woodin with respect to all $A \in \mathfrak{Q}$; the A witnessing non–Woodinness may only be definable over $\mathfrak{Q}$. Similarly, b may drop, and thus $\overline{\delta}$ may only be definably non–Woodin over $\mathfrak{M}_b^{\overline{\mathfrak{T}}}$.

the rest of this paper a measurable cardinal Ω, and a normal measure μ_0 on Ω.

Let us also assume henceforth that *every premouse occurring in a* $\mathbf{K}^c$–*construction is tame.*[74] It follows from Theorem 2.34 and Theorem 2.23 that there is a $\mathbf{K}^c$–construction of length $\Omega + 1$. Let us fix such a construction $\langle \mathfrak{N}_\alpha : \alpha \leq \Omega \rangle$.[75]

Definition 3.1. $\mathbf{K}^c$ is the Ωth model of our designated $\mathbf{K}^c$–construction, that is, $\mathbf{K}^c = \mathfrak{N}_\Omega$.

We shall not need to refer to premice of ordinal height $> \Omega$, and so from now on we let "premouse" stand for "premouse of ordinal height $\leq \Omega$". We call premice which are as long as possible "weasels":

Definition 3.2. A *weasel* is a premouse of ordinal height Ω. A weasel is *universal* just in case it is Ω–universal (*cf.* Definition 2.12).

Our next result, which is sometimes called "cheapo covering", shows that we have been sufficiently liberal about putting extenders on the $\mathbf{K}^c$–sequence. The evidence of this is that we get a weasel which is close enough to $\mathbf{V}$ to compute many successor cardinals correctly, and therefore by Lemma 2.13 is universal.

Theorem 3.3. Suppose every premouse occurring in the $\mathbf{K}^c$ construction is tame; then for μ_0–almost every α we have $(\alpha^+)^{\mathbf{K}^c} = \alpha^+$.

Proof : We use the ultrapower embedding $j : \mathbf{V} \to \mathrm{Ult}(\mathbf{V}, \mu_0)$ that we get from the measurability of Ω. If cheapo covering is false, then $\{\alpha : (\alpha^+)^{\mathbf{K}^c} = \alpha^+\}$ has measure zero, and hence by Łos's theorem $(\Omega^+)^{j(\mathbf{K}^c)} < \Omega^+$. Because $\mathbf{K}^c \models \mathsf{GCH}$ this means that the power set of Ω in $j(\mathbf{K}^c)$ has cardinality Ω.

Let E_j be the extender coding j. If we restrict E_j to $j(\mathbf{K}^c)$, then the resulting extender is in $\mathrm{Ult}(\mathbf{V}, \mu_0)$: By the assumption, we can enumerate $\mathfrak{P}(\Omega) \cap j(\mathbf{K}^c)$ as $\{M_\alpha : \alpha < \Omega\}$. Then $\langle j(M_\alpha) : \alpha < \Omega \rangle$ lies in the ultrapower and therefore we can define the restricted extender by

$$M_\alpha \in (E_j)_a :\Longleftrightarrow a \in j(M_\alpha)$$

in the ultrapower.

We can now show that every proper initial segment of this extender is on the $j(\mathbf{K}^c)$–sequence. These initial segments witness that Ω is Woodin in $j(\mathbf{K}^c)$. It follows by reflection that Ω is a limit of Woodins in $\mathbf{K}^c$, and thus $j(\Omega)$ is a limit of Woodins in $j(\mathbf{K}^c)$. Hence one of the initial segments of E_j overlaps a local Woodin cardinal, and there is a wild mouse on the $j(\mathbf{K}^c)$–construction. Therefore there is a wild mouse on the $\mathbf{K}^c$–construction. □

[74] We shall strengthen this smallness assumption dramatically in a moment.

[75] As we said before, one can show that there is exactly one such construction.

The proof of Theorem 3.3 works in greater generality: if we generalize the $\mathbf{K}^c$–construction so as to allow extenders representing *e.g.* superstrong embeddings, then (granted sufficient iterability[76]), either $\mathbf{K}^c$ satisfies "there is a superstrong cardinal", or $\mathbf{K}^c$ computes successor cardinals correctly μ_0–almost everywhere.

Other evidence that we have put enough extenders into $\mathbf{K}^c$ comes from the fact that various large cardinal properties of $\mathbf{V}$ are inherited by $\mathbf{K}^c$. For example, if one modifies the $\mathbf{K}^c$–construction by omitting the requirement that $\mathfrak{N}_{\alpha+1}$ leave gaps, then either this modified construction reaches a wild mouse, or it converges to a weasel W such that for any Woodin cardinal δ, $W \models$"δ is Woodin".

If $\mathbf{K}^c$ is $\Omega + 1$–iterable, then Theorem 3.3 and Lemma 2.13 immediately imply that $\mathbf{K}^c$ is universal. We have obtained the $\Omega + 1$–iterability of $\mathbf{K}^c$ by restricting ourselves to the situation in which there is no proper class model with a Woodin cardinal (*cf.* Theorem 2.35). So we have

Theorem 3.4. If there is no proper class model with a Woodin cardinal, then $\mathbf{K}^c$ is a universal weasel.

In fact, if there is no proper class model with a Woodin cardinal, then a weasel W is universal if and only if $(\alpha^+)^W = \alpha^+$ for stationarily many $\alpha < \Omega$. This can be seen by coiterating W with $\mathbf{K}^c$.

3.2. **The true core model K.** Any iterate of a universal weasel is itself universal, so there are many universal weasels if there are any. $\mathbf{K}^c$ has some claim to distinction among universal weasels, but it depends too heavily on the universe in which it is constructed to serve some purposes; that is, its definition is not sufficiently generically absolute. This is easy to see:

Lemma 3.5. The first measurable cardinal in $\mathbf{K}^c$, if it exists, is inaccessible in $\mathbf{V}$.

Proof : This is immediate from the fact that the levels where we added new extenders in our $\mathbf{K}^c$–construction were required to leave gaps, and one of the conditions for that was that the critical point of the last extender is inaccessible (*cf.* Definition 2.21). □

Proposition 3.6. Suppose that there is a measurable cardinal in $\mathbf{K}^c$. Then there is a forcing extension $\mathbf{V}[G]$ of $\mathbf{V}$ such that

$$(\mathbf{K}^c)^{\mathbf{V}[G]} \neq (\mathbf{K}^c)^{\mathbf{V}}.$$

Proof : Let κ be the first measurable in $\mathbf{K}^c$, and let G be $\mathrm{Col}(\kappa, \omega)$–generic for $\mathbf{V}$. Then in $\mathbf{V}[G]$, κ is countable and hence by Lemma 3.5, κ can't be the first measurable in $(\mathbf{K}^c)^{\mathbf{V}[G]}$. Hence $(\mathbf{K}^c)^{\mathbf{V}[G]} \neq (\mathbf{K}^c)^{\mathbf{V}}$. □

[76]We would need $\omega_1 + 1$–iterability for countable elementary substructures.

If there is no proper class model with a strong cardinal, then there is a privileged universal weasel, *i.e.* one which generates all other universal weasels by iteration. This canonical universal weasel $\mathbf{K}$ is *maximal*, in the sense that every extender which could be added to its sequence (subject to an iterability constraint) is already on its sequence, and *generically absolute*, in that it has a definition which defines it not just in $\mathbf{V}$, but in every set generic extension of $\mathbf{V}$.[77]

One can construct such a canonical universal weasel under the weaker hypothesis that there is no proper class model with a Woodin cardinal. In fact, the construction we shall present here is much more general than that. We use the assumption that there is no proper class model with a Woodin cardinal only to obtain the $\Omega + 1$–iterability of $\mathbf{K}^c$ via Theorem 2.35. As we remarked after presenting it, the proof of Theorem 2.35 goes through under less restrictive hypotheses, and under such hypotheses our construction of $\mathbf{K}$ will go through as well. To keep the exposition as simple as possible, we shall work here, and for the rest of the paper, under the hypothesis that *there is no proper class model with a Woodin cardinal*. Like the measurability of Ω, we shall not always display this hypothesis in the statements of the results to come which use it.

The idea of our construction is that $\mathbf{K}^c$ is an iterate of $\mathbf{K}$, and therefore we can obtain $\mathbf{K}$ from $\mathbf{K}^c$ by taking a Skolem hull of $\mathbf{K}^c$ which "undoes" the iteration.[78] We now introduce some concepts useful in isolating this Skolem hull.

Definition 3.7. Let

$$\begin{aligned} A_0 := \{\alpha < \Omega \ : \ & \alpha \text{ is inaccessible}, (\alpha^+)^{\mathbf{K}^c} = \alpha^+, \\ & \{\beta < \alpha : \beta \text{ is inaccessible}, (\beta^+)^{\mathbf{K}^c} = \beta^+\} \text{ is not} \\ & \text{stationary in } \alpha\}. \end{aligned}$$

Proposition 3.8. *A_0 is stationary in Ω; moreover, for any $\alpha \in A_0$, α is not the critical point of an extender on the $\mathbf{K}^c$–sequence which is total on $\mathbf{K}^c$.*

Proof : The first condition follows directly from Theorem 3.3, since A_0 is the "first Mahlo derivative" of a set which has μ_0-measure one, and is therefore stationary. For the second condition let $\kappa = \text{crit}(E)$ for some E on the $\mathbf{K}^c$–sequence which is total on $\mathbf{K}^c$. At some step $\mathfrak{N}_{\alpha+1}$ in the construction an extender F was added to the sequence, and eventually, perhaps through taking some cores, F collapsed to E. Since κ is a cardinal of $\mathbf{K}^c$, we get $\text{crit}(F) = \kappa$. Since E is total on $\mathbf{K}^c$, $\mathfrak{N}_{\alpha+1}$ and $\mathbf{K}^c$ have the same subsets of κ. But the requirement that $\mathfrak{N}_{\alpha+1}$ leave gaps gives us at once that

[77] These are results of Dodd, Jensen, and Mitchell; *cf.* [Ze9?] or [St96, §8].

[78] Many of the ideas and arguments behind this construction are due to Mitchell, and can be found in an informally circulated paper which was to be a successor to [Mi84], but was never published. Mitchell's work was transformed into the form it takes here by the second author in [St96].

$\kappa \notin A_0$.[79] □

Definition 3.9. Let W be a weasel and let $\Gamma \subseteq W$; then Γ is *thick* in W iff for all but nonstationarily many $\alpha \in A_0$

- $\alpha \in \Gamma$,
- Γ contains a set which is α–club in α^+, and
- α is not the critical point of an extender from the W–sequence which is total on W.

So by Proposition 3.8

Lemma 3.10. Ω is thick in $\mathbf{K}^c$.

The following elementary lemmas are easy to prove:

Lemma 3.11. If Ω is thick in W, then the class of sets which are thick in W is an Ω-complete filter on Ω.

Lemma 3.12. Let $\pi\colon H \to W$ be elementary, where W is a weasel and $\mathrm{ran}(\pi)$ is thick in W; then $\{\alpha\colon \pi(\alpha) = \alpha\}$ is thick in both H and W.

Lemma 3.13. Let Ω be thick in W, and suppose $\mathfrak{T}$ is an iteration tree on W. Let $\lambda \leq \Omega$, and suppose there is no dropping in $\mathfrak{T}$ along $[0,\lambda]_T$ (and $\lambda < \mathrm{lh}(\mathfrak{T})$), so that $i^{\mathfrak{T}}_{0,\lambda}$ is defined. Suppose $i^{\mathfrak{T}}_{0,\lambda}{}''\Omega \subseteq \Omega$; then $\{\alpha < \Omega\colon i^{\mathfrak{T}}_{0,\lambda}(\alpha) = \alpha\}$ is thick in both W and $\mathfrak{M}^{\mathfrak{T}}_\lambda$.

Proof : This is clear if $\lambda < \Omega$, so let $\lambda = \Omega$. Then $[0,\lambda]_T$ is club in Ω, so for the typical $\alpha \in A_0$, $\alpha \in [0,\lambda]_T$ and $i_{0,\alpha}{}''\alpha \subseteq \alpha$. As α is inaccessible, $i_{0,\alpha}(\alpha) = \alpha$. As α is not the critical point of an extender from the W–sequence which is total on W, α is not the critical point of an extender from the $\mathfrak{M}^{\mathfrak{T}}_\alpha$–sequence which is total on $\mathfrak{M}^{\mathfrak{T}}_\alpha$. Thus α is not the critical point of an extender from the $\mathfrak{M}^{\mathfrak{T}}_\beta$–sequence which is total on $\mathfrak{M}^{\mathfrak{T}}_\alpha$, for any $\beta \geq \alpha$. Thus $\alpha^+ < \mathrm{crit}(i_{\alpha,\lambda})$, so that $i_{0,\lambda}(\alpha) = \alpha$ and $\{\gamma\colon i_{0,\lambda}(\gamma) = \gamma\}$ contains an α–club subset of α^+. □

We shall define $\mathbf{K}$ to be the transitive collapse of the intersection of all thick hulls of $\mathbf{K}^c$. At the moment, it is not even clear that this intersection is a weasel, much less that it is universal.[80] The following concepts are the key to showing $\mathbf{K}$ is universal.

Definition 3.14. Let Ω be thick in M; then M has the *definability property at* α iff for all Γ such that Γ is thick in M, we have

$$\alpha \in \mathrm{Hull}^M(\alpha \cup \Gamma),$$ [81]

[79] This was the entire purpose of the leaving gaps requirement in the $\mathbf{K}^c$–construction.

[80] Every thick hull of $\mathbf{K}^c$ is universal, and hence contains all the reals of $\mathbf{K}^c$, so the intersection is nonempty.

[81] Here $\mathrm{Hull}^M(X)$ denotes the uncollapsed hull of X inside M.

i.e. α is first–order definable over M with parameters from Γ and ordinals $\beta < \alpha$.

Definition 3.15. Let Ω be thick in M; then M has *the hull property at* α iff for all Γ such that Γ is thick in M, we have

$$\mathfrak{P}(\alpha)^M \subseteq \mathcal{H}^M(\alpha \cup \Gamma)$$

where $\mathcal{H}^M(X)$ is the transitive collapse of $\mathrm{Hull}^M(X)$.

Lemma 3.16. There is an $\Omega + 1$–iterable weasel M such that Ω is thick in M and M has the hull property at all $\alpha < \Omega$.

Proof : Let W be any $\Omega + 1$–iterable weasel such that Ω is thick in W; for example, we could take $W = \mathbf{K}^c$. We shall construct inductively decreasing classes $N_\alpha \prec W$ such that N_α is thick in W. The idea is to proceed along the cardinals of the hull of W we are determining, successively discarding all counterexamples to the hull property, so that the transitive collapse of the limit class $N_\Omega := \bigcap_{\xi \in \Omega} N_\xi$ will be the desired weasel M.

We start with $N_0 := W$, and for limit ordinals λ we set $N_\lambda := \bigcap_{\xi \in \lambda} N_\xi$. Now let N_α be given, let

$$\pi \colon N \to N_\alpha$$

be the transitive collapse, and let κ be the αth cardinal of N. By Lemma 3.12, Ω is thick in N. We now discard all counterexamples to the hull property at κ for N: for each $A \subseteq \kappa$ such that $A \in N$ pick a thick class Γ_A such that

$$A \notin \text{ transitive collapse of } \mathrm{Hull}^N(\kappa \cup \Gamma_A),$$

if there is such a thick class, and set $\Gamma_A = \Omega$ otherwise; then set

$$N_{\alpha+1} := \pi\text{"} \bigcap_A \mathrm{Hull}^N(\kappa \cup \Gamma_A).$$

We only have to check that N_Ω is thick in W. By Fodor's lemma, for all but nonstationarily many $\lambda \in A_0$, λ is the λth cardinal of N_λ, and for such λ, $\lambda \in N_{\lambda+1} \Leftrightarrow \lambda \in N_\Omega$. Thus is suffices to show that for all limit ordinals λ, $N_\lambda = N_{\lambda+1}$.

The proof of this illustrates well a typical use of the hull property. Let N be the transitive collapse of N_λ, and let κ be the λth cardinal of N. We must show that N has the hull property at κ. Our construction guarantees that κ is a limit cardinal of N, and that N has the hull property at all $\alpha < \kappa$. Let H be the transitive collapse of $\mathrm{Hull}^N(\kappa \cup \Gamma)$ for some Γ thick in N; it will be enough to show that $\mathfrak{P}(\kappa) \cap N \subseteq H$. For this we compare H with N; let $\mathfrak{T}$ on H and $\mathfrak{U}$ on N be the resulting iteration trees. Since H and N are universal, $\mathfrak{T}$ and $\mathfrak{U}$ have a common last model Q, which is itself a universal weasel. Let

$$i \colon H \to Q \text{ and } j \colon N \to Q$$

be the iteration maps given by $\mathfrak{T}$ and $\mathfrak{U}$. It will be enough to show that $\mathrm{crit}(j) \geq \kappa$, for then if $A \in \mathfrak{P}(\kappa) \cap N$, we get $A = j(A) \cap \kappa \in Q$, which implies $A \in H$. (For the last assertion, notice that H and N agree below κ, so all extenders used in $\mathfrak{T}$ and $\mathfrak{U}$ have length $> \kappa$, and in particular, Q agrees with H below $(\kappa^+)^Q$.)

Suppose then that $\text{crit}(j) < \kappa$. Let E be the first extender used along the branch from N to Q of $\mathfrak{U}$, so that $\text{crit}(E) = \text{crit}(j) := \mu < \kappa$. One can easily show using Lemma 3.13 that Q retains the hull property at μ. The sets $E \restriction \nu$ for $\mu < \nu < \kappa$ witness that Q does not have the hull property at any ν such that $\mu < \nu < \kappa$.[82] But then, by considering how the hull property is passed from H to Q, we see that μ is also the critical point of i, and thus of the first extender F used on the branch from H to Q in $\mathfrak{U}$.

We claim that E is compatible with F, contrary to the fact that they arose in a coiteration. For if $A \subseteq \mu$ and $A \in N$, then by Lemma 3.13 and Lemma 3.11 we can find a finite set $a \subseteq \mu$, a finite set b of common fixed points of i and j, and a Skolem term τ such that $A = \tau^N[a,b] \cap \mu$. Then $j(A) = \tau^Q[a,b] \cap j(\mu)$, and since $A = j(A) \cap \mu$, this implies that $A = \tau^H[a,b] \cap \mu$, and hence $i(A) = \tau^Q[a,b] \cap i(\mu)$. Thus $j(A)$ and $i(A)$ agree below $\inf\{j(\mu), i(\mu)\}$. Because generators are not moved along the branches of an iteration tree, this implies that E is compatible with F, the desired contradiction. □

From Lemma 3.16 we easily get

Theorem 3.17. Let W be an $\Omega+1$–iterable weasel such that Ω is thick in W; then for μ_0–almost every $\alpha < \Omega$, W has the hull property at α.

Proof : Let M be a weasel having the hull property everywhere, as provided by Lemma 3.16. We compare M with W, and obtain iteration maps $i\colon M \to Q$ and $j\colon W \to Q$ into a common iterate. For μ_0–almost every α, $i''\alpha \subseteq \alpha$ and $j''\alpha \subseteq \alpha$. Using this and Lemma 3.13 we can transfer the hull property from M to Q at μ_0–almost every α, and then pull it back from Q to W at μ_0-almost every α (cf. [St96, Lemma 4.6]). □

Corollary 3.18. $\mathbf{K}^c$ has the hull property at μ_0–almost every $\alpha < \Omega$.

Proof : By Proposition 3.8 and Theorem 3.17. □

We turn now to the definability property.

Theorem 3.19. For μ_0–almost every $\alpha < \Omega$, $\mathbf{K}^c$ has the definability property at α.

Proof : The proof proceeds along the lines of the proof of Theorem 3.3. We assume that the set of α such that $\mathbf{K}^c$ has the definability property at α

[82] *Cf.* [St96, Example 4.3] for a proof of this.

has measure zero. Let $j\colon \mathbf{V} \to M = \mathrm{Ult}(\mathbf{V}, \mu_0)$ be the canonical embedding. Since $j(\mathbf{K}^c)$ does not have the definability property at Ω in M, we can in M find a thick hull of $j(\mathbf{K}^c)$ omitting Ω, and hence a map $\pi\colon H \to j(\mathbf{K}^c)$ such that $\mathrm{ran}(\pi)$ is thick in $j(\mathbf{K}^c)$. With some care, one can arrange that E_π is compatible with E_j, so that the fragments of E_j which are in M can be used as background extenders to certify E_π for addition to the $j(\mathbf{K}^c)$–sequence in M. Notice that E_π is total on $j(\mathbf{K}^c)$ by the hull property of $j(\mathbf{K}^c)$ at Ω. But then, as in the proof of Theorem 3.3, Ω is Woodin in $j(\mathbf{K}^c)$[83], contrary to our assumption that there is no proper class model with a Woodin cardinal. □

We proceed to the definition of **K**.

Definition 3.20. **K** is the transitive collapse of

$$D_{\mathbf{K}^c} := \{x : \forall\Gamma(\Gamma \text{ is thick in } \mathbf{K}^c \to x \in \mathrm{Hull}^{\mathbf{K}^c}(\Gamma))\}.$$

Theorem 3.21. **K** is a universal weasel, and $(\alpha^+)^{\mathbf{K}} = \alpha^+$ for μ_0–almost every $\alpha < \Omega$.

Proof : We have to show that $D_{\mathbf{K}^c}$ is unbounded in Ω (this shows that **K** is a weasel) and that $D_{\mathbf{K}^c}$ is unbounded in $(\nu^+)^{\mathbf{K}^c}$ for a μ–measure one set of $\nu < \Omega$ (together with Theorem 3.3 this shows Cheapo Covering for **K**).

Both parts are similar, the unboundedness in Ω uses the definability property of $\mathbf{K}^c$ and the second part uses the hull property. We shall sketch the first part.

If $D_{\mathbf{K}^c}$ is bounded then we easily construct a decreasing sequence of thick classes Γ_ξ such that, letting b_ξ be the least ordinal in $\mathrm{Hull}^{\mathbf{K}^c}(\Gamma_\xi) \setminus D_{\mathbf{K}^c}$, the sequence $\langle b_\xi : \xi < \Omega\rangle$ is strictly increasing and $D_{\mathbf{K}^c} \subseteq b_0$. Now by Theorem 3.19 we can choose ν such that $\nu = \sup\{b_\xi : \xi < \nu\}$ and $\mathbf{K}^c$ has the definability property at ν. Since $\Gamma_{\nu+1}$ is thick, we can find a finite subset a of ν, a finite subset d of $\Gamma_{\nu+1}$, and a Skolem term τ such that $\nu = \tau^{\mathbf{K}^c}[a, d]$. Let $\xi < \nu$ be such that $a \subseteq b_\xi$. We have

$$\mathbf{K}^c \models \exists a \in [b_\xi]^{<\omega}(b_\xi < \tau[a, d] < b_{\nu+1}),$$

and since the parameters $b_\xi, d, b_{\nu+1}$ in this statement all belong to $\mathrm{Hull}^{\mathbf{K}^c}(\Gamma_\xi)$, we can find a^* in $\mathrm{Hull}^{\mathbf{K}^c}(\Gamma_\xi)$ which is a witness to its existential quantifier. By the definition of b_ξ, this means $a^* \in D_{\mathbf{K}^c}$. But that implies

$$\tau^{\mathbf{K}^c}[a^*, d] \in \mathrm{Hull}^{\mathbf{K}^c}(\Gamma_{\nu+1}) \text{ and } b_\xi < \tau^{\mathbf{K}^c}[a^*, d] < b_{\nu+1},$$

which contradicts the definition of $b_{\nu+1}$. □

We lack the space to go much further into the pure theory of **K** here, but we wish to state some basic theorems. First and foremost, there is a full weak covering theorem for **K**.

[83]In fact, the proof would give a wild mouse.

Theorem 3.22. (Mitchell, Schimmerling) Let $\mu < \Omega$ be a singular cardinal; then $(\mu^+)^{\mathbf{K}} = \mu^+$.

This theorem is definitely no cheapo. The central idea is due to Mitchell, and was brought to fruition in the case $\alpha < \mu \Rightarrow \alpha^\omega < \mu$ in [MiSchSt97]. The full theorem was then proved by Mitchell and Schimmerling in [MiSch95].

Concerning embeddings of **K**, the situation is a bit more complicated than it was below a strong cardinal, in that it is consistent with our assumptions[84] that there is an $\Omega+1$-iterable universal weasel which is not an iterate of **K** (*cf.* [St96, §8]). However, it is shown in [St96] that **K** elementarily embeds in every $\Omega + 1$–iterable universal weasel, and that there is no nontrivial embedding of **K** into itself. This leads to a nice characterization of **K**:

Theorem 3.23. **K** is the unique $\Omega + 1$–iterable universal weasel which elementarily embeds into every $\Omega + 1$–iterable universal weasel.

This characterization is Theorem 8.10 of [St96].

Although there may be iterable universal weasels which are not iterates of **K**, it turns out that $\mathbf{K}^c$ itself *is* an iterate of **K**. This is proved in [SchSt9?]. A key ingredient in the proof is a natural *maximality* property of **K** and its iterates. For **K** itself, this result states that every countably certified extender which "could be added" to the **K**–sequence (in that the resulting structure would be a premouse) is already on the **K**–sequence.

Finally, Jensen's Σ^1_3–correctness theorem (*cf.* [Do82]) has an extension to our situation.

Theorem 3.24. Suppose there is a measurable cardinal $\mu < \Omega$; then the Martin-Solovay tree at μ is in **K**, and therefore **K** is Σ^1_3–correct.

This is Theorem 7.9 of [St96]. It is very probably only a provisional result, as the second measurable cardinal μ should not be necessary.

3.3. **The definability of K and generic absoluteness.** We now sketch the application we promised in the introduction. For this we need some further basic results about **K**. We wish to produce a formula defining **K** whose logical form is as simple as possible. We shall then show that the class defined by this formula interpreted in **V** is the same as the class it defines when interpreted in $\mathbf{V}[G]$, for any G which is $\mathbb{P}$–generic over **V** where $\mathbb{P} \in \mathbf{V}_\Omega$.

We have already defined **K**; the only parameter entering into its definition is Ω. This definition involves quantification over $\mathbf{V}_{\Omega+1}$, however, and is therefore too complicated for our purposes. Moreover, it is not clear that this definition is absolute, as we used $\mathbf{K}^c$ to define **K**, and $(\mathbf{K}^c)^{\mathbf{V}[G]} \neq (\mathbf{K}^c)^{\mathbf{V}}$ in general. We show now that, nevertheless, our definition of **K** as the intersection of all thick hulls of $\mathbf{K}^c$ is generically absolute.

[84] *I.e.*, that Ω is measurable and there is no proper class model with a Woodin cardinal.

Definition 3.25. Let M be a weasel, and $S, \Gamma \subseteq M$. We say that Γ *is S–thick in M* iff

1. S is stationary in Ω, and
2. for all but nonstationarily many $\alpha \in S$:
   (a) α is inaccessible, $(\alpha^+)^M = \alpha^+$, and α is not the critical point of an extender from the M–sequence which is total on M, and
   (b) $\alpha \in \Gamma$, and $\Gamma \cap \alpha^+$ contains an α–club.

So Ω is A_0–thick in $\mathbf{K}^c$. We can now relativise our definitions of the hull and definability properties to an arbitrary S in the obvious way: for example, W has the S–definability property at κ iff Ω is S–thick in W, and $\kappa \in \mathrm{Hull}^W(\kappa \cup \Gamma)$ for all Γ which are S–thick in W.

Definition 3.26. A premouse $\mathfrak{M}$ is called *S–very sound*, if there is an $\Omega+1$–iterable weasel W such that $\mathfrak{M} \trianglelefteq W$, Ω is S–thick in W and W has the S–definability property at all $\beta \in \mathrm{Ord} \cap \mathfrak{M}$. We say $\mathfrak{M}$ is *very sound* just in case it is A_0–very sound.

It is easy to see that every proper initial segment of $\mathbf{K}$ is very sound. By the following lemma, this fact characterizes $\mathbf{K}$.

Lemma 3.27. Let $\mathfrak{M}$ be S–very sound and let $\mathfrak{N}$ be T–very sound; then $\mathfrak{M} \trianglelefteq \mathfrak{N}$ or $\mathfrak{N} \trianglelefteq \mathfrak{M}$.

Proof : Let W and R be weasels witnessing the soundness of $\mathfrak{M}$ and $\mathfrak{N}$ respectively. Assume without loss of generality that $\mathrm{Ord}^{\mathfrak{M}} \leq \mathrm{Ord}^{\mathfrak{N}}$.

The proof of Lemma 2.13 shows that for all but nonstationarily many $\alpha \in S \cup T$, $(\alpha^+)^W = (\alpha^+)^R = \alpha^+$. Now let W^* be the (linear) iterate of W obtained by taking an ultrapower by the order zero total measure on α from W, for each $\alpha \in T \setminus \mathrm{Ord}^{\mathfrak{M}}$ such that $W \models$ "α is measurable". Similarly, let R^* be obtained from R by taking an ultrapower by the order zero measure on α at each $\alpha \in S \setminus \mathrm{Ord}^{\mathfrak{N}}$ such that $R \models \alpha$ is measurable. Then W^* and R^* still witness the S and T soundness of $\mathfrak{M}$ and $\mathfrak{N}$, respectively; moreover, Ω is $S \cup T$ thick in each of W^* and R^*.

Let $i\colon W^* \to Q$ and $j\colon R^* \to Q$ come from coiteration, and let

$$\kappa = \inf\{\mathrm{crit}(i), \mathrm{crit}(j)\}.$$

It is enough to show $\mathrm{Ord}^{\mathfrak{M}} \leq \kappa$, for then $\mathfrak{M} \trianglelefteq \mathfrak{N}$ as desired, so assume that $\kappa < \mathrm{Ord}^{\mathfrak{M}}$.

Suppose that $\kappa = \mathrm{crit}(i) < \mathrm{crit}(j)$. Since Ω is T thick in R^* and W^*, and R^* has the T-definability property at κ, we can find a finite set a of common fixed points of i and j such that $\kappa = \tau^{R^*}[a]$ for some term τ. But then $\kappa = j(\kappa) = \tau^Q[a] = i(\tau^{W^*}[a])$, so $\kappa \in \mathrm{ran}(i)$, a contradiction. Similarly, we get $\mathrm{crit}(i) \leq \mathrm{crit}(j)$, and hence $\mathrm{crit}(i) = \mathrm{crit}(j) = \kappa$.

An argument using the S and T hull properties at κ now shows that the first extenders used along the branches giving rise to i and j in our coiteration are compatible with one another, a contradiction. This argument is essentially the same as the "typical use of the hull property" at the end of the proof of Lemma 3.16. □

Corollary 3.28. $\mathfrak{M}$ is a proper initial segment of $\mathbf{K}$ if and only if $\mathfrak{M}$ is very sound if and only if $\mathfrak{M}$ is S–very sound for some S.

These results imply at once that our Definition 3.20 of $\mathbf{K}$ using using thick hulls is generically absolute.

Theorem 3.29. Let G be $\mathbb{P}$–generic over $\mathbf{V}$, where $\mathbb{P} \in \mathbf{V}_\Omega$; then $\mathbf{K}^{\mathbf{V}} = \mathbf{K}^{\mathbf{V}[G]}$.

Proof : We claim that if W is a weasel witnessing that $\mathfrak{M}$ is S–very sound in $\mathbf{V}$, then W continues to witness that $\mathfrak{M}$ is S–very sound in $\mathbf{V}[G]$. For it is clear that S remains stationary in $\mathbf{V}[G]$, and that every S–thick class in $\mathbf{V}[G]$ contains an S-thick class in $\mathbf{V}$. It is less obvious that W remains $\Omega+1$–iterable in $\mathbf{V}[G]$, but this is provable (*cf.* [St96, Lemma 5.12]).

Thus all the proper initial segments of $\mathbf{K}^{\mathbf{V}}$ remain $A_0^{\mathbf{V}}$–very sound in $\mathbf{V}[G]$, as witnessed by the appropriate hulls of $(\mathbf{K}^c)^{\mathbf{V}}$. By Lemma 3.27 we then have $\mathbf{K}^{\mathbf{V}} = \mathbf{K}^{\mathbf{V}[G]}$. □

Unfortunately, the generically absolute definition of $\mathbf{K}$ we have just given is logically too complicated for our application. We now sketch an inductive definition of $\mathbf{K}$ which has the optimal logical complexity.[85]

Definition 3.30. A premouse $\mathfrak{M}$ is α–strong if there is a weasel W witnessing that $\mathcal{J}_\alpha^{\mathfrak{M}}$ is very sound, and an iteration tree $\mathfrak{T}$ on W played according to its unique $\Omega+1$–iteration strategy which uses only extenders with length $\geq \alpha$, and an elementary embedding from $\mathfrak{M}$ to an initial segment of the last model of $\mathfrak{T}$ such that $\pi \restriction \alpha$ is the identity.

This definition of α–strongness above involves quantification over $\mathbf{V}_{\Omega+1}$. We now give a definition by induction on α which is of essentially optimal complexity. The key is a certain iterability property: the α–strong mice are those which are "jointly iterable" with all mice which are β–strong for all $\beta < \alpha$.[86] We now explain this property further.

Definition 3.31. A triple $\langle \mathfrak{M}, \mathfrak{N}, \alpha \rangle$ is called a *phalanx* if

- $\mathfrak{M}$ and $\mathfrak{N}$ are premice with $\mathcal{J}_\alpha^{\mathfrak{M}} = \mathcal{J}_\alpha^{\mathfrak{N}}$,
- α is a cardinal in both models $\mathfrak{M}$ and $\mathfrak{N}$, and
- the projecta $\varrho(\mathfrak{M})$ and $\varrho(\mathfrak{N})$ are $\geq \alpha$

[85] Hugh Woodin has shown that there are universes which satisfy "there is a measurable cardinal and no proper class model with a Woodin cardinal" within which $\mathbf{K}$ has no definition logically simpler than the one we shall sketch.

[86] This idea traces back to Dodd's proof that GCH holds in the models of [Do81].

In building an iteration tree on a phalanx $\langle \mathfrak{M}, \mathfrak{N}, \alpha \rangle$, we pretend that the phalanx is already an iteration tree of length 1, where $\mathfrak{M}_0^{\mathfrak{T}} = \mathfrak{M}$, $\mathfrak{M}_1^{\mathfrak{T}} = \mathfrak{N}$ and α is interpreted, for the purpose of deciding which model to take an ultrapower of, as the length of a "virtual extender" used to get from $\mathfrak{M}$ to $\mathfrak{N}$. Note that $\mathfrak{N}$ need not be an ultrapower of $\mathfrak{M}$ — we simply play the iteration game as if it were.

The notions of ϑ–iterability and iterability for phalanxes are easily derived from this description.

The following theorem yields an inductive characterization of strongness.

Theorem 3.32. Let $\mathfrak{M}$ be a premouse, and suppose that α is a cardinal in $\mathbf{K}$ such that $\mathcal{J}_\alpha^{\mathbf{K}} = \mathcal{J}_\alpha^{\mathfrak{M}}$; then the following are equivalent:

1. $\mathfrak{M}$ is α–strong,
2. If $\mathfrak{N}$ is β–strong for all $\mathbf{K}$–cardinals $\beta < \alpha$, then $\langle \mathfrak{N}, \mathfrak{M}, \alpha \rangle$ is $\Omega + 1$–iterable.
3. If $\mathfrak{N}$ has cardinality $\leq \alpha$, and $\mathfrak{N}$ is β–strong for all $\mathbf{K}$–cardinals $\beta < \alpha$, then $\langle \mathfrak{N}, \mathfrak{M}, \alpha \rangle$ is $\Omega + 1$–iterable.

Proof: For the direction "(2.)⇒(1.)" let W witness that $\mathcal{J}_\alpha^{\mathfrak{M}}$ is very sound. We will show that W has an iterate as demanded in the definition of α–strong. Clearly, W is β–strong for all $\beta < \alpha$, so by (2.) the phalanx $\langle W, \mathfrak{M}, \alpha \rangle$ is $\Omega+1$–iterable. We now compare this phalanx with W in the obvious way, and obtain thereby iteration trees $\mathfrak{T}$ on $\langle W, \mathfrak{M}, \alpha \rangle$ and $\mathfrak{U}$ on W with last models $\mathfrak{R}$ and $\mathfrak{S}$ respectively.[87] The key observation is that $\mathfrak{R}$ is above $\mathfrak{M}$ in the tree $\mathfrak{T}$. This is because otherwise it is above W, in which case we get $\mathfrak{R} = \mathfrak{S}$ by universality, and can argue using the hull and definability properties as in Lemma 3.27 that the first extenders used on the branches from W to $\mathfrak{R}$ and from W to $\mathfrak{S}$ of the $\mathfrak{T}$ and $\mathfrak{U}$ are compatible. It is now easy to see that $\mathfrak{R}$ is an initial segment of $\mathfrak{S}$, all extenders used in $\mathfrak{U}$ have length $> \alpha$, and the embedding $\pi \colon \mathfrak{M} \to \mathfrak{R}$ given by $\mathfrak{T}$ is the identity below α.

For "(1.)⇒(2.)", we use the embeddings given by the fact that our premice are strong at various α and β to lift iteration trees on $\langle \mathfrak{N}, \mathfrak{M}, \alpha \rangle$ to something like iteration trees on $\mathbf{K}^c$, and then use some simple generalizations of our iterability results for $\mathbf{K}^c$.

The equivalence of (2.) with (3.) is a simple Löwenheim–Skolem argument. □

We can now use Theorem 3.32 to give an inductive definition of $\mathbf{K}$. First note the following simple consequence of the definition of α–strength.

Proposition 3.33. Let α be a cardinal of $\mathbf{K}$; then the following are equivalent:

[87] The comparison proceeds by iterating the least disagreement between the current last models as usual, beginning with $\mathfrak{M}$ *versus* W. The rules of iteration trees on premice and phalanxes respectively determine the rest.

1. $\mathfrak{N}$ is some $\mathcal{J}^{\mathbf{K}}_{\beta}$ for $\beta < (\alpha^+)^{\mathbf{K}}$
2. There is an α–strong premouse $\mathfrak{M}$ with projectum α such that $\mathfrak{N} = \mathcal{J}^{\mathfrak{M}}_{\beta}$ for some $\beta < (\alpha^+)^{\mathfrak{M}}$.

We can now define the class of **K**–cardinals α, α–strength for **K**–cardinals α, and $\mathcal{J}^{\mathbf{K}}_{\alpha}$ for **K**–cardinals α, by induction on α. Given a **K**–cardinal α, given $\mathcal{J}^{\mathbf{K}}_{\alpha}$, and given the class of α–strong premice, let $\mathfrak{M}$ be the union of all α–strong mice projecting to α; then the ordinal height of $\mathfrak{M}$ is $(\alpha^+)^{\mathbf{K}}$ and $\mathfrak{M}$ is $\mathcal{J}^{\mathbf{K}}_{(\alpha^+)^{\mathbf{K}}}$. We can then use (3.) of Theorem 3.32 to identify the class of $(\alpha^+)^{\mathbf{K}}$–strong mice. The limit steps in the induction are easy, using (3.) of Theorem 3.32 again to determine the α–strong mice from $\mathcal{J}^{\mathbf{K}}_{\alpha}$.

For our application, we want to measure the complexity of this inductive definition descriptive–set–theoretically, so we shall restrict ourselves to the definition of $\mathcal{J}^{\mathbf{K}}_{\alpha}$, for $\alpha < \omega_1^{\mathbf{V}}$. In this case the mice over which one quantifies in (3.) of Theorem 3.32 are all countable, so one easily obtains

Theorem 3.34. There is a formula $\varphi(v_0, v_1)$ such that the following are equivalent for all $\alpha < \omega_1$:

1. $\mathbf{L}_{\alpha+1}(\mathbb{R}) \models \varphi[x, y]$
2. x is a code for some $\delta < \alpha$ and y codes $\mathcal{J}^{\mathbf{K}}_{\delta}$

As **K** is invariant under forcing of size $< \Omega$, the formula φ defines **K** up to $\omega_1^{\mathbf{V}[G]}$ in $\mathbf{V}[G]$ whenever G is $\mathbb{P}$–generic for a poset $\mathbb{P}$ of size $< \Omega$.

We can now give the application we promised in the introduction:

Theorem 3.35. Suppose Ω is measurable, and for all partial orderings $\mathbb{P} \in \mathbf{V}_{\Omega}$ and all G $\mathbb{P}$–generic over **V** we have

$$(\mathbf{L}_{\omega_1}(\mathbb{R}))^{\mathbf{V}} \equiv (\mathbf{L}_{\omega_1}(\mathbb{R}))^{\mathbf{V}[G]};$$

then there is a proper class model satisfying "there is a Woodin cardinal".

Proof : We suppose toward contradiction that there is no such proper class model. This supposition puts the theory of **K** we have developed at our disposal. In particular, we can use the formula from Theorem 3.34 get a sentence saying that $\omega_1^{\mathbf{V}}$ is a successor cardinal in **K**. We shall call this sentence σ:

$$\mathbf{L}_{\omega_1}(\mathbb{R}) \models \sigma \iff \exists \alpha < (\omega_1)^{\mathbf{V}}((\alpha^+)^{\mathbf{K}} = (\omega_1)^{\mathbf{V}})$$

Our hypotheses guarantee that **K** computes some successor correctly, *i.e.* there is some β such that $\beta^+ = (\beta^+)^{\mathbf{K}}$. Because Ω is measurable, we can assume that β is inaccessible.

We have two different cases:

Case 1 : σ is true in **V**, *i.e.* ω_1 is a successor in **K**. Let G be $\mathrm{Col}(\omega, < \beta)$–generic for **V**. (This is the Levy collapse for collapsing β to ω_1.) Then the generic extension can't satisfy σ anymore.

Case 2 : σ is false in $\mathbf{V}$, *i.e.* ω_1 is a limit in $\mathbf{K}$. Then we collapse β to be countable. By Theorem 3.29 we know that in the generic extension, the successor of β is still computed correctly. But the successor is $(\omega_1)^{\mathbf{V}[G]}$, so $\mathbf{V}[G] \models \sigma$.

In either case, $(\mathbf{L}_{\omega_1}(\mathbb{R}))^{\mathbf{V}}$ and $(\mathbf{L}_{\omega_1}(\mathbb{R}))^{\mathbf{V}[G]}$ disagree as to the truth of σ, a contradiction.[88] □

4. Beyond one Woodin cardinal

We close with some brief remarks on the progress which has been made in extending the theory we have described to core models satisfying stronger large cardinal hypotheses.

The most important obstacle is that we have no proof of the ω_1+1–iterability of the countable elementary submodels of the premice occurring in $\mathbf{K}^c$–constructions. Andretta, Neeman, and the second author (unpublished) have been able to extend the proof for tame mice (Theorem 2.34) to slightly stronger mice, mice strong enough that their existence yields an inner model with a cardinal λ which is both a limit of Woodin cardinals and a limit of cardinals which are $<\lambda$–strong.[89] They obtain thereby

Theorem 4.1. If there is a measurable cardinal and the proper forcing axiom PFA holds, then there is an inner model with a cardinal λ which is both a limit of Woodin cardinals and a limit of cardinals which are $<\lambda$–strong.

This extends earlier work of Schimmerling, who showed that the same hypotheses imply the existence of a wild mouse (*cf.* [Sch95] and [Sch9?]).

It is important in the proof of Theorem 4.1 that one can make do with $\mathbf{K}^c$ in the relative consistency proof; one does not need a generically absolute core model like $\mathbf{K}$. This is important because the full $\Omega+1$–iterability of $\mathbf{K}^c$, on which the theory of $\mathbf{K}$ rests, is another kettle of fish. We obtained full iterability of $\mathbf{K}^c$ under the hypothesis that there is no proper class model with a Woodin cardinal in Theorem 2.35. The argument used collapsing and absoluteness to reflect a failure of iterability to the countable; the level of generic absoluteness required is closely connected to that of the complexity of the ω_1–iteration strategies for countable elementary submodels of $\mathbf{K}^c$.[90] Once our mice are no longer 1–small, this level of generic absoluteness is not provable in ZFC, and consequently, not every model of ZFC is a suitable environment in which to construct $\mathbf{K}$.

---

[88] This argument is due to Woodin, and comes from [Woo82].

[89] This large cardinal hypothesis is known as the $\mathsf{AD}_{\mathbb{R}}$–*hypothesis*, because Woodin (unpublished) has shown that its consistency implies the consistency of $\mathsf{AD}_{\mathbb{R}}$.

[90] We used in Theorem 2.35 the fact that every properly 1–small countable mouse $\mathfrak{M}$ has a $\Delta^1_2(\mathfrak{M})$ ω_1–iteration strategy. For mice which are not 1–small, the iteration strategies are necessarily more complicated.

Here are some examples which illustrate this. The first is due to Jensen (*cf.* [He97]). Let us call a weasel W a *strongly local core model* just in case there is a formula Φ which locally defines W both in W and in all its set generic extensions in the sense that for any (possibly trivial) G which is $\mathbb{P}$–generic over W for some $\mathbb{P} \in W$, and any inaccessible cardinal κ of $W[G]$,

$$W \cap \mathbf{V}_\kappa = \{x \colon W[G] \cap \mathbf{V}_\kappa \models \Phi[x]\}.$$

Then there is no strongly local core model W which satisfies "there is a Woodin cardinal". For let κ be Woodin in W, and let

$$j \colon W \to M \subseteq W[G], \text{ with } \mathrm{crit}(j) = \omega_1^W$$

be a generic elementary embedding coming from Woodin's full stationary tower forcing.[91] Since $\mathrm{crit}(j) = \omega_1$, there is a real $x \in W[G] \setminus W$. But κ remains inaccessible in $W[G]$, so if Φ is our formula locally defining W,

$$W[G] \cap \mathbf{V}_\kappa \models \neg\Phi[x].$$

On the other hand, since j is elementary

$$M \cap \mathbf{V}_\kappa \models \Phi[x].$$

But Woodin's forcing has the property that $M \cap \mathbf{V}_\kappa = W[G] \cap \mathbf{V}_\kappa$, a contradiction.

The second example is due to Woodin. Suppose W is the minimal $\Omega+1$–iterable weasel satisfying "there is a Woodin cardinal".[92] Let κ be the Woodin cardinal of W; then W satisfies "I am not $\kappa^+ + 1$–iterable".[93] Thus, although W really is fully iterable, it doesn't know how to iterate itself.

The third example is the following: let W be the minimal $\Omega + 1$–iterable weasel such that for some α, $\mathcal{J}_\alpha^W$ is not 1–small. It is easy to see that $W = \mathbf{L}[\mathfrak{M}]$, where $\mathfrak{M}$ is essentially the sharp of the minimal fully iterable model with one Woodin. Then $\mathfrak{M}$ is not ordinal definable in W, so that we have a core model which fails to satisfy $\mathbf{V} = \mathrm{HOD}$. As in the previous example, the problem here is that although $\mathfrak{M}$ is iterable, W does not know how to iterate it; in fact, $\mathfrak{M}$ is not even $\omega_1 + 1$–iterable in W.

These examples show that the absoluteness of the definition of $\mathbf{K}$, and what underlies that, the absoluteness of iterability, no longer hold good in the setting of arbitrary models of ZFC once one gets past the 1–small mice. Nevertheless, one can get a useful theory of $\mathbf{K}$ generalizing the one presented in Section 3 to stronger mice by working only in a sufficiently good background universe, a universe closed under building the "lower level" mice over arbitrary sets. In a relative consistency application of this theory of $\mathbf{K}$, one may then have to construct this sufficiently good background universe, and hence the lower level mice, as part of an induction. Here is one theorem proved by such a technique:

---

[91]For a reference, *cf.* the forthcoming monograph [Ma9?].

[92]Reasonable large cardinal hypotheses imply that there is such a weasel.

[93]Neeman sharpened the argument to show that W satisfies "I am not $\kappa + 1$–iterable".

Theorem 4.2 (Woodin). Suppose that for every G which is set generic over $\mathbf{V}$, $\mathbf{L}(\mathbb{R}) \equiv (\mathbf{L}(\mathbb{R}))^{\mathbf{V}[G]}$; then there is an inner model with ω Woodin cardinals.[94]

This result strengthens an earlier result of the second author which gave the same conclusion under the stronger hypothesis that every set of reals in $\mathbf{L}(\mathbb{R})$ is weakly homogeneous.

It is not known how to extend the bootstrapping technique which yields the theorem above, and others like it which involve a full theory of $\mathbf{K}$, so as to obtain models with more than a proper class of Woodin cardinals. Thus there are still basic open problems concerning the iterability of $\mathbf{K}^c$ for uncountable trees within the tame mice. These are connected to a number of very interesting potential relative consistency strength applications. The reader who started with only what we claimed were the prerequisites for this paper and has now made it to the end should have no trouble conquering these problems, although perhaps he would like to take a brief rest first.

References

[Bal83] Stewart **Baldwin**, Generalizing in the Mahlo hierarchy, with applications to the Mitchell models, Annals of Pure and Applied Logic 25 (1983), p. 103–127

[Bal86] Stewart **Baldwin**, Between strong and superstrong. Journal of Symbolic Logic 51 (1986), p. 547–559

[Dev84] Keith **Devlin**, Constructibility, Berlin 1984 [Perspectives of Mathematical Logic]

[DevJen75] Keith **Devlin**, Ronald B. **Jensen**, Marginalia to a theorem of Silver, *in :* Gert H. Müller, Arnold Oberschelp, Klaus Potthoff (*eds.*), ISILC Logic Conference, Proceedings of the International Summer Institute and Logic Colloquium, Kiel, 1974, Berlin 1975 [Lecture Notes in Mathematics 499], p. 115–142

[Do81] Anthony **Dodd**, Strong Cardinals, *handwritten notes* 1981

[Do82] Anthony **Dodd**, The Core Model, Cambridge 1982 [London Mathematical Society Lecture Notes Series 61]

[DoJen81] Anthony **Dodd**, Ronald B. **Jensen**, The Core Model, Annals of Mathematical Logic 20 (1981), p. 43–75

[DoJen82] Anthony **Dodd**, Ronald B. **Jensen**, The covering lemma for $\mathbf{L}[U]$, Annals of Mathematical Logic 22 (1982), p. 127–135

[Gö40] Kurt F. **Gödel**, The Consistency of the Axiom of Choice and the Generalized Continuum Hypothesis with the Axioms of Set Theory, Annals of Mathematical Studies # 3, Princeton 1940

[Gö47] Kurt F. **Gödel**, What is Cantor's Continuum Problem, American Mathematical Monthly 54 (1947), p. 515-525

[He97] Annekathrin **Hegewald**, Generische Ultraprodukte und Woodin–Forcing, Diplomarbeit, Technische Universität Berlin, Berlin 1997

[Jen72] Ronald B. **Jensen**, The Fine Structure of the Constructible Hierarchy, Annals of Mathematical Logic 4 (1972), p. 229–308

[Jen88] Ronald B. **Jensen**, Measures of order zero, *handwritten notes* 1988

---

[94] This result is unpublished. A slightly weaker version of it will appear in [St9?b].

[Jen90] Ronald B. **Jensen**, Nonoverlapping Extenders, *handwritten notes without date*, around 1990

[Jen95] Ronald B. **Jensen**, Inner Models and Large Cardinals, Bulletin of Symbolic Logic 1 (1995), p. 393-407

[Kan94] Akihiro **Kanamori**, The Higher Infinite, Berlin 1994 [Perspectives in Mathematical Logic]

[Ku70] Kenneth **Kunen**, Some applications of iterated ultrapowers in set theory, Annals of Mathematical Logic 1 (1970), p. 179–227

[Ku71] Kenneth **Kunen**, Elementary Embeddings and infinitary combinatorics, Journal of Symbolic Logic 36 (1971), p. 407–413

[Ma9?] Donald A. **Martin**, Determinacy, *monograph in preparation*

[MaSt94] Donald A. **Martin**, John R. **Steel**, Iteration Trees, Journal of the American Mathematical Society 7 (1994), p. 1–73

[Mi74] William J. **Mitchell**, Sets constructible from sequences of ultrafilters, Journal of Symbolic Logic 39 (1974), p. 57–66

[Mi78] William J. **Mitchell**, Hypermeasurable Cardinals, *in :* Maurice Boffa, Dirk van Dalen, Kenneth McAloon (*eds.*), Logic Colloquium '78, Proceedings of the Colloquium held in Mons, August 1978, Amsterdam 1979, p. 303–316

[Mi84] William J. **Mitchell**, The core model for sequences of measures I, Mathematical Proceedings of the Cambridge Philosophical Society 95 (1984), p. 229–260

[MiSch95] William J. **Mitchell**, Ernest **Schimmerling**, Weak covering without countable closure, Mathematical Research Letters 2 (1995), p. 595–609

[MiSchSt97] William J. **Mitchell**, Ernest **Schimmerling**, John R. **Steel**, The weak covering lemma up to a Woodin cardinal, Annals of Pure and Applied Logic 84 (1997), p. 219-255

[MiSt94] William J. **Mitchell**, John R. **Steel**, Fine Structure and Iteration Trees, Berlin 1994 [Lecture Notes in Logic 3]

[Nee98] Itay **Neeman**, Games of countable length, *this volume*

[NeeSt9?] Itay **Neeman**, John R. **Steel**, A Weak Dodd–Jensen Lemma, *submitted to* Journal of Symbolic Logic

[Si71a] Jack H. **Silver**, The consistency of the GCH with the existence of a measurable cardinal, in: Dana S. Scott (*ed.*), Axiomatic Set Theory, Proceedings of Symposia in Pure Mathematics of the American Mathematical Society 13.1 (1971), p. 391–395

[Si71b] Jack H. **Silver**, Measurable cardinals and Δ^1_3 well–orderings, Annals of Mathematics 94 (1971), p. 414–446

[Sch95] Ernest **Schimmerling**, Combinatorial Principles in the core model for one Woodin cardinal, Annals of Pure and Applied Logic 74 (1995), p. 153–201

[Sch9?] Ernest **Schimmerling**, A finite family weak square principle, *to appear in* Journal of Symbolic Logic

[SchSt96] Ernest **Schimmerling**, John R. **Steel**, Fine structure for tame inner models, Journal of Symbolic Logic 61 (1996), p. 621–639

[SchSt9?] Ernest **Schimmerling**, John R. **Steel**, The maximality of the core model, *to appear in* Transactions of the American Mathematical Society

[St82] John R. **Steel**, A classification of jump operators, Journal of Symbolic Logic 47 (1982), p. 347–358

[St93] John R. **Steel**, Inner models with many Woodin cardinals, Annals of Pure and Applied Logic 65 (1993), p. 185–209

[St96] John R. **Steel**, The Core Model Iterability Problem, Berlin 1996 [Lecture Notes in Logic 8]

[St9?a] John R. **Steel**, An Outline of Inner Model Theory, *to appear in :* Matthew Foreman, Akihiro Kanamori, Menachem Magidor (*eds.*), Handbook of Set Theory, *in preparation*

[St9?b] John R. **Steel**, Core models with more Woodin cardinals, *in preparation*

[Woo82] W.Hugh **Woodin**, On the consistency strength of projective absoluteness, *in:* Jacques Stern (*ed.*), Proceedings of the Herbrand Symposium, Logic Colloquium '81, held in Marseille, France, July 1981, Amsterdam 1982 [Studies in Logic and the Foundations of Mathematics 107], p. 365–384

[Ze9?] Martin **Zeman**, Inner Models and Strong Cardinals, *monograph in preparation*

Department of Mathematics, University of California, Berkeley CA 94720, USA

E-mail address, Benedikt Löwe: `loewe@math.berkeley.edu`

E-mail address, John Steel: `steel@math.berkeley.edu`

Games of Countable Length

Itay Neeman
Harvard University

Abstract

We survey the connections between large cardinals and the determinacy of long games, and present a particular calculation (due to John Steel and the author) demonstrating that the large cardinal strength of continuously coded determinacy is somewhere in the region of strong past a Woodin cardinal.

The Axiom of Determinacy (AD) is by now a well established hypothesis in Set Theory, and yields many results concerning definable sets of reals. It is known by results of Martin, Steel, and Woodin (see [MS89] and [MWH]) that AD holds in the model $L(\mathbb{R})$, assuming enough large cardinals in V. Specifically, it was shown in [MS89] that the existence of n Woodin cardinals and a measurable cardinal above them implies the determinacy of all games with a $\mathbf{\Pi}^1_{n+1}$ payoff set. It was later shown by Woodin that the existence of ω Woodin cardinals and a measurable cardinal above them suffices to prove the determinacy of all games with payoff in $L(\mathbb{R})$.

This paper concerns the natural generalizations of these results, first to standard length ω games on ω with payoff sets in bigger pointclasses, and secondly to longer games on reals. These longer games are our main concern. Such games were previously considered, for example in [Bla75], [Mar83], [Ste88], and [Woo]. There are various results regarding the propagation of scales under long game quantifiers assuming the determinacy of the corresponding long game, and various results regarding the determinacy of long games. In this last direction, it was shown by Martin and independently by Woodin that ZF+AD+DC+ "every set admits a scale" proves that for all $\alpha < \omega_1$, games of length α are determined. It was shown in [Ste88] that under the additional assumption that ω_1 is $P(\mathbb{R})$–supercompact, games of continuously coded length (see below) are determined.

Those results antedate the recent progress concerning Woodin cardinals and use strong assumptions on models of AD rather than large cardinals. With the knowledge available today it is possible to replace those assumptions with large cardinal assumptions, and obtain stronger results which in some cases are optimal. Section 1 surveys some of these results. Section 2 attempts to outline the close connections between long games and Woodin cardinals. Long games are used in inner model theory when constructing inner models for large cardinals in the region of Woodin cardinals. Much like L, the inner models constructed come with a definable well-order of their elements. The complexity of the well-order is best expressed using a long game quantifier, and the length of the game depends on the large cardinals considered. Section 2 includes one example of this, connecting a particular long game quantifier with a particular large cardinal assumptions.

1 The Games

For a set $B \subset \mathbb{R}$ [1] we use $G(B)$ to denote the standard length ω game with payoff B (displayed in Figure 1). Each move a_i is an element of ω and a run $a = \langle a_i \mid i < \omega \rangle$ of $G(B)$ is won by player I just in case that $a \in B$. Games of this kind play a central role in modern Descriptive Set Theory. Their importance stems from the fact that under certain assumptions on B, one of the two players has a winning strategy in $G(B)$: σ is called a *strategy* for I if it is a function with domain $\{ s \in \omega^{<\omega} \mid \text{lh}(s) \text{ is even} \}$ and range ω. We say that a run a is *according to* σ if for all n, $a_{2n} = \sigma(a \restriction 2n)$ so that the "even half" of a is completely determined by σ and the "odd half." σ is a *winning strategy* for I in $G(B)$ if for all $a \in \mathbb{R}$, $[a \text{ is according to } \sigma] \Longrightarrow a \in B$. Strategies for II are defined in the same way with "even" replaced by "odd." We say that $G(B)$ is *determined* if one of the two players has a winning strategy.

I	a_0		a_2		a_4		$\ldots$
II		a_1		a_3		$\ldots$	

Figure 1. The game $G(B)$

While using the axiom of choice one can easily construct a set $B \subset \mathbb{R}$ so that $G(B)$ is not determined, it is a commonly accepted principle that for *definable* sets $B \subset \mathbb{R}$ the associated game $G(B)$ *is* determined. For many years this principle has been used as an axiom, with "definable" replaced by various pointclasses such as $\mathbf{\Pi}^1_n$, Inductive, and $\{$ all sets of reals in $\text{L}(\mathbb{R}) \}$. As an axiom, determinacy led to many interesting results in the study of definable sets of reals, which the reader may find in [Mos80] and [Kec95]. Even when determinacy was just taken on faith it was generally expected that it should be possible to *prove* determinacy using large cardinals. This expectation was realized slowly over the years. From the perspective of this

[1]Throughout this paper $\mathbb{R}$ stands for Baire space ω^ω. This space is easily seen to be isomorphic to the irrational numbers with the usual topology.

paper the main break-through came in the mid 1980's with the following theorem of Martin and Steel [MS89]:

Theorem 1.1 (Martin–Steel, [MS89]) *Assume that there exist n Woodin cardinals and a measurable cardinal above them. Then $G(B)$ is determined for all B in $\mathbf{\Pi}^1_{n+1}$.*

This was later extended by Woodin, who in [Woo] showed (a result somewhat stronger than):

Theorem 1.2 (Woodin) *Assume that there exist ω Woodin cardinals and a measurable cardinal above them. Then $G(B)$ is determined for all B in $\mathrm{L}(\mathbb{R})$.*

Our interest here lies in proofs of determinacy for pointclasses larger than $\{$ all sets of reals in $\mathrm{L}(\mathbb{R})$ $\}$. These pointclasses are themselves defined using games, this time of length greater than ω and played on reals rather than natural numbers. Games on reals are already implicit in the results quoted above. For a set $D \subset \mathbb{R}^n$ consider the real game of length n, displayed in Figure 2.

I	x_0		x_2		$\dots$	
II		x_1		$\dots$		x_{n-1}

Figure 2: $G_n(D)$ (for n even).

The two players alternate playing *reals* and I wins just in case that

$$\langle x_0, \dots, x_{n-1} \rangle \in D.$$

Associated to this game is the game quantifier $\Game_n$: For a set $C \subset \mathbb{R}^{n+1}$, $\Game_n C$ is defined to be equal to the set

$$\{ a \in \mathbb{R} \mid \text{I has a winning strategy in the game } G_n(C_a) \}$$

where $C_a = \{ \langle x_0, \dots, x_{n-1} \rangle \in \mathbb{R}^n \mid \langle a, x_0, \dots, x_{n-1} \rangle \in C \}$. This game quantifier may be considered as an operation on pointclasses: For a pointclass Γ we let $\Game_n \Gamma$ be the collection of all sets which are equal to $\Game_n D$ for some $D \in \Gamma$.

In this specific case it is clear that $a \in \Game_n C$ just in case that

$$\exists x_0 \forall x_1 \cdots {}^{\forall}/_{\exists} x_{n-1} [\langle a, x_0, \dots, x_{n-1} \rangle \in C]$$

where the last quantifier is $\forall$ if n is even, and $\exists$ if n is odd. Thus the pointclass $\mathbf{\Sigma}^1_{n+1}$ is $\Game_n \mathbf{\Pi}/\mathbf{\Sigma}^1_1$ ($\mathbf{\Pi}^1_1$ for n odd and $\mathbf{\Sigma}^1_1$ for n even), and Martin–Steel's theorem above states the determinacy of the games $G(B)$ for all B in $\Game_n \mathbf{\Pi}/\mathbf{\Sigma}^1_1$.

One can similarly re-write Woodin's theorem. For $D \subset \mathbb{R}^\omega$ let $G_\omega(D)$ denote the length ω game on reals in which the players alternate playing reals x_i, and I wins just in case that $\langle x_i \mid i < \omega \rangle \in D$. Let $\Game_\omega$ be the associated game quantifier: For $C \subset \mathbb{R}^{1+\omega}$ define

$$\Game_\omega C = \{ a \in \mathbb{R} \mid \text{I has a winning strategy in } G_\omega(C_a) \}.$$

By results of Martin and Steel [MS83] the pointclass $\Game_\omega \mathbf{\Pi}^1_1$ consists of all sets of reals which are definable over L($\mathbb{R}$) via a Σ_1 formula with real parameters only. The conclusion of Woodin's theorem is equivalent to the determinacy of the games $G(B)$ for all B in $\Game_\omega \mathbf{\Pi}^1_1$. (Superficially the latter may seem weaker, but the two are in fact the same — if $G(B)$ is not determined for some $B \in$ L($\mathbb{R}$) then any such B which is minimal with respect to the order of constructibility will be Σ_1 over L($\mathbb{R}$).)

$$\begin{array}{c|cccccccc} I & x_0 & & \dots & & x_\omega & & \dots & \\ \hline II & & x_1 & & \dots & & x_{\omega+1} & & \dots \end{array} \qquad (\alpha \text{ moves})$$

Figure 3: $G_\alpha(D)$.

It is natural then to consider extensions of the above results to pointclasses defined by longer games, and the corresponding stronger game quantifiers. For a countable ordinal α and $D \subset \mathbb{R}^\alpha$ let $G_\alpha(D)$ denote the real game of length α, displayed in Figure 3. The two players alternate playing reals x_ι for $\iota < \alpha$ with player I playing x_ι for even $\iota < \alpha$ (including all limit ι). The run $\langle x_\iota \mid \iota < \alpha \rangle$ is won by I just in case that

$$\langle x_\iota \mid \iota < \alpha \rangle \in D.$$

For a set $C \subset \mathbb{R}^{1+\alpha}$ define

$$\Game_\alpha C = \{ a \in \mathbb{R} \mid \text{I has a winning strategy in } G_\alpha(C_a) \}.$$

Theorem 1.3 *Let $\alpha < \omega_1$. Assume that there exists an increasing sequence of α Woodin cardinals and a measurable cardinal above them. Then $G(B)$ is determined for all B in $\Game_\alpha \mathbf{\Pi}/\mathbf{\Sigma}^1_1$. Further, the games $G_\alpha(D)$ are determined for all $D \subset \mathbb{R}^{1+\alpha}$ in $\mathbf{\Pi}/\mathbf{\Sigma}^1_1$.*

For limit α this was proved by Woodin and myself independently; I established the general case. Observe the distinction between the short games $G(B)$ with payoff in $\Game_\alpha \mathbf{\Pi}/\mathbf{\Sigma}^1_1$, and the long games $G_\alpha(D)$. For very small α this distinction is worth keeping. Once $\alpha \geq \omega^2$ though, the determinacy of the long games by itself imply the determinacy of the short games. Figure 4 illustrates how to simulate the game $G(\Game_\alpha C)$ as the first ω moves of a game $G_{\omega+\alpha}(\tilde{C})$.

$$
\begin{array}{c|cccc}
I & a_0 \quad \ldots & & x_0 \quad \ldots & \\
\hline
II & & a_1 \quad \ldots & & x_1 \quad \ldots
\end{array}
$$

$$\underbrace{\qquad\qquad\qquad}_{\omega \text{ moves}}\underbrace{\qquad\qquad\qquad}_{\alpha \text{ moves}}$$

$$\tilde{C} = \{ \langle a_0, \ldots, x_0, \ldots \rangle \in \mathbb{R}^{\omega+\alpha} \mid \langle a \rangle^\frown \langle x_\iota \mid \iota < \alpha \rangle \in C \text{ where } a = \langle a_i(0) \mid i < \omega \rangle \}$$

Figure 4: Simulating $G(\Game_\alpha C)$ with $G_{\omega+\alpha}(\tilde{C})$.

Each a_i is a real, but only its first digit $a_i(0)$ is important for the payoff. The determinacy of $G_{\omega+\alpha}(\tilde{C})$ clearly implies the determinacy of $G(\Game_\alpha C)$. Now $\tilde{C}$ is $\mathbf{\Pi}/\mathbf{\Sigma}^1_1$ if C is, and $\omega + \alpha = \alpha$ if $\alpha \geq \omega^2$. Thus for $\alpha \geq \omega^2$ the determinacy of $G_\alpha(D)$ for all $D \subset \mathbb{R}^\alpha$ in $\mathbf{\Pi}/\mathbf{\Sigma}^1_1$ implies the determinacy of $G(B)$ for all B in $\Game_\alpha \mathbf{\Pi}^1_1$. In considering games and game quantifiers of length ω^2 or greater, our only concern shall therefore be with the determinacy of the *long game.*

A further distinction is the distinction between *real* moves and *natural number* moves.

Theorem 1.4 (Neeman) *Let $\alpha < \omega_1$ and assume that there exists an increasing sequence of α Woodin cardinals and a measurable cardinal above them. Then all games of length $\omega \cdot (1 + \alpha)$ on* natural numbers *with $\mathbf{\Pi}^1_1$ payoff are determined.*

Changing to play natural numbers rather than reals thus allows an increase in the length of the game. The difference disappears though if $\alpha = \omega \cdot \alpha$.

The above results provide a roughly precise picture for the large cardinal strength of the determinacy of games of *fixed countable length.* Games of uncountable length are beyond the scope of this paper — indeed beyond the scope of current knowledge — but there are many games of length which lies strictly between ω_1 and fixed countable length. Those are the games of *variable countable length* where the actual length of a run of the game (while always countable) depends on the run itself and what the players choose to play. Let us establish some general terminology for such games. A *length condition* L is a subset of $\mathbb{R}^{<\omega_1}$ with the property that for any $\langle x_\iota \mid \iota < \omega_1 \rangle \in \mathbb{R}^{\omega_1}$ there exists some $\alpha < \omega_1$ so that $\langle x_\iota \mid \iota < \alpha \rangle \in L$. L will be used to define the game of "length L." A payoff set D for this game is any subset of L. The game $G_L(D)$ is then played as follows: The two players alternate playing reals x_ι as in Figure 3, and continue playing until they reach the first α such that $\langle x_\iota \mid \iota < \alpha \rangle$ belongs to L. (Note that our restriction on L guarantees that such α will be reached.) At this point the game ends, and I wins just in case that

$$\langle x_\iota \mid \iota < \alpha \rangle \in D.$$

There are many different games of variable countable length. We list below the ones which appear to be the most natural.

GAMES ENDING AT THE FIRST ADMISSIBLE: Let L consist of all sequences $\langle x_\iota \mid \iota < \alpha\rangle$ so that $\mathrm{L}_\alpha[x_\iota \mid \iota < \alpha]$ is admissible and $\mathrm{L}_\beta[x_\iota \mid \iota < \beta]$ is not admissible for $\beta < \alpha$. Games with this length condition are said to end at the first admissible, and are denoted by G_{adm}. The corresponding game quantifier is denoted by $\Game_{adm}$. A payoff set $D \subset L$ is said to be $\mathbf{\Pi}^1_1$ (or in general Γ) in the codes if it is $\mathbf{\Pi}^1_1$ (or Γ) when coded as a subset of $\mathbb{R}$. Note that since α is the first admissible relative to the play $\langle x_\iota \mid \iota < \alpha\rangle$, the play itself induces a canonical enumeration of α in order type ω. Use this enumeration to code the play as a real.

THE CONTINUOUSLY CODED GAME: This game is played according to Figure 5. In round $2 \cdot \iota$ I's moves include a natural number n_ι in addition to the real $x_{2\cdot\iota}$. We require I to never repeat the same natural number twice. Sooner or later I must then run out of natural numbers to play, and at this point the game ends. We denote this game by G_{cont} and the corresponding game quantifier by $\Game_{cont}$. Officially, the length condition L for this game consists of all sequences $\langle x_\iota \mid \iota < \alpha\rangle$ so that $\{\, x_\iota(0) \mid \iota < \alpha \text{ even}\,\}$ is equal to $\mathbb{N}$ or else for $\iota \neq \iota' < \alpha$ both even we have $x_\iota(0) = x_{\iota'}(0)$.

I	n_0, x_0		n_1, x_2		$\dots$		n_ω, x_ω		$\dots$
II		x_1		x_3		$\dots$		$x_{\omega+1}$	$\dots$

Figure 5: $G_{cont}(D)$.

A payoff set is any $D \subset L$ which includes only sequences which satisfy the first of the two conditions above. Note that any sequence $\vec{x} = \langle x_\iota \mid \iota < \alpha\rangle \in L$ comes complete with an injection of α into $\mathbb{N}$. We use this injection to code $\vec{x}$ as a real, and say that D is $\mathbf{\Pi}^1_1$ (or Γ in general) if it is $\mathbf{\Pi}^1_1$ (or Γ) in the codes.

f–CONTINUOUSLY CODED GAMES: The game G_{f-cont} is played in a manner similar to that of G_{cont}, except that both players participate in choosing the integer code n_ι. To be precise: fix a function $f\colon \mathbb{R}^2 \to \mathbb{N}$. In round ι both players collaborate in playing the real x_ι (as in Figure 1). We then let $n_\iota = f(z, x_\iota)$ where z is the canonical code for the sequence $\langle x_\zeta \mid \zeta < \iota\rangle$ played so far. The game ends once the n_ι's exhaust all natural numbers, or if ever a natural number is repeated twice. Positions $\langle x_\iota \mid \iota < \alpha\rangle$ in the game are coded as before, using the injection of α into $\mathbb{N}$ given by $\iota \mapsto n_\iota$. As before a payoff set D is $\mathbf{\Pi}^1_1$ (or in general Γ) if it is $\mathbf{\Pi}^1_1$ (or Γ) in the codes. Note that this game is stronger than G_{cont}. It is easy to simulate G_{cont} as a run of G_{f-cont} by placing appropriate rules on the payoff D and the function f.

GAMES ENDING AT ω_1 IN L[PLAY]: L consists of all sequences $\langle x_\iota \mid \iota < \alpha\rangle$ so that $\mathrm{L}[x_\iota \mid \iota < \alpha] \models$ "α is the first uncountable cardinal". It is not clear (and indeed not always true) that L satisfies our requirement for being a length condition. We need a certain amount of reflection to make sure α is always countable. Assume then that every subset of ω_1 has a sharp. Under this assumption it is certainly true that for any $\langle x_\iota \mid \iota < \omega_1 \rangle$ there exists a countable α for which $\langle x_\iota \mid \iota < \alpha \rangle \in L$. Games with this length condition are denoted by $G_{\omega_1^{L[p]}}$ and the corresponding game quantifier is $\Game_{\omega_1^{L[p]}}$. As before some remarks are in order on how we code a payoff set $D \subset L$ in this case. One way is to note that for $\vec{x} = \langle x_\iota \mid \iota < \alpha \rangle \in L$, the sharp of $\vec{x}$ projects to ω and therefore defines a canonical enumeration of α. Use this enumeration to code $\vec{x}$ as a real.

General clopen length ω_1 games: This class of games consists of $G_L(D)$ for all length conditions L. We denote these games simply as G_{ω_1} and use $\Game_{\omega_1}$ for the game quantifier. It is not clear how to code a run of the game by a real. The simplest way is to require as part of the rules of the game that one of the players plays this code. Clopen length ω_1 games were considered by Steel [Ste88] who proved a general theorem on the propagation of scales under the game quantifier $\Game_{\omega_1}$, assuming the determinacy of the G_{ω_1} games involved.

The long games presented above are arranged in the order of increasing strength. The reader can easily verify that for a pointclass Γ

The determinacy of the games $G_{\omega_1}(D)$ for all D in Γ

$\Downarrow$

the determinacy of the games $G_{\omega_1^{L[p]}}(D)$ for all D in Γ

$\Downarrow$

the determinacy of the games $G_{f-cont}(D)$ for all D, f in Γ

$\Downarrow$

the determinacy of the games $G_{cont}(D)$ for all D in Γ

$\Downarrow$

the determinacy of the games $G_{adm}(D)$ for all D in Γ

$\Downarrow$

the determinacy of the games $G_\alpha(D)$ for all D in Γ, and any *fixed* $\alpha < \omega_1$.

One would expect these levels of determinacy to correspond to levels of the large cardinal hierarchy. For games with $\mathbf{\Pi}^1_1$ payoff, considerations arising from inner model theory seem to suggest that:

- The determinacy of G_{ω_1} should roughly correspond to a cardinal which is both measurable and Woodin.
- The determinacy of $G_{\omega_1^{L[p]}}$ should roughly correspond to a cardinal which is both Woodin and a limit of Woodin cardinals.

- The determinacy of $G_{f\text{-}cont}$ should roughly correspond to a "strong past a Woodin cardinal" (see below).
- The determinacy of G_{adm} should roughly correspond to a (measurable) cardinal κ which is a limit of Woodin cardinals and has $o(\kappa) = \kappa^{++}$, where $o(\kappa)$ is the order type of the Mitchell order of measures on κ.

At the moment none of these rough expectation is known to be true, but we can approximate the third:

Theorem 1.5 (Neeman, [Nee96]) *Assume that there are cardinals $\kappa < \delta$ and an extender E so that* $\text{crit}(E) = \kappa$, *E is δ-strong, and δ is a Woodin cardinal in* $\text{Ult}(V, E)$ *(E is said to be* strong past a Woodin cardinal*). Assume further that there is a measurable cardinal above δ. Then the games $G_{cont}(D)$ are determined for all D which are $\mathbf{\Pi}^1_1$ in the codes.*

It is not clear whether the conclusion of Theorem 1.5 can be strengthened to give the determinacy of $G_{f\text{-}cont}(D)$ for D, f in $\mathbf{\Pi}^1_1$. The inner model considerations sketched in the next section seem to suggest that this should be possible.

2 Large cardinals

Some of the most interesting long games are *iteration games* on models which contain a rich enough collection of extenders. Such games arise naturally in the study of inner models for large cardinals. Iteration games provide the key not only to the proofs of Theorems 1.3, 1.4, and 1.5 but also to the proof that the assumptions used in these theorem are (roughly) optimal. The proof of this latter direction is sketched in this section.

We shall need a few concepts from inner model theory. A naive (but we hope sufficient) descriptions is included below. The reader may consult [MiSt94] for the Full Monty.

A *pre-mouse* M is any model of the form $M = \mathcal{J}_\eta[\vec{E}]$ [2] where $\vec{E}$ is a sequence of extenders E_γ, $\gamma < \eta$ which satisfies certain first order properties. A pre-mouse M may have an additional predicate E, which is an extender over $\mathcal{J}_\eta[\vec{E}]$. We shall abuse notation and suppress this predicate as E_η. The extenders E_γ may be partial — each E_γ is only required to measure subsets of its critical points which belong to $\mathcal{J}_\gamma[\vec{E} \restriction \gamma]$. $E_\gamma = \emptyset$ is also allowed, and is regarded as the extender of the identity map. The precise first order properties which $\vec{E}$ must satisfy can be found in [MiSt94]. For our purposes the

[2]The $\mathcal{J}$–hierarchy is a convenient rearrangement of the L–hierarchy. The reader who is not familiar with it may think of $\mathcal{J}$ as being L, although this will make some of the results quoted false.

most relevant property of a pre-mouse is the *soundness* of its initial segments. More specifically, we are interested in a particular consequence of soundness which states that for any $\kappa < \alpha < \eta$, if a new subset of κ appears in $\mathcal{J}_{\alpha+1}[\vec{E}]$ which did not appear in $\mathcal{J}_\alpha[\vec{E}]$, then the cardinality of α is the same as that of κ and this is witnessed in $\mathcal{J}_{\alpha+1}[\vec{E}]$; in fact in $\mathcal{J}_{\alpha+1}[\vec{E}]$ there is a surjection of κ onto $\mathcal{J}_\alpha[\vec{E}]$. Our own concern here lies with *reals*, namely subsets of ω, and from soundness we know that if a real x appears in $\mathcal{J}_{\alpha+1}[\vec{E}] - \mathcal{J}_\alpha[\vec{E}]$ then α is countable and this is known to the model $\mathcal{J}_{\alpha+1}[\vec{E}]$. Let us say that a pre-mouse $M = \mathcal{J}_\eta[\vec{E}]$ is *short* if η is a successor ordinal, $\alpha + 1$ say, and in M there exists a surjection of ω onto $\mathcal{J}_\alpha[\vec{E}]$. By the above, if a real x appears in any pre-mouse, it must appear in a short pre-mouse [3].

What makes pre-mice so important is our ability to compare them with each other (or at least try). For two pre-mice $M = \mathcal{J}_\eta[\vec{E}]$ and $N = \mathcal{J}_\theta[\vec{F}]$ we say that $M \leq^* N$ (M is an initial segment of N) just in case that $\eta \leq \theta$ and $\vec{F} \restriction \eta + 1 = \vec{E}$. If neither of the two pre-mice is an initial segment of the other, there must exist some ordinal $\gamma \leq min\{\eta, \theta\}$ so that $\vec{E} \restriction \gamma = \vec{F} \restriction \gamma$ and $E_\gamma \neq F_\gamma$. This γ is called the *least disagreement* between M and N. Both E_γ and F_γ measure sets which belong to $\mathcal{J}_\gamma[\vec{E} \restriction \gamma] = \mathcal{J}_\gamma[\vec{F} \restriction \gamma]$. For the two extenders to differ it must be that either $\text{crit}(E_\gamma) \neq \text{crit}(F_\gamma)$ or else there exists a particular set $A \subset \text{crit}(E_\gamma) = \text{crit}(F_\gamma)$ in $\mathcal{J}_\gamma[\vec{E} \restriction \gamma] = \mathcal{J}_\gamma[\vec{F} \restriction \gamma]$ which the two extenders measure differently. This A is said to *witness the disagreement.*

It is one of the miracles of inner model theory that disagreements can be overcome, by what is known as the process of comparison. In this process we form a sequence of models M_ξ and N_ξ, starting with $M_0 = M$ and $N_0 = N$, until eventually reaching a ξ for which $M_\xi \leq^* N_\xi$ or $N_\xi \leq^* M_\xi$. The sequences are formed in such a way that the M_ξ's and N_ξ's preserve enough of the information contained in M_0 and N_0. The new models obtained are *iterates* of M, N via an *iteration tree.* The construction of such a tree is best viewed as a game known as the *iteration game.* The definition below is taken from [Ste95b] and is based on the ideas of [MS94] and [MiSt94].

Definition 2.1 (Iteration Game) *For a pre-mouse M and an ordinal τ, the full iteration game on M of length τ, $\mathcal{G}(M, \tau)$, is played as follows: Set $M_0 = M$. Through the game the players shall define a sequence of pre-mice $M_\xi = \mathcal{J}_{\eta_\xi}[\vec{E^\xi}]$, $\xi < \tau$, a tree order $\prec$ on τ, and a particular set $D \subset \tau$ called the set of* drops. *Player* I *plays only in successor stages $\xi + 1 < \tau$, choosing $\gamma_\xi \leq \eta_\xi$ so that ${E^\xi}_{\gamma_\xi}$ is an extender on the sequence of M_ξ. For notational*

[3] The concept of "short" as we use it here is not related to the concepts of short defined by Steel or by Koepke.

convenience call this extender E_ξ. I *must also choose some $\zeta_\xi \le \xi$ so that*

$$\forall \epsilon [\zeta_\xi \le \epsilon < \xi \Longrightarrow \text{crit}(E_\xi) < \text{lh}(E_\epsilon)]$$

This last condition implies that the models M_{ζ_ξ} and M_ξ are sufficiently in agreement so that the extender E_ξ can be applied to the model M_{ζ_ξ}. Specifically, we let $M_{\xi+1} = \text{Ult}(P, E_\xi)$ where $P = \mathcal{J}_{\bar\eta}[\vec{E}^{\zeta_\xi} \restriction \bar\eta + 1]$ for the largest $\bar\eta \le \eta_{\zeta_\xi}$ such that E_ξ measures all subsets of its critical point belonging to P. P is therefore the largest initial segment of M_{ζ_ξ} to which we can apply E_ξ.

We put $\alpha + 1 \in D$ and say that $\mathcal{T}$ drops *at $\alpha + 1$ if P is a* proper *initial segment of M_{ζ_ξ}. Otherwise we let $i_{\zeta_\xi,\xi}: M_{\zeta_\xi} \to M_\xi$ be the ultrapower embedding. In any case we extend the tree order by putting $\zeta_\xi \prec \xi$ (and of course $\zeta' \prec \xi$ for any ζ' such that $\zeta' \prec \zeta$).*

At limit stage $\xi < \tau$ player II *plays.* II *must pick a cofinal branch of the iteration tree produced so far, namely some unbounded $b_\xi \subset \xi$ which is a branch of $\prec \restriction \xi$. $D \cap b_\xi$ must be finite (or else* II *loses). We then have embeddings $i_{\zeta',\zeta}$ for cofinally many $\zeta' < \zeta$ in b_ξ. Let M_ξ be the direct limit of the models $M_\zeta, \zeta \in b_\xi$ under these embeddings. For $\zeta \in b_\xi - sup(D \cap b_\xi)$ we let $i_{\zeta,\xi}: M_\zeta \to M_\xi$ be the direct limit embeddings, and set $\zeta \prec \xi$ for all $\zeta \in b_\xi$.*

The game is closed for player II *— if for any $\xi < \tau$ the model M_ξ is ill-founded then* II *loses. Otherwise* II *wins.*

For $\xi < \tau$, we say that "there are no drops on the branch leading to M_ξ" if $D \cap \{\zeta \le \xi \mid \zeta \preceq \xi\} = \emptyset$. If there are no drops on the branch leading to M_ξ then the tree gives rise to an embedding $i_{0,\xi}: M_0 \to M_\xi$. A run $\mathcal{T} = \langle M_\xi^{\mathcal{T}}, \xi < \tau \rangle, \prec^{\mathcal{T}}, D^{\mathcal{T}}$ of the above game is called an *aspiring iteration tree.* An *iteration tree* is a run in which II has not lost. Iteration trees look a bit like

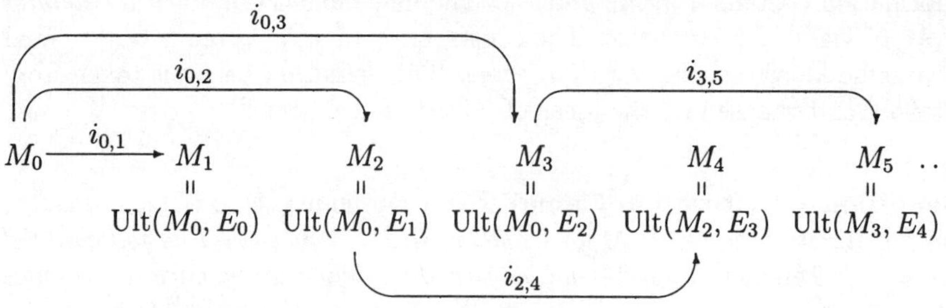

Figure 6: An iteration tree, with $0 \prec^{\mathcal{T}} 1$, $0 \prec^{\mathcal{T}} 2 \prec^{\mathcal{T}} 4$, and $0 \prec^{\mathcal{T}} 3 \prec^{\mathcal{T}} 5$.

An (aspiring) iteration tree $\mathcal{T}$ is called *non-overlapping* if for each ξ, (given γ_ξ) ζ_ξ is taken to be the *least* legal move. The reason for the name non-overlapping is that for such trees if $\xi + 1 \prec \xi'$ and $i_{\xi+1,\xi'}$ exists, then

$\text{crit}(i_{\xi+1,\xi'}) \geq \text{lh}(E_\xi)$. Thus on any branch b of a non-overlapping iteration tree $\mathcal{T}$, the critical points of extenders used do not overlap the lengths of previous extenders.

In trying to compare two pre-mice M and N we simply form two *non-overlapping* iteration trees $\mathcal{T}$ and $\mathcal{U}$ on M and N respectively, using always the least disagreement as our next extender. To be slightly more precise, in round $\xi+1$ of the construction we play round $\xi+1$ of the iteration games on M and N. If $M_\xi^{\mathcal{T}} \leq^* N_\xi^{\mathcal{U}}$ or $N_\xi^{\mathcal{U}} \leq^* M_\xi^{\mathcal{T}}$ then the comparison is successfully over and our job is done. Otherwise we let γ_ξ be the least disagreement between these two models, and let I play γ_ξ in both iteration games. This description of course misses II's moves in the iteration game, and for a very good reason — it is not always possible to play for II.

Definition 2.2 (Iterability) *A pre-mouse M is said to be τ–*iterable *if* II *has a winning strategy in the game $\mathcal{G}(M, \tau)$.*

Iterable pre-mice are called *mice.* This terminology hides τ, and we shall use it only when τ is clear from the context.

If M and N are τ–iterable we can follow the process of comparison, at limit stages using II's winning strategies to pick branches of the iteration trees constructed. Our hope is to finish the comparison at some ξ *before* τ.

Lemma 2.3 (Comparison lemma) *Let M and N be pre-mice. Let $\tau = max(|M|, |N|)^+$ and assume that M, N are $\tau+1$–iterable. Let Σ and Λ be winning strategies for* II *which witness this. Then there are non-overlapping iteration trees $\mathcal{T}$ on M and $\mathcal{U}$ on N, played according to Σ and Λ respectively, having last models $M_\xi^{\mathcal{T}}$ and $N_\xi^{\mathcal{U}}$ for some $\xi < \tau$ so that either*

- *$M_\xi^{\mathcal{T}} \leq^* N_\xi^{\mathcal{U}}$ and there are no drops on the branch of $\mathcal{T}$ leading to $M_\xi^{\mathcal{T}}$, or else*

- *$N_\xi^{\mathcal{U}} \leq^* M_\xi^{\mathcal{T}}$ and there are no drops on the branch of $\mathcal{U}$ leading to $N_\xi^{\mathcal{U}}$.*

If the conclusion of Lemma 2.3 holds for two pre-mice M and N we say that M and N are *comparable.* Lemma 2.3 then states that any two iterable pre-mice are comparable. The key to its proof is the fact that both trees are non-overlapping. This implies that the same set cannot witness a disagreement at two different ξ's on the same branch, which in turn means that before reaching τ the comparison must terminate as all possible witnesses for disagreements were used. The curious reader may wish to look up the proof in [MiSt94, section 7]. The ability to compare mice is essential and lies at the heart of inner model theory. The reader is referred to [MiSt94, section 8] for one use (out of many) of the comparison process. For our own purpose the essential use of Lemma 2.3 is given by the next corollary.

Corollary 2.4 *Let* $M = \mathcal{J}_{\alpha+1}[\vec{E}]$ *and* $N = \mathcal{J}_{\beta+1}[\vec{F}]$ *be two short pre-mice which are* comparable. *Then either* $M \leq^* N$ *or else* $N \leq^* M$.

Proof. Without loss of generality the first of the two possible options in the conclusion of 2.3 holds so that $M_\xi^{\mathcal{T}} \leq^* N_\xi^{\mathcal{U}}$ and there are no drops on the branch of $\mathcal{T}$ leading to $M_\xi^{\mathcal{T}}$. But $M \models$ "all ordinals are countable" and so in particular no extender on the M sequence could possibly be total. Any use of an extender $E_\gamma \neq \emptyset$ would therefore necessitate dropping. Since there are no drops on the branch of $\mathcal{T}$ leading to $M_\xi^{\mathcal{T}}$ we conclude that no extenders were used and $M_\xi^{\mathcal{T}} = M_0^{\mathcal{T}} = M$. On the N–side, let κ be the smallest critical point used in the tree $\mathcal{U}$. κ must then be a cardinal of $N_\xi^{\mathcal{U}}$. Now M is an initial segment of $N_\xi^{\mathcal{U}}$ so $N_\xi^{\mathcal{U}} \models$ "α is countable". Hence $\alpha < \kappa$ and M is already an initial segment of N. □

A short mouse is a short pre-mouse which is $\omega_1 + 1$–iterable. An iteration strategy for a short mouse M is any winning strategy for II in the game $\mathcal{G}(M, \omega_1 + 1)$. By Lemma 2.3 any two short mice are comparable. By 2.4 then short mice are well ordered by $\leq^*$, and this allows us to well-order all the reals which belong to mice. Let

$$\mathbb{R}^{mice} = \{ x \in \mathbb{R} \mid \text{There exists a short mouse } M \text{ with } x \in M \}$$

For $x, y \in \mathbb{R}^{mice}$ put

$$x \trianglelefteq y \implies \text{There exists a short mouse } M \text{ with } x, y \in M \text{ and } x \leq_L^M y \text{ where } \leq_L^M \text{ is the order of constructibility in } M.$$

Since $M \leq^* N$ implies that $\leq_L^N$ end-extends $\leq_L^M$, it follows from Corollary 2.4 that $x \trianglelefteq y$ just in case that $x \leq_L^M y$ *for all* short mice M which construct both x and y. This in turn immediately implies that $\trianglelefteq$ is a well-order. The key to our estimates of determinacy strength is a calculation of the complexity of $\trianglelefteq$ given particular restrictions on the mice considered. As a simple example: Say that a pre-mouse $M = \mathcal{J}_\eta[\vec{E}]$ is trivial if $E_\alpha = \emptyset$ for all $\alpha \leq \eta$. Clearly

$$\mathbb{R} \cap \mathrm{L} = \{ x \in \mathbb{R} \mid \text{There exists a trivial short mouse } M \text{ with } x \in M \}$$

Let $\trianglelefteq^{\mathrm{L}}$ be the restriction of $\trianglelefteq$ to reals in L. Then $\trianglelefteq^{\mathrm{L}}$ is a well-order of $\mathbb{R} \cap \mathrm{L}$ and

$$x \trianglelefteq^{\mathrm{L}} y \iff \text{There exists a trivial short mouse } M \text{ with } x, y \in M \text{ and } x \leq_L^M y \quad (1)$$

$$\iff \text{For all trivial short mice } M, \text{ if } x, y \in M \text{ then } x \leq_L^M y. \quad (2)$$

Being a pre-mouse is a a first-order property of M and contributes little to the complexity of (1) and (2). The order $\leq_L^M$ is first-order definable over M

and similarly does not contribute to the complexity. *Iterability is therefore the main factor affecting the complexity of* $\trianglelefteq$. Now clearly a trivial pre-mouse is iterable if and only if it is well-founded. "M is a trivial short mouse" is therefore equivalent to "M is a well-founded short pre-mouse." Being well-founded is a Π^1_1 statement about M. Thus the statements (1) and (2) above are Σ^1_2 and Π^1_2 respectively so that $\trianglelefteq$ is a Δ^1_2 well-order of $\mathbb{R} \cap \mathrm{L}$.

Our simple example is of course nothing more than Gödel's proof that the canonical well-order of the reals in L is Δ^1_2. By absoluteness L knows its reals are well-ordered by a Δ^1_2 relation and this allows us to conclude that

$$\mathrm{L} \not\models \text{"The games } G(B) \text{ are determined for all } B \in \boldsymbol{\Pi}^1_1\text{"}$$

whence ZFC does *not* prove $\boldsymbol{\Pi}^1_1$ determinacy.

Replacing L with M_n — the canonical minimal model for n Woodin cardinals — Steel in [Ste95a] showed that

$$M_n \models \text{"There is a well-order of the the reals which is } \Delta^1_{2+n}\text{"}$$

whence this model does *not* satisfy $\boldsymbol{\Pi}^1_{n+1}$ determinacy and so the existence of n Woodin cardinals by itself does not suffice for Theorem 1.1. As in our simple example, Steel defines a particular class of pre-mice, called n–small, so that the reals of M_n are exactly those which belong to n–small short mice. He then computes the complexity of the statement "M is iterable" for short n–small pre-mice, proving that it is Π^1_{n+1} in a code for M. The relationship between the complexity of "being iterable" and the number of Woodin cardinals stems from the following theorem of [Ste93] (a proof is given in [MiSt94], and essentially traces back to a result of Martin and Steel in [MS94]):

Theorem 2.5 *Let M be a pre-mouse, $\mathcal{T}$ an iteration tree of length τ on M, and assume that b, c are two* distinct *cofinal branches through $\mathcal{T}$. Let M_b, M_c be the direct limit models along the branches b, c.*

Set $\delta(\mathcal{T})$ (below referred to as δ) to be the supremum of $\mathrm{lh}(E_\xi)$ *for $\xi < \tau$. Suppose that $f\colon \delta \to \delta$ is a function which belongs to both M_b and M_c. Then*

$$M_b \models \text{"}\delta \text{ is Woodin with respect to } f\text{."}$$

Note that we do not assume M_b, M_c to be well-founded. (In the applications f will belong to the well-founded part of both models, but even this is not necessary.)

Theorem 2.5 provides a connection between *actual* moves in the iteration games, namely stages in the iteration where there is more than one possible branch that player II could choose and still win the game, and Woodin cardinals. It is the key to many results which assign a certain amount of Woodin cardinals to the consistency strength of a given statement.

At the low levels of finitely many Woodin cardinals the complexity arguments outlined above are rather crude. Much finer results exist providing equivalences of determinacy (within the projective hierarchy) and large cardinals (at the level of finitely many Woodin cardinals). For example, it is well known not only that $\mathbf{\Pi}^1_1$ determinacy cannot be proved in ZFC, but that in fact it implies the existence of $0^\sharp$. Unfortunately such tight results do not yet exist higher up, and all we have to rely on for the time being are the estimates given by calculating the complexity of "being iterable."

What follows is one such calculation, due to John Steel and the author, which indicates that Theorem 1.5 is not too far from optimal. Following [Ste93] call a pre-mouse $M = \mathcal{J}_\eta[\vec{E}]$ *tame* iff for any $\alpha \leq \eta$ and any δ

$$\text{crit}(E_\alpha) \leq \delta < \alpha \Longrightarrow \mathcal{J}_\alpha[\vec{E} \restriction \alpha] \models \text{"}\delta \text{ is not a Woodin cardinal."}$$

Thus a pre-mouse is tame if its sequence contains no extenders which appear to be strong past a Woodin cardinal when listed. Let $\mathbb{R}^{tame}$ be the set of reals which belong to short tame mice, and let $\trianglelefteq^{tame}$ be the restriction of $\trianglelefteq$ to $\mathbb{R}^{tame}$.

Lemma 2.6 (Neeman–Steel) *Assuming the determinacy of all the relevant games, there exists a function $f\colon \mathbb{R}^2 \to \mathbb{N}$ with Δ^1_2 graph so that the well-order $\trianglelefteq^{tame}$ is $\Game_{f\text{-}cont}\Delta^1_2$.*

Proof. The idea is to try and re-package the iteration game used for the comparison of two tame pre-mice M, N, and present it in such a way that it becomes f–continuously coded for some f. On the face of it we need $\omega_1 + 1$ moves in the iteration game to make sure that M and N are comparable (after all this is what Lemma 2.3 requires). The fact that M and N are tame though allows us to make do with fewer moves, and in fact play natural number codes for each round in the comparison without repeating the same number twice. These natural numbers essentially code a witness for the least disagreement in the round. To be (for a moment) naive, in each round (ι) of the game we have the models M_ι, N_ι left by the previous rounds, and the two players produce iteration trees $\mathcal{T}_\iota$ on M_ι (player I) and $\mathcal{U}_\iota$ on N_ι (player II) which begin to compare M_ι and N_ι. We let $M_{\iota+1}$ be the last model of $\mathcal{T}_\iota$ and $N_{\iota+1}$ the last model of $\mathcal{U}_\iota$, and pass to the next round where we continue the comparison. The natural number n_ι should somehow code the least disagreement between the first extender used on the main branch of $\mathcal{T}_\iota$, and the first extender used on the main branch of $\mathcal{U}_\iota$. If we successfully arrange that the models M_ι, $M_{\iota+1}$ and N_ι, $N_{\iota+1}$ all lie on the main branches of the two trees produced by the comparison (once the game is over), then the usual comparison argument for showing that the same disagreement is never repeated twice on the same branch will show that the same natural number is never used twice in the game.

Let $\delta_{\iota+1}$ be the supremum of indices of extenders used in the tree $\mathcal{T}_\iota$ (which should be the same as the supremum of indices used in the tree $\mathcal{U}_\iota$). $\delta_{\iota+1}$ then represents the level of agreement between the models $M_{\iota+1}$ and $N_{\iota+1}$. To make sure that $M_{\iota+1}, N_{\iota+1}$ lie on the main branches of the full comparison trees it is enough to arrange that no future extenders in the comparison overlap $\delta_{\iota+1}$. This in turn can be secured using the *tameness* of M and N and Theorem 2.5. If one of the trees $\mathcal{T}_\iota$, $\mathcal{U}_\iota$ has a second *distinct* cofinal branch, we can add the direct limit along this branch to the comparison, thereby making sure that $\delta_{\iota+1}$ appears to be Woodin in the lined-up part of all future models, so that by tameness no future extender used in the comparison can have a critical point below $\delta_{\iota+1}$. It then follows that $M_{\iota+1}$ and $N_{\iota+1}$ lie on the main branches of the full comparison trees (produced at the end of the game) as desired.

By far the most serious difficulty which this naive description ignores is making sense of the phrase "iteration trees .... which begin to *compare* M_ι and N_ι." Let's view things from player I's side. Player I must play $\mathcal{T}_\iota$ which is the tree arising from a comparison of M_ι and N_ι. The tree must be fairly long (for after all it may have to reach a level with two distinct cofinal branches). Being the tree of a comparison process means that at each level ξ, I must use the extender given by the least disagreement between the ξ-th model on her tree and the ξ-th model on the tree $\mathcal{U}_\iota$. Here precisely lies the difficulty. The two players play their trees in the same round at the same time, and I does *not* have advance knowledge of the tree $\mathcal{U}_\iota$ and its ξ-th model. I knows of course that this tree too arises in a comparison, but I does *not* know which branches this tree picks at limits, and so has no idea what the ξ-th model of the tree is for $\xi \geq \omega$.

Player I must therefore *guess* the structure of the tree $\mathcal{U}_\iota$, and play $\mathcal{T}_\iota$ accordingly, so that the two trees form a comparison. We use $\mathcal{U}_{\text{I}}$ and $\mathcal{T}_{\text{I}}$ to refer to I's guess of $\mathcal{U}_\iota$ and her own tree $\mathcal{T}_\iota$ played according to this guess. Similarly player II will have to guess the structure of the tree $\mathcal{T}_\iota$ and play $\mathcal{U}_\iota$ which arises from a comparison with her guess of $\mathcal{T}_\iota$. We use $\mathcal{T}_{\text{II}}$ and $\mathcal{U}_{\text{II}}$ to refer to II's trees.

Of course, we may end up with two pairs $\langle \mathcal{T}_{\text{I}}, \mathcal{U}_{\text{I}} \rangle$ and $\langle \mathcal{T}_{\text{II}}, \mathcal{U}_{\text{II}} \rangle$ which are *not the same*. Each of the pairs represents a comparison of M_ι with N_ι and this dictates the choice of extenders used on the trees. If the pairs differ, it can only be because at some limit stage ξ they pick different branches. If ξ is chosen to be minimal then the trees $\mathcal{T}_{\text{I}}$, $\mathcal{T}_{\text{II}}$ are the same up to ξ. We let $\mathcal{T}_\iota$ be these trees cut at ξ, with the last branch of the tree $\mathcal{T}_{\text{I}}$. Similarly $\mathcal{U}_{\text{I}}$, $\mathcal{U}_{\text{II}}$ are the same up to ξ and we let $\mathcal{U}_\iota$ be these trees cut at ξ, with the last branch of the tree $\mathcal{U}_{\text{II}}$.

Our choice of ξ is such that the pairs $\langle \mathcal{T}_{\text{I}}, \mathcal{U}_{\text{I}} \rangle$ and $\langle \mathcal{T}_{\text{II}}, \mathcal{U}_{\text{II}} \rangle$ differ at ξ. For example it may be that $\mathcal{T}_{\text{I}}$ and the guess $\mathcal{T}_{\text{II}}$ pick different branches at ξ. In this case the branch that II guessed in $\mathcal{T}_{\text{II}}$ is *distinct* from the branch that

I played, and this gives us the second distinct branch through $\mathcal{T}_\iota$. We can add the direct limit along this branch to the comparison and thereby make sure that no future extenders overlap $\delta_{\iota+1}$. Similarly if $\mathcal{U}_{\mathrm{I}}$ and $\mathcal{U}_{\mathrm{II}}$ differ at ξ we see that the branch player I guessed in $\mathcal{U}_{\mathrm{I}}$ is distinct from the branch player II played in $\mathcal{U}_{\mathrm{II}}$. Adding the direct limit along the guessed branch to the comparison will make sure that no future extenders overlap $\delta_{\iota+1}$.

This roughly describes the idea of the game $\mathcal{G}_{tame}(M,N)$ which we use to simulate the comparison of M and N. Note that in each round we may end up adding models to the comparison. Thus in round ι we have to compare not only the iterates M_ι and N_ι of M and N, but also many other iterates of models which were added in previous rounds. This complicates somewhat the notation in our proof. In pages 175–177 we declare in advance our notation for all the objects which will come up during the game. We then (pages 177–180) present the rules governing round ι of the game and describe how the moves played in this round determine the natural number code n_ι. In page 181 we state the pay-off condition which determines the winner once the game reaches its end.

We define $\mathcal{G}_{tame}(M,N)$ so that if at some point after round ι an extender overlaps a $\delta_{\iota+1}$ then the game ends, and one of the players loses. Roughly speaking, the player who played the shorter tree in round ι loses, and both players lose if the trees have equal length. Sub-Claim 2.8.1 shows that if the models M and N are tame, then we (playing for I say) can avoid losing on account of an overlap, essentially by making sure in round ι that either our tree is longer than II's tree, or if the trees have the same length, that a model was added to the comparison (so that $\delta_{\iota+1}$ appears to be Woodin in future models). Claim 2.7 shows (using the fact that the game is defined to never allow overlaps) that $\iota \neq \iota' \implies n_\iota \neq n_{\iota'}$ so that the f–continuously coded game $\mathcal{G}_{tame}(M,N)$ is indeed long enough to complete the comparison.

The most involved part of the proof is Claim 2.8 which attempts to show that if M is an *iterable* tame pre-mouse then player I wins $\mathcal{G}_{tame}(M,N)$. However we cannot show this directly. Suppose for a moment that we are player I and that we have at our disposal an iteration strategy Σ for the model M. We wish to use Σ to win the game $\mathcal{G}_{tame}(M,N)$. Using Σ we can pick branches for the tree $\mathcal{T}_{\mathrm{I}}$ played in round ι, but unfortunately this is not enough. As part of our move in round ι we must also play the guess $\mathcal{U}_{\mathrm{I}}$ which is an iteration tree on N_ι. We must therefore be able to iterate N_ι, and more than that, *predict* in our guess which branches II is going to play in her tree on N_ι.

Because of this additional burden, what we in fact show in Claim 2.8 is that if Λ is a strategy for II in the game $\mathcal{G}_{tame}(M,N)$ and M is iterable then we (player I) can play to defeat Λ. We use an iteration strategy Σ for M to play the tree $\mathcal{T}_{\mathrm{I}}$, and use Λ to predict II's moves and so correctly guess $\mathcal{U}_{\mathrm{I}}$.

One final complication in the proof of Claim 2.8 is presented by the fact that in round ι we (player I) have to play trees not only on M_ι and N_ι, but also on iterates of all the models which were added to the comparison in previous rounds. To handle this complication we are forced to play $\mathcal{G}_{tame}(M, N)$ against Λ on many different boards simultaneously (see pages 184–193).

Once done with the proof of Claim 2.8 we use the game $\mathcal{G}_{tame}(M, N)$ and its properties to finally prove Lemma 2.6.

Let us begin with the actual definition of the function f, and the f–continuously coded game $\mathcal{G}_{tame}(M, N)$, intended to simulate the comparison process of short tame pre-mice M, N (and many additional pre-mice created on the way). As presented the game is not quite an f–continuously coded game. What we present below as round ι is in fact two rounds (the rounds $2 \cdot \iota$ and $2 \cdot \iota + 1$). The natural number code n_ι will be determined at the end of the first of these two rounds, and we do not bother giving a code for the second. Formally the code for the first round at ι should be $2 \cdot n_\iota$ and the code for the second round should be $2 \cdot n_\iota + 1$. Below we refer to (what formally are) the rounds $2 \cdot \iota$ and $2 \cdot \iota + 1$ as the first and second "halves" of round ι.

At the beginning of round ι we shall have the models M_ι and N_ι which are the "main" iterates of M and N we are to compare. We shall also have a (possibly finite) countable collection of models M_ι^k defined for $k \in X_\iota \subset \omega \times \omega$ and models N_ι^k defined for $k \in Y_\iota \subset \omega \times \omega$ [4]. The M–models are controlled by I (meaning that I will be responsible for finding branches through the iteration trees on these models) and the N–models are controlled by II.

For $\iota < \iota'$ the model $M_{\iota'}$ will be an iterate of the model M_ι, and we use $\pi_{\iota,\iota'}$ to denote the iteration embedding. Similarly for $k \in X_\iota$ the model $M_{\iota'}^k$ will be an iterate of M_ι^k and we use $\pi_{\iota,\iota'}^k$ for the embedding. The same will hold for the models controlled by II. We use $\sigma_{\iota,\iota'}: N_\iota \to N_{\iota'}$ and $\sigma_{\iota,\iota'}^k: N_\iota^k \to N_{\iota'}^k$ (for $k \in Y_\iota$) to denote the embeddings. For $l \in \omega \times \omega$ we let $\iota(l)$ be the least such that $l \in X_{\iota+1}$ or $l \in Y_{\iota+1}$. Then l will not belong to $X_\iota \cup Y_\iota$ and $M_{\iota+1}^l$ [or $N_{\iota+1}^l$] will be an iterate of one of the models in round ι, namely one of M_ι, M_ι^k for $k \in X_\iota$, N_ι, or N_ι^k for $k \in Y_\iota$. For notational convenience we shall denote this model by M_ι^l [or N_ι^l] even though it isn't. Similarly for $\zeta < \iota$ we shall use M_ζ^l [or N_ζ^l] to denote the model in round ζ of which $M_{\iota+1}^l$ [or $N_{\iota+1}^l$] is an iterate.

In addition to the models and embeddings we shall have enumerations witnessing that the models are all countable. Specifically, $d_\iota: M_\iota \to \omega^2$, $d_\iota^k: M_\iota^k \to \omega^2$ (for $k \in X_\iota$), $e_\iota: N_\iota \to \omega^2$ and $e_\iota^k: N_\iota^k \to \omega^2$ (for $k \in Y_\iota$) all injective. These embeddings will cohere with each other in the sense that for $\zeta < \iota$ all the diagrams in Figure 7 will commute.

[4] It is worthwhile emphasizing that k is not a natural number, but an element of $\omega \times \omega$.

This coherence allows us to define the models and embeddings for limit ι: M_ι is defined to be the direct limit of M_ζ, $\zeta < \iota$. $\pi_{\zeta,\iota}$ are defined to be the direct limit maps, and d_ι is defined to be the (unique) map which will make the first diagram of Figure 7 commute for all $\zeta < \iota$. It follows easily that d_ι is injective. We similarly define M^k_ι, d^k_ι, N_ι, e_ι, N^k_ι and e^k_ι. Finally for limit ι we let $X_\iota = \bigcup_{\zeta<\iota} X_\zeta$ and $Y_\iota = \bigcup_{\zeta<\iota} Y_\zeta$.

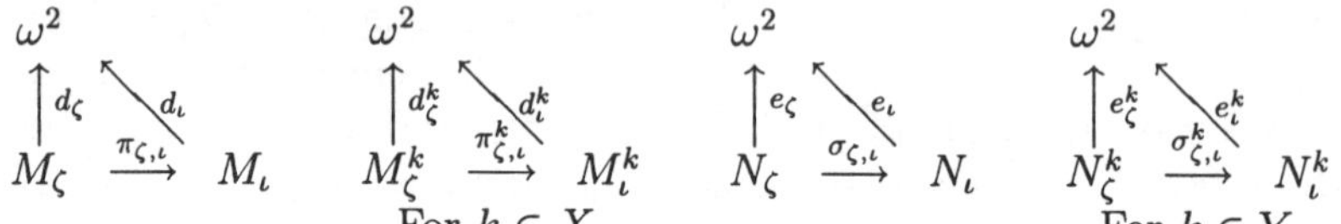

Figure 7: Coherence of the enumerations

Remark: It is of course possible that there was a drop on the iteration from e.g. M_ι to $M_{\iota+1}$, and that *there is no* embedding of M_ι into $M_{\iota+1}$. In this case we take $\pi_{\iota,\iota+1}$ to be a *partial* embedding on M_ι, defined (from the iteration tree) on as large a part of this model as possible. If the embedding $\pi_{\zeta,\iota}$ is partial, we still require the coherence given by the first diagram of Figure 7 to hold. The same of course goes for the remaining embeddings and diagrams.

To add more letters to our menagerie we shall have at the beginning of round ι also a countable ordinal δ_ι. These ordinals will keep track of the level of agreement between all the models compared. The index of the first disagreement between any two models at the start of round ι will be at least δ_ι. The δ's will form an increasing sequence and for limit ι we let δ_ι be the supremum of δ_ζ, $\zeta < \iota$.

At the end of the first half of round ι the moves played by the two players will determine a natural number n_ι in a manner explained below. We shall use these numbers in keeping track of the above injections. Specifically, the ranges of e_ι, e^k_ι, d_ι and d^k_ι will all be subsets of

$$[\{0\} \cup \{1 + n_\zeta \mid \zeta < \iota\}] \times \omega.$$

We also use the n_ι's in defining the set X_ι and Y_ι. We shall maintain through the game that

$$X_\iota, Y_\iota \subset \{n_\zeta \mid \zeta < \iota\} \times \omega$$

and moreover, if $l = \langle n_\zeta, i \rangle$ for some $\zeta < \iota$ and $i \in \omega$ then $\iota(l) = \zeta$.

Again at the end of the first half, the players' moves will determine a value for the variable $G(\iota)$. $G(\iota)$ indicates which player was *guilty* in round ι, and can be one of I,II, and N which stands for "neither." In later rounds when ending the game, we may look back at the value of G for a previous round, to decide which player wins.

Finally, a matter of notation: Fix a bijection $\ulcorner\urcorner: \omega \times \omega \longleftrightarrow \omega$. We let the letter ∂ range over $2 \times \omega$, and use $M_\iota^\partial/d_\iota^\partial/\pi_{\iota,\iota'}^\partial$ to denote

- $M_\iota/d_\iota/\pi_{\iota,\iota'}$ if $\partial = \langle 0, 0\rangle$.
- $M_\iota^k/d_\iota^k/\pi_{\iota,\iota'}^k$ if $\partial = \langle 0, 1 + \ulcorner k \urcorner\rangle$ (here $k \in \omega \times \omega$),
- $N_\iota/e_\iota/\sigma_{\iota,\iota'}$ if $\partial = \langle 1, 0\rangle$, and
- $N_\iota^k/e_\iota^k/\sigma_{\iota,\iota'}^k$ if $\partial = \langle 1, 1 + \ulcorner k \urcorner\rangle$.

Before starting the game let's fix another bijection

$$\phi\colon [(2 \times \omega) \times \omega^2] \cup [(2 \times \omega) \times \omega^2]^2 \longleftrightarrow \omega$$

which will be used in determining the codes n_ι for each round.

For round 0, we start with $M_0 = M$, $N_0 = N$, and $X_0 = Y_0 = \emptyset$ so that no additional M– or N–models are involved (yet). M and N are assumed to be countable, and we fix $d_0\colon M \to \omega^2$, $e_0\colon N \to \omega^2$ injective with range contained in $\{0\} \times \omega$. We let $\delta_0 = 0$.

In the first half of round ι the players suggest their versions of a simultaneous comparison of *all* models accumulated so far. I plays non-overlapping aspiring iteration trees $\mathcal{T}_{\mathrm{I}}$, $\mathcal{U}_{\mathrm{I}}$, $\mathcal{T}_{\mathrm{I}}^k$ (for $k \in X_\iota$) and $\mathcal{U}_{\mathrm{I}}^k$ (for $k \in Y_\iota$) on the models M_ι, N_ι, M_ι^k and N_ι^k respectively, all of countable lengths. (We omit the ι on the trees, for notational convenience.) Later rules will require these to be actual iteration trees (namely trees whose models are all *well-founded*) which arise from a comparison process, in other words for each ξ the index γ_ξ used to determine the ξ-th extender will have to be *the same* on all trees, and be furthermore the least index which witnesses a disagreement between some (not necessarily all) of the ξ-th models on the various trees. For the trees controlled by I the player must also play final branches cofinal in the length of the tree. Denote these branches by b (through $\mathcal{T}_{\mathrm{I}}$) and b^k (through $\mathcal{T}_{\mathrm{I}}^k$). (If a tree does not have limit length then the branch played must be the final branch of the tree.)

Player II must similarly play her version of the comparison, given by non-overlapping aspiring iteration trees $\mathcal{T}_{\mathrm{II}}$, $\mathcal{U}_{\mathrm{II}}$, $\mathcal{T}_{\mathrm{II}}^k$ (for $k \in X_\iota$) and $\mathcal{U}_{\mathrm{II}}^k$ (for $k \in Y_\iota$) on the models M_ι, N_ι, M_ι^k and N_ι^k respectively, all of countable lengths. II must further play final branches c (through $\mathcal{U}_{\mathrm{II}}$) and c^k (through $\mathcal{U}_{\mathrm{II}}^k$) which are cofinal in the lengths of the respective trees.

Once the trees were played, we let ϵ be the *least* countable ordinal satisfying one of the following conditions:

1. One or more of the trees played has length ϵ
2. For some ∂, either $\mathcal{T}_{\mathrm{I}}^\partial$ or $\mathcal{T}_{\mathrm{II}}^\partial$ picks at ϵ a branch with an *ill-founded* direct limit.

3. For all ∂, the trees $\mathcal{T}_{\mathrm{I}}^{\partial}$ and $\mathcal{T}_{\mathrm{II}}^{\partial}$ are the same up to ϵ, and have the same ϵ-th model (denoted here by Q_{ϵ}^{∂}), but for some $\tilde{\partial}$ the index γ_{ξ} for the ξ-th extender used on the tree $\mathcal{T}_{\mathrm{I}}^{\tilde{\partial}}$ or $\mathcal{T}_{\mathrm{II}}^{\tilde{\partial}}$ is *not* the index of the least disagreement between the models Q_{ϵ}^{∂}.

4. For all ∂, the trees $\mathcal{T}_{\mathrm{I}}^{\partial}$ and $\mathcal{T}_{\mathrm{II}}^{\partial}$ are the same up to ϵ, but for one or more ∂, the trees $\mathcal{T}_{\mathrm{I}}^{\partial}$ and $\mathcal{T}_{\mathrm{II}}^{\partial}$ have different (well-founded) branches at stage ϵ. (And hence have different direct limit models at stage ϵ.)

5. Conditions (1-4) all fail for ϵ. Let E_{ϵ}^{∂} denote the extender used on the trees $\mathcal{T}_{\mathrm{I}}^{\partial}$ and $\mathcal{T}_{\mathrm{II}}^{\partial}$. For some ∂, the extender E_{ϵ}^{∂} has critical point below δ_{ι}.

If condition (5) holds for ϵ we end the game immediately. In this case we look at the least ζ so that any of the extenders E_{ϵ}^{∂} has critical point below $\delta_{\zeta+1}$. If $G(\zeta)$ is I then I loses. If $G(\zeta)$ is II then II loses. If $G(\zeta)$ is N then I wins, but we shall also call this a *draw*.

If conditions (2) or (3) hold for ϵ we again end the game. The rules for deciding the winner in this case will be explained later.

Otherwise we proceed as follows:

For each (valid) ∂ let $\bar{\mathcal{T}} = \mathcal{T}_{\mathrm{I}}^{\partial} \upharpoonright \epsilon = \bar{\mathcal{T}}_{\mathrm{II}}^{\partial} \upharpoonright \epsilon$. Let $\bar{b}^{\partial}$ be the cofinal branch through $\bar{\mathcal{T}}^{\partial}$ which is

- the ϵ-th branch through $\mathcal{T}_{\mathrm{I}}^{\partial}$ if $\partial = \langle 0, a\rangle$ (meaning that the tree is controlled by I), or

- the ϵ-th branch through $\mathcal{T}_{\mathrm{II}}^{\partial}$ if $\partial = \langle 1, a\rangle$ (meaning that the tree is controlled by II).

Note that above if by chance ϵ is equal to the length of the tree $\mathcal{T}_{\mathrm{I/II}}^{\partial}$ then we take the branch b^{∂} or c^{∂} played by the appropriate player.

As defined, the trees $\bar{\mathcal{T}}^{\partial}$ all have length ϵ, use only extenders with critical points which are greater than or equal to δ_{ι}, and are trees which arise from a comparison of the models M_{ι}^{∂}.

Set $\delta_{\iota+1}$ to be the supremum of the γ's used on any (all) of the trees $\bar{\mathcal{T}}^{\partial}$. We set $M_{\iota+1}^{\partial}$ to be the direct limit along the branch $\bar{b}^{\partial}$ of $\bar{\mathcal{T}}^{\partial}$ and define the corresponding iteration embeddings $\pi_{\iota,\iota+1}^{\partial}\colon M_{\iota}^{\partial} \to M_{\iota+1}^{\partial}$.

If any of these embeddings has critical point *strictly greater* then δ_{ι} we let n_{ι} be the least n which is equal to $\phi(\partial, d_{\iota}^{\partial}(\delta_{\iota}))$ for a $\partial \in 2 \times \omega$ so that $\mathrm{crit}(\pi_{\iota,\iota+1}^{\partial}) > \delta_{\iota}$.

Otherwise, let E^{∂} be the first extender used on the branch $\bar{b}^{\partial}$. These extenders must all have critical points equal to δ_{ι}. It follows by the usual comparison argument that there is a subset A of δ_{ι} so that two of these extenders disagree on A. We let n_ι be the least n which is equal to $\phi(\partial, d_\iota^\partial(A), \tilde{\partial}, d_\iota^{\tilde{\partial}}(A))$ for some ∂ and $\tilde{\partial}$ so that E^∂ and $E^{\tilde{\partial}}$ disagree on a subset A of δ_ι.

n_ι therefore represents a *witness A for the least disagreement*, essentially by pairing together the codes for A in the two relevant models ($d_\iota^\partial(A)$ and $d_\iota^{\tilde{\partial}}(A)$), and enough information to say which models these are (namely ∂ and $\tilde{\partial}$).

Having defined n_ι, we define the maps $d_{\iota+1}^\partial$ by setting

$$d_{\iota+1}^\partial(\pi_{\iota,\iota+1}^\partial(z)) = d_\iota^\partial(z)$$

for $z \in M_\iota^\partial$, thus securing the coherence described in Figure 7, and on $M_{\iota+1}^\partial - \text{range}(\pi_{\iota,\iota+1}^\partial)$ we let $d_{\iota+1}^\partial$ map injectively into $\{1 + n_\iota\} \times \omega$.

Finally we must define $G(\iota)$, which is supposed to indicate which of the two players restricted the lengths of the trees in this round.
Let $D \subset 2 \times \omega$ be the set of ∂'s which are "valid" in round ι, namely

$$\begin{aligned}\{\, \partial \mid M_\iota^\partial \text{ exists} \,\} \;=\; & \{\, \langle 0,0\rangle, \langle 1,0\rangle \,\} \cup \\ & \{\, \langle 0, 1 + \ulcorner k \urcorner\rangle \mid k \in X_\iota \,\} \cup \\ & \{\, \langle 1, 1 + \ulcorner k \urcorner\rangle \mid k \in Y_\iota \,\}\end{aligned}$$

Let $\ll$ well-order D with $\langle 0,0\rangle$ first, $\langle 1,0\rangle$ second, and then — arranged lexicographically first by $\zeta < \iota$ and then by $i < \omega$ — the pair $\langle 0, 1 + \ulcorner n_\zeta, i \urcorner\rangle$ followed by $\langle 1, 1 + \ulcorner n_\zeta, i \urcorner\rangle$.

- If the trees $\mathcal{T}_{\mathrm{I}}^\partial$ and $\mathcal{T}_{\mathrm{II}}^\partial$ *all* had length ϵ then both players are equally guilty and we set $G(\iota) = N$.

- Otherwise: If condition (1) above holds for ϵ, we look for the $\ll$-least $\partial \in D$ such that one of the trees $\mathcal{T}_{\mathrm{I}}^\partial$, $\mathcal{T}_{\mathrm{II}}^\partial$ has length ϵ. If $\mathcal{T}_{\mathrm{I}}^\partial$ has length ϵ we set $G(\iota) = \mathrm{I}$, and otherwise we set $G(\iota) = \mathrm{II}$.

- Otherwise: If condition (4) above holds for ϵ, we look for the $\ll$-least $\partial \in D$ such that the trees $\mathcal{T}_{\mathrm{I}}^\partial$ and $\mathcal{T}_{\mathrm{II}}^\partial$ have distinct branches at ϵ. If $\partial = \langle 0, a\rangle$ (meaning that I controls this tree) we hold II responsible for playing a wrong branch through a tree controlled by I, and set $G(\iota) = \mathrm{II}$. If $\partial = \langle 1, a\rangle$ then the tree is controlled by II, and we hold I responsible for playing the wrong branch, setting $G(\iota) = \mathrm{I}$.

During the second half round we give the players an opportunity to add models to the comparison. The two players play sequences $\langle \partial_i^{\mathrm{I}}, b_i^{\mathrm{I}} \mid i < \omega\rangle$ and $\langle \partial_i^{\mathrm{II}}, b_i^{\mathrm{II}} \mid i < \omega\rangle$.

Each ∂_i^{I} must either be empty, or be the index of an existing tree $\bar{\mathcal{T}}^\partial$ (either a tree controlled by I or a tree controlled by II) in which case b_i^{I} must be a cofinal branch through this tree and be *distinct from* $\bar{b}^\partial$. We then *add* the direct limits along these branches to the game. Specifically, we let $M_{\iota+1}^{\langle n_\iota, i\rangle}$ be the direct limit along the branch b_i^{I} and set

$$X_{\iota+1} = X_\iota \cup \{\, \langle n_\iota, i\rangle \mid \partial_i^{\mathrm{I}} \text{ is not empty} \,\}$$

$M_{\iota+1}^{\langle n_\iota, i\rangle}$ is then an iterate of the model $M_\iota^{\partial_i^{\mathrm{I}}}$ (and by convention we may now use $M_\iota^{\langle n_\iota, i\rangle}$ to denote the model $M_\iota^{\partial_i^{\mathrm{I}}}$, and $\pi_{\iota,\iota+1}^{\langle n_\iota, i\rangle}$ for the iteration embedding). We define $d_{\iota+1}^{\langle n_\iota, i\rangle}$ so as to make the coherence of Figure 7 hold and keep $d_{\iota+1}^{\langle n_\iota, i\rangle}{}''(M_{\iota+1}^{\langle n_\iota, i\rangle} - \mathrm{range}(\pi_{\iota,\iota+1}^{\langle n_\iota, i\rangle}))$ a subset of $\{\, 1 + n_\iota \,\} \times \omega$.

We work similarly with $\langle \partial_i^{\mathrm{II}}, b_i^{\mathrm{II}} \mid i < \omega\rangle$. Each ∂_i^{II} must either be empty, or be the index of an existing tree $\bar{\mathcal{T}}^{\partial}$ (again this tree may be controlled by either player) in which case b_i^{II} must be a cofinal branch through this tree and be *distinct from* $\bar{b}^{\partial}$. We let $N_{\iota+1}^{\langle n_\iota, i\rangle}$ be the direct limit along the branch b_i^{II} and set

$$Y_{\iota+1} = Y_\iota \cup \{\, \langle n_\iota, i\rangle \mid \partial_i^{\mathrm{II}} \text{ is not empty} \,\}$$

$N_{\iota+1}^{\langle n_\iota, i\rangle}$ is then an iterate of the model $M_\iota^{\partial_i^{\mathrm{II}}}$ (and by convention we may now use $N_\iota^{\langle n_\iota, i\rangle}$ to denote the model $M_\iota^{\partial_i^{\mathrm{II}}}$, and $\sigma_{\iota,\iota+1}^{\langle n_\iota, i\rangle}$ for the iteration embedding). We define $e_{\iota+1}^{\langle n_\iota, i\rangle}$ so as to make the coherence of Figure 7 hold and keep $e_{\iota+1}^{\langle n_\iota, i\rangle}{}''(N_{\iota+1}^{\langle n_\iota, i\rangle} - \mathrm{range}(\sigma_{\iota,\iota+1}^{\langle n_\iota, i\rangle}))$ a subset of $\{\, 1 + n_\iota \,\} \times \omega$.

Note that the players are *not required* to add new models to the comparison, and are at liberty to simply play empty ∂_i for all i. In this case we shall have $X_{\iota+1} = X_\iota$ and $Y_{\iota+1} = Y_\iota$, and we say that "no new models were added during round ι." If at least one of the players did add a new model (i.e., either $X_{\iota+1} \neq X_\iota$ or $Y_{\iota+1} \neq Y_\iota$ or both) we say that "a new model was added during round ι."

This completes our description of the game $\mathcal{G}_{tame}(M, N)$. The game ends once one of the six conditions below is satisfied:

2. Condition (2) above (one of the trees picked a branch leading to an ill-founded model).

3. Condition (3) above (one of the trees used an extender which was not the least disagreement).

5. Condition (5) above (one of the trees used an extender with critical point below δ_ι).

6. The n_ι's exhaust all natural numbers,

7. n_ι as defined above is equal to n_ζ for some $\zeta < \iota$.

8. All the models $M_\iota, M_\iota^k, N_\iota, N_\iota^k$ agree, in the sense that for any two models M_ι^∂ and $M_\iota^{\tilde\partial}$ either $M_\iota^\partial \leq^* M_\iota^{\tilde\partial}$ or $M_\iota^{\tilde\partial} \leq^* M_\iota^\partial$.

The actual moves played in $\mathcal{G}_{tame}(M, N)$ are reals x_ι rather than aspiring iteration trees. $\langle x_\iota(2n) \mid n < \omega \rangle$ codes I's trees while $\langle x_\iota(2n+1) \mid n < \omega \rangle$ codes II's trees. The particular method of coding used is immaterial, except that both players must be allowed to procrastinate (even indefinitely) with any tree. This introduces the following additional way for the game to end:

0. The tree $\mathcal{T}_{\mathrm{I}}^\partial$ or $\mathcal{T}_{\mathrm{II}}^\partial$ was not played (meaning that the player responsible for this tree procrastinated indefinitely).

If the game ends with condition (8) then I wins, but we shall also call this situation a *draw*. If the game ends with condition (5) we use G to decide on the winner (see above). If the game ends with one of the conditions (0), (2) or (3):
If condition (0) holds for one of the trees $\mathcal{T}_{\mathrm{I}}^\partial$ or $\mathcal{T}_{\mathrm{II}}^\partial$, let $\partial \in D$ be $\ll$-least witnessing this. If condition (0) holds for $\mathcal{T}_{\mathrm{I}}^\partial$ then I loses. Otherwise condition (0) holds for $\mathcal{T}_{\mathrm{II}}^\partial$ and II loses.
If condition (0) does not hold for any of the trees, but condition (2) does, let $\partial \in D$ be $\ll$-least so that condition (2) holds for one of the trees $\mathcal{T}_{\mathrm{I}}^\partial$ or $\mathcal{T}_{\mathrm{II}}^\partial$. If condition (2) holds for $\mathcal{T}_{\mathrm{I}}^\partial$ then I loses, otherwise II loses.
Finally, if conditions (0) and (2) do not hold for any of the trees, but condition (3) does, let $\partial \in D$ be $\ll$-least so that condition (3) holds for one of the trees $\mathcal{T}_{\mathrm{I}}^\partial$ or $\mathcal{T}_{\mathrm{II}}^\partial$. If condition (3) holds for $\mathcal{T}_{\mathrm{I}}^\partial$ then I loses, otherwise II loses.

We shall show below that the game cannot end because of conditions (6) and (7).

It is clear that the game $\mathcal{G}_{tame}(M, N)$ is (essentially) an f continuously coded game. The payoff is Δ_2^1 as it is clearly definable over $\mathrm{L}[z]$ (where z is a real coding the play) via a Δ_1 formula. Similarly f has Δ_2^1 graph. In both cases Δ_2^1 is an over-kill.

The next claim shows that the game we defined is long enough to reach the end of a comparison. More precisely, the natural number codes n_ι above are defined in such a way that the same code is *never repeated twice.* The reason for this is that our natural numbers essentially code the least disagreement in each round, and since all trees are non-overlapping, a disagreement cannot be repeated twice.

Claim 2.7 *The game $\mathcal{G}_{tame}(M, N)$ never reaches its end through condition (7) above.*

Proof. Assume for contradiction that in round ι trees $\mathcal{T}_{\mathrm{I}}^\partial$ and $\mathcal{T}_{\mathrm{II}}^\partial$ were played in such a way that n_ι is equal to n_ζ for some $\zeta < \iota$.

Case 1: $n_\zeta = \phi(\partial, d^\partial_\zeta(\delta_\zeta))$ for some ∂. Then n_ι must equal $\phi(\partial, d^\partial_\iota(\delta_\iota))$ and $d^\partial_\iota(\delta_\iota)$ must be equal to $d^\partial_\zeta(\delta_\zeta)$. On the other hand, our rules for determining n_ζ also imply that $\mathrm{crit}(\pi^\partial_{\zeta,\zeta+1}) > \delta_\zeta$. Since $\mathrm{crit}(\pi^\partial_{\zeta+1,\iota}) \geq \delta_{\zeta+1}$ it follows that $\mathrm{crit}(\pi^\partial_{\zeta,\iota}) > \delta_\zeta$ and in particular $\pi^\partial_{\zeta,\iota}(\delta_\zeta) = \delta_\zeta \neq \delta_\iota$. But this means that $d^\partial_\zeta(\delta_\zeta) = d^\partial_\iota(\pi^\partial_{\zeta,\iota}(\delta_\zeta)) = d^\partial_\iota(\delta_\zeta) \neq d^\partial_\iota(\delta_\iota)$, contradiction.
Case 2: $n_\zeta = \phi(\partial, d^\partial_\zeta(A), \tilde{\partial}, d^{\tilde{\partial}}_\zeta(A))$ for some $\partial, \tilde{\partial}$ and $A \subset \delta_\zeta$ which belongs to $M^\partial_\zeta \cap M^{\tilde{\partial}}_\zeta$. Then n_ι must be equal to $\phi(\partial, d^\partial_\iota(A'), \tilde{\partial}, d^{\tilde{\partial}}_\iota(A'))$ for some $A' \subset \delta_\iota$ which belongs to $M^\partial_\iota \cap M^{\tilde{\partial}}_\iota$. Furthermore, $d^\partial_\iota(A')$ and $d^{\tilde{\partial}}_\iota(A')$ must be equal (respectively) to $d^\partial_\zeta(A)$ and $d^{\tilde{\partial}}_\zeta(A)$. By the coherence requirement of Figure 7 it follows that $A' = \pi^\partial_{\zeta,\iota}(A)$ and similarly on the other side $A' = \pi^{\tilde{\partial}}_{\zeta,\iota}(A)$. In particular $\pi^\partial_{\zeta,\iota}(A) = \pi^{\tilde{\partial}}_{\zeta,\iota}(A)$. On the other hand, our rules for determining n_ζ also imply that A is a witness for the disagreement of the extenders E^∂ and $E^{\tilde{\partial}}$ in round ζ, whence $\pi^\partial_{\zeta,\zeta+1}(A) \neq \pi^{\tilde{\partial}}_{\zeta,\zeta+1}(A)$ and in fact

$$\pi^\partial_{\zeta,\zeta+1}(A) \cap \delta_{\zeta+1} \neq \pi^{\tilde{\partial}}_{\zeta,\zeta+1}(A) \cap \delta_{\zeta+1}.$$

Since $\pi^\partial_{\zeta+1,\iota}$ and $\pi^{\tilde{\partial}}_{\zeta+1,\iota}$ both have critical point at least $\delta_{\zeta+1}$ it follows that $\pi^\partial_{\zeta,\iota}(A) \neq \pi^{\tilde{\partial}}_{\zeta,\iota}(A)$, contradiction. □

A moment's reflection shows that $\mathcal{G}_{tame}(M, N)$ can also never reach its end because of condition (6). Thus the game can only end with conditions (0), (2), (3), (5), or (8). Condition (8) is the one we strive for, as it indicates a successful comparison of M and N. Condition (2) can be avoided if the model we are responsible for is *iterable*, and conditions (0), (3) can be avoided if we play trees which arise through a comparison. Avoiding condition (5) is the most delicate task, and will require heavy use of our assumption that M and N are tame. Note the importance of condition (5) — it is because of condition (5) that generators are not moved in our comparison (in other words $\mathrm{crit}(\pi^\partial_{\iota,\iota+1}) \geq \delta_\iota$) and this was essential for the proof of Claim 2.7.

Claim 2.8 *Let M and N be countable tame pre-mice. Assume further that M is ω_1+1-iterable. Then* II *does not have a winning strategy in the game $\mathcal{G}_{tame}(M, N)$.*

Proof. Let us fix an iteration strategy Σ for M, and assume for contradiction that Λ is a winning strategy for II in the game $\mathcal{G}_{tame}(M, N)$. We describe below how to play so as to defeat Λ.

The idea (for round ζ say) is to use Λ in predicting the final branches that II will choose on the trees for which she is responsible ($\mathcal{U}$ and $\mathcal{U}^k$), and then re-play the same round with longer trees, using these branches in I's version of the comparison ($\mathcal{U}_\mathrm{I}$ and $\mathcal{U}^k_\mathrm{I}$). As long as Λ uses the same branches in all the "re-plays" we continue extending the trees, still in round ζ. When finally we reach a point where two re-plays pick distinct branches through the same tree we move to the next round, adding both direct limit model to the comparison. $\delta_{\zeta+1}$ will then be a *Woodin cardinal* in the lined-up part of all future models by Theorem 2.5. By tameness of M and N no future extender used in the comparison can overlap $\delta_{\zeta+1}$ and this guarantees that the critical points of all future embeddings are above $\delta_{\zeta+1}$.

Sub-Claim 2.8.1 *Let P represent a play according to Λ of length $\iota+1$. Assume that for any $\zeta < \iota$ either $G(\zeta) = \mathrm{II}$ or else a new model was added to the comparison in round ζ.*

Then all the extenders used in the trees $\bar{\mathcal{T}}^{\partial}$ in round ι have critical points which are greater than or equal to δ_ι.

Proof. Assume otherwise, and let ζ be least such that one of the extenders has critical point below $\delta_{\zeta+1}$. Note that by definition the game $\mathcal{G}_{tame}(M, N)$ ends in round ι, and if $G(\zeta) = \mathrm{II}$ then II loses. But Λ is a winning strategy for II so $G(\zeta) = \mathrm{II}$ is impossible.

By hypothesis then a new model was added to the comparison in round ζ. This model must be either $M_{\zeta+1}^{\langle n_\zeta, i\rangle}$ or $N_{\zeta+1}^{\langle n_\zeta, i\rangle}$ for some $i < \omega$. We work below assuming that it is $M_{\zeta+1}^{\langle n_\zeta, i\rangle}$. The proof in the other case is similar.

The model $M_{\zeta+1}^{\langle n_\zeta, i\rangle}$ added in round ζ must be the direct limit along a cofinal branch of some tree $\bar{\mathcal{T}}_\zeta^{\partial_0}$ played in round ζ and must furthermore be distinct from the branch $\bar{b}_\zeta^{\partial_0}$ that gave rise to the model $M_{\zeta+1}^{\partial_0}$. Let us denote these two distinct branches by b_0 (with direct limit $M_{\zeta+1}^{\partial_0}$) and b_1 (with direct limit $M_{\zeta+1}^{\langle n_\zeta, i\rangle}$). Let us further put $\partial_1 = \langle 0, 1 + \ulcorner n_\zeta, i \urcorner\rangle$ so that $M_{\zeta+1}^{\langle n_\zeta, i\rangle} = M_{\zeta+1}^{\partial_1}$. Observe that $\delta(\bar{\mathcal{T}}_\zeta^{\partial_0}) = \delta_{\zeta+1}$

Since one of the extenders used in the trees in round ι has critical point below $\delta_{\zeta+1}$, there must be some $\tilde{\partial}$ for which the extender E_{γ_ϵ} on the tree $\bar{\mathcal{T}}_\iota^{\tilde{\partial}}$ has critical point below $\delta_{\zeta+1}$. Let us use Q^∂ to stand for the ϵ-th model on the tree $\bar{\mathcal{T}}_\iota^{\partial}$, and $\vec{E}^\partial$ for the extender sequence of this model. Let $\gamma = \gamma_\epsilon$, and let E^∂ be the extender indexed by γ_ϵ in the model Q^∂.

Now by choice of $\tilde{\partial}$, the extender $E^{\tilde{\partial}}$ has critical point below $\delta_{\zeta+1}$. Since $Q^{\tilde{\partial}}$ is *tame* it follows that

$$\mathcal{J}_\gamma[\vec{E}^{\tilde{\partial}} \restriction \gamma] \models \text{“}\delta_{\zeta+1} \text{ is } \textit{not} \text{ a Woodin cardinal.”}$$

There must therefore exist a specific function $f\colon \delta_{\zeta+1} \to \delta_{\zeta+1}$ in the model $\mathcal{J}_\gamma[\vec{E}^{\tilde{\partial}} \restriction \gamma]$ so that

$$\mathcal{J}_\gamma[\vec{E}^{\tilde{\partial}} \restriction \gamma] \models \text{“}\delta_{\zeta+1} \text{ is } \textit{not} \text{ Woodin with respect to } f.\text{”}$$

Since γ represent the least disagreement in the comparison at stage ϵ, all the sequences $\vec{E^\partial}$ are the same up to γ. Thus f belongs to all the models Q^∂, and furthermore we may replace $\tilde{\partial}$ by ∂_0 to conclude that

$$(*) \quad \mathcal{J}_\gamma[\vec{E^{\partial_0}} \restriction \gamma] \models \text{“}\delta_{\zeta+1} \text{ is } not \text{ Woodin with respect to } f.\text{”}$$

Q^{∂_0} is an iterate of the model $M^{\partial_0}_{\zeta+1}$, and since f belongs to Q^{∂_0} it follows that f belongs also to $M^{\partial_0}_{\zeta+1}$. Similarly f belongs to $M^{\partial_1}_{\zeta+1}$. By Theorem 2.5, used for the tree $\bar{\mathcal{T}}^{\partial_0}_\zeta$ and the two branches b_0 and b_1, it follows that

$$(**) \quad M^{\partial_0}_{\zeta+1} \models \text{“}\delta_{\zeta+1} \text{ is Woodin with respect to } f\text{”}$$

Let $\pi_0\colon M^{\partial_0}_{\zeta+1} \to Q^{\partial_0}$ be the iteration embeddings. Note that by the minimality of ϵ this embedding has critical point at least $\delta_{\zeta+1}$. Since the truth of the statement "δ is Woodin with respect to a function f" depends on the existence of an extender with length less than δ, it follows from (**) that

$$Q^{\partial_0} \models \text{“}\delta_{\zeta+1} \text{ is Woodin with respect to } f.\text{”}$$

But this contradicts (*) above. □

Remark: The proof of Sub-Claim 2.8.1 is an adaptation of Steel's observation that weak iterability is sufficient for the comparison of tame mice. See [Ste93, Theorem 1.10] (note that there too M and N are assumed to be tame, though this is not stated).

Our strategy for defeating Λ is to play in each round ι making sure that either $G(\iota) = \text{II}$ or else a new model is added in round ι. By Sub-Claim 2.8.1 it will follow that the game can only end with condition (8) — in which case we win — or one of conditions (0), (2), (3). To avoid condition (2) we must be sure to only pick branches which lead to well-founded direct limits. This is easy to do for trees on M, since M is iterable. During the game however we shall have to add new models to the comparison, possibly models which are iterates of N, and for all we know those models are not iterable. Our solution is to play against Λ simultaneously on many different boards. All boards will have exactly the same models, but we shall make sure that each model M^∂ added is controlled by II on at least one of the boards. This will enable us to iterate M^∂ simply by copying II's moves from a board where II controls this model.

We begin the game on the main board — denoted B. At the start of round ι we shall have additional boards B_η for (some, possibly not all) $\eta < \iota$. The models compared in all these boards will be the same, though the same model may appear as M^∂ on one board and as $M^{\partial'}$ on another. To avoid confusion we shall take M^∂ to mean M^∂ on the main board B, unless otherwise stated. We shall define assignments $\eta \mapsto \partial(\eta)$ and the model M^∂ on the board B

will be equal to the model $M^{\partial(\eta)}$ on the board B_η. Another way to view the assignment is to fix η and look at $\partial \mapsto \partial(\eta)$. Viewed this way the assignment is a bijection between the indices ∂ on the board B at any particular round ι, and the indices ∂' on the board B_η in the same round ι. For notational convenience we regard B as the board B_{-1}, and have $\partial(-1) = \partial$.
Remark: These assignments may change from round to round, and are therefore best regarded as $\partial \mapsto \partial^\iota(\eta)$. For $\iota < \iota'$ we may have $\partial^\iota(\eta) \neq \partial^{\iota'}(\eta)$ (see for example Case 2 below). But even then on the board B_η, $M_{\iota'}^{\partial^{\iota'}(\eta)}$ will be an iterate of $M_\iota^{\partial^\iota(\eta)}$ and the iteration trees witnessing this will be exactly the same trees which on the board B witness that $M_{\iota'}^\partial$ is an iterate of M_ι^∂.

It is essential for the construction that if a model $M^{\partial'}$ is controlled by us (I) on a certain board B_η, then there is another board $B_{\bar\eta}$ on which the same model is controlled by II, and is indexed *earlier*. To be precise, we maintain:

> For ∂ and η, if $\partial(\eta) = \langle 0, 1 + \ulcorner n_\zeta^\eta, i \urcorner\rangle$ then there exists some $\bar\eta$ so that either
>
> - $\partial(\bar\eta) = \langle 1, 0\rangle$, or else
>
> (*) - $\partial(\bar\eta) = \langle 1, \ulcorner n_{\bar\zeta}^{\bar\eta}, \bar i \urcorner\rangle$ and either
>
>   - $\bar\zeta < \zeta$, or
>   - $\bar\zeta = \zeta$ and $\bar i < i$.

(by n_ζ^η we mean the natural number code for round ζ on the board B_η. Note that the same round ζ may have different codes on the different boards.)

We shall need to make frequent reference to the $\bar\eta$ given by (*). For convenience let us denote (one such) $\bar\eta$ as η/∂. Note that η/∂ depends on ∂.

It is time now to begin the first half of round ι. We are playing simultaneously on all boards, and must on *each* board η play trees $\mathcal{T}_{\mathrm{I}}^{\partial(\eta)}$. (We play against Λ, and so the trees $\mathcal{T}_{\mathrm{II}}^{\partial(\eta)}$ on the board B_η are always given as Λ's response to our moves.) To begin with we describe how to play $\mathcal{T}_{\mathrm{I}}^{\partial(\eta)}$ on B_η when $\partial(\eta) = \langle 0, 1 + \ulcorner k \urcorner\rangle$. In this case we simply *copy* the tree $\mathcal{T}_{\mathrm{II}}^{\partial(\eta/\partial)}$ from the board $B_{\eta/\partial}$. Note that on the board B_η we control this tree and must therefore also play a final branch $b^{\partial(\eta)}$. We simply copy the final branch $c^{\partial(\eta/\partial)}$ which II plays on the board $B_{\eta/\partial}$ — where by (*) the tree is controlled by II. Note that in copying we of course have to procrastinate until II begins to play her tree, and only then we can start to play ours.

The copying takes care of all trees except for $\mathcal{T}_{\mathrm{I}}$ (on M_ι) and those trees which are controlled by II (on the board B_η the trees $\mathcal{T}_{\mathrm{I}}^{\partial(\eta)}$ when $\partial(\eta) = \langle 1, a\rangle$). We shall play and re-play these trees over many stages in the same round ι. The stages are indexed by the ordinal θ, and during stage θ all the trees we play will have length θ. We use $\mathcal{T}_{\mathrm{I}}^{\partial(\eta)}(\theta)$ to denote the tree played in stage θ on the board B_η. We shall use the copying procedure only for limit θ. For successor θ there is no need to play final branches and so the copying is not necessary.

We shall maintain the following for stage θ:

i On each board the ϵ defined by conditions (1-5) is equal to θ.

ii For any (valid) ∂, and any η, η', the trees $\mathcal{T}_{\mathrm{I}}^{\partial(\eta)}(\theta)$ and $\mathcal{T}_{\mathrm{I}}^{\partial(\eta')}(\theta)$ are the same up to θ (note that by (i) they are also equal to the trees $\mathcal{T}_{\mathrm{II}}^{\partial(\eta)}(\theta)$ and $\mathcal{T}_{\mathrm{II}}^{\partial(\eta')}(\theta)$ up to θ, and must furthermore be the trees which arise from a comparison process of all models involved).

iii For $\theta' < \theta$ and any η, the tree $\mathcal{T}_{\mathrm{I}}^{\partial(\eta)}(\theta)$ restricted to θ' is exactly equal to the tree $\mathcal{T}_{\mathrm{I}}^{\partial(\eta)}(\theta')$ restricted to θ'. Note that this second tree is (by (i)) equal to the tree $\mathcal{T}_{\mathrm{II}}^{\partial(\eta)}(\theta')$ up to θ'.

iv For $\theta' < \theta$ and any η: If $\partial(\eta) = \langle 1, a\rangle$ (meaning that the corresponding tree is controlled by II on B_η) then the θ'-th branch in the tree $\mathcal{T}_{\mathrm{I}}^{\partial(\eta)}(\theta)$ is equal to the θ'-th branch in the tree $\mathcal{T}_{\mathrm{II}}^{\partial(\eta)}(\theta')$. Note that the latter may be the *final* branch $\bar{b}^{\partial(\eta)}(\theta')$ through the tree $\mathcal{T}_{\mathrm{II}}^{\partial(\eta)}(\theta')$ in stage θ'. Such final branch *was* played by II since II controls the tree.

v The trees on M_ι all use the iteration strategy Σ to pick branches at limit stages.

Conditions (i-v) in fact specify precisely how we must play. We begin with $\theta = 1$, and let γ_0 be the index of a least disagreement between the models M_ι^∂. On the board B_η then, we play the iteration tree $\mathcal{T}_{\mathrm{I}}^{\partial(\eta)}(1)$, of length 1, which begins by using the extender indexed at γ_0.

Having played the trees for stage θ, conditions (i-v) specify precisely what we must do in stage $\theta+1$. On the board B_η we must play $\mathcal{T}_{\mathrm{I}}^{\partial(\eta)}(\theta+1)$ which extends the tree $\mathcal{T}_{\mathrm{II}}^{\partial(\eta)}(\theta)$ restricted to θ.

If θ is a limit ordinal we must first extend the tree by picking a final branch which is cofinal in θ. We use the branches played in stage θ, by I if $\partial(\eta) = \langle 0, a\rangle$, and by II if $\partial(\eta) = \langle 1, a\rangle$.

Once the branches were picked (or immediately if θ is a successor ordinal) we have a θ-th model $Q_\theta^{\partial(\eta)}$ for the tree $\mathcal{T}_{\mathrm{I}}^{\partial(\eta)}(\theta+1)$ on the board B_η. We continue to stage $\theta+1$ only if

vi For (valid) ∂ and any η, η', the models $Q_\theta^{\partial(\eta)}$ and $Q_\theta^{\partial(\eta')}$ are the same. In other words, at stage θ the final branches $\bar{b}^{\partial(\eta)}(\theta)$ on the board B_η and $\bar{b}^{\partial(\eta')}(\theta)$ on the board $B_{\eta'}$ were the same.

In this case we let γ_θ be the index of the least disagreement between any of the models Q^{∂}_{θ}. On the board B_η we then extend the tree $\mathcal{T}_{\mathrm{I}}^{\partial(\eta)}(\theta)$ by using γ_θ as the index of the θ-th extender.

Conditions (i-vi), together with our copying procedure, also specify precisely what we must play in stage θ for limit θ. If $\partial(\eta) = \langle 0, 1 + \ulcorner k \urcorner \rangle$ then on the board B_η we play $\mathcal{T}_{\mathrm{I}}^{\partial(\eta)}(\theta)$ by copying the tree $\mathcal{T}_{\mathrm{II}}^{\partial(\eta/\partial)}(\theta)$ from the board $B_{\eta/\partial}$. If $\partial(\eta) = \langle 0, 0 \rangle$ or $\langle 1, a \rangle$ we let $\mathcal{T}_{\mathrm{I}}^{\partial(\eta)}(\theta)$ be the union of the trees $\mathcal{T}_{\mathrm{I}}^{\partial(\eta)}(\theta') \restriction \theta'$, $\theta' < \theta$. In the case $\partial(\eta) = \langle 0, 0 \rangle$ we must also play a final branch, which we pick using Σ.

There are several mis-haps that can cause our construction to end at stage θ. First, condition **(vi)** may fail at limit θ. Another option is for condition **(i)** to fail at θ. Note that condition (vi) combined with the fact that all trees arise from comparison (itself guaranteed by condition (i)) are enough to give condition (ii). Conditions (iii-v) are always satisfied by our construction. (For limit θ, note that if $\partial(\eta) = \langle 0, 1 + \ulcorner k \urcorner \rangle$ then our copied trees $\mathcal{T}_{\mathrm{I}}^{\partial(\eta)}(\theta)$ still satisfy conditions (iii) and (iv), since the "originals" $\mathcal{T}_{\mathrm{II}}^{\partial(\eta/\partial)}(\theta)$ are equal to $\mathcal{T}_{\mathrm{I}}^{\partial(\eta/\partial)}(\theta)$ (or else condition (i) fails at θ) and these latter trees satisfy conditions (iii) and (iv).) Thus the only other potential problem is that the trees played during stage θ of round ι cause the game on at least one of the boards to **end**, because of one of the conditions (8), (0), (2), or (3). If the game ends with condition (8) then we are done, having defeated Λ. The next three claims show that on no board can the game end with one of the conditions (0), (2), or (3).

Sub-Claim 2.8.2 *In stage θ,* all *trees on all boards were played. In other words there does* not *exists an η and some ∂ so that player* I/II *on the board B_η procrastinated indefinitely in playing the tree $\mathcal{T}_{\mathrm{I/II}}^{\partial(\eta)}(\theta)$.*

Proof. Assume otherwise. Observe first that θ must be a limit ordinal, for otherwise we (player I) do not procrastinate at all. Thus if a player procrastinated indefinitely it can only be player II, but then II loses, which is impossible since Λ is a winning strategy for II.

Even for limit θ, we (player I) do not procrastinate at all on the trees $\mathcal{T}_{\mathrm{I}}^{\partial(\eta)}(\theta)$ for $\partial = \langle 0, 0 \rangle$ and $\partial = \langle 1, a \rangle$ for any a. In particular, we do not procrastinate on $\mathcal{T}_{\mathrm{I}}^{\partial(\eta)}(\theta)$ when $\partial = \langle 0, 0 \rangle$ or $\langle 1, 0 \rangle$. It follows that II does not procrastinate indefinitely on the trees $\mathcal{T}_{\mathrm{II}}^{\partial(\eta)}(\theta)$ when $\partial = \langle 0, 0 \rangle$ or $\langle 1, 0 \rangle$, for otherwise the game ends and II loses, which is impossible.

Thus there must exist some $\partial = \langle 0/1, 1 + \ulcorner k \urcorner \rangle$ and some η so that on the board B_η a player procrastinated indefinitely with the tree $\mathcal{T}_{\mathrm{I/II}}^{\partial(\eta)}(\theta)$.

Fix the least $\zeta < \iota$, then least $i < \omega$, and finally least $j \in \{0, 1\}$ so that for some η one of the players on B_η procrastinated indefinitely with $\mathcal{T}_{\mathrm{I/II}}^{\partial(\eta)}(\theta)$

where $\partial = \langle j, 1 + \ulcorner n^{\eta}_{\zeta}, i \urcorner\rangle$.
Case 1: If $j = 1$, then by construction we (player I) do not procrastinate at all on the tree $\mathcal{T}_{\mathrm{I}}^{\partial(\eta)}(\theta)$. Thus it must be that on the board B_η player II procrastinated indefinitely with the tree $\mathcal{T}_{\mathrm{II}}^{\partial(\eta)}(\theta)$. But then by minimality of ζ, i, j it follows that on the board B_η player II loses, which is impossible.
Case 2: If $j = 0$. If we (player I) did not procrastinate indefinitely with the tree $\mathcal{T}_{\mathrm{I}}^{\partial(\eta)}(\theta)$ then just as in case 1 it follows that II loses, which is impossible. Thus it must be that we procrastinated indefinitely with the tree $\mathcal{T}_{\mathrm{I}}^{\partial(\eta)}(\theta)$. But by construction we played this tree by copying the tree $\mathcal{T}_{\mathrm{II}}^{\partial(\eta/\partial)}(\theta)$ which II played on the board $B_{\eta/\partial}$. It must therefore be the case that on the board $B_{\eta/\partial}$ player II procrastinated indefinitely with the tree $\mathcal{T}_{\mathrm{II}}^{\partial(\eta/\partial)}(\theta)$. Fix $\bar{\eta}, \bar{\zeta}, \bar{i}$ such that $\eta/\partial = \bar{\eta} = \langle 1, 1 + \ulcorner n^{\bar{\eta}}_{\bar{\zeta}}, \bar{i} \urcorner\rangle$. By (*) then $\bar{\zeta} < \zeta$ or else $\bar{\zeta} = \zeta$ and $\bar{i} < i$. But this contradicts the minimality in our choice of ζ, i. □

Let $\epsilon(\theta)$ be the least ordinal which on one or more of the boards B_η is equal to the ϵ determined by the play at stage θ. (We do not yet assume $\epsilon(\theta) = \theta$.)

Sub-Claim 2.8.3 *On no board does the game end at stage θ because condition (2) holds for $\epsilon(\theta)$.*

Proof. Similar to the previous claim, replacing "procrastinate" with "play an ill-founded branch." Note that for $\partial(\eta) = \langle 0, 0\rangle$ or $\langle 1, a\rangle$ the trees $\mathcal{T}_{\mathrm{I}}^{\partial(\eta)}(\theta)$ which we play on the board B_η always pick well-founded branches. For $\partial(\eta) = \langle 0, 0\rangle$ this is because we use Σ — an iteration strategy for M — to pick branches. For $\partial(\eta) = \langle 1, a\rangle$ it is because we play in stage θ branches that were previously played by II in the tree $\mathcal{T}_{\mathrm{II}}^{\partial(\eta)}(\theta')$ $(\theta' < \theta)$. □

Sub-Claim 2.8.4 *On no board does the game end at stage θ because condition (3) holds for $\epsilon(\theta)$.*

Proof. Similar to the previous two claims. In the case of successor θ observe that the trees $\mathcal{T}_{\mathrm{I}}^{\partial(\eta)}(\theta)$ which we play on the board B_η *do* use the least disagreement to determine the next extender. □

The only potential reasons to end our construction at stage θ are thus given by conditions **(8)**, **(vi)**, and the possibility of $\epsilon(\theta) < \theta$ which can result only from conditions **(1)** or **(4)**. The next claim shows that one of these *must happen* for a countable θ (on at least one of the boards).

Sub-Claim 2.8.5 *The construction in round ι cannot go on for ω_1 stages.*

Proof. Assume otherwise. Let $\mathcal{T}^\partial$ be the length ω_1 iteration tree given by the union of the trees $\mathcal{T}_{\mathrm{I}}^{\partial}(\theta) \restriction \theta$. We use M^∂_ξ for the ξ-th model on the tree $\mathcal{T}^\partial$, and $\mathcal{T}, M_\xi$ to stand for $\mathcal{T}^{\langle 0,0\rangle}, M_\xi^{\langle 0,0\rangle}$.

Since Σ is an $\omega_1 + 1$ iteration strategy for M, and since the branches in $\mathcal{T}$ were picked according to Σ, there exists a cofinal branch b through the tree $\mathcal{T}$. Let M_b denote the direct limit along this b. Note that b is a closed unbounded subset of ω_1. A standard argument shows that $\omega_1 \in M_b$, and that there must therefore be some $\xi \in B$ and some κ so that $\pi_{\xi,b}(\kappa) = \omega_1$. As κ is the critical point of $\pi_{\xi,b}$ it must be measurable in the ξ-th model of $\mathcal{T}$. By elementarity then ω_1 is measurable in M_b.

M_b, being tame, certainly does not have measurable Woodin cardinals. Thus

$$M_b \models \text{“}\omega_1^V \text{ is not a Woodin cardinal.”}$$

Using $\vec{E}_b$ for the extender sequence of M_b, there must therefore exists some μ so that

$$\mathcal{J}_{\mu+1}[\vec{E}_b \restriction \mu + 1] \models \text{“}\omega_1^V \text{ is not a Woodin cardinal”}$$

Let $Q = \mathcal{J}_{\mu+1}[\vec{E}_b \restriction \mu + 1]$ for the least such μ, and fix $f \in Q$ so that, in Q, ω_1^V is not Woodin with respect to f. By the minimality of μ, f belongs to $\mathcal{J}_{\mu+1}[\vec{E}_b \restriction \mu + 1] - \mathcal{J}_{\mu}[\vec{E}_b \restriction \mu]$.

Working in $V^{\text{col}(\omega,\omega_1)}$, let $\mathcal{R}^\partial$ be the tree of attempts to construct a cofinal branch c through $\mathcal{T}^\partial$ in such a way that Q belongs to the direct limit along c. (Note we do not require the direct limit to be well-founded — well-foundedness cannot be witnessed by a "tree of attempts.") By Theorem 2.5, for each $\partial \neq \langle 0, 0 \rangle$ there can be at most one branch of $\mathcal{R}^\partial$.

Case 1: If for each $\partial \neq \langle 0, 0 \rangle$ there is (in $V^{\text{col}(\omega,\omega_1)}$) *exactly* one branch of $\mathcal{R}^\partial$. Denote this branch by c^∂. Note that the uniqueness of c^∂ implies that $c^\partial \in V$. But then working in V we have cofinal branches through all the trees $\mathcal{T}^\partial$. The usual argument for showing that the comparison of countable mice ends *before* ω_1 now gives a contradiction. (See for example the last claim of [MiSt94, pp.72].)

Case 2: There must therefore be some $\partial \neq \langle 0, 0 \rangle$ so that (in $V^{\text{col}(\omega,\omega_1)}$) there are *no* branches through the tree $\mathcal{R}^\partial$. In other words this tree is well-founded. Let H be a countable elementary sub-structure of a sufficiently large rank initial segment of V, which contains all relevant objects. Let $\bar{V}$ be the transitive collapse of H and let $j: \bar{V} \longleftrightarrow H$ be the collapse embedding. Let $\bar{\omega_1}, \bar{Q}, \bar{\mu}, \bar{f}, \bar{b}, \bar{\mathcal{R}}^\partial$ be the collapse of $\omega_1, Q, \mu, f, b, \mathcal{R}^\partial$ respectively. For notational convenience let $\xi = \bar{\omega_1}$. Note that since b is closed unbounded, $\xi \in b$ and so $\bar{b}$ is equal to the ξ-th branch of $\mathcal{T}$. Let M_ξ be the ξ-th model of $\mathcal{T}$.Then M_ξ is equal to the direct limit model along $\bar{b}$. Now in the model $\bar{V}^{\text{col}(\omega,\bar{\omega_1})}$, the tree $\bar{\mathcal{R}}^\partial$ is well-founded. By absoluteness then $\bar{\mathcal{R}}^\partial$ is also well-founded in V. Let c be the $\bar{\omega_1}$-th branch of the iteration tree $\mathcal{T}^\partial$, and let $\bar{M}_c^\partial$ be the direct limit along this branch (namely the ξ-th model on the tree $\mathcal{T}^\partial$, M_ξ^∂). Then c cannot represent a branch through $\bar{\mathcal{R}}^\partial$ whence $\bar{Q} \notin \bar{M}_c^\partial = M_\xi^\partial$. But this means that the least disagreement between M^{∂}_{ξ} and M_{ξ} must be $\bar{\mu}$ or less. In particular the index used in the comparison at stage ξ is at most $\bar{\mu}$. But then — if a non-empty extender was used in $\mathcal{T}$ at stage ξ — $\bar{f}$ does not belong to $M_{\xi+1}$, and indeed $\bar{f} \notin M_{\xi'}$ for any $\xi' > \xi$. This is a contradiction since $\bar{f} = f \cap \xi$ certainly belongs to $M_b = M_{\omega_1}$. Even if the extender used in $\mathcal{T}$ at stage ξ was empty — at least for some ∂' a non-empty extender was used in $\mathcal{T}^{\partial'}$ so $\bar{f} \notin M^{\partial'}_{\xi+1}$. But this is enough to conclude that $\bar{f} \notin M_b$. □

Our tactic for playing I's moves in round ι centers on the observation that a break-down of construction provides us with moves in round ι which are according to Λ and satisfy one of the three conditions below:

A On at least one of the boards $G(\iota) = \mathrm{II}$.

B A new model was added (on all boards).

C On at least one of the boards the game ended with condition (8).

(C) represent victory. (A) and (B) are enough for Claim 2.8.1 to apply (as required, to ensure that the game does not end with condition (5)).

Suppose the construction breaks-down at θ.

Case 1: The construction broke-down because $\epsilon(\theta) < \theta$ and condition (1) holds.

In the first half of round ι play the trees $\mathcal{T}^{\partial(\eta)}(\theta)$ on the board B_η. Λ then responds with the trees $\mathcal{T}^{\partial(\eta)}_{\mathrm{II}}(\theta)$ on the board B_η.

We claim that on at least one board $G(\iota) = \mathrm{II}$. Fix ∂ so that for some η one of the trees $\mathcal{T}^{\partial(\eta)}_{\mathrm{I}}(\theta)$, $\mathcal{T}^{\partial(\eta)}_{\mathrm{II}}(\theta)$ has length $\epsilon(\theta)$. If $\partial = \langle j, 1 + \ulcorner k \urcorner \rangle$ — where $k = \langle n^{\eta}_{\zeta}, i \rangle$ — choose ∂ so as to minimize first ζ, then i, and finally j, over all possible η.

Note now that $\mathcal{T}^{\partial(\eta)}_{\mathrm{I}}(\theta)$ has length *equal to* θ. This is clear if $\partial = \langle 0/1, 0 \rangle$ for in this case in stage θ of the construction we played a tree of length θ. The same holds if $\partial = \langle 1, 1 + \ulcorner k \urcorner \rangle$, and for all ∂ if θ is not a limit ordinal. Finally, if $\partial = \langle 0, 1 + \ulcorner k \urcorner \rangle$ and θ is a limit ordinal then the tree $\mathcal{T}^{\partial(\eta)}_{\mathrm{I}}(\theta)$ was copied from the tree $\mathcal{T}^{\partial(\eta/\partial)}_{\mathrm{II}}(\theta)$ on the board $B_{\eta/\partial}$, which by the minimality of ζ, i, j must have length θ.

Thus $\mathcal{T}^{\partial(\eta)}_{\mathrm{II}}(\theta)$ must have length $\epsilon(\theta)$, and $G(\iota) = \mathrm{II}$ on the board B_η.

This completes the first half of round ι, and condition (A) holds. We therefore have no need for adding models during the second half. Of course, player II may decide to add models on some of the boards. Since it is essential for our construction that all boards have *exactly the same models* we must make sure, if II adds a particular model on a particular board, that this model is added to *all* boards. this can be achieved as follows:

If on board $B_{\bar{\eta}}$ II plays non empty $\partial^{\mathrm{II}}_{\bar{\imath}}$ — say $\partial^{\mathrm{II}}_{\bar{\imath}} = \bar{\partial}(\bar{\eta})$ — and a branch $b^{\mathrm{II}}_{\bar{\imath}} = \bar{b}$ then we must for each $\eta \neq \bar{\eta}$ fix some $\mathbf{i} > \bar{\mathbf{i}}$ and play $\partial^{\mathrm{I}}_{i} = \bar{\partial}(\eta)$ and

$b_i^{\mathrm{I}} = \bar{b}$ on the board B_η. This guarantees that the model $N_{\iota+1}^{\langle n_\iota^{\bar{\eta}}, \bar{i}\rangle}$ on the board $B_{\bar{\eta}}$ is equal to the model $M_{\iota+1}^{\langle n_\iota^{\eta}, i\rangle}$ on the board B_η.

We do this for each $\eta \neq \bar{\eta}$. Denote the i used by $i(\eta)$. Let

$$\hat{\partial}(\eta) = \begin{cases} \langle 0, 1 + \ulcorner n_\iota^\eta, i(\eta)\urcorner\rangle & \text{If } \eta \neq \bar{\eta} \\ \langle 1, 1 + \ulcorner n_\iota^{\bar{\eta}}, \bar{i}\urcorner\rangle & \text{If } \eta = \bar{\eta} \end{cases}$$

Then $\hat{\partial}(\eta)$ is the index of our new model on the board B_η. We then set $\eta/\hat{\partial} = \bar{\eta}$. This secures (*), since $\mathbf{i}(\eta) > \bar{\mathbf{i}}$.

It is a matter of simple book-keeping to do this for all the models that II adds on any board, making sure that $\eta/\hat{\partial}$ is defined for all new $\hat{\partial}$'s and satisfies (*).

Case 2: The construction broke down because of condition (vi).

In the first half of round ι play $\mathcal{T}_{\mathrm{I}}^{\partial(\eta)}(\theta)$ for I on the board B_η. Λ then responds with the tree $\mathcal{T}_{\mathrm{II}}^{\partial(\eta)}(\theta)$. We use $\bar{\mathcal{T}}^{\partial(\eta)}$ for the trees then determined by the rules of $\mathcal{G}_{tame}(M, N)$ on the board B_η, and $\bar{b}^{\partial(\eta)}$ for the final branches.

In this case we *add* models to the comparison during the second half of round ι. Particularly, suppose ∂ is a witness that condition (vi) holds. In other words, there exists η and η' so that on two boards B_η and $B_{\eta'}$ different final branches were played trough the trees $\bar{\mathcal{T}}^{\partial(\eta)}$ and $\bar{\mathcal{T}}^{\partial(\eta')}$. Note that $\partial \neq \langle 0, 0\rangle$ since we (player I) play the same branch through the tree $\mathcal{T}$ on all boards. In the case that $\partial(\eta) = \langle 0, 1 + \ulcorner n_\zeta^\eta, i\urcorner\rangle$ we may replace η with η/∂ and still (because of our copying) have the same tree and final branch. Thus we may without loss of generality assume that the trees $\bar{\mathcal{T}}^{\partial(\eta)}$ on the board B_η and $\bar{\mathcal{T}}^{\partial(\eta')}$ on the board $B_{\eta'}$ are both *controlled by player* II.

Let b be the branch $\bar{b}^{\partial(\eta)}$ through $\bar{\mathcal{T}}^{\partial(\eta)}$ played on the board B_η and let b' be the branch $\bar{b}^{\partial(\eta')}$ through $\bar{\mathcal{T}}^{\partial(\eta')}$ played on the board $B_{\eta'}$. Note that the trees $\bar{\mathcal{T}}^{\partial(\eta)}$ and $\bar{\mathcal{T}}^{\partial(\eta')}$ are *equal.* Thus b and b' are two distinct branches through the same tree. We (as player I) will add the direct limit along b as a model in *all* boards $B_{\dot{\eta}}$ for which $\bar{b}^{\partial(\dot{\eta})} \neq b$. One such $\dot{\eta}$ certainly exists (e.g., η') and so certainly models are added and (B) is satisfied. Of course we also add the direct limit along b' to all boards $B_{\dot{\eta}}$ such that $\bar{b}^{\partial(\dot{\eta})} \neq b'$. This serves a double purpose. It guarantees that a new model was added on *all* boards (as for all $\dot{\eta}$ either $\bar{b}^{\partial(\dot{\eta})} \neq b$ or else $\bar{b}^{\partial(\dot{\eta})} \neq b'$) and also serves to make sure that we end up with the same models on all boards.

To be precise: Fix $\bar{\partial}$ and $\bar{\eta}$ so that $\bar{\partial}(\bar{\eta}) = \langle 1, a\rangle$. Let $\bar{b}$ be the branch $\bar{b}^{\bar{\partial}(\bar{\eta})}$ played on the board $B_{\bar{\eta}}$. Let $\eta \neq \bar{\eta}$ be such that $\bar{b}^{\bar{\partial}(\eta)}$ played on the board B_η is *not* equal to $\bar{b}$. We then pick some $i = i(\eta)$ and play $\partial_i^{\mathrm{I}} = \bar{\partial}(\eta)$, $b_i^{\mathrm{I}} = \bar{b}$.

For each η let

$$\hat{\partial}(\eta) = \begin{cases} \bar{\partial}(\eta) & \text{if } \bar{b}^{\bar{\partial}(\eta)} = \bar{b} \\ \langle 0, 1 + \ulcorner n_\iota^\eta, i(\eta)\urcorner\rangle & \text{if } \bar{b}^{\bar{\partial}(\eta)} \neq \bar{b} \end{cases}$$

$\hat{\partial}$ then is the index of the direct limit along $\bar{b}$ on all boards. Set for each η

$$\eta/\hat{\partial} = \begin{cases} \eta/\bar{\partial} & \text{if } \bar{b}^{\bar{\partial}(\eta)} = \bar{b} \\ \bar{\eta} & \text{if } \bar{b}^{\bar{\partial}(\eta)} \neq \bar{b} \end{cases}$$

This secures (*). If $\bar{b}^{\bar{\partial}(\eta)} = \bar{b}$ then (*) holds here because it held in previous rounds. If $\bar{b}^{\bar{\partial}(\eta)} \neq \bar{b}$ then (*) holds here because either $\bar{\partial}(\bar{\eta}) = \langle 1, 0\rangle$ or else $\bar{\partial}(\bar{\eta}) = \langle 1, 1 + \ulcorner n_\zeta^{\bar{\eta}}, \bar{\imath}\urcorner\rangle$ for some $\zeta < \iota$.

Remark: Note that $\eta \mapsto \hat{\partial}(\eta)$ represent the new assignment which takes effect from round $\iota + 1$ onwards (and should formally be $\eta \mapsto \hat{\partial}^{\iota+1}(\eta)$), while $\eta \mapsto \bar{\partial}(\eta)$ represent the old assignment which was in effect until round ι (and should formally be $\eta \mapsto \bar{\partial}^{\iota}(\eta)$). If $\bar{b}^{\bar{\partial}(-1)} = \bar{b}$ then $\hat{\partial}(-1) = \bar{\partial}(-1)$ (in other words $\hat{\partial} = \bar{\partial}$) but the assignments $\eta \mapsto \hat{\partial}(\eta)$ and $\eta \mapsto \bar{\partial}(\eta)$ may be different.

We repeat this addition for *every* possible $\bar{\partial}$ and $\bar{\eta}$. We also repeat the additions described in Case 1 to make sure that any models II adds on one board are added on all boards. It is again a matter of simple book-keeping to add all these models.

Remark: The same model may well be added many times, appearing as $M_{\iota+1}^{\partial}$ for many distinct new ∂'s. This makes no difference to the construction.

Case 3: The construction broke down in stage θ because $\epsilon(\theta) < \theta$ and condition (4) holds.

Fix $\dot{\eta} < \iota$ so that *on the board* $B_{\dot{\eta}}$ the construction broke down because of condition (4). In this case we *open a new board*, B_ι, which will be a continuation of the board $B_{\dot{\eta}}$.

Let $\dot{\theta} = \epsilon(\theta)$. On the board B_η (for $\eta < \iota$) play $\mathcal{T}_{\mathrm{I}}^{\partial(\eta)}(\dot{\theta})$ in the first half round. Λ then responds with $\mathcal{T}_{\mathrm{II}}^{\partial(\eta)}(\dot{\theta})$. Observe that we use the trees from stage $\dot{\theta}$ of the construction. Thus the trees $\bar{\mathcal{T}}^{\partial(\eta)}$ on the board B_η all have length $\dot{\theta}$.

As advertised, we open a new board B_ι. For rounds $\iota' < \iota$ we copy to B_ι the moves played on the board $B_{\dot{\eta}}$. For round ι we play $\mathcal{T}_{\mathrm{I}}^{\partial(\dot{\eta})}(\theta)$ on the board B_ι. Λ then responds with $\mathcal{T}_{\mathrm{II}}^{\partial(\dot{\eta})}(\theta)$. Note that for the board B_ι we take the moves played in stage θ (rather than $\dot{\theta}$) on the board $B_{\dot{\eta}}$. Since $\epsilon(\theta) = \dot{\theta}$, on this board too the length of the trees $\bar{\mathcal{T}}^{\partial(\iota)}$ is equal to $\dot{\theta}$.

Since condition (4) is assumed to holds there must be some ∂ so that $\mathcal{T}_{\mathrm{I}}^{\partial(\dot{\eta})}(\theta)$ and $\mathcal{T}_{\mathrm{II}}^{\partial(\dot{\eta})}(\theta)$ pick different branches at $\dot{\theta}$. Pick $\dot{\partial}$ so that the above holds for $\dot{\partial}(\dot{\eta})$, and $\dot{\partial}(\dot{\eta})$ is $\ll$-least such, on the board $B_{\dot{\eta}}$.

If $\dot{\partial}(\dot{\eta}) = \langle 0, a\rangle$ then on the board B_ι, $G(\iota) = \mathrm{II}$. Thus (A) holds, and we proceed directly to the second half round, playing as we did in Case 1.

If $\dot{\partial}(\dot{\eta}) = \langle 1, a\rangle$ then the tree $\mathcal{T}_{\mathrm{I}}^{\dot{\partial}(\dot{\eta})}(\theta)$ played on the board B_ι is by construction an extension of the tree $\bar{\mathcal{T}}^{\dot{\partial}(\dot{\eta})}$ played on the board $B_{\dot{\eta}}$. The $\dot{\theta} = \epsilon(\theta)$-th branch in $\mathcal{T}_{\mathrm{I}}^{\dot{\partial}(\dot{\eta})}(\theta)$ is by construction equal to the branch $\bar{b}^{\partial(\dot{\eta})}$

which II played on the board $B_{\dot\eta}$. By assumption the trees $\mathcal{T}_{\mathrm{I}}^{\partial(\dot\eta)}(\theta)$ and $\mathcal{T}_{\mathrm{II}}^{\partial(\dot\eta)}(\theta)$ pick different branches at $\dot\theta$. Since II controls this tree, the branch $\bar b^{\dot\partial(\iota)}$ on the board B_ι is equal to the $\dot\theta$-th branch through $\mathcal{T}_{\mathrm{II}}^{\partial(\dot\eta)}(\theta)$.

It follows than that the branches $\bar b^{\dot\partial(\dot\eta)}$ on the board $B_{\dot\eta}$ and $\bar b^{\dot\partial(\iota)}$ on the board B_ι are distinct branches through the same tree. Moreover, on both boards II controls the tree. This precisely is the situation we had in Case 2, and we proceed to play in the second half precisely as we did in Case 2, adding models to the comparison so that (B) holds.

Case 4: The last remaining possibility is for condition (8) to hold in stage θ.

In this case play $\mathcal{T}_{\mathrm{I}}^{\partial(\eta)}(\theta)$ for I on the board B_η during the first half round. Λ will respond with $\mathcal{T}_{\mathrm{II}}^{\partial(\eta)}(\theta)$. The game then *ends* as required for (C).

The above 4 cases describe how we can play to secure conditions (A-C). Claim 2.8.1 easily adapts to show that on no board does the game end with condition (5) (observe that if on one board the game ends with condition (5), then since on all boards the trees played are the same it follows that on all boards the game ends with condition (5) and that this is witnessed by the same extender on all boards).

When finally on one of the boards the game ends, it must then be due to condition (8). On that board we Λ. □(Claim 2.8)

Corollary 2.9 *Let M and N be countable, tame pre-mice. Assume that M is ω_1+1–iterable. If* II *has a draw strategy in $\mathcal{G}_{tame}(M,N)$ then M and N are comparable.*

Proof. Let Λ be II's draw strategy and follow the proof of Claim 2.8. The game must end in a draw, and since the models are tame it cannot end with condition (5). Thus it must end with condition (8). Let P represent the complete run of the game. Let ι be the length of the play. Let $\mathcal{T}$ be the linear composition of all the trees $\bar{\mathcal{T}}_\zeta^{\langle 0,0\rangle}$, $\zeta<\iota$ played in P and let $\mathcal{U}$ be the linear composition of all the trees $\bar{\mathcal{T}}_\zeta^{\langle 1,0\rangle}$, $\zeta<\iota$ played in P. Let $\xi+1$ be the length of these trees. Let b and c be the last branches of $\mathcal{T}$ and $\mathcal{U}$ respectively. From condition (8) it follows that either $M_\xi^{\mathcal{T}} \leq^* N_\xi^{\mathcal{U}}$ or else $N_\xi^{\mathcal{U}} \leq^* M_\xi^{\mathcal{T}}$. Observe that by condition (5) both $\mathcal{T}$ and $\mathcal{U}$ are non-overlapping. The usual comparison argument then shows that at least on one of b, c there were no drops, and that if one side ends shorter then on that side there were no drops (see for example [MiSt94, pp.64]). □

Let $\mathcal{G}'_{tame}(M,N)$ the the game which is identical to $\mathcal{G}_{tame}(M,N)$ except that II (rather than I) wins in the case of a *draw*. The game $\mathcal{G}_{tame}(M,N)$ is symmetric enough so that the proof of Claim 2.8 can be modified to yield

Claim 2.10 *Let M and N be countable, tame pre-mice. Assume that N is ω_1+1–iterable. Then* I *does not have a winning strategy in the game $\mathcal{G}'_{tame}(M,N)$.*

Corollary 2.11 *Let M and N be countable, tame pre-mice. Assume that N is ω_1+1–iterable. If* I *has a draw strategy in the game $\mathcal{G}_{tame}(M, N)$ then M and N are comparable.*

For reals $x, y \in \mathbb{R}^{tame}$, let $\mathcal{G}_{tame}(x, y)$ be the game played as follows: During the first round players I and II play (reals coding) short tame pre-mice M and N respectively. It must be that $x, y \in M$ and $x \leq_L^M y$ — or else I loses. If indeed $x \leq_L^M y$ then it must be that $x, y \in N$ and $y <_L^N x$ — or else II loses. The two players then proceed to play the game $\mathcal{G}_{tame}(M, N)$. Observe that by Corollary 2.4 and the argument following it, M and N are *not* comparable.

Claim 2.12 *If $x \trianglelefteq^{tame} y$ then* II *does not have a winning strategy in the game $\mathcal{G}_{tame}(x, y)$. If $x \triangleright^{tame} y$ then* I *does not have a winning strategy in the game $\mathcal{G}_{tame}(x, y)$.*

Proof. Assume $x \trianglelefteq^{tame} y$. Then there exists an ω_1+1–iterable, short, tame pre-mouse M so that $x \leq_L^M y$. Let I play this M in the first round of $\mathcal{G}_{tame}(x, y)$. Then use Claim 2.8 to defeat II. Note that II's strategy cannot even reach a draw in this case, since M and N are not comparable.

Assume $x \triangleright^{tame} y$. Then there exists an ω_1+1–iterable, short, tame pre-mouse N so that $y <_L^N x$. Let II play this N in the first round of $\mathcal{G}_{tame}(x, y)$. Then use Claim 2.10 to defeat I. Note that again I cannot reach a draw since M and N are not comparable. □

Clearly $\mathcal{G}_{tame}(x, y)$ has the form of an f–continuously coded game with Δ^1_2 payoff. Thus

Corollary 2.13 *If for all $x, y \in \mathbb{R}^{tame}$ the game $\mathcal{G}_{tame}(x, y)$ is determined, then $\trianglelefteq^{tame}$ is $\Game_{f\text{-}cont}\Delta^1_2$, where f has Δ^1_2 graph.*

□(Lemma 2.6)

The determinacy of f–continuously coded games follows from results of the author. Let $G_{\gamma\text{-}cont}$ denote the games which are played just like the continuously coded game, except that the coding is postponed by γ moves. For example in the game $G_{1\text{-}cont}$ player II plays the natural number n_ι during round $2\cdot\iota+1$.

Theorem 2.14 *Assume that there exist cardinals $\kappa < \delta_0 < \delta_1$ and an extender E so that* crit$(E) = \kappa$, *E is δ_1–strong, and δ_0, δ_1 are both Woodin cardinals in* Ult(V, E) *(E is said to be* strong past 2 Woodin cardinals*). Assume further that there is a measurable cardinal above δ_1. Then the games $G_{1\text{-}cont}(D)$ are determined for all D which are $\mathbf{\Pi}^1_1$ in the codes.*

Remark: Similar results are known, connecting "strong past $1+\gamma$ Woodin cardinals" with the determinacy of γ–continuously coded games.

It is easy to see that 1–continuously coded games are stronger than f–continuously coded games, and that the conclusion of Theorem 2.14 implies (more than) the determinacy of $G_{f-cont}(D)$ for f and D which are Δ^1_2. Thus Theorem 2.14 and Lemma 2.6 together imply that the large cardinal strength of f–continuously coded determinacy is somewhere between the existence of a non-tame mouse, and the existence of a cardinal strong past 2 Woodin cardinals.

Lemma 2.6 seems to suggest that in fact "strong past (one) Woodin cardinal" should be enough to prove the determinacy of f–continuously coded games. Indeed, it seems likely that the proof in [Nee96] of Theorem 1.5 combined with the techniques of [Nee95] should allow proving the determinacy of f–continuously coded games, using only a strong past one Woodin cardinal. At the moment however we do not know how to do this.

References

[Bla75] Andreas Blass. The equivalence of two strong forms of determinacy. *Proc. Amer. Math. Soc.*, 52:373–376, 1975.

[Kec95] Alexander S. Kechris. *Classical Descriptive Set Theory.* Springer-Verlag, 1995.

[Mar83] D.A. Martin. The real game quantifier propagates scales. In *Cabal seminar 1979-81*, number 1019 in Lecture notes in Mathematics, pages 157–171. Springer–Verlag, 1983.

[Mos80] Yiannis N. Moschovakis. *Descriptive Set Theory.* North-Holland, 1980.

[MS83] D.A. Martin and John Steel. The extent of scales in L($\mathbb{R}$). In *Cabal seminar 1979-81*, number 1019 in Lecture notes in Mathematics, pages 86–96. Springer-Verlag, 1983.

[MS89] D.A. Martin and John Steel. A proof of projective determinacy. *J. Amer. Math. Soc.*, 2(1):71–125, 1989.

[MS94] D.A. Martin and John Steel. Iteration trees. *J. Amer. Math. Soc.*, 7(1):1–73, 1994.

[MiSt94] W.J. Mitchell and John Steel. *Fine Structure and Iteration Trees*, volume 3 of *Lecture notes in Logic.* Springer, 1994.

[MWH] Adrian Mathias, W.H. Woodin, and Kai Hauser. *The axiom of determinacy.* Forthcoming.

[Nee95] Itay Neeman. Optimal proofs of determinacy. *Bull. Symbolic Logic*, 1(3):327–339, 1995.

[Nee96] Itay Neeman. *Determinacy and Iteration trees.* PhD thesis, UCLA, June 1996.

[Ste88] John Steel. Long games. In *Cabal seminar 1981-85*, number 1333 in Lecture notes in Mathematics, pages 56–97. Springer-Verlag, 1988.

[Ste93] John Steel. Inner models with many Woodin cardinals. *Ann. Pure. Appl. Logic*, 65(2):185–209, 1993.

[Ste95a] John Steel. Projectively well-ordered inner models. *Ann. Pure. Appl. Logic*, 74(1):77–104, 1995.

[Ste95b] John Steel. $\mathrm{HOD}^{L(\mathbb{R})}$ is a core model below θ. *Bull. Symbolic Logic*, 1(1):75–84, 1995.

[Woo] W.H. Woodin. Unpublished notes.

On the Complexity of the Propositional Calculus

Pavel Pudlák*
Mathematical Institute, Prague

Abstract

We show that research into the complexity of propositional proofs is related to various problems in other branches of mathematics. A nondeterministic procedure for a $co\mathcal{NP}$-complete can be viewed as a propositional proof system. We survey several proof systems which were proposed for problems in integer linear programming, algebra and graph theory. Then we consider a general method of proving lower bounds on the lengths of propositional proofs which is based on a property of some systems called *effective interpolation*. This means that the circuit complexity of interpolating boolean functions $f(x)$ of an implication $\phi(x, y) \rightarrow \psi(x, z)$ can be bounded by a polynomial in the length of a proof of the implication. We shall prove such a theorem for a system for integer linear programming proposed by Lovász and Schrijver [34] and show a relation between a problem on the complexity of linear programming and a problem of proving lower bounds on the lengths of proofs in this system. We shall consider the lengths of proofs in nonclassical logics, in particular monotone and intuitionistic logics, and, in a similar way, show a relation to the complexity of sorting networks.

1 Introduction

It is well-known that the theory of $\mathcal{NP}$-completeness and the problem $\mathcal{P} = \mathcal{NP}$? originated in the research of the complexity of propositional calculus. The computational content of the concept of a proof is a nondeterministic computation. Also, the satisfiability of a boolean formula has this nature.

*pudlak@math.cas.cz, this work was partially supported by grant A1019602 of AV ČR and by cooperative research grant INT-9600919/ME-103 from the NSF (USA) and the MŠMT (Czech republic)

Thus the question $co\mathcal{NP} = \mathcal{NP}$?, closely related to the famous problem $\mathcal{P} = \mathcal{NP}$?, can be stated as a problem on the lengths of proofs in propositional calculus. To be more precise, we shall talk about propositional *proof systems.* In the general definition of a proof system (equivalent to the one of [14]), a proof can be any object, we only require that the relation *"x is a proof of y"* in the proof system is checkable in polynomial time. Of course, we also assume that the provable propositions are just tautologies. In this definition we do not specify the class of formulas from which we take the tautologies. It is assumed that it is sufficiently universal in the sense that it is possible to code in this class all formulas in some complete basis of connectives. In this setting the problem $co\mathcal{NP} = \mathcal{NP}$? is equivalent to the question whether there is a polynomially bounded proof system, ie., a system in which every tautology has a proof of polynomial length.

The usual textbook proof systems are the so-called *Frege systems* or *sequent calculi.* One may naively expect that they are not particularly efficient and it would be easy to prove that they are not polynomially bounded, but the contrary is true. At present this problem seems hopelessly difficult. In order to get some experience with proving lower bounds on the lengths of proofs in propositional calculus we have to consider weaker proof systems. In practice there is a demand for solving even $\mathcal{NP}$-hard problems; in some cases a solution is considered satisfactory even if the algorithm is nondeterministic and gives short "proofs" for a large subset of instances. Such systems fall under the general concept of the propositional proof system, if we think of the objects as a suitable class of formulas. So a proof system is just a nondeterministic algorithm for a $co\mathcal{NP}$-complete problem. Several such procedures have been investigated mainly in combinatorial optimization, algebra and graph theory. We shall survey some systems in section 2.

The $co\mathcal{NP} = \mathcal{NP}$? and practical proof systems are not the only motivation for the research into the lengths of proofs in propositional calculus. It has been shown that superpolynomial lower bounds for certain proof systems can be used to prove the independence of true sentences from first order theories. In this way one could get explicit independent Π_1 sentences different from Gödel sentences. The theories for which these relations have been studied are weak subsystems of Peano arithmetic with the generic name *Bounded Arithmetic.* In some special cases independence results have been achieved. Theoretically this approach to independence results can work for any theory (containing a fragment of arithmetic such as Robinson's Q). However, for strong theories this leads to combinatorial problems of such complexity that there is little hope of making any progress in this way. We shall not survey this connection here and refer the reader to Krajíček's book [25].

Several methods of proving lower bounds on the lengths of proofs have been developed. In section 3 we shall explain the most appealing one, which is based on the well-known Craig interpolation theorem. The idea is to bound the complexity of the interpolant in terms of the complexity of the proof of the implication. In this way the lower bound problem on the length of proofs is reduced to a lower bound problem on the circuit size of particular circuits. Methods of complexity theory can then be applied directly. We shall prove such an interpolation theorem for one of the systems for integer linear programming.

Very little is known about the lengths of proofs in proof systems for non-classical logics. Some nonclassical calculi are important for Artificial Intelligence, thus the question of the efficiency is of a practical interest. Here we show in section 4 that some restricted systems for *classical* propositional logic can be viewed as parts of systems for some *nonclassical* logics. In particular the resolution system is a part of the monotone sequent calculus which in turn is included in the intuitionistic sequent calculus.

The interesting feature of the research of such "practical" proof systems is that it connects with important mathematical problems. In particular our Problems 1 to 4 are related to two important subjects in theoretical computer science: the complexity of linear programming and the complexity of sorting.

The reader interested in the subject treated here can learn more in the book [25] and surveys [8, 28, 30, 38].

2 Mathematical proof systems

In this section we shall review some systems from mathematical practice. This means that they were proposed to give an explicit or efficient solution of a mathematical problem, or are motivated by such a result. We start, however, with one which is rather logical.

2.1 Resolution System

This is the propositional part of Robinson's resolution procedure for first order logic. We can view it as a refutation system for sets of disjunctions of *literals,* which are propositional variables and their negations. We think of disjunctions of literals as *sets* of literals, hence we do not need any structural rules. The only rule is the *resolution rule,* or the *cut,*

$$\frac{\bigvee A, \quad \bigvee B}{\bigvee (A \cup B) \setminus \{p, \neg p\}},$$

provided that $p \in A$ and $\neg p \in B$. A set of disjunctions is refuted by deriving the empty disjunction. We can look at it also dually; a resolution proof is then a proof of a formula in DNF. The complexity of a proof is measured by the number of disjunctions in the proof. A more restricted version is when one thinks of proofs as trees and counts the number of nodes. Exponential lower bounds are known for the unrestricted resolution system [19].

2.2 Proof systems for integer linear programming

An integer linear programming problem is given by a set of linear inequalities with real coefficients. In order to be able to consider the complexity of this problem, we shall assume that the coefficients are rational. The problem is to find integers satisfying these inequalities. The problem is known to be $\mathcal{NP}$-complete. A restricted version, where we look only for 0–1 solutions, is equally hard. In this case we can think of the variables as being propositional variables and inequalities as being a special kind of formulas, *threshold formulas.* Several proof systems have been proposed for this problem. Though they can also be used to derive inequalities from a given set of inequalities, it is more convenient to treat them as refutation systems, in a similar manner as resolution. Thus a proof starts with a given set of inequalities, proceeds with deriving new ones and ends with deriving a contradiction which is a false inequality, such as $0 \geq 1$.

Since we consider only 0–1 solutions, we shall assume that for each propositional variable x, the two inequalities $x \leq 1$ and $x \geq 0$ are among the initial ones. All systems contain rules making linear combinations from derived inequalities. We shall also assume that $1 \geq 0$ is given for free. Recall that we allow only rational coefficients in the inequalities.

2.2.1 Cutting Planes

The additional rule in this system is the *rounding rule:*

$$\frac{a_1 x_1 + \cdots + a_n x_n + b \geq 0}{a_1 x_1 + \cdots + a_n x_n + \lfloor b \rfloor \geq 0}$$

provided that $a_1, \ldots, a_n$ are integers. The completeness was proved by Gomory [17]; for 0–1 solutions it can be proved by simulating resolution. Exponential lower bounds for this system have been proved [21, 37].

2.2.2 Lovász and Schrijver [34]

The idea here is to allow also quadratic inequalities. The new rule is

$$\frac{F \geq 0}{xF \geq 0, \quad (1-x)F \geq 0}$$

where F is a linear term and x is a variable. This rule is sound for $0 \leq x \leq 1$. Note that we can apply this rule only to linear inequalities, thus in order to be able to apply it again, we have to eliminate the quadratic terms using linear combinations of the quadratic inequalities. Furthermore one can use the inequality $x^2 - x \geq 0$, for every variable, which gives, using the rule, $x^2 - x = 0$. The completeness can again be proved by simulating resolution.

Let us prove the following lemma as an example of a proof in this calculus.

Lemma 1 *The inequality $x+y+z \leq 1$ is derivable from $x+y \leq 1$, $x+z \leq 1$ and $y+z \leq 1$.*

Proof. By multiplying $x+y \leq 1$ by x and using $x^2 = x$ we get $xy \leq 0$. In the same way we get also $xz \leq 0$ and $yz \leq 0$. Adding $x+y \leq 1$ and $y+z \leq 1$, we get $x+2y+z \leq 2$. Multiplying it by $1-x$ and using $x^2 = x$, we get $2x+2y+z-2xy-xz \leq 2$. Using $xy \leq 0$ and $xz \leq 0$ it reduces to $2x+2y+z \leq 2$. Multiplying by $1-z$ and using $z^2 = z$ we get $2x+2y+2z-2xz-2yz \leq 2$, which reduces to $2x+2y+2z \leq 2$. This gives the conclusion by dividing it by 2. □

The following proposition implies that the calculus is stronger than resolution, as the *Pigeon Hole Principle* tautologies PHP do not have subexponential proofs in resolution.

Proposition 1 *The PHP tautologies have polynomial size proofs in the Lovasz-Schrijver calculus.*

Proof. The n-th PHP tautology is stated as the unsatisfiability of the following system of equations:

$$\sum_j x_{ij} \geq 1, \quad \text{for } i = 1, \ldots, n+1,$$
$$x_{ij} + x_{i'j} \leq 1, \quad \text{for } i, i' = 1, \ldots, n+1,\ i \neq i' \text{ and } j = 1, \ldots n.$$

By adding the first set of inequalities, we get $\sum_{ij} x_{ij} \geq n+1$. To get a contradiction, we shall derive $\sum_{ij} x_{ij} \leq n$ from the second set of inequalities. For each j we shall show, gradually for all $r = 1, \ldots, n$, $\sum_{t=i}^{i+r} x_{t,j} \leq 1$, for all $i \leq n+1-r$. For $r = 1$, these are just inequalities contained in the initial ones. To go from r to $r+1$, take $\sum_{t=i}^{i+r} x_{t,j} \leq 1$ and $\sum_{t=i+1}^{i+r+1} x_{t,j} \leq 1$ and apply the lemma to $x = x_{i,j}$, $y = \sum_{t=i+1}^{i+r} x_{t,j}$ and $z = x_{i+r+1,j}$. This gives $\sum_{t=i}^{i+r+1} x_{t,j} \leq 1$. The number of inequalities that we have to consider in this proof is, clearly, polynomial. □

Furthermore Lovász and Schrijver considered extensions of this system obtained by adding the *multiplication rule* (multiplying two linear inequalities) and the axiom schema $F^2 \geq 0$ for every F linear.

2.2.3 Chvátal [unpublished]

This system resembles natural deduction, as one works with assumptions. In a proof we can pick an arbitrary linear inequality $F \geq a$ with *integer coefficients* and we branch the proof according to the two possibilities $F \geq a$ and $F \leq a-1$ (since we consider only integral solutions, we do not have to consider the values between $a-1$ and a). Such assumptions can be nested arbitrarily many times. Otherwise we use only linear combinations. Formally we can represent such a proof as a rooted binary tree, where the edges are labelled by linear inequalities in such a way that for each node the two edges have labels $F \geq a$ and $F \leq a-1$. At each leaf there is a linear combination of the initial inequalities and those on the branch which leads to that leaf which produces a contradiction. In passing, note that the contradictory linear combinations can be found using polynomial time algorithms for linear programming, thus the proof is determined by the branching inequalities. It is an easy exercise to simulate Cutting Planes in this system. The relation to the previous system is not known.

Chvátal proposed also an apparently stronger version where the proofs have the structure of general DAGs. An exponential lower bound has been obtained for this system with the restriction that the coefficients are integers bounded by a polynomial in the input length [29].

2.3 Ideal generation systems

Let us now consider general algebraic equations

$$f_1 = 0, \ldots, f_n = 0$$

over some field. Again, the existence of a solution in the field is $\mathcal{NP}$-hard (and $\mathcal{NP}$-complete, if the field is finite). A well-known general criterion on unsolvability is the Hilbert Nullstellensatz. It says that the system is unsolvable in the algebraic closure of the field iff 1 is in the ideal generated by the polynomials of the equations. This leads to the following two systems. In both systems we assume that the set of equations includes $x^2 - x = 0$ for every variable, which ensures that solutions can be only 0–1. (In this case the Nullstellensatz has an easy proof by induction on the number of variables.)

2.3.1 Nullstellensatz proof system

A proof in this system is a sequence of polynomials $g_1, \ldots, g_n$ such that

$$g_1 f_1 + \cdots + g_n f_n = 1$$

(in the ring of polynomials). A natural measure of complexity of such a proof is the sum of the numbers of monomials in the polynomials $g_1, \ldots, g_n$, as this is roughly the length of a sequence encoding them. Another natural measure is the maximal degree of the polynomials $g_1, \ldots, g_n$. Clearly, if the degree is bounded by a constant, then the number of monomials is bounded by a polynomial. The converse is, of course, not true, but at least for some nice sets of polynomials $f_1, \ldots, f_n$, one can obtain a constant degree proof from a polynomial size proof by applying a random restriction. The restricted proof uses a "similar" set of polynomials. An example of such a set of polynomials are polynomials expressing the pigeon hole principle.

In general (when the equations $x^2 = x$ are not present) the minimal degree of the polynomials $g_1, \ldots, g_n$ can be exponential. Here, it can be at most the number of variables plus 1, as the equations $x^2 - x = 0$ ensure that multilinear polynomials are sufficient.

This system was investigated in connection with another proof system [2, 5] (the so called *bounded depth Frege system*). An exponential lower bound on the lengths of proofs of some particular tautologies in that system was shown by reducing the question to a nonconstant lower bound on the degree of Nullstellensatz proofs.

An interesting feature of this system is that for a constant degree the proof search is easy; namely, a proof can be found in polynomial time by solving linear equations for the coefficients of the polynomials $g_1, \ldots, g_n$.

2.3.2 Polynomial calculus (also: Groebner proof system)[9]

Here we generate members of the ideal sequentially, using the basic closure conditions. Thus a proof is a sequence of polynomials $h_1, \ldots, h_m$ where each polynomial either belongs to the initial ones, or is obtained by the addition of two previously derived polynomials, or by multiplying a previously derived polynomial by an arbitrary one. Again, the most interesting parameter is the maximal degree of the polynomials $h_1, \ldots, h_m$. Proving lower bounds on the degree is harder than for the previous system; the first nonconstant lower bounds were found only recently ([24] for the field of reals, [43] for all fields, [27] for finite fields).

The proof search problem is in $\mathcal{P}$ also for polynomial proofs of constant degree. This follows from the Groebner basis algorithm [9].

2.3.3 Ideals in graph theory

Given a graph G on $\{1, \ldots, n\}$, we associate to it the polynomial

$$f_G = \prod_{i<j \text{ adjacent}} (x_i - x_j).$$

Some monotone families of graphs can be determined by ideals. E.g. the independence number of a graph is less than k iff f_G belongs to the ideal

$I(k,n)$ of the polynomials which vanish whenever k variables are set equal. In some cases it is possible to find a set of generators such that the membership in this set can be decided in polynomial time. In the example above such a set of generators consists of f_H where H ranges over graphs which are unions of k vertex disjoint complete graphs [33]. Hence, for a graph G with independence number $\leq k$, the polynomial f_G can be expressed as

$$f_G = \sum_H g_H f_H,$$

where H are unions of k vertex disjoint complete graphs and g_H are some polynomials. As in the Nullstellensatz proof system, we can think of this expression as a proof (of G having the independence number $\leq K$). The natural measures of complexity are the size of this expression and the number of generators used. We do not know about any bounds on the complexity of such proofs.

2.4 Hajós calculus

This is a calculus for proving that a given graph G is not three-colorable. The initial axiom is the graph K_4, the complete graph on four vertices. The rules are: *edge/vertex introduction,* which allows to add vertices and edges arbitrarily, *contraction,* which allows to contract two nonadjacent vertices into one, and the following *join rule.* Suppose two constructed graphs share a vertex a and suppose (a,b) is an edge in one graph and (a,c) is in the other one. Then we deduce that the graph obtained from them by deleting $(a,b),(a,c)$ and adding (b,c) is also not three-colorable. Hájos introduced this system and proved its completeness [18].

The problem of proving lower bounds seems very hard, as this system simulates even *extended Frege systems* [36]. Surprisingly, for the version with proof trees an exponential lower bound was found [32].[1]

3 Lower bounds based on effective interpolation

Craig's interpolation theorem, restricted to propositional calculus, states that for a true implication $\phi(\bar{u},\bar{x}) \rightarrow \psi(\bar{u},\bar{y})$ where $\bar{u},\bar{x},\bar{y}$ are disjoint sets of variables, there exists a formula $\chi(\bar{u})$ containing only the common variables

[1] In the case of the extended Frege system the versions with trees and DAGs are equivalent, so this shows that the tree structure is not preserved in the simulation of the Hájos calculus.

of the antecedent and succedent such that both $\phi(\bar{u},\bar{x}) \to \chi(\bar{u})$ and $\chi(\bar{u}) \to \psi(\bar{u},\bar{y})$ are true [15, 16]. Let

$$X = \{\bar{a};\ \exists \bar{b}\ \phi(\bar{a},\bar{b})\}, \quad Y = \{\bar{a};\ \exists \bar{c}\ \neg\psi(\bar{a},\bar{c})\}, \quad Z = \{\bar{a};\ \chi(\bar{a})\}.$$

Then the validity of the implication $\phi(\bar{u},\bar{x}) \to \psi(\bar{u},\bar{y})$ means that $X \cap Y = \emptyset$. The interpolation property of $\chi(\bar{u})$ means that $X \subseteq Z$ and $Z \cap Y = \emptyset$, ie., Z separates the sets X and Y. It was first observed by Mundici [35] that the question of how small interpolants χ can one find for given ϕ and ψ is essentially the question, how difficult is to separate two disjoint $\mathcal{NP}$ sets. It is a generally accepted conjecture, very important for cryptography, that the separating set cannot be in $\mathcal{P}$ for some $\mathcal{NP}$ disjoint pairs. This conjecture implies that for the corresponding implication there is no polynomial size interpolant formula.

The idea of proving a lower bound on the lengths of proofs using interpolation, as proposed by Krajíček [26], is to show a bound on the complexity of the interpolant not in terms of the complexity of ϕ and ψ, but in terms of the complexity of the *proof* of $\phi(\bar{u},\bar{x}) \to \psi(\bar{u},\bar{y})$. Then, if we succeed in showing that for some implication of this form the interpolant cannot be simple, we can conclude that the proof of $\phi(\bar{u},\bar{x}) \to \psi(\bar{u},\bar{y})$ cannot be simple either. Such a relation between the size of a proof of $\phi(\bar{u},\bar{x}) \to \psi(\bar{u},\bar{y})$ and the complexity of an interpolant is known for some systems, eg. for Resolution and Cutting Planes. Namely, the set Z can be computed by a boolean circuit whose size is polynomial in the size of the proof. This property of a proof system is called *effective interpolation.* An effective interpolation does not suffice for proving superpolynomial lower bounds on the lengths of proofs. We can derive such bounds only using some conjectures from cryptography. These conjectures are quite strong, thus one may wonder what we gain by proving an effective interpolation theorem. Isn't it better to use just the conjecture $\mathcal{NP} \neq co\mathcal{NP}$? But we do get more. Firstly, $\mathcal{NP} \neq co\mathcal{NP}$ is not known to be comparable with such conjectures, however strong they seem. Secondly, using effective interpolation and an assumption that a particular function is one way, we get explicit tautologies which should not have short proofs. Furthermore, having effective interpolation may be the first step to getting an unconditional lower bound, as was the case eg. for cutting planes.

3.1 Interpolation for the calculus of Lovász and Schrijver

Here we prove such effective interpolation for the calculus of Lovász and Schrijver described above. Our proof actually gives a stronger result – effective interpolation for the union of Cutting Planes and the Lovász – Schrijver calculus with the general multiplication rule. As we have to use the rounding rule anyway, we shall consider this stronger calculus. (Actually, we do not know if the following lemma holds for the Lovász – Schrijver calculus alone.)

Lemma 2 *Let $\bar{x}$ and $\bar{y}$ be two disjoint sets of variables and $E_{\bar{x}}$ and $E_{\bar{y}}$ sets of linear inequalities containing only the corresponding variables. If we have a proof of a contradiction from the union of the two sets, then we can construct in polynomial time a proof of contradiction from one of the sets.*

Proof. We shall use an idea of some previous proofs of effective interpolation [37]: we will split the proof into two proofs each using only one kind of variables, either $\bar{x}$, or $\bar{y}$. An inequality in the original proof is the sum of the corresponding pair of the new inequalities except that the constant term may be larger in the original proof than the sum of the constant terms in the new inequalities. Hence the new inequalities are at least as strong as the original one. Thus at least one of the proofs must contain a contradiction.

First we redefine the concept of a proof. A line in a proof will be only a *linear* inequality. To get a new linear inequality we may use a more complicated construction, namely, take some positive linear combinations of the given inequalities, multiply some of them by variables and take their sum. This is allowed only if all quadratic terms cancel in the sum. In other words, we do not consider the intermediate quadratic inequalities as separate steps. Furthermore we shall use only inequalities with integer coefficients. Thus the rounding rule will be interpreted as taking a linear combination and rounding in one step.

An inequality $F_1(\bar{x}) + F_2(\bar{y}) + c \geq 0$, with F_1, F_2 linear forms and constant c, of the original proof will correspond to inequalities $F_1(\bar{x}) + c_1 \geq 0$ and $F_2(\bar{y}) + c_2 \geq 0$ in the two new proofs. We shall prove by induction that $c_1 + c_2 \leq c$, so the two inequalities are at least as strong as the original one.

It is trivial to show that the rounding rule, if done in parallel in the two proofs, preserves this property. Thus it remains to show this for the multiplication rule of the Lovász – Schrijver calculus. An inequality derived by this rule has the following general form

$$L^{(1)}(\bar{x}) + L^{(2)}(\bar{y}) +$$

$$\sum_i a_i(x_i^2 - x_i) + \sum_k L_k^{(3)}(\bar{x}) L_k^{(4)}(\bar{x}) +$$

$$\sum_j b_j(y_j^2 - y_j) + \sum_l L_l^{(5)}(\bar{y}) L_l^{(6)}(\bar{y}) +$$

$$\sum_m L_m^{(7)}(\bar{x}) L_m^{(8)}(\bar{y}) \geq 0$$

where $L^{(1)}(\bar{x}), L^{(2)}(\bar{y}), L_k^{(3)}(\bar{x}) \geq 0, \ldots, L_m^{(8)}(\bar{y}) \geq 0$ are some linear inequalities derived before. Let us note that in each of the following expressions

$$L^{(1)}(\bar{x}), \quad L^{(2)}(\bar{y}),$$

$$\sum_i a_i(x_i^2 - x_i) + \sum_k L_k^{(3)}(\bar{x})L_k^{(4)}(\bar{x}),$$

$$\sum_j b_j(y_j^2 - y_j) + \sum_l L_l^{(5)}(\bar{y})L_l^{(6)}(\bar{y}),$$

$$\sum_m L_m^{(7)}(\bar{x})L_m^{(8)}(\bar{y}),$$

all quadratic terms cancel. So we can assume w.l.o.g. that we actually first derive the inequalities with these linear terms and then we sum them. The first four are all right as they use only one type of variable, so we only need to consider the last one. Let us write it as $F_1(\bar{x}) + F_2(\bar{y}) + c \geq 0$, with F_1, F_2 linear forms and a constant c. The crucial observation is that this inequality follows from $L_m^{(7)}(\bar{x}) \geq 0$, $L_m^{(8)}(\bar{y}) \geq 0$ without using the assumption that x_i and y_j are integers. (Recall that $x_i^2 - x_i \geq 0$ and $y_j^2 - y_j \geq 0$ are the only axioms which express that fact and we do not use them here.) Hence $F_1(\bar{x}) + F_2(\bar{y}) + c \geq 0$ must be a linear combination of these inequalities. Taking separately the parts of this linear combination with $\bar{x}$ resp. $\bar{y}$ we get inequalities $F_1(\bar{x})+c_1 \geq 0$ and $F_2(\bar{y})+c_2 \geq 0$ with $c_1+c_2 = c$ as consequences of the previous inequalities. The constants c_1, c_2 need not be integers, so we round them down.

Now, we have to find the constants c_1, c_2 in polynomial time. This is done using a polynomial time algorithm for linear programming. To find $-c_k$, for $k = 1, 2$, we maximize F_k using the inequalities above as constraints.

In the two new proofs the coefficients of the variables come from the original proof, thus we only have to watch if the constant terms are not too small (they are always less than or equal to the original ones). But if the constant is less than the negative of the sum of the coefficients of the variables, the inequality cannot be satisfied and we can stop. □

Theorem 2 *The calculus of Lovász and Schrijver combined with Cutting Planes has effective interpolation.*

Proof. Suppose we have a proof of contradiction from the union of the two sets of linear inequalities $E_{\bar{u},\bar{x}}$ and $E_{\bar{u},\bar{y}}$, with $\bar{x}$ and $\bar{y}$ two disjoint sets of variables. We want to construct a polynomial size circuit which for a given input $\bar{a}$ tells which of the two sets $E_{\bar{a},\bar{x}}$, $E_{\bar{a},\bar{y}}$ is inconsistent. It is enough to have a polynomial algorithm, as it can be turned into a polynomial size circuit using standard methods. Such an algorithm was shown in the previous lemma. □

We do not know if effective interpolation holds for stronger versions of the calculus, in particular, if it holds when inequalities $F^2 \geq 0$ are added for all linear functions F.

An interesting observation, due to J. Sgall, is that *interpolating functions for this system is as hard as solving an instance of linear programming* (the one with arbitrary rationals). Namely, take $\sum_i -a_{ij}x_i + d_j \geq 0$, $\sum_i c_i x_i \geq u$ and $\sum_j a_{ij}y_j + c_i \geq 0$, $\sum_j d_j y_j \geq -u + \varepsilon$ for a small $\epsilon > 0$. Here u is the single common variable. To determine for which u which part is contradictory, means to locate the minimax e with ε precision. (Here we take u to be an arbitrary real, but instead we can take several 0–1 variables multiplied by powers of 2 to approximate it.) Similar characterizations of the complexity of the interpolation are known for other systems: interpolants for Resolution are characterized by boolean circuits [26], interpolants for Nullstellensatz by span programs [39].

3.2 Monotone interpolation

As we have observed above, an effective interpolation does not suffice to prove a superpolynomial lower bound on the lengths of proofs. However, in some cases (Resolution, Cutting Planes, Nullstellensatz) it is possible to get a reduction to some monotone circuits for which it is possible to prove exponential lower bounds. In particular for Cutting Planes, assuming that the coefficients with the variables $\bar{x}$ are all positive, an interpolant can be computed by a *monotone real circuit.* Such circuits are generalizations of monotone boolean circuits in the sense that they are allowed to use arbitrary reals in computations and the gates are arbitrary binary nondecreasing real functions. The lower bound techniques for monotone boolean circuits can be easily extended to this kind of circuits [21, 37].

In order to prove a lower bound by a reduction to monotone real circuits for the calculus of Lovász and Schrijver, we need to show that a suitably presented version of linear programming can be computed by monotone real circuits, namely:

Problem 1 *Let a matrix $\{a_{ij}\}$ and a vector $\vec{d}$ of rational numbers be fixed. Does there exist a polynomial size monotone real circuit which for given input* *vector $\vec{c}$ of rational numbers computes*

$$\max\{\sum_j d_j y_j;\ \forall i\ \sum_j a_{ij} y_j \le c_i\}\ ?$$

Note that the maximum as a function of the variables c_i is nondecreasing, since by increasing a c_i we increase the set over which we maximize. This problem is interesting independently of its application to the bounds on the length of proofs. A negative answer would show that all polynomial algorithms for linear programming must be, in some sense, indirect. Namely, any polynomial size arithmetic circuit for linear programming would have to compute some values which do not depend on the input in a monotone way. For the negative solution to this problem, it suffices to find a counterexample to the monotone version of the interpolation theorem above, namely a short proof of a "monotone" tautology for which we know that it does not have an interpolant computable by small monotone real circuits. We know, for instance, that there is no subexponential size real monotone circuit which separates graphs with m cliques from $m-1$-colorable graphs [37]. Thus, if one can solve negatively the above problem by finding subexponential size proofs of the following tautology expressing that for every graph (coded by the variables p_{ij}) either it does not have a clique of size m (coded by the variables q_{ki}), or it is not $m-1$-colorable (the coloring coded by the variables r_{il}).

$$\sum_i q_{ki} \ge 1, \qquad \text{for } k = 1, \ldots, m;$$

$$\sum_k q_{ki} \le 1, \qquad \text{for } i = 1, \ldots, n;$$

$$\sum_l r_{il} \ge 1, \qquad \text{for } i = 1, \ldots, n;$$

$$q_{ki} + q_{k'j} - p_{ij} \le 1 \qquad \text{for } 1 \le i < j \le n,\ 1 \le k, k' \le m,\ k \ne k';$$

$$p_{ij} + r_{il} + r_{jl} \le 2, \qquad \text{for } 1 \le i < j \le n, l = 1, \ldots, m-1.$$

Let us note that these are the tautologies for which exponential lower bounds were shown in [37], so it would also show that Cutting Planes are weaker than Lovász and Schrijver. By adding corresponding inequalities in the last two sets we can eliminate the variables p_{ij} and get a stronger system where the last two sets are replaced by

$$q_{ki} + q_{k'i} + r_{il} + r_{jl} \leq 3$$
for $1 \leq i < j \leq n,\ 1 \leq k, k' \leq m,\ k \neq k',\ 1 \leq i < j \leq n,\ l = 1, \ldots, m-1.$

These equations express that there are two mappings, one defined on an m element set, the other having domain an $m-1$ element set, such that the composition of the two is a one-to-one mapping. The unsatisfiability of this set is a sort of a stronger pigeon hole principle.

3.3 Existential interpolation

It has been shown that effective interpolation is unlikely for some strong systems [31]. Therefore one has to look for other methods or one has to generalize the concept of effective interpolation. There is one particularly natural generalization, called *existential interpolation.* We say that a proof system P has this property, if there exists a polynomial p such that the following holds. If $\phi(\bar{x}) \vee \psi(\bar{y})$, with $\bar{x}$ and $\bar{y}$ disjoint sets of variables, has a proof of length n in P, then either $\phi(\bar{x})$ or $\psi(\bar{y})$ has a proof of length at most $p(n)$ in P. (In a more general version we allow the latter proof to be in some stronger proof system P'.) All the systems for which we know that effective interpolation holds, also satisfy this condition. On the other hand, there is a proof system for which only existential interpolation has been shown. This is the system of Chvátal described above (with proofs as trees), see [22]. However, for this system a slightly less effective interpolation theorem has been shown where the circuits are of size $n^{O(\log n)}$ [22] and effective interpolation was proved for the case of polynomially bounded integer coefficients [29] (the last result is actually stronger: it works for proofs as DAGs and there is a monotone version of it).

For using existential interpolation to prove superpolynomial lower bounds we need two disjoint $\mathcal{NP}$ sets A, B which cannot be separated using a pair of disjoint $co\mathcal{NP}$ sets A', B' in such a way that $A \subseteq A'$ and $B \subseteq B'$. Let us note that the pairs of disjoint $\mathcal{NP}$ sets A, B which one gets from the functions used in cryptography do not satisfy this condition, since they are actually complements.

4 The lengths of proofs in nonclassical logics

In artificial intelligence we need to automate logical deduction and quite often nonclassical logics seem the better alternative. Hence one should study the lengths of proofs also for nonclassical calculi. But even if we stick to classical logic (say, we are interested in the complexity theory and aim at $co\mathcal{NP} \neq$

$\mathcal{NP}$), proof systems for some nonclassical logics may help. The aim of this section is to show that.

The basic observation is: *Propositional Resolution system is the sequent system restricted to propositional variables.* In more detail, we represent the disjunction $\neg x_1 \vee \ldots \vee \neg x_m \vee y_1 \vee \ldots \vee y_n$ by the sequent $x_1, \ldots, x_m \longrightarrow y_1, \ldots, y_m$. If we use only propositional variables as formulas, the only rules that can be used in the classical sequent calculus LK are structural rules and *cut.* The structural rules correspond to the fact that in resolution we think of disjunctions as sets, the cut is the resolution rule.

4.1 Monotone sequent calculus

Now we can embed Resolution in other calculi by relaxing the condition on the formulas used in proofs. One such calculus that is natural but has not been much investigated is the *monotone sequent calculus.* This is LK restricted to formulas in the monotone basis $\wedge, \vee$. The only monotone tautological *formula* is the constant for truth $\top$, thus we are interested rather in proving tautological *sequents.* The interpretation of monotone sequents is an implication $\phi \rightarrow \psi$ with ϕ and ψ monotone formulas.

While for monotone boolean circuits exponential lower bounds were proved long ago, we do not have any nontrivial lower bounds on the lengths of proofs in the monotone sequent calculus.

There is a trivial and useless interpolation theorem here; namely, for a monotone sequent $\Phi(\bar{u}, \bar{x}) \longrightarrow \Psi(\bar{u}, \bar{y})$, we have interpolants $\Phi(\bar{u}, \top, \ldots, \top)$ and $\Psi(\bar{u}, \bot, \ldots, \bot)$. Note that this is different from the translation of the interpolation theorem for Resolution. The form we get in this way is the following. Two *sets* of sequents are given, one in variables $\bar{u}, \bar{x}$ and the other in variables $\bar{u}, \bar{y}$. Furthermore we have a proof of contradiction from the union of the two sets. In the case of Resolution, formulas occurring in the sequents are only propositional variables. Then, there exists a polynomial size circuit which for a given assignment $\bar{a}$ for the variables $\bar{u}$ decides which of the sets is contradictory. It is rather unlikely that this generalizes to sets of sequents with arbitrary monotone formulas, as it would imply effective interpolation for classical (ie., with negations) Frege systems.

In several cases the first sequence of tautologies for which an exponential lower bound was shown was the Pigeon Hole Principle (PHP). Also here it is important as several other tautologies (eg. those used for exponential lower bounds for cutting planes) can be reduced to PHP. Note that PHP is a monotone tautology ("every pigeon sits in a hole $\longrightarrow$ there is a hole with two pigeons"). We do not know, if PHP has polynomial proofs in the monotone sequent calculus. This is, a bit surprisingly, related to the question on the complexity of sorting networks. The usual proofs of PHP go by induction counting the pigeons assigned to a segment of holes. To construct such a proof in monotone calculus we need to formalize the concept of counting. There are constructions of polynomial size monotone counting formulas. The first is a probabilistic construction of Valiant [44], the second follows from the famous sorting network of Ajtai, Komlós and Szemeredi [1]. However, having small monotone counting formulas does not suffice for simulating the inductive proof of the PHP; we need also short proofs of the basic properties of the counting formulas. There is little hope of getting it for the first kind of monotone formulas, as they are not explicit, their existence is shown by a probabilistic argument. In the case of the formulas obtained from the sorting network the problem is that the proof of the properties of the circuit is highly nontrivial and, moreover, relies on the construction of expander graphs. All the proofs of the expanding properties of the known explicit constructions of expanders are nonelementary. Thus the following seems to be a difficult problem:

Problem 2 *Are there polynomial size monotone counting formulas with polynomial size proofs of the basic properties?*

The counting formulas are formulas which are true iff exactly k variables are true; k ranges from 1 to the number of variables; the basic properties express that if exactly k variables are true and we flip a false variable, we get $k + 1$ true variables. A negative solution to the problem would be an explanation why there are no "simple" monotone counting formulas, hence no simple sorting networks of logarithmic depth. Let us also state the related problem on PHP explicitly.

Problem 3 *Does PHP have proofs of polynomial length in the monotone sequent calculus?*

4.2 Intuitionistic propositional calculus

This is the most famous among nonclassical calculi. We arrive at it by observing that monotone sequents which are classical tautologies are also intuitionistic tautologies. Strictly speaking, the monotone sequent calculus is not a subsystem of the intuitionistic sequent system, as in the intuitionistic sequent calculus only sequents with the empty or one-element consequents are allowed, but there is a simple simulation. Consequents with more than one formulas are simulated by disjunctions of the formulas.

No superpolynomial lower bounds are known on the lengths of intuitionistic sequent proofs. This problem is not necessarily harder than the same for the monotone sequent calculus, since the language is richer. It seems to be easier than proving superpolynomial lower bounds for the classical sequent calculus, since the set of intuitionistically valid tautologies is $\mathcal{PSPACE}$-complete.

We do not have a proof of a superpolynomial lower bound for proofs in the intuitionistic propositional calculus, but we can at least prove an effective interpolation theorem. This follows from a recent result of Buss and Mints, which answers a problem posed in an earlier version of this paper.

Theorem 3 ([6]) *There is a polynomial time algorithm which for every proof of $\longrightarrow \phi \vee \psi$ in the intuitionistic propositional sequent calculus constructs a proof, in the same calculus, of $\longrightarrow \phi$ or $\longrightarrow \psi$?*

Note that we do not assume that the formulas have disjoint sets of variables here. This theorem is a computationally efficient version of the well-known theorem on intuitionistically derivable disjunctions in the propositional calculus.

Corollary 4 *Let a proof D of $\chi(\bar{u}) \longrightarrow \phi(\bar{u},\bar{x}) \vee \psi(\bar{u},\bar{y})$ in the intuitionistic propositional calculus be given. Then it is possible to construct a circuit C whose size is polynomial in the size of D and which, for every vector $\bar{a}$ of 0's and 1's which satisfies $\chi(\bar{a})$, tells one of the formulas $\phi(\bar{a},\bar{x})$, $\psi(\bar{a},\bar{y})$ which is a tautology.*

Proof. Let $\bar{a}$ be given. Think of $\bar{a}$ as a vector of symbols for the truth and the falsehood and substitute it for the variables $\bar{u}$ in the proof D. If $\chi(\bar{a})$ is true, then it is derivable by a proof whose size is polynomial in the size of χ since the formula does not contain variables. Thus we get a proof of $\longrightarrow \phi(\bar{a},\bar{x}) \vee \psi(\bar{a},\bar{y})$. Then we apply the algorithm of the theorem above to get a proof of $\phi(\bar{a},\bar{x})$ or $\psi(\bar{a},\bar{y})$. □

Corollary 5 *Suppose two disjoint sets $X, Y \in \mathcal{NP}$ are given such that X and Y cannot be separated by a set in $\mathcal{P}/poly$. Then it is possible to construct a sequence of propositional intuitionistic tautologies which do not have polynomial size proofs.*

Proof. Let X and Y be given. Furthermore, let a natural number n be given. The disjoint sets $X \cap \{0,1\}^n$ and $Y \cap \{0,1\}^n$ can be defined by $\{\bar{a};\ \exists \bar{b}\ \neg\phi(\bar{a},\bar{b})\}$ and $\{\bar{a};\ \exists \bar{c}\ \neg\psi(\bar{a},\bar{c})\}$ with ψ, ϕ propositional formulas of polynomial size in n. Hence $\phi(\bar{u},\bar{x}) \vee \psi(\bar{u},\bar{y})$ is a classical tautology. Let $\chi(\bar{u})$ be $(u_1 \vee \neg u_1) \wedge \ldots \wedge (u_n \vee \neg u_n)$. Consider the following formula.

$$\chi(\bar{u}) \longrightarrow \neg\neg\phi(\bar{u},\bar{x}) \vee \neg\neg\psi(\bar{u},\bar{y}) \tag{1}$$

If it had a polynomial size proof, then, by the previous corollary, we could construct a polynomial size circuit which separates the sets $X \cap \{0,1\}^n$ and $Y \cap \{0,1\}^n$. That would be a contradiction with the assumption about the sets X and Y. Hence it remains to show that the formula (1) is an intuitionistic tautology. The reason is that we have *excluded middle* for the common variables, thus we can prove it "by cases". Each case reduces to a classical formula.

Arguing more precisely, first we observe that $\chi(\bar{u})$ is equivalent to $\bigvee_{\bar{a}}(u_1^{a_1} \wedge \ldots \wedge u_n^{a_n})$, where the superscript indicates whether we take the variable or its negation. Thus we only need to show that, for every $\bar{a}$, either

$$u_1^{a_1} \wedge \ldots \wedge u_n^{a_n} \longrightarrow \neg\neg\phi(\bar{u}, \bar{x}) \tag{2}$$

or

$$u_1^{a_1} \wedge \ldots \wedge u_n^{a_n} \longrightarrow \neg\neg\psi(\bar{u}, \bar{y}) \tag{3}$$

is intuitionistically valid. To this end we first prove by induction on subformulas α of ϕ and ψ that

$$u_1^{a_1} \wedge \ldots \wedge u_n^{a_n} \longrightarrow \alpha(\bar{u}) \equiv \alpha(\bar{a}) \tag{4}$$

Since either $\phi(\bar{a}, \bar{x})$ or $\psi(\bar{a}, \bar{y})$ is a classical tautology, we have that either $\neg\neg\phi(\bar{a}, \bar{x})$ or $\neg\neg\psi(\bar{a}, \bar{y})$ is an intuitionistic tautology. Thus (4) gives (2) or (3), hence (1) is an intuitionistic tautology. □

In order to get an unconditional lower bound, one may try to prove a monotone version of effective interpolation for the intuitionistic calculus. This should give a monotone circuit C in Corollary 4 for suitable formulas ϕ, ψ, χ. Thus we might get an exponential lower bound on the lengths of proofs of the tautologies on cliques and colorings of graphs, as we did in case of Resolution and Cutting Plane proofs.

The clique-coloring tautology says that it is not possible to have two mappings, one is from an m element set in the graph such that the image of the m element set is a clique, the other is a mapping from the graph such that no two vertices connected by an edge are mapped to the same element. Hence the composition of the two alleged mappings violates the PHP. Though the composition is definable in the intuitionistic calculus by a polynomial size formula and the properties of the two mappings ensure that it is one-to-one, it seems that the clique-coloring tautologies, in the form required by the interpolation theorem above, are not derivable from PHP tautologies using polynomial size intuitionistic proofs. If that was possible we could reduce proving a lower bound on PHP in the monotone sequent calculus to a lower bound on the clique-coloring tautologies in the intuitionistic propositional calculus.

Problem 4 *Does PHP have proofs of polynomial size in the intuitionistic sequent calculus?*

It is well-known that PHP has polynomial size proofs in the classical sequent calculus [7].

There are many more logics that are interesting from the point of view of the complexity of proofs. For instance, there is a calculus called *Positive Logic*, which is between monotone and intuitionistic logic. It uses the connectives $\wedge, \vee, \rightarrow$, thus it generalizes the monotone calculus by allowing more than one implication in a formula; see [40] for more information.

Acknowledgment

This paper resulted to a large extent from discussions with Vašek Chvátal, Jan Krajíček, László Lovász, Jiří Sgall and Gaisi Takeuti, which I appreciate very much.

References

[1] M. Ajtai, J. Komlós and E. Szemerédi, *An $O(n \log n)$ sorting network,* Proc. 15-th ACM Symp. on Theory of Computing, 1983

[2] P. Beame, R. Impagliazzo, J. Krajíček, T. Pitassi and P. Pudlák, *Lower bounds on Hilbert's Nullstellensatz and propositional proofs,* Proc. London Math. Soc. (3) 73, 1996, 1-26

[3] Bonet, M., Pitassi, T. and Raz, R., *Lower bounds for Cutting Planes proofs with small coefficients,* Proc. 27-th STOC, 1995, 575-584.

[4] Bonet, M., Pitassi, T. and Raz, R., *No feasible interpolation for TC^0-Frege proofs,* Proc. 38-th FOCS, 1997, 254-263

[5] S. Buss, R. Impagliazzo, J.Krajíček, P. Pudlák, A.A. Razborov and J. Sgall, *Proof complexity in algebraic systems and constant depth Frege systems with modular counting,* Computational Complexity 6, (1996/1997), 256-298

[6] S.R. Buss, G. Mints, *The complexity of the disjunction end existence properties in intuitionistic logic*, preprint, 1997

[7] S.R. Buss, *The propositional pigeonhole principle has polynomial size Frege proofs,* JSL 52, 916-27

[8] S. R. Buss, *Propositional proof complexity*, preprint, 1997

[9] M. Clegg, J. Edmonds, R. Impagliazzo, *Using the Groebner basis algorithm to find proofs of unsatisfiability,* Proc. 28-th ACM STOC, 1996, 174-183

[10] V. Chvátal, *Edmonds polytopes and a hierarchy of combinatorial problems,* Discrete Math., 4 (1973), 305-337.

[11] V. Chvátal, *Some linear programming aspects of combinatorics,* in Proc. of the Conf. on Algebraic Aspects of Combinatorics, Toronto 1975, Utilitas Math. Publishing Inc., Winnipeg, 2-30.

[12] V. Chvátal, W. Cook, M. Hartmann, *On cutting plane proofs in combinatorial optimization,* Linear Algebra and Its Applications 114/115 (1989), 455-499.

[13] W. Cook, C.R. Coullard, and Gy. Turán, *On the complexity of Cutting Plane proofs*, Discrete Applied Mathematics 18, (1987), 25-38

[14] S.A.Cook and R.A. Reckhow, *The relative efficiency of propositional proof systems,* JSL 44, 1987, 25-38

[15] W. Craig, *Linear reasoning: A new form of the Herbrand-Gentzen theorem,* Journ. of Symb. Logic 22(3), (1957), 250-287

[16] W. Craig, *Three uses of the Herbrand-Gentzen theorem in relating model theory and proof theory,* Journ. of Symb. Logic 22(3), (1957), 269-285

[17] R.E. Gomory, *An algorithm for integer solutions of linear programs,* in Recent Advances in Mathematical Programming, eds. R.L. Graves and P. Wolfe, McGraw-Hill, 1963, 269-302

[18] G. Hajós, *Über eine Konstruktion nicht n-färbener Graphen,* Wiss. Zeitschr. M. Luther Univ. Halle-Wittenberg, A 10, 1961, 116-117

[19] A. Haken, *The intractability of resolution,* Theoretical Computer Science, 39, (1985), 297-308

[20] A. Haken, *Counting bottlenecks to show monotone $P \neq NP$,* Proc. 36-th Annual IEEE Symp. on Foundations of Computer Science, Milwaukee, 1995, 36-40

[21] A. Haken and S.A.Cook, *An exponential lower bound for the size of monotone real circuits*, J. of Computer and System Sciences, to appear

[22] R. Impagliazzo, T. Pitassi, *Interpolation for generalized cutting planes system,* preprint, 1996

[23] R. Impagliazzo, T. Pitassi, A. Urquhart, *Upper and lower bounds for tree-like cutting planes proofs,* Proc. 9-th Annual IEEE Symp. on Logic in Computer Science, 1994, 220-228

[24] R. Impagliazzo, P. Pudlák and J. Sgall, *Simplified lower bounds for the polynomial calculus,* Computational Complexity, to appear

[25] J. Krajíček, *Bounded Arithmetic, Propositional Logic and Complexity Theory,* Cambridge University Press 1995.

[26] J. Krajíček, *Interpolation theorems, lower bounds for proof systems, and independence results for bounded arithmetic,* Journ. of Symb. Logic 62/2, 1997, 457-486

[27] J. Krajíček, *On the degree of ideal membership proofs from uniform families of polynomials over a finite field,* submitted, 1997

[28] J. Krajíček, *A fundamental problem of mathematical logic,* Collegium Logicum, Annals of Kurt Gödel Society Vol. 2, 1996, 56-64

[29] J. Krajíček, *Discretely ordered modules as a first-order extension of the cutting planes proof system,* JSL, to appear

[30] J. Krajíček, *On methods for proving lower bounds in propositional logic,* M.L. Dalla Chiara et al. (eds.), Logic and Scientific Methods, Kluwer Acad. Publ., 69-83

[31] J. Krajíček and P. Pudlák, *Some consequences of cryptographical conjectures for S_2^1 and EF,* Information and Computation 140, 1998, 82-94.

[32] K. Iwama and T. Pitassi, *Exponential lower bound for the Tree-like Hajós Calculus,* 1997, preprint

[33] S-Y. R. Li and W-C. W. Li, *Independence numbers of graphs and generators of ideals,* Combinatorica 1,1981, 55-61

[34] L. Lovász and A. Schrijver, *Cones of matrices and set-functions and 0-1 optimization,* SIAM J. Optimization 1(2), 166-190

[35] D. Mundici, *NP and Craig's interpolation theorem,* Logic Colloquium 1982, North-Holland, 345-358

[36] T. Pitassi and A. Urquhart, *The complexity of the Hajós calculus,* Proc. 33-th Symp. on Foundations of Computer Science, IEEE, 1992, 187-196

[37] P. Pudlák, *Lower bounds for resolution and cutting planes proofs and monotone computations,* JSL 62/3, 1997, 981-998

[38] P. Pudlák, *The lengths of proofs,* in *Handbook of Proof Theory,* S. Buss ed., North-Holland, 1998, 547-637

[39] P. Pudlák and J. Sgall, *Algebraic models of computation and interpolation for algebraic proof systems,* in Proof Complexity and Feasible Arithmetics, P.W. Beame and S.R. Buss eds., DIMACS Series in Discrete Mathematics and Theoretical Computer Science 39, 279-295

[40] H. Rasiowa, *An algebraic approach to non-classical logics,* PWN-Polish Scientific Publishers and North-Holland Publishing Company, 1974

[41] A.A. Razborov, *Lower bounds on the monotone complexity of some boolean functions,* Doklady Akad. Nauk SSSR 282, (1985), 1033-1037

[42] A.A. Razborov, *Unprovability of lower bounds on the circuit size in certain fragments of Bounded Arithmetic,* Izvestiya of the R.A.N. 59(1), (1995), 201-222; see also Izvestiya: Mathematics 59:1, 205-227.

[43] A.A. Razborov, *Lower bound for the polynomial calculus,* Computational Complexity, to appear

[44] L.G. Valiant, *Short monotone formulae for the majority function,* J. of Algorithms 5, 1984, 363-366

The Realm of Ordinal Analysis

Michael Rathjen
Department of Pure Mathematics, University of Leeds
Leeds LS2 9JT, United Kingdom

Denn die Pioniere der Mathematik hatten sich von gewissen Grundlagen brauchbare Vorstellungen gemacht, aus denen sich Schlüsse, Rechnungsarten, Resultate ergaben, deren bemächtigten sich die Physiker, um neue Ergebnisse zu erhalten, und endlich kamen die Techniker, nahmen oft bloß die Resultate, setzten neue Rechnungen darauf und es entstanden Maschinen. Und plötzlich, nachdem alles in schönste Existenz gebracht war, kamen die Mathematiker - jene, die ganz innen herumgrübeln, - darauf, daß etwas in den Grundlagen der ganzen Sache absolut nicht in Ordnung zu bringen sei; tatsächlich, sie sahen zuunterst nach und fanden, daß das ganze Gebäude in der Luft stehe. Aber die Maschinen liefen! Man muß daraufhin annehmen, daß unser Dasein bleicher Spuk ist; wir leben es, aber eigentlich nur auf Grund eines Irrtums, ohne den es nicht entstanden wäre.

ROBERT MUSIL: Der mathematische Mensch (1913)

1 Introduction

A central theme running through all the main areas of Mathematical Logic is the classification of sets, functions or theories, by means of transfinite hierarchies whose ordinal levels measure their 'rank' or 'complexity' in some sense appropriate to the underlying context. In Proof Theory this is manifest in the assignment of 'proof theoretic ordinals' to theories, gauging their 'consistency strength' and 'computational power'. Ordinal-theoretic proof theory came into existence in 1936, springing forth from Gentzen's head in the course of his consistency proof of arithmetic. To put it roughly, ordinal analyses attach ordinals in a given representation system to formal theories. Though this area of mathematical logic has is roots in Hilbert's "Beweistheorie" - the aim of which was to lay to rest all worries[1] about the foundations of mathematics once and for all by securing mathematics via an absolute proof of consistency - technical results in proof theory are not different from those in any other

[1]As, in a rather amusing way, described by Musil in the above quote.

branch of mathematics, inasmuch as they can be understood in a way that does not at all refer to any kind of (modified) Hilbert programme. In actuality, most proof theorists do not consider themselves pursuing consistency proofs.

The present paper is based on the lectures that I gave at the LC '97. The lectures were an attempt to give an overview of results that have been achieved by means of ordinal analyses, and to explain the current rationale and goals of ordinally informative proof theory as well as its salient technical tools of analysis. They were aimed at a general logic audience, assuming very little knowledge of proof theory, basically cut elimination for Gentzen's sequent calculus.

The paper is divided into three parts. In Section 1 I try to explain the nature of the connection between ordinal representation systems and theories established in ordinal analyses. Furthermore, I gather together some general conclusions that can be drawn from an ordinal analysis. In the literature, the result of an ordinal analysis of a given theory T is often stated in a rather terse way by saying that the supremum of the provable recursive well-orderings of T (hereafter called $|T|_{\mathrm{sup}}$) is a certain ordinal α. This is at best a shorthand for a much more informative statement. From questions that I've been asked over the years, I know that sloppy talk about proof-theoretic ordinals has led to misconceptions about ordinal-theoretic proof theory. One of the recurring questions is whether it is always possible, given a decent theory T, to cook up a well-ordering $\prec$ such that the order-type of $\prec$ amounts to $|T|_{\mathrm{sup}}$ and T proves all initial segments of $\prec$ to be well-founded, thereby making a mockery of the task of performing an ordinal analysis of T. This time I decided, I'd better take such questions seriously. Section 1 will scrutinize the norm $|.|_{\mathrm{sup}}$, compare it with other scales of strengths and also attend to the above and related questions.

Section 2 is devoted to results that have been achieved through ordinal analyses. They fall into four groups: (1) Consistency of subsystems of classical second order arithmetic and set theory relative to constructive theories, (2) reductions of theories formulated as conservation theorems, (3) combinatorial independence results, and (4) classifications of provable functions and ordinals.

As an introduction to the techniques used in ordinal analysis and in order to illustrate its more subtle features, Section 3 provides sketches of ordinal analyses for two theories. These theories are Kripke-Platek set theory and an extension of the latter, called **KPM**, which formalizes a recursively Mahlo universe of sets. **KPM** is considerably stronger than than the fragment of second order arithmetic with Δ^1_2 comprehension. It is distinguished by the fact that it is essentially the 'strongest' classical theory for which a consistency proof in Martin-Löf type theory can be carried out.

2 Measures in proof theory

2.1 Gentzen's result

Gentzen showed that transfinite induction up to the ordinal

$$\varepsilon_0 = \sup\{\omega, \omega^\omega, \omega^{\omega^\omega}, \ldots\} = \text{least } \alpha.\, \omega^\alpha = \alpha$$

suffices to prove the consistency of Peano Arithmetic, **PA**. To appreciate Gentzen's result it is pivotal to note that he applied transfinite induction up to ε_0 solely to primitive recursive predicates and besides that his proof used only finitistically justified means. Hence, a more precise rendering of Gentzen's result is

$$\mathbf{F} + \text{PR-TI}(\varepsilon_0) \vdash \mathsf{Con}(\mathbf{PA}), \tag{1}$$

where **F** signifies a theory that is acceptable in finitism (e.g. **F** = **PRA** = *Primitive Recursive Arithmetic*) and PR-TI(ε_0) stands for transfinite induction up to ε_0 for primitive recursive predicates. Gentzen also showed that his result is best possible in that **PA** proves transfinite induction up to α for arithmetic predicates for any $\alpha < \varepsilon_0$. The compelling picture conjured up by the above is that the non-finitist part of **PA** is encapsulated in PR-TI(ε_0) and therefore "measured" by ε_0, thereby tempting one to adopt the following definition of *proof-theoretic ordinal* of a theory T:

$$|T|_{Con} = \text{least } \alpha.\, \mathbf{PRA} + \text{PR-TI}(\alpha) \vdash \mathsf{Con}(T). \tag{2}$$

The foregoing definition of $|T|_{Con}$ is, however, inherently vague because the following issues have not been addressed:

- How are ordinals to be represented in **PRA**?
- (2) is definitive only with regard to a prior choice of *ordinal representation system.*
- Different ordinal representation systems may provide different answers to (2).

Notwithstanding that, for "natural" theories T and with regard to a "natural" ordinal representation system, the ordinal $|T|_{Con}$ encapsulates important information about the proof strength of T. To demonstrate the serious deficiencies of the above concept it might be illuminating to exhibit a clearly pathological ordinal representation system for the ordinal ω which underscores the dependence of $|T|_{Con}$ on the choice of the ordinal representation system. This example is due to Kreisel [52]. For a given theory T, it shows how to cook up an ordinal representation system which trivializes the determination of $|T|_{Con}$ by coding the proof predicate for T, Proof_T, into the ordinal representation system. Suppose T is a consistent (primitive recursively axiomatized) extension of **PRA**. Define

$$n <_T m \Leftrightarrow \begin{cases} n < m & \text{if } \forall i < n \, \neg \mathsf{Proof}_T(i, \ulcorner \bot \urcorner) \\ m < n & \text{if } \exists i < n \, \mathsf{Proof}_T(i, \ulcorner \bot \urcorner) \end{cases}$$

where $\bot$ is $\bar{0} = \bar{1}$. If T were inconsistent, then there would exists a least natural number k_0 such that $\mathsf{Proof}_T(k_0, \ulcorner \bot \urcorner)$ and the ordering $<_T$ would look like

$$k_0 \,_T\!> k_0 - 1 \,_T\!> \cdots \,_T\!> 0 \,_T\!> k_0 + 1 \,_T\!> k_0 + 2 \,_T\!> k_0 + 3 \,_T\!> \cdots .$$

Otherwise, $<_T$ is just the standard ordering on the natural numbers. At any rate, $<_T$ is a linear ordering (provably so in **PRA**). However, by assumption, T is consistent and thus $n <_T m \Leftrightarrow n < m$. Consequently, the order-type of $<_T$ is ω. In view of its definition, $<_T$ is primitive recursive. Furthermore,

$$\mathbf{PRA} + \mathrm{PR} - \mathrm{TI}(<_T) \vdash \mathsf{Con}(T). \tag{3}$$

Let $A(x) := \forall u \le x \, \neg \mathsf{Proof}_T(u, \ulcorner \bot \urcorner)$. To see that (3) holds, it suffices to prove **PRA** $\vdash \forall x \, [\forall y <_T x \, A(y) \rightarrow A(x)]$. So assume $\forall y <_T a \, A(y)$. We have to show $A(a)$. But $\neg A(a)$ would imply $a + 1 <_T a$ and thus yield $A(a+1)$ which implies $A(a)$. Therefore, $A(a)$ must hold.

2.2 How natural ordinal representation systems arise

Natural ordinal representation systems are frequently derived from structures of the form

$$\mathfrak{A} \;=\; \langle \alpha, f_1, \ldots, f_n, <_\alpha \rangle \tag{4}$$

where α is an ordinal, $<_\alpha$ is the ordering of ordinals restricted to elements of α and the f_i are functions

$$f_i : \underbrace{\alpha \times \cdots \times \alpha}_{k_i \text{ times}} \longrightarrow \alpha$$

for some natural number k_i.

$$\mathbb{A} \;=\; \langle A, g_1, \ldots, g_n, \prec \rangle \tag{5}$$

is a *recursive representation of* $\mathfrak{A}$ if the following conditions hold:

1. $A \subseteq \mathbb{N}$
2. A is a recursive set.
3. $\prec$ is a recursive total ordering on A.
4. The functions g_i are recursive.
5. $\mathfrak{A} \cong \mathbb{A}$, i.e. the two structures are isomorphic.

Gentzen's ordinal representation system for ε_0 is based on the *Cantor normal form*, i.e. for any ordinal $0 < \alpha < \varepsilon_0$ there exist uniquely determined ordinals $\alpha_1, \dots, \alpha_n < \alpha$ such that

- $\alpha_1 \geq \cdots \geq \alpha_n$
- $\alpha = \omega^{\alpha_1} + \cdots \omega^{\alpha_n}$.

To indicate the Cantor normal form we write $\alpha =_{CNF} \omega^{\alpha_1} + \cdots \omega^{\alpha_n}$. Now define a function

$$\ulcorner . \urcorner : \varepsilon_0 \longrightarrow \mathbb{N}$$

by

$$\ulcorner \alpha \urcorner = \begin{cases} 0 & \text{if } \alpha = 0 \\ \langle \ulcorner \alpha_1 \urcorner, \dots, \ulcorner \alpha_n \urcorner \rangle & \text{if } \alpha =_{CNF} \omega^{\alpha_1} + \cdots \omega^{\alpha_n} \end{cases}$$

where $\langle n_1, \cdots, k_n \rangle := 2^{n_1+1} \cdot \ldots \cdot p_n^{k_n+1}$ with p_i being the ith prime number (or any other coding of tuples). Further define

$$\begin{array}{rcl} A_0 & := & \mathbf{ran}(\ulcorner . \urcorner) \\ \ulcorner \alpha \urcorner \prec \ulcorner \beta \urcorner & :\Leftrightarrow & \alpha < \beta \\ \ulcorner \alpha \urcorner \hat{+} \ulcorner \beta \urcorner & := & \ulcorner \alpha + \beta \urcorner \\ \ulcorner \alpha \urcorner \hat{\cdot} \ulcorner \beta \urcorner & := & \ulcorner \alpha \cdot \beta \urcorner \\ \hat{\omega}^{\ulcorner \alpha \urcorner} & := & \omega^{\alpha}. \end{array}$$

Then

$$\langle \varepsilon_0, +, \cdot, \delta \mapsto \omega^\delta, < \rangle \;\cong\; \langle A_0, \hat{+}, \hat{\cdot}, x \mapsto \hat{\omega}^x, \prec \rangle.$$

$A_0, \hat{+}, \hat{\cdot}, x \mapsto \hat{\omega}^x, \prec$ are recursive, in point of fact, they are all elementary recursive.

2.3 Elementary ordinal representation systems

The next definition garners some features (following [32]) that ordinal representation systems used in proof theory always have, and collectively calls them *"elementary ordinal representation system"*. One reason for singling out this notion is that it leads to an elegant characterization of the provably recursive functions of theories equipped with transfinite induction principles for such ordinal representation systems (cf. Propositions 3.19, 3.20).

Definition 2.1 *Elementary recursive arithmetic,* **ERA**, is a weak system of number theory, in a language with $0, 1, +, \times, E$ (exponentiation), $<$, whose axioms are:

1. the usual recursion axioms for $+, \times, E, <$.

2. induction on Δ_0-formulae with free variables.

ERA is referred to as elementary recursive arithmetic since its provably recursive functions are exactly the Kalmar *elementary functions,* i.e. the class of functions which contains the successor, projection, zero, addition, multiplication, and modified subtraction functions and is closed under composition and bounded sums and products (cf. [89]).

Definition 2.2 For a set X and and a binary relation $\prec$ on X, let $\mathrm{LO}(X, \prec)$ abbreviate that $\prec$ linearly orders the elements of X and that for all u, v, whenever $u \prec v$, then $u, v \in X$.

A *linear ordering* is a pair $\langle X, \prec \rangle$ satisfying $\mathrm{LO}(X, \prec)$.

Definition 2.3 An *elementary ordinal representation system* (EORS) for a limit ordinal λ is a structure $\langle A, \lhd, n \mapsto \lambda_n, +, \times, x \mapsto \omega^x \rangle$ such that:

(i) A is an elementary subset of $\mathbb{N}$.

(ii) $\lhd$ is an elementary well-ordering of A.

(iii) $|\lhd| = \lambda$.

(iv) Provably in **ERA**, $\lhd \restriction \lambda_n$ is a proper initial segment of $\lhd$ for each n, and $\bigcup_n \lhd \restriction \lambda_n = \lhd$. In particular, $\mathbf{ERA} \vdash \forall y\, \lambda_y \in A \wedge \forall x \in A \exists y\, [x \lhd \lambda_y]$.

(v) $\mathbf{ERA} \vdash \mathrm{LO}(A, \lhd)$

(vi) $+, \times$ are binary and $x \mapsto \omega^x$ is unary. They are elementary functions on elementary initial segments of A. They correspond to ordinal addition, multiplication and exponentiation to base ω, respectively. The initial segments of A on which they are defined are maximal.

$n \mapsto \lambda_n$ is an elementary function.

(vii) $\langle A, \lhd, +, \times, \omega^x \rangle$ satisfies "all the usual algebraic properties" of an initial segment of ordinals. In addition, these properties of $\langle A, \lhd, +, \times, \omega^x \rangle$ can be proved in **ERA**.

(viii) Let $\tilde{n}$ denote the nth element in the ordering of A. Then the correspondence $n \leftrightarrow \tilde{n}$ is elementary.

(ix) Let $\alpha = \omega^{\beta_1} + \cdots \omega^{\beta_k}, \beta_1 \geq \cdots \geq \beta_k$ (Cantor normal form). Then the correspondence $\alpha \leftrightarrow \langle \beta_1, \ldots, \beta_k \rangle$ is elementary.

Elements of A will often be referred to as *ordinals*, and denoted $\alpha, \beta, \ldots$.

2.4 The conceptual problem of characterizing natural ordinal representation systems

It is an empirical fact that ordinal representation systems emerging in proof theory are always elementary recursive and their basic properties are provable in weak fragments of arithmetic like **ERA**. Sommer has investigated the question of complexity of ordinal representation systems at great length in [104, 105]. His case studies revealed that with regard to complexity measures considered in complexity theory the complexity of ordinal representation systems involved in ordinal analyses is rather low. It appears that they are always Δ_0-representable (cf. [104]) and that computations on ordinals in actual proof-theoretic ordinal analyses can be handled in the theory $I\Delta_0 + \Omega_1$, where Ω_1 is the assertion that the function $x \mapsto x^{\log_2(x)}$ is total.

Sommer's findings clearly underpin the fact that the naturalness of ordinal representation systems involved in proof-theoretic ordinal analyses cannot be described in terms of the computational complexity of the representations of these ordinals. Intuitively, computational complexity is inadequate because it says nothing about how ordinals are built up. It has been suggested (cf. [51], [25]) that it is important to address the broader question *"What is a natural well-ordering?"* A criterion for naturalness put forward in [51] is uniqueness up to recursive isomorphism. Furthermore, in [51], Kreisel seems to seek naturalness in algebraic characterizations of ordered structures. Feferman, in [22], discerns the properties of completeness, repleteness, relative categoricity and preservation of these under iteration of the critical process as significant features of systems of natural representation. Girard [35] appears to propose dilators to capture the abstract notion of a notation system for ordinals. However, in my opinion, similar attempts in the Philosophy of Science of defining *'natural properties'* and the complete failure of these attempts show that it is futile to look for a formal definition of *'natural well-ordering'* that will exclude every pathological example. Moreover, it is rather unlikely that such a definition would be able to discern and explain an important feature of EORSs found in proof theory, namely their versatility in establishing equivalences between classical non-constructive theories and intuitionistic constructive theories based on radically different ontologies. To obtain the reductions of classical (non-constructive) theories to constructive ones (as related, for instance, in [26], [83],§2) it appears to be pivotal to work with very special and well-structured ordinal representation systems.

"Natural" well-orderings have arisen using several sources of inspiration:

Set-theoretical (*Cantor, Veblen, Gentzen, Bachmann, Schütte, Feferman, Pfeiffer, Isles, Bridge, Buchholz, Pohlers, Jäger, Rathjen*)

- Define hierarchies of functions on the ordinals.
- Build up terms from function symbols for those functions.
- The ordering on the values of terms induces an ordering on the terms.

Reductions in proof figures (*Takeuti, Yasugi, Kino, Arai*)

- Ordinal diagrams; formal terms endowed with an inductively defined ordering on them.

Patterns of partial elementary substructurehood (*Carlson*, cf. [16])

- Finite structures with Σ_n-elementary substructure relations .

Category-theoretical (*Aczel, Girard, Vauzeilles*)

- Functors on the category of ordinals (with strictly increasing functions) respecting direct limits and pull-backs.

Examples for the set-theoretical approach to ordinal representation systems, in particular the use and role of large cardinals therein, will be presented in section 4.

2.5 Proof-theoretical reductions

Ordinal analyses of theories allow one to compare the strength of theories. This subsection defines the notions of *proof-theoretic reducibility* and *proof-theoretic strength* that will be used henceforth.

All theories T considered in the following are assumed to contain a modicum of arithmetic. For definiteness let this mean that the system **PRA** of Primitive Recursive Arithmetic is contained in T, either directly or by translation.

Definition 2.4 Let T_1, T_2 be a pair of theories with languages $\mathcal{L}_1$ and $\mathcal{L}_2$, respectively, and let Φ be a (primitive recursive) collection of formulae common to both languages. Furthermore, Φ should contain the closed equations of the language of **PRA**.

We then say that T_1 is *proof-theoretically Φ-reducible to* T_2, written $T_1 \leq_\Phi T_2$, if there exists a primitive recursive function f such that

$$\mathbf{PRA} \vdash \forall \phi \in \Phi \, \forall x \, [\mathsf{Proof}_{T_1}(x, \phi) \ \rightarrow \ \mathsf{Proof}_{T_2}(f(x), \phi)]. \tag{6}$$

T_1 and T_2 are said to be *proof-theoretically Φ-equivalent*, written $T_1 \equiv_\Phi T_2$, if $T_1 \leq_\Phi T_2$ and $T_2 \leq_\Phi T_1$.

The appropriate class Φ is revealed in the process of reduction itself, so that in the statement of theorems we simply say that T_1 is *proof-theoretically reducible* to T_2 (written $T_1 \leq T_2$) and T_1 and T_2 are *proof-theoretically equivalent* (written $T_1 \equiv T_2$), respectively. Alternatively, we shall say that T_1 and T_2 have the *same proof-theoretic strength* when $T_1 \equiv T_2$.

As observed above (23), by Kreisel's theorem (cf. [102], Theorem 5.2.1), we could replace **PRA** in (6) with T_2.

Remark 2.5 Feferman's notion of proof-theoretic reducibility in [26] is more relaxed in that he allows the reduction to be given by a T_2-recursive function f, i.e.

$$T_2 \vdash \forall \phi \in \Phi \, \forall x \, [\mathsf{Proof}_{T_1}(x, \phi) \ \rightarrow \ \mathsf{Proof}_{T_2}(f(x), \phi)]. \tag{7}$$

The disadvantage of (7) is that one forfeits the transitivity of the relation $\leq_\Phi$. Furthermore, in practice, proof-theoretic reductions always come with a primitive recursive reduction, so nothing seems to be lost by using the stronger notion of reducibility.

2.6 The general form of ordinal analysis

In this section I attempt to say something general about all ordinal analyses that have been carried out thus far. One has to bear in mind that these concern "natural" theories. Also, to circumvent countless and rather boring counter examples, I will only address theories that have at least the strength of **PA** and and always assume the pertinent ordinal representation systems are closed under $\alpha \mapsto \omega^\alpha$.

Before delineating the general form of an ordinal analysis, we need several definitions.

Definition 2.6 Let T be a framework for formalizing a certain part of mathematics. T should be a true theory which contains a modicum of arithmetic.

Let A be a subset of $\mathbb{N}$ ordered by $\prec$ such that A and $\prec$ are both definable in the language of T. If the language of T allows for quantification over subsets of $\mathbb{N}$, like that of second order arithmetic or set theory, *well-foundedness* of $\langle A, \prec \rangle$ will be formally expressed by

$$\mathrm{WF}(A, \prec) \ := \ \forall X \subseteq \mathbb{N} \, [\forall u{\in}A(\forall v \prec u \, v{\in}X \to u{\in}X) \ \to \ \forall u{\in}A \, u{\in}X.] \tag{8}$$

If, however, the language of T does not provide for quantification over arbitrary subsets of $\mathbb{N}$, like that of Peano arithmetic, we shall assume that it contains a new unary predicate $\mathbf{U}$. $\mathbf{U}$ acts like a free set variable, in that no special properties of it will ever be assumed. We will then resort to the following formalization of well-foundedness:

$$\mathrm{WF}(A, \prec) \ := \ \forall u{\in}A(\forall v \prec u \, \mathbf{U}(v) \to \mathbf{U}(u)) \ \to \ \forall u{\in}A \, \mathbf{U}(u), \tag{9}$$

where $\forall v \prec u \ldots$ is short for $\forall v(v \prec u \to \ldots)$.

We also set

$$\mathrm{WO}(A, \prec) \ := \ \mathrm{LO}(A, \prec) \wedge \mathrm{WF}(A, \prec). \tag{10}$$

If $\langle A, \prec \rangle$ is well-founded, we use $|\prec|$ to signify its set-theoretic order-type. For $a{\in}A$, the ordering $\prec\restriction a$ is the restriction of $\prec$ to $\{x{\in}A : \ x \prec a\}$.

The ordering $\langle A, \prec \rangle$ is said to be *provably well-founded in* T if

$$T \vdash \mathrm{WO}(A, \prec). \tag{11}$$

The supremum of the provable well-orderings of T, $|T|_{\mathrm{sup}}$, is defined as follows:

$$|T|_{\mathrm{sup}} \ := \ \sup \big\{ \alpha : \ \alpha \text{ provably recursive in } T \big\} \tag{12}$$

where an ordinal α is said to be provably recursive in T if there is a recursive well–ordering $\langle A, \prec \rangle$ with order–type α such that

$$T \vdash \mathrm{WO}(A, \prec)$$

with A and $\prec$ being provably recursive in T. Note that, by definition, $|T|_{\mathrm{sup}} \leq \omega_1^{CK}$, where ω_1^{CK} is the supremum of the order-types of all recursive well-orderings on $\mathbb{N}$. Another characterization of ω_1^{CK} is that it is the least admissible ordinal $> \omega$.

Definition 2.7 Suppose $\mathrm{LO}(A, \lhd)$ and $F(u)$ is a formula. Then $\mathrm{TI}_{\langle A, \lhd \rangle}(F)$ is the formula

$$\forall n{\in}A \, [\forall x \lhd n F(x) \to F(n)] \ \to \ \forall n{\in}A \, F(n). \tag{13}$$

$\mathrm{TI}(A, \lhd)$ is the schema consisting of $\mathrm{TI}_{\langle A, \lhd \rangle}(F)$ for all F.

Given a linear ordering $\langle A, \lhd \rangle$ and $\alpha \in A$ let $A_\alpha = \{\beta \in A : \beta \lhd \alpha\}$ and $\lhd_\alpha$ be the restriction of $\lhd$ to A_α.

In what follows, quantifiers and variables are supposed to range over the natural numbers. When n denotes a natural number, $\bar{n}$ is the canonical name in the language under consideration which denotes that number.

Observation 2.8 *Every ordinal analysis of a classical or intuitionistic theory* $\mathbf{T}$ *that has ever appeared in the literature provides an EORS* $\langle A, \lhd, \dots \rangle$ *such that* $\mathbf{T}$ *is proof-theoretically reducible to* $\mathbf{PA} + \bigcup_{\alpha \in A} \mathrm{TI}(A_{\bar{\alpha}}, \lhd_{\bar{\alpha}})$.

Moreover, if T *is a classical theory, then* T *and* $\mathbf{PA} + \bigcup_{\alpha \in A} \mathrm{TI}(A_{\bar{\alpha}}, \lhd_{\bar{\alpha}})$ *prove the same arithmetic sentences, whereas if* T *is based on intuitionististic, then* T *and* $\mathbf{HA} + \bigcup_{\alpha \in A} \mathrm{TI}(A_{\bar{\alpha}}, \lhd_{\bar{\alpha}})$ *prove the same arithmetic sentences.*

Furthermore, $|T|_{\mathrm{sup}} = |\lhd|$.

Remark 2.9 There is a lot of leeway in stating the latter observation. For instance, instead of **PA** one could take **PRA** or **ERA** as the base theory, and the scheme of transfinite induction could be restricted to Σ^0_1 formulae as it follows from Proposition 3.20 that $\mathbf{PA} + \bigcup_{\alpha \in A} \mathrm{TI}(A_{\bar{\alpha}}, \lhd_{\bar{\alpha}})$ and $\mathbf{ERA} + \bigcup_{\alpha \in A} \Sigma^0_1\text{-}\mathrm{TI}(A_{\bar{\alpha}}, \lhd_{\bar{\alpha}})$ have the same proof-theoretic strength, providing that A is closed under exponentiation $\alpha \mapsto \omega^\alpha$.

Observation 2.8 lends itself to a formal definition of the notion of *proof-theoretic ordinal* of a theory T. Of course, before one can go about determining the proof-theoretic ordinal of T, one needs to be furnished with representations of ordinals. Not surprisingly, a great deal of ordinally informative proof theory has been concerned with developing and comparing particular ordinal representation systems. Assuming that a sufficiently strong EORS $\langle A, \lhd, \dots \rangle$ has been provided, we define

$$|T|_{\langle A, \lhd, \dots \rangle} \;:=\; \text{least } \rho \in A.\ T \equiv \mathbf{PA} + \bigcup_{\alpha \lhd \rho} \mathrm{TI}(A_{\bar{\alpha}}, \lhd_{\bar{\alpha}}) \qquad (14)$$

and call $|T|_{\langle A, \lhd, \dots \rangle}$, providing this ordinal exists, the *proof-theoretic ordinal* of T with respect to $\langle A, \lhd, \dots \rangle$.

Since, in practice, the ordinal representation systems used in proof theory are comparable, we shall frequently drop mentioning of $\langle A, \lhd, \dots \rangle$ and just write $|T|$ for $|T|_{\langle A, \lhd, \dots \rangle}$.

Note, however, that $|T|_{\langle A, \lhd, \dots \rangle}$ might not exist even if the order-type of $\lhd$ is bigger than $|T|_{\mathrm{sup}}$. A simple example is provided by the theory $\mathbf{PA} + \mathsf{Con}(\mathbf{PA})$ (where $\mathsf{Con}(\mathbf{PA})$ expresses the consistency of **PA**) when we take $\langle A, \lhd, \dots \rangle$ to be a standard EORS for ordinals $> \varepsilon_0$; the reason being that $\mathbf{PA} + \mathsf{Con}(\mathbf{PA})$ is proof-theoretically strictly stronger than $\mathbf{PA} + \bigcup_{\alpha \lhd \varepsilon_0} \mathrm{TI}(A_{\bar{\alpha}}, \lhd_{\bar{\alpha}})$ but also strictly weaker than $\mathbf{PA} + \bigcup_{\alpha \lhd \varepsilon_0 + 1} \mathrm{TI}(A_{\bar{\alpha}}, \lhd_{\bar{\alpha}})$. Therefore, as opposed to

$|\cdot|_{\sup}$, the norm $|\cdot|_{\langle A,\lhd,\ldots\rangle}$ is only partially defined and does not induce a prewellordering on theories T with $|T|_{\sup} < |\lhd|$.

The remainder of this subsection expounds on important consequences of ordinal analyses that follow from Observation 2.8.

Proposition 2.10 $\mathbf{PA} + \bigcup_{\alpha\in A} \mathrm{TI}(A_{\bar\alpha}, \lhd_{\bar\alpha})$ *and* $\mathbf{HA} + \bigcup_{\alpha\in A} \mathrm{TI}(A_{\bar\alpha}, \lhd_{\bar\alpha})$ *prove the same sentences in the negative fragment.*

Proof: $\mathbf{PA} + \bigcup_{\alpha\in A} \mathrm{TI}(A_{\bar\alpha}, \lhd_{\bar\alpha})$ can be interpreted in $\mathbf{HA} + \bigcup_{\alpha\in A} \mathrm{TI}(A_{\bar\alpha}, \lhd_{\bar\alpha})$ via the Gödel–Gentzen $\neg\neg$-translation. Observe that for an instance of the schema of transfinite induction we have

$$\begin{aligned}&\big(\forall u\,\big[\forall x\,(\forall y\,[y \prec x \to \phi(y)] \to \phi(x)) \;\to\; \phi(u)\big]\big)^{\neg\neg} \;\equiv \\ &\quad\big(\forall u\,\big[\forall x\,(\forall y\,[\neg\neg y \prec x \to \neg\neg\phi(y)] \to \neg\neg\phi(x)) \;\to\; \neg\neg\phi(u)\big]\big).\end{aligned}$$

Thus for primitive recursive $\prec$ the $\neg\neg$-translation is $\mathbf{HA}$ equivalent to an instance of the same schema. □

Corollary 2.11 $\mathbf{PA} + \bigcup_{\alpha\in A} \mathrm{TI}(A_{\bar\alpha}, \lhd_{\bar\alpha})$ *and* $\mathbf{HA} + \bigcup_{\alpha\in A} \mathrm{TI}(A_{\bar\alpha}, \lhd_{\bar\alpha})$ *prove the same* Π^0_1 *sentences.*

Since many well-known and important theorems as well as conjectures from number theory are expressible in Π^0_1 form (examples: the quadratic reciprocity law, Wiles' theorem, also known as Fermat's conjecture, Goldbach's conjecture, the Riemann hypothesis), Π^0_1 conservativity ensures that many mathematically important theorems which turn out to be provable in S will be provable in T, too.

However, Π^0_1 conservativity is not always a satisfactory conservation result. Some important number-theoretic statements are Π^0_2 (examples are: the twin prime conjecture, miniaturized versions of Kruskal's theorem, totality of the van der Waerden function), and in particular, formulas that express the convergence of a recursive function for all arguments. Consider a formula $\forall n\,\exists m\,P(n,m)$, where $P(n,m)$ is a primitive recursive formula expressing that *"m codes a complete computation of algorithm A on input n."* The $\neg\neg$-translation of this formula is $\forall n\,\neg\forall m\,\neg P(n,m)$, conveying the convergence of the algorithm A for all inputs only in a weak sense. Fortunately, Proposition 2.11 can be improved to hold for sentences of Π^0_2 form.

Proposition 2.12 $\mathbf{PA} + \bigcup_{\alpha\in A} \mathrm{TI}(A_{\bar\alpha}, \lhd_{\bar\alpha})$ *and* $\mathbf{HA} + \bigcup_{\alpha\in A} \mathrm{TI}(A_{\bar\alpha}, \lhd_{\bar\alpha})$ *prove the same* Π^0_2 *sentences.*

The missing link to get from Proposition 2.10 to Proposition 2.12 is usually provided by *Markov's Rule* for primitive recursive predicates, $\mathbf{MR}_{PR}$: if

$\neg\forall n\neg Q(n)$ (or, equivalently, $\neg\neg\,\exists n\, Q(n)$) is a theorem, where Q is a primitive recursive relation, then $\exists n\, Q(n)$ is a theorem. Kreisel [49] showed that $\mathbf{MR}_{PR}$ holds for $\mathbf{HA}$. A variety of intuitionistic systems have since been shown to be closed under $\mathbf{MR}_{PR}$, using a variety of complicated methods, notably Gödel's dialectica interpretation and normalizability. A particularly elegant and short proof for closure under $\mathbf{MR}_{PR}$ is due to Friedman [28] and, independently, to Dragalin [18]. However, though the Friedman–Dragalin argument works for a host of systems, it doesn't seem to work in the case of $\mathbf{HA} + \bigcup_{\alpha\in A}\mathrm{TI}(A_{\bar\alpha}, \lhd_{\bar\alpha})$.

Proof of Proposition 2.12: We will give a direct proof, i.e. without using Proposition 2.10. So suppose

$$\mathbf{PA} + \bigcup_{\alpha\in A}\mathrm{TI}(A_{\bar\alpha}, \lhd_{\bar\alpha}) \vdash \forall x\, \exists y\, \phi(x,y),$$

where ϕ is Δ_0. Then there already exists a $\delta \in A$ such that

$$\mathbf{PA} + \mathrm{TI}(A_{\bar\delta}, \lhd_{\bar\delta}) \vdash \forall x\, \exists y\, \phi(x,y). \tag{15}$$

We now use the coding of infinitary $\mathbf{PA}_\infty$ derivations presented in [99], section 4.2.2. Let $d \,|\!\frac{\beta}{\rho}\, \ulcorner\psi\urcorner$ signify that d is the code of a $\mathbf{PA}_\infty$ derivation with length $\leq \beta$, cut-rank ρ and end formula ψ. (15) implies that there is a d_0 and $n < \omega$ such that

$$\mathbf{HA} + \bigcup_{\alpha\in A}\mathrm{TI}(A_{\bar\alpha}, \lhd_{\bar\alpha}) \vdash d_0 \,|\!\frac{\delta\cdot\omega}{n}\, \ulcorner\forall x\, \exists y\, \phi(x,y)\urcorner. \tag{16}$$

To obtain a cut-free proof of $\forall x\, \exists y\, \phi(x,y)$ in $\mathbf{PA}_\infty$ one needs transfinite induction up to the ordinal $\omega_n^{\delta\cdot\omega}$, where $\omega_0^\gamma := \gamma$ and $\omega_{m+1}^\gamma := \omega^{\omega_m^\gamma}$. This amount of transfinite induction is available in our background theory $\mathbf{HA} + \bigcup_{\alpha\in A}\mathrm{TI}(A_{\bar\alpha}, \lhd_{\bar\alpha})$ as A is closed under $\xi \mapsto \omega^\xi$. Also note that the cut-elimination procedure is completely effective. Thus from (16) we obtain, for some d^*,

$$\mathbf{HA} + \bigcup_{\alpha\in A}\mathrm{TI}(A_{\bar\alpha}, \lhd_{\bar\alpha}) \vdash d^* \,|\!\frac{\omega_n^{\delta\cdot\omega}}{0}\, \ulcorner\forall x\, \exists y\, \phi(x,y)\urcorner, \tag{17}$$

and further

$$\mathbf{HA} + \bigcup_{\alpha\in A}\mathrm{TI}(A_{\bar\alpha}, \lhd_{\bar\alpha}) \vdash \forall x \exists d\, d \,|\!\frac{\omega_n^{\delta\cdot\omega}}{0}\, \ulcorner\exists y\, \phi(\dot x,y)\urcorner \tag{18}$$

(where Feferman's dot convention has been used here). Let Tr_{Σ_1} be a truth predicate for Gödel numbers of disjunctions of Σ_1 formulae (cf. [109], section 1.5, in particular 1.5.7). We claim that

$$\mathbf{HA} + \bigcup_{\alpha\in A}\mathrm{TI}(A_{\bar\alpha}, \lhd_{\bar\alpha}) \vdash \forall d\, \forall\beta \leq \omega_n^{\delta\cdot\omega}\, \forall\Gamma \subseteq \Sigma_1 \left[d \,|\!\frac{\beta}{0}\, \Gamma \;\rightarrow\; \mathsf{Tr}_{\Sigma_1}(\textstyle\bigvee\Gamma)\right], \tag{19}$$

where $\forall \Gamma \subseteq \Sigma_1$ is a quantifier ranging over Gödel numbers of finite sets of Σ_1 formulae and $\bigvee \Gamma$ stands for the Gödel number corresponding to the disjunction of all formulae of Γ. (19) is proved by induction on β by observing that all formulae occurring in a cut-free $\mathbf{PA}_\infty$ proof of a set of Σ_1 formulae are Σ_1 themselves and the only inferences therein are either axioms or instances of the ($\exists$) rule or improper instances of the ω rule. Combining (18) and (19) we obtain

$$\mathbf{HA} + \bigcup_{\alpha \in A} \mathrm{TI}(A_{\bar{\alpha}}, \lhd_{\bar{\alpha}}) \vdash \forall x \, \mathsf{Tr}_{\Sigma_1}(\ulcorner \exists y \, \phi(\dot{x}, y) \urcorner). \tag{20}$$

As

$$\mathbf{HA} \vdash \forall x \, [\, \mathsf{Tr}_{\Sigma_1}(\ulcorner \exists y \, \phi(\dot{x}, y) \urcorner) \leftrightarrow \exists y \, \phi(x, y)]$$

(cf. [109], Theorem 1.5.6), we finally obtain

$$\mathbf{HA} + \bigcup_{\alpha \in A} \mathrm{TI}(A_{\bar{\alpha}}, \lhd_{\bar{\alpha}}) \vdash \forall x \, \exists y \, \phi(x, y).$$

□

In section 2 we considered the ordinal $|T|_{Con}$. What is the relation between $|T|_{Con}$ and $|T|_{\langle A, \lhd, \ldots \rangle}$? First we have to delineate the meaning of $|T|_{Con}$, though. The latter is only determined with respect to a given ordinal representation system $\langle B, \prec, \ldots \rangle$. Thus let

$$|T|_{Con} \;=\; \text{least } \alpha \in B. \, \mathbf{PRA} + \text{PR-TI}(\alpha) \vdash \mathsf{Con}(T).$$

It turns out that the two ordinals are the same when T is proof-theoretically reducible to $\mathbf{PA} + \bigcup_{\alpha \in A} \mathrm{TI}(A_{\bar{\alpha}}, \lhd_{\bar{\alpha}})$, A is closed under $\alpha \mapsto \omega^\alpha$ and $\langle B, \prec, \ldots \rangle$ is a proper end extension of $\langle A, \lhd, \ldots \rangle$. The reasons are as follows:

Proposition 2.13 *The consistency of* $\mathbf{PA} + \bigcup_{\alpha \in A} \mathrm{TI}(A_{\bar{\alpha}}, \lhd_{\bar{\alpha}})$ *can be proved in the theory* $\mathbf{PRA} + \text{PR-TI}(A, \lhd)$, *where* PR-TI$(A, \lhd)$ *stands for transfinite induction along* $\lhd$ *for primitive recursive predicates.*

Hint of proof. First note that $\mathbf{PRA} + \text{PR-TI}(A, \lhd) \vdash \Pi^0_1\text{-TI}(A, \lhd)$. The key to showing this is that for each $\alpha \in A$ and each $x \in \omega$ we can code α and x by the ordinal $\omega \cdot \alpha + x$ which is less than $\omega \cdot \alpha$ and therefore in A.

Secondly, one has to show that an ordinal analysis of $\mathbf{PA} + \bigcup_{\alpha \in A} \mathrm{TI}(A_{\bar{\alpha}}, \lhd_{\bar{\alpha}})$ can be carried out in $\mathbf{PRA} + \Pi^0_1\text{-TI}(A, \lhd)$. The main tool to achieve this is to embed $\mathbf{PA} + \bigcup_{\alpha \in A} \mathrm{TI}(A_{\bar{\alpha}}, \lhd_{\bar{\alpha}})$ into a system of Peano arithmetic with an infinitary rule, the so-called ω-rule, and a *repetition rule*, Rep, which simply repeats the premise as the conclusion. The ω-rule allows one to infer $\forall x \phi(x)$ from the infinitely many premises $\phi(\bar{0}), \phi(\bar{1}), \phi(\bar{2}), \ldots$ (where $\bar{n}$ denotes the nth numeral); its addition accounts for the fact that the infinitary system enjoys cut-elimination. The addition of the Rep rule enables one to carry out a *continuous cut elimination*, due to Mints [60], which is a continuous operation in the usual tree topology on prooftrees. A further pivotal step consists in making the ω-rule more constructive by assigning codes to proofs, where codes for applications of finitary rules contain codes for the proofs of the premises, and codes for applications of the ω-rule contain Gödel numbers for primitive recursive functions enumerating codes of the premises. Details can be found in [99]. The main idea here is that we can do everything with primitive recursive proof–trees instead of arbitrary derivations. A proof–tree is a tree, with each node labelled by: A sequent, a rule of inference or the designation "Axiom", two sets of formulas specifying the set of principal and minor formulas,respectively, of that inference, and two ordinals (length and cut–rank) such that the sequent is obtained from those immediately above it through application of the specified rule of inference. The well-foundedness of a proof–tree is then witnessed by the (first) ordinal "tags" which are in reverse order of the tree order. As a result, the notion of being a (code of a) proof tree is Π^0_1. The cut elimination for infinitary proofs with finite cut rank (as presented in [99]) can be formalized in $\mathbf{PRA} + \Pi^0_1\text{-TI}(A, \lhd)$. The last step consists in recognizing that every endformula of Π^0_1 form of a cut free infinitary proof is true. The latter employs $\Pi^0_1\text{-TI}(A, \lhd)$. For details see [99]. □

2.7 The orderings of consistency-strength and Π^0_1 conservativity

Two orderings figure prominently among orderings that have been suggested for comparing the strength of theories. These are the orderings of consistency-strength ($\leq_{\mathsf{Con}}$) and Π^0_1 conservativity ($\subseteq_{\Pi^0_1}$) (cf. [19, 113]). If one has ordinal analyses for two theories S, T such that $|S| \leq |T|$, then $S \leq_{\mathsf{Con}} T$ and $S \subseteq_{\Pi^0_1} T$. The latter, however, need not obtain if one merely knows that $|S|_{\mathsf{sup}} \leq |T|_{\mathsf{sup}}$ as will be shown in the subsequent subsection.

I consider the results of this section folklore, though I have no references.

Definition 2.14 T_2 is *conservative over* T_1 *for* Φ, written $T_1 \subseteq_\Phi T_2$ if

$$\forall \phi\, [\phi \in \Phi \wedge T_1 \vdash \phi \;\rightarrow\; T_2 \vdash \phi].$$

Definition 2.15 Let $\mathsf{Proof}_T(x, y)$ express that x is the code of a proof in T such that y is the code of its endformula. We use $\mathrm{Pr}_T(y)$ for $\exists x \mathsf{Proof}_T(x, y)$. The sentence $\mathsf{Con}(T)$ expressing the consistency of T can be taken as $\neg\, \mathrm{Pr}_T(\ulcorner 0 = 1 \urcorner)$.

The ordering of *consistency strength* between theories is defined by

$$S \leq_{\mathsf{Con}} T \quad :\Leftrightarrow \quad \text{the consistency of } T \text{ implies the consistency of } S. \tag{21}$$

One point needs to be attended to: Where should relative consistency be proven? If one is actually interested in the consistency of S relative to T it would suffice to prove the relative consistency result in T:

$$T \vdash \mathsf{Con}(T) \rightarrow \mathsf{Con}(S). \tag{22}$$

However, the provability within T of such an implication might be rather meaningless, as is the case for $T := \mathbf{PA} + \neg\mathsf{Con}(\mathbf{PA})$, and the relation $\leq_{\mathsf{Con}}$ wouldn't even be transitive.[2]

In practice, one shows the relative consistency result in a sound base theory like **PRA**. Moreover, as Kreisel noted, if the proof of $\mathsf{Con}(T) \rightarrow \mathsf{Con}(S)$ in T provides an effective transformation, i.e.

$$T \vdash \forall n \left[\mathsf{Proof}_S(n, \ulcorner 0 = 1 \urcorner) \rightarrow \mathsf{Proof}_T(f(n), \ulcorner 0 = 1 \urcorner)\right], \tag{23}$$

where f is primitive recursive, then

$$\mathbf{PRA} + \mathsf{Con}(T) \vdash \forall n \left[\mathsf{Proof}_S(n, \ulcorner 0 = 1 \urcorner) \rightarrow \mathsf{Proof}_T(f(n), \ulcorner 0 = 1 \urcorner)\right] \tag{24}$$

(cf. [102], Theorem 5.2.1), and therefore

$$\mathbf{PRA} \vdash \mathsf{Con}(T) \rightarrow \mathsf{Con}(S). \tag{25}$$

The upshot of the above is that if the proof of $\mathsf{Con}(T) \rightarrow \mathsf{Con}(S)$ is done at all nicely, i.e. in the sense of (23), it automatically follows that (23) can be proven in the weaker theory **PRA**. On the strength of the latter we adopt (24) as our official definition of $S \leq_{\mathsf{Con}} T$.

Remark 2.16 By definition, $T_1 \leq_\Phi T_2$ implies $T_1 \leq_{\mathsf{Con}} T_2$ and $T_1 \subseteq_\Phi T_2$. The converses are by no means true. (For a trivial counterexample, take $T_1 := \mathbf{ZF}$, $T_2 := \mathbf{PRA}$, and let Φ be the closed equations $\mathcal{L}(\mathbf{PRA})$. Then $T_1 \subseteq_\Phi T_2$ but not $T_1 \leq_\Phi T_2$, assuming T_1 is consistent.)

It is a striking empirical fact that many "natural" theories, i.e. theories which have something like an "idea" to them, are comparable with regard to consistency strength. This has actually been proved in many cases, for theories whose ideas and motivations have nothing at all to do with one

[2]Let $T_0 := \mathbf{ACA}$, $T_1 := \mathbf{PA} + \neg\mathsf{Con}(\mathbf{PA})$, and $T_2 := \mathbf{PA} + \mathsf{Con}(\mathbf{PA})$. Then $T_1 \vdash \mathsf{Con}(T_1) \rightarrow \mathsf{Con}(T_0)$ simply because $T_1 \vdash \neg\mathsf{Con}(T_1)$. Also (cf. [102], Corollary 2.2.4) $T_2 \vdash \mathsf{Con}(T_2) \rightarrow \mathsf{Con}(T_1)$. But surely we don't have $T_2 \vdash \mathsf{Con}(T_2) \rightarrow \mathsf{Con}(T_0)$ since T_0 is proof-theoretically stronger than $T_2 + \mathsf{Con}(T_2)$.

another. A plethora of results in proof theory and set theory seems to provide compelling evidence that $\leq_{\mathsf{Con}}$ is a linear ordering on "natural" theories. To illustrate this by way of examples from set theory, with a few exceptions, large cardinal axioms have been shown to form a well-ordered hierarchy when ordered as follows:

$$\phi \leq_{\mathsf{Con}} \psi \ := \ \mathbf{ZFC} + \phi \leq_{\mathsf{Con}} \mathbf{ZFC} + \psi,$$

where ϕ and ψ are large cardinal axioms. This has not been established for all of the large cardinal axioms which have been proposed to date; but there is strong conviction among set theorists that this will eventually be accomplished (cf. [19, 113]).

The mere fact of linearity of $\leq_{\mathsf{Con}}$ is remarkable. But one must emphasize "natural" here, because one can construct a pair of self-referential sentences which yield incomparable theories. We first give an example of true theories which are not ordered by $\subseteq_{\Pi_1^0}$.

Proposition 2.17 *There are true arithmetic statements ψ_0 and ψ_1 such that the theories $\mathbf{PA} + \psi_0$ and $\mathbf{PA} + \psi_1$ are not comparable with regard to $\subseteq_{\Pi_1^0}$.*

Proof: For existential, arithmetical formulae $\phi \equiv \exists x \phi_0(x)$ and $\psi \equiv \exists u \psi_0(u)$ define

$$\begin{aligned} \phi \preceq \psi \ &:\Leftrightarrow \ \exists x \left[\phi_0(x) \wedge \forall u < x \, \neg\psi_0(u)\right] \\ \phi \prec \psi \ &:\Leftrightarrow \ \exists x \left[\phi_0(x) \wedge \forall u \leq x \, \neg\psi_0(u)\right]. \end{aligned} \tag{26}$$

For T an extension of $\mathbf{PRA}$, let $\Box_T(\phi) := \exists x \, \mathsf{Proof}_T(x, \ulcorner \phi \urcorner)$. By the Diagonalization Lemma (cf. [102], Theorem 2.2.1) we find a sentence θ so that

$$\mathbf{PA} \vdash \theta \ \leftrightarrow \ \Box_{\mathbf{PA}}(\neg\theta) \preceq \Box_{\mathbf{PA}}(\theta). \tag{27}$$

Let

$$\begin{aligned} \psi_0 \ &:= \ \neg\theta \\ \psi_1 \ &:= \ \neg(\Box_{\mathbf{PA}}(\theta) \prec \Box_{\mathbf{PA}}(\neg\theta)). \end{aligned} \tag{28}$$

We claim that ψ_0 and ψ_1 are both true. To see that ψ_0 is true, note that because of (27), θ implies $\mathbf{PA} \vdash \mathrm{Pr}_{\mathbf{PA}}(\ulcorner \neg\theta \urcorner)$, yielding $\neg\theta$, so ψ_0 is true. As a result we get $\theta \rightarrow \neg\theta$, which implies $\neg\theta$ and hence ψ_0.

To see that ψ_1 is true, note that $\Box_{\mathbf{PA}}(\theta) \prec \Box_{\mathbf{PA}}(\neg\theta)$ implies $\mathrm{Pr}_{\mathbf{PA}}(\ulcorner \theta \urcorner)$, which yields θ, and by the foregoing arguments also $\neg\theta$; thus $\neg(\Box_{\mathbf{PA}}(\theta) \prec \Box_{\mathbf{PA}}(\neg\theta))$ must hold.

Let $T_0 := \mathbf{PA} + \psi_0$ and $T_1 := \mathbf{PA} + \psi_1$. We claim that

$$T_0 \nvdash \psi_1, \tag{29}$$
$$T_1 \nvdash \psi_0. \tag{30}$$

For a contradiction, assume $T_0 \vdash \psi_1$. Then $\mathbf{PA} \vdash \neg\psi_0 \vee \psi_1$, whence, using (27),

$$\begin{aligned} \mathbf{PA} \ \vdash \ & \exists x[\mathsf{Proof}_{\mathbf{PA}}(x, \ulcorner\neg\theta\urcorner) \wedge \forall u < x \neg\mathsf{Proof}_{\mathbf{PA}}(u, \theta)] \\ & \vee \forall x[\mathsf{Proof}_{\mathbf{PA}}(x, \ulcorner\theta\urcorner) \rightarrow \exists u < x \mathsf{Proof}_{\mathbf{PA}}(u, \neg\theta)]. \end{aligned}$$

The latter yields

$$\mathbf{PA} \vdash \forall x\,[\mathsf{Proof}_{\mathbf{PA}}(x, \ulcorner\theta\urcorner) \rightarrow \exists u < x \mathsf{Proof}_{\mathbf{PA}}(u, \neg\theta)]$$

and hence $\mathbf{PA} \vdash \psi_1$. Further, $\mathbf{PA} \vdash \psi_1$ implies $\mathbf{PA} \vdash \mathrm{Pr}_{\mathbf{PA}}(\ulcorner\theta\urcorner) \rightarrow \theta$, and thus $\mathbf{PA} \vdash \theta$ by Löb's theorem (cf. [102], 4.1.1). The latter yields $\mathbf{PA} \vdash \exists x \, \mathsf{Proof}_{\mathbf{PA}}(x, \ulcorner\neg\theta\urcorner)$ and hence $\mathbf{PA} \vdash \neg\theta$. But then we have $\mathbf{PA} \vdash \theta$ as well as $\mathbf{PA} \vdash \neg\theta$ and $\mathbf{PA}$ would be inconsistent.

To show $T_1 \nvdash \psi_0$, we assume $T_1 \vdash \psi_0$, working towards a contradiction. We then get $\mathbf{PA} \vdash \neg\psi_1 \vee \neg\theta$ which yields $\mathbf{PA} \vdash \neg(\Box_{\mathbf{PA}}(\theta) \prec \Box_{\mathbf{PA}}(\neg\theta))$ and hence $\mathbf{PA} \vdash \neg\theta$ by (27). Thus we have $\mathbf{PA} \vdash \mathrm{Pr}_{\mathbf{PA}}(\ulcorner\neg\theta\urcorner)$. $\mathbf{PA} \vdash \neg\theta$ and $\mathbf{PA} \vdash \mathrm{Pr}_{\mathbf{PA}}(\ulcorner\neg\theta\urcorner)$ together imply $\mathbf{PA} \vdash \mathrm{Pr}_{\mathbf{PA}}(\ulcorner\theta\urcorner)$, and hence the contradiction $\mathbf{PA} \vdash \neg\mathsf{Con}(\mathbf{PA})$. □

Next, we give an example of a pair of true theories S_0, S_1 which cannot be compared with regard to $\leq_{\mathsf{Con}}$.

Proposition 2.18 *There is a pair of sound theories S_0, S_1 which are extensions of* **PRA** *such that*

$$\mathbf{PA} \nvdash \mathsf{Con}(S_0) \rightarrow \mathsf{Con}(S_1), \tag{31}$$
$$\mathbf{PA} \nvdash \mathsf{Con}(S_1) \rightarrow \mathsf{Con}(S_0).$$

Proof: By [103], chap. 7, Corollary 2.6, one can construct Π^0_1 sentences χ, η satisfying

$$\mathbf{PRA} + \mathsf{Con}(\mathbf{PRA}) \vdash \psi_0 \leftrightarrow \mathsf{Con}(\mathbf{PRA} + \chi), \tag{32}$$
$$\mathbf{PRA} + \mathsf{Con}(\mathbf{PRA}) \vdash \psi_1 \leftrightarrow \mathsf{Con}(\mathbf{PRA} + \eta),$$

where ψ_0, ψ_1 are from Example 2.17. Since ψ_0 and ψ_1 are true, $\mathsf{Con}(\mathbf{PRA}+\chi)$ and $\mathsf{Con}(\mathbf{PRA} + \eta)$ must be true, too. Therefore, as χ is Π^0_1 it must be true as well, for otherwise $\mathbf{PRA}$ would prove $\neg\chi$, yielding that $\mathbf{PRA} + \chi$ is inconsistent, colliding with $\mathsf{Con}(\mathbf{PRA} + \chi)$ being true. By the same token, η is true.

Now set $S_0 := \mathbf{PRA}+\chi$ and $S_1 := \mathbf{PRA}+\eta$. Note that $\mathbf{PA} \vdash \mathsf{Con}(\mathbf{PRA})$. Thus, using (32), $\mathbf{PA} \vdash \mathsf{Con}(S_0) \rightarrow \mathsf{Con}(S_1)$ would imply $\mathbf{PA} \vdash \psi_0 \rightarrow \psi_1$ and $\mathbf{PA} \vdash \mathsf{Con}(S_1) \rightarrow \mathsf{Con}(S_0)$ would imply $\mathbf{PA} \vdash \psi_1 \rightarrow \psi_0$, both contradicting the results of Proposition 2.17. □

2.8 The proof-theoretic ordinal of a theory and the supremum of its provable recursive well-orderings

In several papers and books the calibration of $|T|_{\text{sup}}$ has been called *ordinal analysis of T*. The definition of $|T|_{\text{sup}}$ has the advantage that it is not notation-sensitive. But as to the activity named "ordinal analysis" it is left completely open what constitutes such an analysis. One often encounters this kind of sloppy talk of ordinals in proof theory, though it is mostly a shorthand for conveying a far more interesting result.

In this subsection the norm $|\cdot|_{\text{sup}}$ will be compared with the other previously introduced norms. It will also become clear that, in general, the mere knowledge of $|T|_{\text{sup}}$ is not the goal of an ordinal analysis of T.

First, it should be mentioned that, in general, $|T|_{\text{sup}}$ has several equivalent characterizations; though some of these hinge upon the mathematical strength of T. As the next the result below will show, the concept $|T|_{\text{sup}}$ is very robust.

Proposition 2.19 *(i) Suppose that for every elementary well-ordering* $\langle A, \prec\rangle$, *whenever* $T \vdash \mathrm{WO}(A, \prec)$, *then*

$$T \vdash \forall u\,[A(u) \rightarrow (\forall v \prec u P(v)) \rightarrow P(u)] \rightarrow \forall u\,[A(u) \rightarrow P(u)]$$

holds for all provably recursive predicates P of T. Then

$$\begin{aligned} |T|_{\text{sup}} &= \sup\{\alpha : \ \alpha \textit{ is provably elementary in } T\} \qquad (33)\\ &= \sup\{\alpha : \ \alpha \textit{ is provably } \Sigma^0_1 \textit{ in } T\}. \end{aligned}$$

Moreover, if $T \vdash \mathrm{WO}(A, \prec)$ and $A, \prec$ *are provably recursive in T, then one can find an elementary well-ordering* $\langle B, \lessdot\rangle$ *and a recursive function f such that* $T \vdash \mathrm{WO}(B, \lessdot)$, *$f$ is provably recursive in T, and T proves that f supplies an order isomorphism between* $\langle B, \lessdot\rangle$ *and* $\langle A, \prec\rangle$.

Examples for (i) are the theories $\mathbf{I\Sigma_1}$, $\mathbf{WKL_0}$ *and* $\mathbf{PA}$.

(ii) If T comprises $\mathbf{ACA_0}$, *then*

$$|T|_{\text{sup}} = \sup\{\alpha : \ \alpha \textit{ is provably arithmetic in } T\}. \qquad (34)$$

(iii) If T comprises $\mathbf{\Sigma^1_1 - AC_0}$, *then*

$$|T|_{\text{sup}} = \sup\{\alpha : \ \alpha \textit{ is provably analytic in } T\}, \qquad (35)$$

where a relation on $\mathbb{N}$ *is called* analytic *if it is lightface* Σ^1_1.

Proof: **(i)**: Suppose $T \vdash \mathrm{WO}(A, \lhd)$, where A and $\lhd$ are defined by Σ^0_1 arithmetic formulae. We shall reason informally in T. We may assume that A contains at least two elements since there are elementary well-orderings for any finite order-type. Without loss of generality we may also assume $0 \notin A$ as $\langle A, \lhd \rangle$ could be replaced by $\langle \{n+1 : n \in A\}, \{(n+1, m+1) : n \lhd m\} \rangle$. A crucial observation is now that there are elementary R and f such that $x \lhd y \leftrightarrow \exists z R(x,y,z)$ and f enumerates A, i.e. $A = \{f(n) : n \in \mathbb{N}\}$. It is wellknown that such A and f can be chosen among the primitive recursive ones; the usual proof actually furnishes this stronger result (cf. [89], p. 30).

Next, define a function h by $h(0) = 0$, and $h(v+1) = f(i)$ if i is the smallest integer $\leq v+1$ such that $f(i) \neq h(0), \ldots, f(i) \neq h(v-1), f(i) \neq h(v)$ and

$$\forall u \leq v \, \exists w \leq v \, [h(u) \neq 0 \rightarrow R(f(i), h(u), w) \vee R(h(u), f(i), w)];$$

let $h(v+1) = 0$ if there is no such $i \leq v+1$. Clearly, $h(v) \leq \Pi_{u \leq v} f(u)$. Thus h is a primitive recursive function bounded by an elementary function. As the auxiliary functions entering the definition of h are elementary, h is elementary too (cf. [89], Theorem 3.1). Obviously, h enumerates $\{0\} \cup A$, moreover, for each $a \in A$ there is exactly one v such that $h(v) = a$.

Define the elementary relation $\lessdot$ via

$$x \lessdot y \quad \text{iff} \quad \exists w \leq \max(x,y) \, R(h(x), h(y), w). \tag{36}$$

We want to show that $\lessdot$ linearly orders the elementary set $B := \{n : h(n) \neq 0\}$. If x is in the field of $\lessdot$, i.e. $\exists y \, (x \lessdot y \vee y \lessdot x)$, then clearly $x \in B$ by definition of $\lessdot$ and h. Conversely, if $h(x) \neq 0$, then $h(x) \in A$, and thus $h(x) \lhd a \vee a \lhd h(x)$ for some a since A has at least two elements. Pick y such that $a = h(y)$. By definition of h, $\exists w \leq \max(x,y) \, [R(h(x), h(y), w) \vee R(h(y), h(x), w)]$. Hence $x \lessdot y \vee y \lessdot x$.

As $\lessdot$ is clearly irreflexive, to verify $\mathrm{LO}(B, \lessdot)$ it remains to be shown that $\lessdot$ is transitive. Assume $x \lessdot y \wedge y \lessdot z$. Then $h(x) \lhd h(z)$, and, by definition of h, if $a < z$ then $\exists w \leq z \, R(h(x), h(z), w)$, whereas $z < x$ implies $\exists w \leq x \, R(h(x), h(z), w)$; thus $x \lessdot z$.

To prove $\mathrm{WF}(B, \lessdot)$, assume

$$\forall x \in B \, [\forall y \lessdot x U(y) \rightarrow U(x)]. \tag{37}$$

We want to show $\forall x \in B \, U(x)$. Define

$$g(x) \;=\; \begin{cases} \text{least } v.h(v) = x & \text{if } v \in A \\ 0 & \text{otherwise.} \end{cases}$$

Notice that g is provably recursive in T. Let $G(u)$ be the formula $U(g(u))$, and assume $v \in A$ and $\forall u \lhd v \, G(u)$. Then $\forall y \lessdot g(v) \, F(y)$ as $y \lessdot g(v)$ yields

$h(y) \lhd h(x) = v$. So (37) yields $U(g(v))$; thus $G(v)$. We then get $\forall v{\in}A\, G(v)$ employing $\mathrm{WO}(A, \lhd)$. Hence $\forall x{\in}B\, U(x)$. The upshot of the foregoing is that

$$T \vdash \mathrm{WO}(B, \lessdot). \tag{38}$$

The desired result now follows by noticing that h furnishes an order preserving mapping from $\langle B, \lessdot\rangle$ onto $\langle A, \lhd\rangle$ (provably in T), thereby yielding $|\lessdot| = |\lhd|$.

(ii): We are going to draw on the notations of Lemma 2.20 below. Let $\lhd$ be a binary Σ^1_1 relation on $\mathbb{N}$ such that $T \vdash \mathrm{WO}(\lhd)$. Let

$$\mathfrak{S} \;:=\; \{e \in Rec:\ \exists f\, \mathsf{Emb}(f, \prec_e, \lhd) \ \wedge\ \mathrm{LO}(\prec_e)\},$$

where $\mathsf{Emb}(f, \prec_e, \lhd)$ stands for $\forall n \forall m\, [n \prec_e m \ \rightarrow\ f(n) \lhd f(m)]$.

The formula "$\exists f\, \mathsf{Emb}(f, \prec_e, \lhd)$" is Σ^1_1. Note that $T \vdash \mathfrak{S} \subseteq \mathfrak{W}_{Rec}$. By Lemma 2.20 below (a formalized, effective version of the Σ^1_1 Bounding Principle) we can find an integer a such that

$$T \vdash \bar{a} \in \mathfrak{W}_{Rec} \setminus \mathfrak{S},$$

in particular, $T \vdash \neg\exists f\, \mathsf{Emb}(f, \prec_{\bar{a}}, \lhd)$, and hence the Π^1_1 faithfulness of T yields $|\lhd| \le |\prec_a|$, thus $|\lhd| < |T|_{\mathsf{sup}}$.

(iii): This time let $\lhd$ be a binary Σ^1_1 relation on $\mathbb{N}$. The proof is basically the same as for (ii) though the formula "$\exists f\, \mathsf{Emb}(f, \prec_e, \lhd)$" may not be strictly Σ^1_1, but it is equivalent to a Σ^1_1 formula provably in T, by using the Σ^1_1 axiom of choice.

Lemma 2.20 *Let* $Rec := \{e \in \mathbb{N} :\ e$ *is an index of a total recursive function*$\}$. *With each* $e \in Rec$ *there is associated a relation* $\prec_d$ *via* $n \prec_d m :\Leftrightarrow \{d\}(\langle n, m\rangle) = 0$, *where* $\langle .,. \rangle$ *is a primitive recursive pairing function. Let*

$$\mathfrak{W}_{Rec} \;:=\; \{e \in \mathbb{N} :\ e \in Rec \ \wedge\ \mathrm{WO}(\prec_e)\}.$$

Suppose $H(x)$ *is a* Σ^1_1 *formula such that*

$$T \vdash \forall n[H(n) \rightarrow x \in \mathfrak{W}_{Rec}].$$

Then there exists $e \in Rec$ *such that*

$$T \vdash \bar{e} \in \mathfrak{W}_{Rec} \ \wedge\ \neg H(\bar{e}).$$

Proof: See [76], Lemma 1.1. □

Theorem 2.21 *Let* T *be a* Σ^1_1 *axiomatizable theory.*

(i) If T *is* Π^1_1*-faithful, then* $|T|_{\mathsf{sup}} < \omega_1^{CK}$.

(ii) If $\mathbf{ACA_0} \subseteq T$ *and* $|T|_{\sup} < \omega_1^{CK}$*, then* T *is* Π^1_1*-faithful.*

(iii) There are consistent primitive recursive theories T *such that* $|T|_{\sup} = \omega_1^{CK}$*.*

Proof: **(i)**: The set $X := \{e \in Rec : T \vdash \mathrm{WO}(\prec_{\bar{e}})\}$ is Σ^1_1. Π^1_1-faithfulness ensures that $X \subseteq \mathfrak{W}_{Rec}$. So by Σ^1_1 bounding there is a recursive well-ordering that has a bigger order-type than all the orderings $\prec_e$ with $e \in \mathrm{Rec}$. Consequently, $|T|_{\sup} < \omega_1^{CK}$.

(ii): For a contradiction, suppose T is not Π^1_1-faithful. Then there is a false Π^1_1-sentence B such that $T \vdash B$. Rendering B in Π^1_1 normal form, one obtains a primitive recursive well-ordering $\prec$ such that $\mathbf{ACA_0} \vdash B \leftrightarrow \mathrm{WF}(\prec)$. As a result, $T \vdash \mathrm{WF}(\prec)$, but $\prec$ is not well-founded. Now let $\lhd$ be an arbitrary recursive well-ordering. Put

$$\mathbb{T} := \{\langle\rangle\} \cup \{\langle\langle x_0, y_0\rangle, \dots, \langle x_i, y_i\rangle\rangle : x_i \lhd \cdots \lhd x_0;\ y_i \prec \cdots \prec y_0;\ i \in \mathbb{N}\}$$

and let $<_{\mathbb{T}}$ be the Kleene-Brouwer linearization of $\mathbb{T}$. Since $T \vdash \mathrm{WF}(\prec)$ it follows $T \vdash \mathrm{WO}(<_{\mathbb{T}})$. Since $\lhd$ is well-founded, $<_{\mathbb{T}}$ is a well-ordering in the "real world". We claim that $<_{\mathbb{T}}$ has at least the order-type of $\lhd$. To this end, let $(f(n))_{n\in\mathbb{N}}$ be an infinite descending $\prec$ sequence, i.e. $f(n+1) \prec f(n)$ for all n. Put

$$\mathbb{S} := \{\langle\rangle\} \cup \{\sigma \in \mathbb{T} : \sigma = \langle\langle x_0, f(0)\rangle, \dots, \langle x_i, f(i)\rangle\rangle\}.$$

Being a subtree of $\mathbb{T}$, the Kleene-Brouwer ordering on $\mathbb{S}$, $<_{\mathbb{S}}$, is also well-founded. Define $g(x)$ be the $<_{\mathbb{S}}$-least $\sigma \in \mathbb{S}$ of the form $\sigma * \langle\langle x, f(i)\rangle\rangle$. Now, if $y \lhd x$, then

$$g(y) \leq_{\mathbb{S}} g(x) * \langle\langle y, f(i+1)\rangle\rangle <_{\mathbb{S}} g(x).$$

This shows that $\mathsf{Emb}(g, \lhd, <_{\mathbb{S}})$ and a fortiori $\mathsf{Emb}(g, \lhd, <_{\mathbb{T}})$, verifying the claim. As $\lhd$ was an arbitrary recursive well-ordering, it follows $|T|_{\sup} = \omega_1^{CK}$, contradicting $|T|_{\sup} < \omega_1^{CK}$.

(iii): Due to (ii), an example is provided by $\mathbf{ACA_0} + \neg\mathsf{Con}(\mathbf{ACA_0})$. □

Remark 2.22 If one considers it worthwhile looking at theories which are not Π^1_1 faithful, though consistent, one can amuse oneself by producing theories S, T such that S and T are equiconsistent but differ with respect to their $|.|_{\sup}$ norms. Just let $T := \mathbf{ACA_0}$ and $S := \mathbf{ACA_0} + \neg\mathsf{Con}(\mathbf{ACA_0})$. Then $T \equiv_{\mathsf{Con}} S$ by [102], Corollary 2.2.4 and $|T|_{\sup} < |S|_{\sup}$ by Theorem 2.21, (ii).

Sloppy talk about what constitutes an ordinal analysis of a theory T is prone to trivialization. Given a faithful theory T, one easily concocts a definition of a well-ordering whose order-type is $|T|_{\sup}$ by simply amalgamating the provable well-orderings of T into one big ordering.

Theorem 2.23 *Let T be a primitive recursive Π^1_1-faithful theory which comprises* $\mathbf{RCA_0}$. *Then there exists a primitive recursive well-ordering $\lhd$ such that*

$$|T|_{\sup} = |\lhd|, \tag{39}$$
$$\mathbf{RCA_0} \vdash \mathrm{WF}(\lhd) \rightarrow \mathsf{Con}(T); \tag{40}$$
$$\textit{For each proper initial segment } \lhd' \textit{ of } \lhd,\ T \vdash \mathrm{WO}(\lhd'). \tag{41}$$

The third assertion probably requires some clarification. For definiteness, by a proper initial segment of $\lhd$ we mean any ordering of the form $\{(n,m) : n \lhd m \wedge m \lhd n_0\}$ such that $n_0 \lhd k$ for some k.

Define

$$\begin{aligned} \phi(n) \ &:\Leftrightarrow\ \exists e\, \exists m \left[n = \langle e, m\rangle \wedge \mathsf{Proof}_T(m, \ulcorner \mathrm{WO}(\prec_{\bar{e}})\urcorner)\right]; \\ x \prec_n y \ &:\Leftrightarrow\ \phi(n) \wedge x \prec_{(n)_0} y; \\ \langle n, x\rangle \lhd \langle n', y\rangle \ &:\Leftrightarrow\ \phi(n) \wedge \phi(n') \wedge \left[n < n' \vee (n = n' \wedge x \prec_{(n)_0} y)\right]. \end{aligned}$$

In view of its definition, $\lhd$ is primitive recursive and $|T|_{\sup} = |\lhd|$.

To verify (40), we reason in $\mathbf{RCA_0}$. Assume $\neg\mathsf{Con}(T)$. Then T is inconsistent and thus proves every statement. In particular, T proves then that the ordering $0 \gtrdot 1 \gtrdot 2 \gtrdot 3 \gtrdot \cdots$ is a well-ordering. But $\lessdot$ is embeddable into $\lhd$; thus $\lhd$ cannot be a well-ordering.

As to (41), let $\lhd\restriction r := \{(n,m) : n \lhd m \wedge m \lhd r\}$ be an initial segment of $\lhd$. Then there exists s such that $r \lhd s$. In particular, $\phi(r)$ and $T \vdash \phi(\bar{r})$. Let $\langle e_0, p_0\rangle, \ldots, \langle e_t, p_t\rangle$ be the list of all pairs $< s$ such that $\mathsf{Proof}_T(e_i, \ulcorner \mathrm{WO}(\prec_{\overline{p_i}}$ $)\urcorner)$. Then

$$\begin{aligned} &T \vdash \forall u \lhd \bar{r}\, [(u)_0 = \langle \overline{e_0}, \overline{p_0}\rangle \vee \cdots \vee (u)_0 = \langle \overline{e_t}, \overline{p_t}\rangle]; \\ &T \vdash \mathrm{WO}(\prec_{\overline{e_0}}) \wedge \cdots \wedge \mathrm{WO}(\prec_{\overline{e_t}}). \end{aligned}$$

The latter implies $T \vdash \mathrm{WO}(\lhd\restriction\bar{r})$. □

Another reason why the ordinal $|T|_{\sup}$, even when presented in the shape of a natural ordinal representation system, does not convey all the information obtained by an ordinal analysis of T is that one can find theories T_1, T_2 of different proof-theoretic strength which satisfy $|T_1|_{\sup} = |T_2|_{\sup}$. More precisely, the ordinal $|T|_{\sup}$ usually doesn't change when one augments T by true Σ^1_1 statements.

In the main, the next result is due to Kreisel. But I couldn't find a reference and don't know how Kreisel proved it.

Proposition 2.24 *Let T be a primitive recursive, Π^1_1-faithful theory of second order arithmetic such that* $\mathbf{PA} \subseteq T$. *Let $\lhd$ be a primitive recursive well-ordering such that $|T|_{\sup} = |\lhd|$ and*

$$\mathbf{PA} + \mathrm{TI}(\lhd) \vdash \mathsf{Proof}_T(\ulcorner F \urcorner) \rightarrow F \tag{42}$$

holds for all arithmetic formulae F which may contain free second order set variables but no free number variables. Then, for any true Σ^1_1 statement B,

$$|T|_{\text{sup}} = |T + B|_{\text{sup}}.$$

Proof: Let $B := \exists X\, C(X)$ be a true Σ^1_1 sentence with $C(X)$ being arithmetic. Let $S := T + B$. Note that S is also Π^1_1 faithful.

We want to show that $|T|_{\text{sup}} = |S|_{\text{sup}}$. So suppose

$$S \vdash \mathrm{WO}(\prec)$$

for some arithmetic well-ordering $\prec$. Then let $E(U)$ be the statement that U is the graph of a function on $\mathbb{N}$ which maps the field of $\lhd$ order-preservingly onto an initial segment (not necessarily proper) of the field of $\prec$. Then

$$S + \exists X\, E(X) \vdash \mathrm{TI}(\lhd).$$

Thus, in view of (42), one gets

$$S \vdash \exists X\, E(X) \rightarrow \mathsf{Proof}_S(\ulcorner \forall X \neg E(X) \urcorner) \rightarrow \forall X \neg E(X). \tag{43}$$

The latter yields (using predicate logic)

$$S \vdash \mathsf{Proof}_S(\ulcorner \forall X \neg E(X) \urcorner) \rightarrow \forall X \neg E(X), \tag{44}$$

and thus, by Löb's Theorem (cf. [102], Theorem 4.1.1),

$$S \vdash \forall X \neg E(X). \tag{45}$$

As a result, since S proves only true statements, $\forall X \neg E(X)$ must be true and therefore the order-type of $\prec$ must be less than the order-type of $\lhd$. In conclusion,

$$|T|_{\text{sup}} = |S|_{\text{sup}}.$$

□

Remark 2.25 In all the examples I know, if T is a subsystem of classical second order arithmetic for which an ordinal analysis has been carried out via an ordinal representation system $(A, \lhd)$, (42) is satisfied.

If one takes, e.g. $B := \mathsf{Con}(T)$, then S is of greater proof-theoretic strength than T.

3 Rewards of ordinal analyses and ordinal representation systems

This section is devoted to results that have been achieved through ordinal analyses. They fall into four groups: (1) Consistency of subsystems of classical second order arithmetic and set theory relative to constructive theories, (2) reductions of theories formulated as conservation theorems, (3) combinatorial independence results, and (4) classifications of provable functions and ordinals.

3.1 Hilbert's programme extended: Constructive consistency proofs

A natural modification of Hilbert's programme consists in loosening the requirement of reduction to finitary methods by allowing reduction to constructive methods more generally.[3]

The point of an extended Hilbert programme (H.P.) is that one wants a constructive conception for which there is an absolute guarantee that, whatever one proves in a sufficiently strong classical theory T, say, a fragment of second order arithmetic or set theory, there would be an interpretation of the proof according to which the theorem is contructively true. Moreover, one wants the theory T to be such as to make the process of formalization of mathematics in T almost trivial, in particular T should be sufficiently strong for all practical purposes. This is a very Hilbertian attitude: show once and for all that non-constructive methods do not lead to false constructive conclusions and then proceed happily on with non-constructive methods.

There are several aspects of an extended H.P. that require clarification. Let's first dispense with the question of how to delineate a sufficiently strong classical theory T as this is an easy one. It was already observed by Hilbert-Bernays [38] that classical analysis can be formalized within second order arithmetic. Further scrutiny revealed that a small fragment is sufficient. Even without knowledge of that program carried out under the rubric of "reverse mathematics", it is easily seen that most of ordinary mathematics can be

[3]Such a shift from the original programme is implicit in Hilbert-Bernays' [38] apparent acceptance of Gentzen's consistency proof for **PA** under the heading "Überschreitung des bisherigen methodischen Standpunktes der Beweistheorie". The need for a modified Hilbert programme has clearly been recognized by Gentzen (cf. [34]) and Bernays [6]: *It thus became apparent that the "finite Standpunkt" is not the only alternative to classical ways of reasoning and is not necessarily implied by the idea of proof theory. An enlarging of the methods of proof theory was therefore suggested: instead of reduction to finitist methods of reasoning it was required only that the arguments be of a constructive character, allowing us to deal with more general forms of inferences.*

formalized in $\Delta^1_2-\mathbf{CA}+\mathbf{BI}$ without effort (**BI** stands for the principle of bar induction, i.e. the assertion that transfinite induction along a well-founded set relation holds for arbitrary classes; this is the pendant of the foundation axiom in set theory). A more convenient framework for formalizing mathematics is set theory. A set theory which proves the same theorems of second order arithmetic is the set theory **KPi** which is an extension of Kripke-Platek set theory via an axiom that asserts the existence of many admissible sets, namely every set is contained in an admissible set.

It may not be clear how ordinal analysis can contribute to an extended H.P. The system of ordinal representations used in consistency proofs of stronger and stronger theories becomes more and more complicated. To say that the consistency proof has been carried out by transfinite induction on a certain complicated ordering tells us nothing about what constructive principles are involved in the proof of its well-ordering. Are we to take transfinite induction with respect to these ordinal representation as a fundamental constructive principle? The answer could hardly be "yes" lest only specialists on ordinal representations should be convinced. Therefore it becomes necessary to give a detailed account of what constructive principles are allowed in any well-ordering proof and to carry out well-ordering proofs for ordinal representations using only these principles. The problem is thus to find some basic constructive principles upon which a coherent system of constructive reasoning may be built. Several frameworks for constructivism that relate to Bishop's constructive mathematics as theories like **ZFC** relate to Cantorian set theory have been proposed by Myhill, Martin–Löf, Feferman and Aczel. Among those are Feferman's "Explicit mathematics", a constructive theory of operations and classes ([23, 24]), and Martin-Löf's intuitionistic type theory of [59] (the latter does not contain Russell's infamous *reducibility axiom*). Type theory is a logic free theory of constructions within which the logical notions can be defined whereas systems of Explicit mathematics leave the logical notions unanalysed. For this reason we consider type theory to be more fundamental.

By employing an ordinal analysis for **KPi** it has been shown that **KPi** and consequently $\Delta^1_2-\mathbf{CA}+\mathbf{BI}$ can be reduced to both these theories.

Theorem 3.1 (Feferman [23], Jäger [43], Jäger and Pohlers [45]) $\Delta^1_2-\mathbf{CA}+\mathbf{BI}$ *(or* **KPi***) and* $\mathbf{T_0}$ *are proof-theoretically equivalent. In particular, these theories prove the same theorems in the negative arithmetic fragment.*

Theorem 3.2 (Rathjen [36]; Setzer [101]) *The consistency of* $\Delta^1_2-\mathbf{CA}+\mathbf{BI}$ *and* **KPi** *is provable in Martin-Löf's 1984 type theory.*

On the part of the intuitionists/constructivists, the following objection could be raised against the significance of consistency proofs: even if it had been constructively demonstrated that the classical theory T cannot lead to mutually contradictory results, the theorems of T would nevertheless be propositions without sense and their investigation therefore an idle pastime. Well, it turns out that the constructive well-ordering proof of the representation system used in the analysis of $\Delta^1_2 - \mathbf{CA{+}BI}$ yields more than the mere consistency of the latter system. For the important class of Π^0_2 statements one obtains a conservativity result.

Theorem 3.3 (Rathjen [36]; Setzer [101])

- *The soundness of the negative arithmetic fragment of* $\Delta^1_2 - \mathbf{CA{+}BI}$ *(or* **KPi***) is provable in Martin-Löf's 1984 type theory.*
- *Every* Π^0_2 *statement provable in* $\Delta^1_2 - \mathbf{CA{+}BI}$ *(or* **KPi***) has a proof in Martin-Löf's 1984 type theory.*

3.2 Reductions of theories formulated as conservation theorems

The motivation for an extended Hilbert programme depends on the conviction that constructive methods are, in some sense, superior. Another way to conceive of the results mentioned in the previous subsection is to simply view them as proof-theoretic reductions and to formulate them as conservation theorems. Ordinal analyses have been used many times to prove that a foundationally interesting theory is in some sense reducible to or equiconsistent with another foundationally interesting theory. Here I'm going to list just a few examples, their selection being a very biased choice. A plethora of further reductions can be found in [70, 44, 15, 73, 72].

1. The proofs that the theories $\mathbf{ATR}_0$ and $\mathbf{KPi}_0$ are reducible to Feferman's system of predicative analysis, **IR**, in the sense that they are conservative over **IR** for Π^1_1 sentences involves the ordinal Γ_0. The foundational significance of these systems is as follows. $\mathbf{ATR}_0$ is a subsystem of second order arithmetic that frequently arises in reverse mathematics and is equivalent to many mathematical statements, e.g. the open Ramsey theorem, the perfect set theorem, Ulm's theorem, the König duality theorem for countable bipartite graphs, etc. $\mathbf{KPi}_0$ is, on the one hand, an extension of Kripke-Platek set theory via an axiom that asserts that any set is contained in an admissible set, but, on the other hand, a weakening of Kripke-Platek set theory in that the foundation axiom is completely missing.

   The reductions can be described as contributing to a foundational program of predicative reductionism. The reduction of $\mathbf{ATR}_0$ to **IR** was obtained in [29]. The reduction of $\mathbf{KPi}_0$ to **IR** is due to [44].

2. The study of formal theories featuring inductive definitions in both single and iterated form was initiated by Kreisel [50]. The immediate stimulus was the question of constructive justification of Spector's 1961 consistency proof for analysis via his interpretation in the so-called bar-recursive functionals of finite type. Let ν denote a fixed ordinal in a given ordinal representation system. Ordinal analysis has shown that the classical theory $\mathbf{ID}^c_\nu$ of ν-times iterated arithmetical inductions is reducible to the intuitionistic theory $\mathbf{ID}^i_\nu(\mathfrak{O})$ of ν-times iterated constructive number classes $\mathfrak{O}$. The history of these results is described in the monograph [14].[4]

3. T_0 is Feferman's system of Explicit mathematics. The results of Feferman [23], Jäger [43], Jäger and Pohlers [45]) yield that $\Delta^1_2 - \mathbf{CA}{+}\mathbf{BI}$, **KPi**, and $\mathbf{T_0}$ are proof-theoretically equivalent. In particular, these theories prove the same theorems in the negative arithmetic fragment.

   No proof of the above result has been found that doesn't use ordinal representations.

4. Inspired by work of Myhill [62] on constructive set theories, Aczel (cf. [1, 2, 3]) proposed an intuitionistic set theory, termed *Constructive Zermelo-Fraenkel set theory* (**CZF**), that bears a close relation to Martin-Löf type theory. The novel ideas were to replace Powerset by the (classically equivalent) Subset Collection Axiom and to discard full Comprehension while strengthening Collection to Strong Collection. Aczel corroborated the constructiveness of **CZF** by interpreting it in Martin-Löf's intuitionistic type theory. A very nice aspect of **CZF** is the fact that one can develop a good theory of large sets with the right consistency strength. Since in intuitionistic set theory $\in$ is not a linear ordering on ordinals the notion of a cardinal does not play a central role. Consequently, one talks about *"large set properties"* instead of *"large cardinal properties"*. Classically though, the large cardinal axioms and the pertinent large set axiom are of the same strength.

   Up to now, the notions of inaccessible, Mahlo and 2-strong sets that classically correspond to inaccessible, Mahlo and weakly compact cardinals, respectively, have been investigated (cf. [85, 86, 87]). As to consistency strength and conservativity the following theories have the same consistency strength and actually prove the same Π^0_2-sentences:

   (i) $\mathbf{CZF} + \forall x \exists I\,[x{\in}I \wedge$ "I inaccessible"$]$ and
   $\mathbf{KP} + \forall \alpha \exists \kappa\,[\alpha{\in}\kappa \wedge$ "κ recursively inaccessible"$]$.

[4]For limit ordinals ν, Sieg (cf. [14]) obtained the reduction $\bigcup_{\alpha<\nu}\mathbf{ID}^c_\alpha \equiv \bigcup_{\alpha<\nu}\mathbf{ID}^i_\alpha(\mathfrak{O})$ without the use of ordinal analysis, but his approach is still proof-theoretic as it employs cut-elimination for infinitary derivations.

(ii) $\mathbf{CZF} + \forall x \exists M\, [x \in M \ \wedge$ "M Mahlo"$]$ and
$\mathbf{KP} + \forall \alpha \exists \kappa\, [\alpha \in \kappa \ \wedge$ "κ recursively Mahlo ordinal"$]$.

(iii) $\mathbf{CZF} + \forall x \exists K\, [x \in K \ \wedge$ "K 2-strong"$]$ and
$\mathbf{KP} + \forall \alpha \exists \kappa\, [\alpha \in \kappa \ \wedge$ "κ Π_3-reflecting"$]$.

The proof that the intuitionistic theory has at least the strength of the classical one requires an ordinal analysis of the classical theories as given in [74, 75, 82] and a proof of the well-foundedness of the pertinent ordinal representation system in the intuitionistic theory.

3.3 Combinatorial independence results and new combinatorial principles

Since 1931, the year Gödel's Incompleteness Theorems were published, logicians have been looking for a strictly mathematical example of an incompleteness in first-order Peano arithmetic, one which is mathematically simple and interesting and does not require the numerical coding of notions from logic. The first such examples were found early in 1977. The most elegant of these is a strengthening of the Finite Ramsey Theorem due to Paris and Harrington (cf. [65]). The original proofs of the independence of combinatorial statements from **PA** all used techniques from non-standard models of arithmetic. Only later on alternative proofs using proof-theoretic techniques were found. However, results from ordinal-theoretic proof theory turned out to be pivotal in providing independence results for stronger theories than **PA**, and even led to a new combinatorial statement. The stronger theories referred to are Friedman's system $\mathbf{ATR}_0$ of *arithmetical transfinite recursion* and the system $\Pi^1_1 - CA$ based on Π^1_1-comprehension. The independent combinatorial statements have their origin in certain embeddability questions in the theory of finite graphs. The first is a famous theorem of Kruskal asserting that every set of finite trees has only finitely many minimal elements.

Definition 3.4 A *finite tree* is a finite partially ordered set $\mathbb{B} = (B, \leq)$ such that:

(i) B has a smallest element (called the *root* of $\mathbb{B}$);

(ii) for each $s \in B$ the set $\{t \in B : t \leq s\}$ is a totally ordered subset of B.

Definition 3.5 For finite trees $\mathbb{B}_1$ and $\mathbb{B}_2$, an *embedding* of $\mathbb{B}_1$ into $\mathbb{B}_2$ is a one-to-one mapping $f : \mathbb{B}_1 \longrightarrow \mathbb{B}_2$ such that $f(a \wedge b) = f(a) \wedge f(b)$ for all $a, b \in \mathbb{B}_1$, where $a \wedge b$ denotes the infimum of a and b.

We write $\mathbb{B}_1 \leq \mathbb{B}_2$ to mean that there exists an embedding $F : \mathbb{B}_1 \longrightarrow \mathbb{B}_2$.

Theorem 3.6 (Kruskal's theorem) *For every infinite sequence of trees* $(\mathbb{B}_k : k < \omega)$, *there exist indices* i *and* j *such that* $i < j < \omega$ *and* $\mathbb{B}_i \leq \mathbb{B}_j$. *(In particular, there is no infinite set of pairwise nonembeddable trees.)*

Theorem 3.7 *Kruskal's Theorem is not provable in* $\mathbf{ATR}_0$ *(*cf. [100]*).*

The proof of the above independence result exploits a connection between finite trees and ordinal representations for ordinals $< \Gamma_0$ and the fact that Γ_0 is the proof-theoretic ordinal of $\mathbf{ATR}_0$. Each ordinal representation $\mathfrak{a}$ is assigned a finite tree $\mathbb{B}_\mathfrak{a}$ to the effect that for two representations $\mathfrak{a}$ and $\mathfrak{b}$, $\mathbb{B}_\mathfrak{a} \leq \mathbb{B}_\mathfrak{b}$ implies $\mathfrak{a} \leq \mathfrak{b}$. Hence Kruskal's theorem implies the well-foundedness of Γ_0 and is therefore not provable in $\mathbf{ATR}_0$. The connection between finite trees and ordinal representations for ordinals $< \Gamma_0$ was noticed by Friedman (cf. [100]) and independently by Diana Schmidt (cf. [93]).

A hope in connection with ordinal analyses is that they lead to new combinatorial principles which encapsulate considerable proof-theoretic strength. Examples are still scarce. One case where ordinal notations led to a new combinatorial result was Friedman's extension of Kruskal's Theorem, EKT, which asserts that finite trees are well-quasi-ordered under gap embeddability (see [100]). The gap condition imposed on the embeddings is directly related to an ordinal notation system that was used for the analysis of Π^1_1 comprehension. The principle EKT played a crucial role in the proof of the graph minor theorem of Robertson and Seymour (see [30]).

Definition 3.8 For $n < \omega$, let $\mathcal{B}_n$ be the set of all finite trees with labels from n, i.e. $(\mathbb{B}, \ell) \in \mathcal{B}_n$ if $\mathbb{B}$ is a finite tree and $\ell : B \to \{0, \ldots, n-1\}$. The set $\mathcal{B}_n$ is quasiordered by putting $(\mathbb{B}_1, \ell_1) \leq (\mathbb{B}_2, \ell_2)$ if there exists an embedding $f : \mathbb{B}_1 \to \mathbb{B}_2$ with the following properties:

1. for each $b \in B_1$ we have $\ell_1(b) = \ell_2(f(b))$;

2. if b is an immediate successor of $a \in \mathbb{B}_1$, then for each $c \in \mathbb{B}_2$ in the interval $f(a) < c < f(b)$ we have $\ell_2(c) \geq \ell_2(f(b))$.

The condition (ii) above is called a *gap condition.*

Theorem 3.9 *For each* $n < \omega$, $\mathcal{B}_n$ *is a well quasi ordering (abbreviated* $\mathrm{WQO}(\mathcal{B}_n)$*), i.e. there is no infinite set of pairwise nonembeddable trees.*

Theorem 3.10 $\forall n < \omega\ \mathrm{WQO}(\mathcal{B}_n)$ *is not provable in* $\Pi^1_1 - \mathbf{CA}_0$.

The proof of Theorem 3.10 employs an ordinal representation system for the proof-theoretic ordinal of $\Pi^1_1 - \mathbf{CA}_0$. The ordinal is $\psi_0(\Omega_\omega)$ in the ordinal representation system of [8] or $\theta\Omega_\omega 0$ in that of [97]. Let $\mathcal{T}(\psi_0(\Omega_\omega))$ denote the ordinal representation system. The connection between $< \omega$ labelled trees and $\mathcal{T}(\psi_0(\Omega_\omega))$ is that $\forall n < \omega$ WQO($\mathcal{B}_n$) implies the wellfoundedness of $\mathcal{T}(\psi_0(\Omega_\omega))$ on the basis of $\mathbf{ACA}_0$. The connection is even closer in that the gap condition imposed on the embeddings between trees is actually gleaned from the ordering of the terms in $\mathcal{T}(\psi_0(\Omega_\omega))$. If one views these terms as labelled trees, then the gap condition is exactly what one needs to ensure that an embedding of two such trees implies that the ordinal corresponding to the first tree is less than the ordinal corresponding to the second tree.

It is also for that reason that criticism had been levelled against the principle EKT for being too contrived or too metamathematical. But this was superseded by the crucial role that EKT played in the proof of the graph minor theorem of Robertson and Seymour (see [30]).

As to the importance attributed to the graph minor theorem, I quote from a book on Graph Theory [17], p. 349.

> *Our goal* [...] *is a single theorem, one which dwarfs any other result in graph theory and may doubtless be counted among the deepest theorems that mathematics has to offer:* in every infinite set of graphs there are two such that one is a minor of the other. *This* minor theorem, *inconspicuous though it may look at first glance, has made a fundamental impact both outside graph theory and within. Its proof, due to Neil Robertson and Paul Seymour, takes well over 500 pages.*

Definition 3.11 Let $e = xy$ be an edge of a graph $G = (V, E)$, where V and E denote its vertex and edge set, respectively. By G/e we denote the graph obtained from G by *contracting* the edge e into a new vertex v_e, which becomes adjacent to all the former neighbours of x and of y. Formally, G/e is a graph (V', E') with vertex set $V' := (V \setminus \{x, y\}) \cup \{v_e\}$ (where v_e is the "new" vertex, i.e. $v \notin V \cup E$) and edge set

$$\begin{aligned} E' \;:=\; & \big\{vw \in E | \{v, w\} \cap \{x, y\} = \emptyset\big\} \\ & \cup \big\{v_e w | xw \in E \setminus \{e\} \;\;\vee\;\; xy \in E \setminus \{e\}\big\}. \end{aligned}$$

If X is obtained from Y by first deleting some vertices and edges, and then contracting some further edges, X is said to be a *minor* of Y. In point of fact, the order in which deletions and contractions are applied is immaterial as any graph obtained from another by repeated deletions and contractions in any order is its minor.

Theorem 3.12 (Robertson and Seymour 1986-1997) *If $G_0, G_1, G_2, \ldots$ is an infinite sequence of finite graphs, then there exist $i < j$ so that G_i is isomorphic to a minor of G_j.*

Corollary 3.13 *(i)* (Vázsonyi's conjecture) *If all the G_k are trivalent, then there exist $i < j$ so that G_i is embeddable into G_j.*

(ii) (Wagner's conjecture) *For any 2-manifold M there are only finitely many graphs which are not embeddable in M and are minimal with this property.*

Theorem 3.14 (Friedman, Robertson, Seymour [30])

(i) GMT *implies* EKT *within, say,* $\mathbf{RCA}_0$.

(ii) GMT *is not provable in* $\Pi^1_1 - \mathbf{CA}_0$.

A further independence result that ensues from ordinal analysis is due to Buchholz [9]. It concerns an extension of the hydra game of Kirby and Paris. It is shown in [9] that the assertion that Hercules has a winning strategy in this game is not provable in the theory $\Pi^1_1 - \mathbf{CA} + \mathbf{BI}$.

3.4 Classifications of provable functions and ordinals

An apt leitmotif for this subsection is provided by Kreisel's question (cf. [49]): *"What more do we know if we have proved a theorem by restricted means than if we merely know that it is true?"*

3.4.1 Provable recursive functions

In the case of **PA** an answer to the foregoing question was provided by Kreisel in [48], where he characterized the provably recursive functions of **PA** as those which are α-recursive for some $\alpha < \varepsilon_0$. However, there is nothing special about **PA** when it comes to extracting the latter kind of information. Indeed, it is a general fact that an ordinal analysis of a theory T yields, as a by-product, a characterization of the provably recursive functions of T. As stated in section 2, an ordinal analysis of T via an ordinal representation system $\langle A, \lhd, \ldots \rangle$ provides a reduction of T to $\mathbf{PA} + \bigcup_{\alpha \in A} \mathrm{TI}(A_{\bar{\alpha}}, \lhd_{\bar{\alpha}})$ and further ensures Π^0_2-conservativity. On the strength of the latter, it suffices to characterize the provably recursive functions of

$$\mathbf{S} := \mathbf{PA} + \bigcup_{\alpha \in A} \mathrm{TI}(A_{\bar{\alpha}}, \lhd_{\bar{\alpha}})$$

for EORSs $\langle A, \lhd, \ldots \rangle$.

Definition 3.15 Let $\alpha \in A$ such that $0 \lhd \alpha$. A number-theoretic function f is called α-recursive if it can be generated by the usual schemes for generating primitive recursive functions plus the following scheme:

$$f(m, \vec{n}) = \begin{cases} h(m, \vec{n}, f(\theta(m, \vec{n}), \vec{n})) & \text{if } 0 \lhd m \lhd \alpha \\ g(m, \vec{n}) & \text{otherwise,} \end{cases}$$

where g, h, θ are α-recursive and θ satisfies $\theta(\beta) \lhd \beta$ whenever $0 \lhd \beta \lhd \alpha$.

Theorem 3.16 *The provably recursive functions of* $\mathbf{PA} + \bigcup_{\alpha \in A} \mathrm{TI}(A_{\bar{\alpha}}, \lhd_{\bar{\alpha}})$ *are excactly the recursive functions which are* α*-recursive for some* $\alpha \in A$.

The technical tool for achieving this characterization is to embed $\mathbf{PA} + \bigcup_{\alpha \in A} \mathrm{TI}(A_{\bar{\alpha}}, \lhd_{\bar{\alpha}})$ into a system of Peano arithmetic with an infinitary rule, the so-called ω-rule, and a *repetition rule*, Rep, which simply repeats the premise as the conclusion. The ω-rule allows one to infer $\forall x \phi(x)$ from the infinitely many premises $\phi(\bar{0}), \phi(\bar{1}), \phi(\bar{2}), \ldots$ (where $\bar{n}$ denotes the nth numeral); its addition accounts for the fact that the infinitary system enjoys cut-elimination. The addition of the Rep rule enables one to carry out a *continuous cut elimination*, due to Mints [60], which is a continous operation in the usual tree topology on prooftrees. A further pivotal step consists in making the ω-rule more constructive by assigning codes to proofs, where codes for applications of finitary rules contain codes for the proofs of the premises, and codes for applications of the ω-rule contain Gödel numbers for partial functions enumerating codes of the premises. The aforementioned enumerating functions can be required to be partial recursive, making the proof trees recursive, or even primitive recursive in the presence of the rule Rep which enables one to stretch recursive trees into primitive recursive trees. Theorem 3.16 can be extracted from Kreisel-Mints-Simpson [53], Lopez-Escobar [54], or Schwichtenberg [99] and was certainly known to these authors. A variant of the characterization of Theorem 3.16 is given in Friedman-Sheard [32], where the provable functions of $\mathbf{PA} + \bigcup_{\alpha \in A} \mathrm{TI}(A_{\bar{\alpha}}, \lhd_{\bar{\alpha}})$ are classified as the *descent recursive functions over A*. But before discussing this and related results, I'd like to draw attention to a more recent approach which has the great advantage over the previous one that one need not bother with codes for infinite derivations. In this approach one adds an extra feature to infinite derivations by which one can exert a greater control on derivations so as to be able to directly read off numerical bounds from cut free proofs of Σ_1^0 statements. This has been carried out by Buchholz-Wainer [11] for the special case of **PA**. In much greater generality and flexibility this approach has been developed by Weiermann [112].

The remainder of this section presents further results about theories of the shape $\mathbf{PA} + \bigcup_{\alpha \in A} \mathrm{TI}(A_{\bar{\alpha}}, \lhd_{\bar{\alpha}})$, thereby providing more information that can be extracted from ordinal analyses. Propositions 3.19, 3.20, and 3.23 are due to Friedman-Sheard [32].

Definition 3.17 For each $\alpha \in A$, $\mathrm{ERWF}(\lhd, \bar{\alpha})$ is the schema

$$\forall \vec{x} \exists y [f(\vec{x}, y) \unlhd f(\vec{x}, y+1) \lor f(\vec{x}, y) \notin A \lor \bar{\alpha} \unlhd f(\vec{x}, y)]$$

for each (definition of an) elementary function f.

ERWF($\lhd$) is the schema

$$\forall \vec{x} \exists y [f(\vec{x}, y) \unlhd f(\vec{x}, y+1) \vee f(\vec{x}, y) \notin A]$$

for each elementary function f.

The schemata PRWF($\lhd, \bar{\alpha}$) and PRWF($\lhd$) are defined identically, except that f ranges over the primitive recursive functions.

Definition 3.18 $\mathbf{DRA}_{\langle A, \lhd \rangle}$ (*Descent Recursive Arithmetic*) is the theory whose axioms are $\mathbf{ERA} + \bigcup_{\alpha \in A} \mathrm{ERWF}(\lhd, \bar{\alpha})$.

$\mathbf{DRA}(\lhd^+)$ is the theory whose axioms are $\mathbf{ERA}$ + ERWF($\lhd$).

The difference is that $\mathbf{DRA}(\lhd)$ asserts only the non-existence of elementary infinitely descending sequences below each $\alpha \in A$, where α is given at the meta-level.

Combined with 2.8 the latter result leads to a neat characterization of the provably recursive functions of $\mathbf{T}$ due to the following observation:

Proposition 3.19 ([32]) *The provably recursive functions of* $\mathbf{DRA}_{\langle A, \lhd \rangle}$ *are all functions* f *of the form*

$$f(\vec{m}) \;=\; g(\vec{m}, \textit{least } n. h(\vec{m}, n) \unlhd h(\vec{m}, n+1)) \tag{46}$$

where g *and* h *are elementary functions and* $\mathbf{ERA} \vdash \forall \vec{x} y\, h(\vec{x}, y) \in A_{\bar{\alpha}}$ *for some* $\alpha \in A$.

The above class of recursive functions will be referred to as the *descent recursive functions over A*.

Proposition 3.20 ([32, 4.4]) $\mathbf{DRA}_{\langle A, \lhd \rangle}$ *and* $\mathbf{PA} + \bigcup_{\alpha \in A} \mathrm{TI}(A_{\bar{\alpha}}, \lhd_{\bar{\alpha}})$ *prove the same* Π^0_2 *sentences.*

From 2.8 and 3.20 we get:

Observation 3.21 *Suppose an ordinal analysis of the formal system* T *has been attained using an EORS* $\langle A, \lhd, \ldots \rangle$. *Then the provably recursive functions of* T *are the descent recursive functions over* A.

We shall list some complementary results.

Definition 3.22 If T is a theory, the 1-*consistency of* T is the schema

$$\forall u [Pr_T(\ulcorner F(\dot{u}) \urcorner) \to F(u)]$$

for Σ^0_1 formulae $F(u)$ with one free variable u.

Proposition 3.23 ([32, 4.5]) *The following are equivalent over* **PRA***:*

(i) *1-consistency of* **PA** $+ \bigcup_{\alpha \in A} \mathrm{TI}(A_{\bar{\alpha}}, \lhd_{\bar{\alpha}})$

(ii) $\mathrm{PRWF}(\lhd^{+})$

(ii) $\mathrm{ERWF}(\lhd^{+})$.

Observation 3.24 *Again, let* T *be a theory for which an ordinal analysis has been carried out via* $\langle A, \lhd \rangle$*. Then the following are equivalent over* **PRA***:*

(i) *1-consistency of* T

(ii) $\mathrm{PRWF}(\lhd^{+})$

(ii) $\mathrm{ERWF}(\lhd^{+})$.

A characterization of the provably recursive functions of a formal system T as the α-recursive functions for $\alpha \in A$ or the descent recursive functions over A is notation-sensitive. None the less, it is sometimes possible to extract further notation-free information, in particular independence results that are not couched in terms of a given ordinal representation system $\langle A, \lhd, \ldots \rangle$. Usually, though, to obtain such results one needs to be furnished with a specific well-structured hierarchy $(F_\alpha)_{\alpha \in A}$ of functions such that every provable function of T is majorized by a function in the hierarchy. An example is Kirby and Paris' result on the unprovability of the termination of all Goodstein sequences in **PA**. A proof-theoretic proof of this result (cf. [11, 108]) employs the fact that for every provable recursive functions of **PA** there is a function H_α for some $\alpha < \varepsilon_0$ that majorizes it. Here H_α is the αth function in the so-called Hardy hierarchy. What complicates matters is that the definition of the functions H_α hinges upon a particular assignment of fundamental sequences to limit ordinals. It appears that only "natural" assignments of fundamental sequences, which take into account their algebraic properties rather than their codes, lead to function hierarchies that can be used for combinatorial independence results. A general discussion about different hierarchies $(F_\alpha)_{\alpha \in A}$ and their relations can be found in [13].

3.4.2 Provable set functions and ordinals

The extraction of classifications of provable functions from ordinal analyses is not confined to recursive functions on natural numbers. In the case of fragments of second order arithmetic, one may also classify the provable hyperarithmetical as well as the provable Δ^1_2 functions on $\mathbb{N}$. In the case of set theories one may classify several kinds of provable set functions.

In the following we will be concerned with norms that can be assigned to set theories. In general, they can also be extracted from an ordinal analysis of a set theory T. Among other results, they lead to a classification of the provable set functions of T.

The first of these norms will be denoted $|T|^E$, where the superscript E signifies *E-recursion*, also termed *set recursion*. E-recursion theory extends the notion of computation from the natural numbers to arbitrary sets. For details see [90].

Definition 3.25 The intent is to assign meaning to $\{e\}(x)$ for every set x via an appropriate notion of computation. E-recursion is defined by the following schemes:

1. $e = \langle 1, n, i \rangle$,
   $\{e\}(x_1, \ldots, x_n) = x_i.$

2. $e = \langle 2, n, i, j \rangle$,
   $\{e\}(x_1, \ldots, x_n) = x_i \backslash x_j.$

3. $e = \langle 3, n, i, j \rangle$,
   $\{e\}(x_1, \ldots, x_n) = \{x_i, x_j\}.$

4. $e = \langle 5, n, m, e', e_1, \ldots, e_n \rangle$,
   $\{e\}(x_1, \ldots, x_n) \simeq \{e'\}(\{e_1\}(x_1, \ldots, x_n), \ldots, \{e_m\}(x_1, \ldots, x_n)).$

5. $e = \langle 6, n, m \rangle$,
   $\{e\}(e_1, x_1, \ldots, x_n, y_1, \ldots, y_m) \simeq \{e_1\}(x_1, \ldots, x_n).$

$\simeq$ is Kleene's symbol for strong equality. If g and f are partial functions, then $f(x) \simeq g(x)$ iff neither $f(x)$ nor $g(x)$ is defined, or $f(x)$ and $g(x)$ are defined and equal.

Recall that L_α, the αth level of Gödel's constructible hierarchy L, is defined by $L_0 = \emptyset$, $L_{\beta+1} = \{X : X \subseteq L_\beta;\ X \text{ definable over } \langle L_\beta, \in \rangle\}$ and $L_\lambda = \bigcup\{L_\beta : \beta < \lambda\}$ for limits λ. So any element of L of level α is definable from elements of L with levels $< \alpha$ and L_α.

Definition 3.26 For a collection of formulae (in the language of set theory), $\mathcal{F}$, we say that L_α is an *$\mathcal{F}$-model of T* if for all $B \in \mathcal{F}$, whenever $T \vdash B$, then $L_\alpha \models B$. Let

$$|T|_{\mathcal{F}} \ := \ \min\{\alpha : \ L_\alpha \text{ is an } \mathcal{F}\text{-model of } T\}.$$

Definition 3.27 The next notions are due to A. Schlüter [91].

$$|T|^E_{\Sigma_1} \;:=\; \min\{\alpha : \text{ for all } e \in \omega,\, T \vdash \{e\}(\omega)\downarrow \text{ implies } \{e\}(\omega) \in L_\alpha\}.$$

$$|T|_{\Pi^E_2} := \\ \min\{\alpha > \omega : \text{ for all } e \in \omega,\, T \vdash \forall x\, \{e\}(x)\downarrow \text{ implies } \forall x{\in}L_\alpha \{e\}(x) \in L_\alpha\}.$$

Definition 3.28 Let $\mathcal{F}$ be a collection of sentences. A set theory T is said to be *$\mathcal{F}$-sound* if for every $\mathcal{F}$ theorem ϕ of T, $L \models \phi$ holds.

For a collection of formulae $\mathcal{F}$, let $\mathcal{F}(L_\alpha)$ consist of all formulae A^{L_α} with $A \in \mathcal{F}$.

The system **PRST** (for *Primitive Recursive Set Theory*) is formulated in the language of set theory augmented by symbols for all primitive recursive set functions. The *axioms of* **PRST** are Extensionality, Pair, Union, Infinity, Δ_0-Separation, the Foundation Axiom (i.e. $x \neq \emptyset \rightarrow (\exists y{\in}x)(\forall z{\in}y)\, z{\notin}x)$ and the defining equations for the primitive recursive set functions.

In the following we shall assume that all set theories contain **PRST** either directly or via interpretation.

Proposition 3.29 *Suppose T is Π_2 sound and comprises Δ_0-collection. Furthermore, suppose that $T \vdash B$ implies $T \vdash \exists\alpha\exists x(x = L_\alpha \wedge B^x)$ for all Σ_1-sentences B. If T has a Σ_1-model then T has a Π_2-model and*

$$|T|_{\Sigma_1} \;=\; |T|_{\Pi_2}. \tag{47}$$

Proof: [77]. Theorem 2.1. □

Proposition 3.30 *If T is a Π_2 sound theory, then*

$$|T|^E_{\Sigma_1} \;=\; |T|^E_{\Pi_2}. \tag{48}$$

Recall that ω_1^{CK} stands for the least admissible ordinal $> \omega$. If, in addition, T proves the existence of ω_1^{CK}, then

$$|T|_{\Sigma_1(L(\omega_1^{CK}))} \;=\; |T|_{\Pi_2(L(\omega_1^{CK}))}. \tag{49}$$

Proof: (49) is an immediate consequence of the proof of [77], Theorem 2.1 and a slight modification of the latter proof yields (48). (48) is stated and proved in [91], 6.14. □

Theorem 3.31 *If T is Π_2-sound and $T \vdash \forall x\, \exists y\, [x{\in}y \;\wedge$ "y is an admissible set"], then*

$$|T|_{\mathsf{sup}} = |T|^E_{\Sigma_1} = |T|^E_{\Pi_2} = |T|_{\Sigma_1(L(\omega_1^{CK}))} = |T|_{\Pi_2(L(\omega_1^{CK}))}.$$

Proof: A detailed proof of $|T|^E_{\Sigma_1} = |T|_{\Sigma_1(L(\omega_1^{CK}))}$ can be found in [91], Satz 6.15. The equality $|T|_{\mathsf{sup}} = |T|_{\Sigma_1(L(\omega_1^{CK}))}$ also follows from the proof of [91], Satz 6.15, but has been observed previously (cf. [75], Theorem 7.14). □

Definition 3.32 Another notion that is closely related to the the norm $|T|_{\Sigma_1}$ is the notion of *good* Σ_1*-definition* from admissible set theory (see [5], II.5.13). Given a set theory T, we say that an ordinal α has a *good* Σ_1*-definition in* T if there is a Σ_1-formula $\phi(u)$ such that

$$L \models \phi[\alpha] \text{ and } T \vdash \exists! x \phi(x).$$

Let

$$\mathbf{sp}_{\Sigma_1}(T) := \{\alpha : \ \alpha \text{ has a good } \Sigma_1 \text{ definition in } T\}.$$

One obviously has $\sup(\mathbf{sp}_{\Sigma_1}(T)) = |T|_{\Sigma_1}$. In many cases the set $\mathbf{sp}_{\Sigma_1}(T)$ bears interesting connections to the ordinals of the representation system that has been used to analyze T. Ordinal representation systems that have been developed via a detour through large cardinals allow for an alternative interpretation wherein the large cardinals are replaced by their recursively large counterparts. The latter interpretation gives rise to a canonical interpretation of the ordinal terms of the representation system in $\mathbf{sp}_{\Sigma_1}(T)$. In general, however, the ordinals of $\mathbf{sp}_{\Sigma_1}(T)$ stemming from the ordinal representation form a proper subset of $\mathbf{sp}_{\Sigma_1}(T)$ with many 'holes'. It would be very desirable to find a 'natural' property which could distinguish the ordinals of the representation system within $\mathbf{sp}_{\Sigma_1}(T)$ so as to illuminate their naturalness. I consider this to be one of the most important problems in the area of strong ordinal representation systems. A more thorough discussion will follow in section 4.

4 Examples of ordinal analyses

In this last section, I'm going to sketch the ordinal analyses of two systems of set theory which are intended to illustrate the main ideas and techniques used in ordinal analysis. Some attempts will be made to explain the role of large cardinals that appear in the definition procedures of so-called *collapsing functions* which then give rise to strong ordinal representation systems.

4.1 A brief history of ordinal analyses

To set the stage for the following, a very brief history of ordinal-theoretic proof theory since Gentzen reads as follows: In the 1950's proof theory flourished in the hands of Schütte: in [94] he introduced an infinitary system for first order number theory with the so-called ω-rule, which had already been proposed by Hilbert [37]. Ordinals were assigned as lengths to derivations and via cut-elimination he re-obtained Gentzen's ordinal analysis for number theory in a particularly transparent way. Further, Schütte extended his approach to systems of ramified analysis and brought this technique to perfection in his monograph "Beweistheorie" [95]. Independently, in 1964 Feferman [21] and Schütte [96], [97] determined the ordinal bound Γ_0 for theories of autonomous ramified progressions.

A major breakthrough was made by Takeuti in 1967, who for the first time obtained an ordinal analysis of a strong fragment of second order arithmetic. In [106] he gave an ordinal analysis of Π^1_1 comprehension, extended in 1973 to Δ^1_2 comprehension in [107] jointly with Yasugi. For this Takeuti returned to Gentzen's method of assigning ordinals (ordinal diagrams, to be precise) to purported derivations of the empty sequent (inconsistency).

The next wave of results, which concerned theories of iterated inductive definitions, were obtained by Buchholz, Pohlers, and Sieg in the late 1970's (see [14]). Takeuti's methods of reducing derivations of the empty sequent ("the inconsistency") were extremely difficult to follow, and therefore a more perspicuous treatment was to be hoped for. Since the use of the infinitary ω-rule had greatly facilitated the ordinal analysis of number theory, new infinitary rules were sought. In 1977 (see [7]) Buchholz introduced such rules, dubbed Ω-rules to stress the analogy. They led to a proof-theoretic treatment of a wide variety of systems, as exemplified in the monograph [15] by Buchholz and Schütte. Yet simpler infinitary rules were put forward a few years later by Pohlers, leading to the *method of local predicativity*, which proved to be a very versatile tool (see [67, 68, 69]). With the work of Jäger and Pohlers (see [41, 42, 45]) the forum of ordinal analysis then switched from the realm of second-order arithmetic to set theory, shaping what is now called *admissible proof theory*, after the models of *Kripke-Platek set theory*, **KP**. Their work culminated in the analysis of the system with Δ^1_2 comprehension plus **BI** [45]. In essence, admissible proof theory is a gathering of cut-elimination techniques for infinitary calculi of ramified set theory with Σ and/or Π_2 reflection rules[5] that lend itself to ordinal analyses of theories of the form **KP**+ *"there are x many admissibles"* or **KP**+ *"there are many admissibles"*. By way of illustration, the subsystem of analysis with Δ^1_2 comprehension and bar induction can be couched in such terms, for it is naturally interpretable in the set theory $\mathbf{KPi} := \mathbf{KP} + \forall y \exists z (y \in z \wedge z \textit{ is admissible})$ (cf. [45]).

After an intermediate step [75], which dealt with a set theory **KPM** that formalizes a recursively Mahlo universe, a major step beyond admissible proof

[5] Recall that the salient feature of admissible sets is that they are models of Δ_0 collection and that Δ_0 collection is equivalent to Σ reflection on the basis of the other axioms of **KP** (see [5]). Furthermore, admissible sets of the form L_α also satisfy Π_2 reflection.

theory was taken in [82]. That paper featured ordinal analyses of extensions of **KP** by Π_n reflection. A generalization of the methods of [82] underlies the treatment of $\Pi^1_2 - \mathbf{CA}$ sketched in [83].

4.2 An ordinal analysis of Kripke-Platek set theory

Until the late 70s the systems treated by ordinal analysis were either fragments of second order arithmetic or theories of iterated inductive definitions. A direct proof-theoretic treatment of systems of set theory was pioneered by Jäger (cf. [41, 42]). A first impression of ordinal analysis will be given by way of the example of Kripke-Platek set theory.

4.2.1 The system KP

Though considerably weaker than **ZF**, a great deal of set theory requires only the axioms of **KP**. The axioms of **KP** are:[6]

Extensionality: $a = b \rightarrow [F(a) \leftrightarrow F(b)]$ for all formulas F.

Foundation: $\exists x G(x) \rightarrow \exists x[G(x) \wedge (\forall y \in x)\neg G(y)]$

Pair: $\exists x\ (x = \{a, b\})$.

Union: $\exists x\ (x = \bigcup a)$.

Infinity: $\exists x\ \big[x \neq \emptyset \wedge (\forall y \in x)(\exists z \in x)(y \in z)\big]$.[7]

Δ_0 *Separation:* $\exists x\ \big(x = \{y \in a : F(y)\}\big)$[8] for all Δ_0–formulas F in which x does not occur free.

Δ_0 *Collection:* $(\forall x \in a)\exists y G(x, y) \rightarrow \exists z(\forall x \in a)(\exists y \in z)G(x, y)$ for all Δ_0–formulas G.

By a Δ_0 formula we mean a formula of set theory in which all the quantifiers appear restricted, that is have one of the forms $(\forall x \in b)$ or $(\exists x \in b)$.

KP arises from **ZF** by completely omitting the power set axiom and restricting separation and collection to absolute predicates (cf. Barwise [1975]), i.e. Δ_0 formulas. These alterations are suggested by the informal notion of 'predicative'.

---

[6]For technical convenience, $\in$ will be taken to be the only predicate symbol of the language of set theory. This does no harm, since equality can be defined by $a = b :\Leftrightarrow (\forall x \in a)(x \in b) \wedge (\forall x \in b)(x \in a)$, provided that we state extensionality in a slightly different form than usually.

[7]This contrasts with Barwise [1975] where Infinity is not included in **KP**.

[8]$x = \{y \in a : F(y)\}$ stands for the Δ_0–formula $(\forall y \in x)[y \in a \wedge F(y)] \wedge (\forall y \in a)[F(y) \rightarrow y \in x]$.

4.2.2 An ordinal representation system for the Bachmann-Howard ordinal

This section introduces an ordinal representation system which encapsulates the strength of **KP**.

Definition 4.1 The *Veblen-function* φ figures prominently in elementary proof theory (cf. [22, 71, 98]). It is defined by transfinite recursion on α by letting $\varphi_0(\xi) := \omega^\xi$ and, for $\alpha > 0$, φ_α be the function that enumerates the class of ordinals

$$\{\gamma : \forall \xi < \alpha\, [\varphi_\xi(\gamma) = \gamma]\}.$$

We shall write $\varphi_\alpha(\beta)$ instead of $\varphi\alpha\beta$. Let Γ_α be the αth ordinal $\rho > 0$ such that for all $\beta, \gamma < \rho$, $\varphi\beta\gamma < \rho$

Corollary 4.2 *(i)* $\varphi 0 \beta = \omega^\beta$.

(ii) $\xi, \eta < \varphi\alpha\beta \Longrightarrow \xi + \eta < \varphi\alpha\beta$.

(iii) $\xi < \zeta \Longrightarrow \varphi\alpha\xi < \varphi\alpha\zeta$.

(iv) $\alpha < \beta \Longrightarrow \varphi\alpha(\varphi\beta\xi) = \varphi\beta\xi$.

The least ordinal (> 0) closed under the function φ is called Γ_0. The proof-theoretic ordinal of **KP**, however, is bigger than Γ_0 and we need another function to obtain a sufficiently large ordinal representation system.

Definition 4.3 Let Ω be a "big" ordinal. By recursion on α we define sets $C^\Omega(\alpha, \beta)$ and the ordinal $\psi_\Omega(\alpha)$ as follows:

$$C^\Omega(\alpha,\beta) \quad = \quad \begin{cases} \textit{closure of } \beta \cup \{0, \Omega\} \\ \textit{under:} \\ +, (\xi \mapsto \omega^\xi) \\ (\xi \longmapsto \psi_\Omega(\xi))_{\xi<\alpha} \end{cases} \tag{50}$$

$$\psi_\Omega(\alpha) \quad \simeq \quad \min\{\rho < \Omega :\ C^\Omega(\alpha,\rho) \cap \Omega = \rho\}. \tag{51}$$

Note that if $\rho = \psi_\Omega(\alpha)$, then $\psi_\Omega(\alpha) < \Omega$ and $[\rho, \Omega) \cap C^\Omega(\alpha, \rho) = \emptyset$, thus the order-type of the ordinals below Ω which belong to the Skolem hull $C^\Omega(\alpha, \rho)$ is ρ . In more pictorial terms, ρ is the αth *collapse* of Ω.

Lemma 4.4 *$\psi_\Omega(\alpha)$ is always defined; in particular $\psi_\Omega(\alpha) < \Omega$.*

Proof: The claim is actually not a definitive statement as I haven't yet said what largeness properties Ω has to satisfy. In the proof below, we assume $\Omega := \aleph_1$, i.e. Ω is the first uncountable cardinal.

Observe first that for a limit ordinal λ,

$$C^{\Omega}(\alpha,\lambda) = \bigcup_{\xi<\lambda} C^{\Omega}(\alpha,\xi)$$

since the right hand side is easily shown to be closed under the clauses that define $C^{\Omega}(\alpha,\lambda)$. Thus we can pick $\omega \leq \eta < \Omega$ such that $\Omega \in C^{\Omega}(\alpha,\eta)$. Now define

$$\begin{aligned} \eta_0 &= \sup C^{\Omega}(\alpha,\eta) \cap \Omega \\ \eta_{n+1} &= \sup C^{\Omega}(\alpha,\eta_n) \cap \Omega \\ \eta^* &= \sup_{n<\omega} \eta_n. \end{aligned} \tag{52}$$

Since the cardinality of $C^{\Omega}(\alpha,\eta)$ is the same as that of η and therefore less than Ω, the regularity of Ω implies that $\eta_0 < \Omega$. By repetition of this argument one obtains $\eta_n < \Omega$, and consequently $\eta^* < \Omega$. The definition of η^* then ensures

$$C^{\Omega}(\alpha,\eta^*) \cap \Omega = \bigcup_n C^{\Omega}(\alpha,\eta_n) \cap \Omega = \eta^* < \Omega.$$

Therefore, $\psi_{\Omega}(\alpha) < \Omega$. □

Let $\varepsilon_{\Omega+1}$ be the least ordinal $\alpha > \Omega$ such that $\omega^{\alpha} = \alpha$. The next definition singles out a subset $\mathcal{T}(\Omega)$ of $C^{\Omega}(\varepsilon_{\Omega+1},0)$ which gives rise to an ordinal representation system, i.e., there is an elementary ordinal representation system $\langle \mathcal{OR}, \lhd, \hat{\Re}, \hat{\psi}, \dots \rangle$, so that

$$\langle \mathcal{T}(\Omega), <, \Re, \psi, \dots \rangle \cong \langle \mathcal{OR}, \lhd, \hat{\Re}, \hat{\psi}, \dots \rangle. \tag{53}$$

"..." is supposed to indicate that more structure carries over to the ordinal representation system.

Definition 4.5 $\mathcal{T}(\Omega)$ is defined inductively as follows:

1. $0, \Omega \in \mathcal{T}(\Omega)$.
2. If $\alpha_1, \dots, \alpha_n \in \mathcal{T}(\Omega)$ and $\omega^{\alpha_1} + \cdots + \omega^{\alpha_n} > \alpha_1 \geq \ldots \geq \alpha_n$, then $\omega^{\alpha_1} + \cdots + \omega^{\alpha_n} \in \mathcal{T}(\Omega)$.
3. If $\alpha \in \mathcal{T}(\Omega)$ and $\alpha \in C^{\Omega}(\alpha, \psi_{\Omega}(\alpha))$, then $\psi_{\Omega}(\alpha) \in \mathcal{T}(\Omega)$.

The side condition in 4.5.2 is is easily explained by the desire to have unique representations in $\mathcal{T}(\Omega)$. The requirement $\alpha \in C^{\Omega}(\alpha, \psi_{\Omega}(\alpha))$ in 4.5.3 also serves the purpose of unique representations (and more) but is probably a bit harder to explain. The idea here is that from $\psi_{\Omega}(\alpha)$ one should be able to retrieve the stage (namely α) where it was generated. This is reflected by $\alpha \in C^{\Omega}(\alpha, \psi_\Omega(\alpha))$.

It can be shown that the foregoing definition of $\mathcal{T}(\Omega)$ is deterministic, that is to say every ordinal in $\mathcal{T}(\Omega)$ is generated by the inductive clauses of 4.5 in exactly one way. As a result, every $\gamma \in \mathcal{T}(\Omega)$ has a unique representation in terms of symbols for $0, \Omega$ and function symbols for $+, (\alpha \mapsto \omega^\alpha), (\alpha \mapsto \psi_\Omega(\alpha))$. Thus, by taking some primitive recursive (injective) coding function $\ulcorner \cdots \urcorner$ on finite sequences of natural numbers, we can code $\mathcal{T}(\Omega)$ as a set of natural numbers as follows:

$$\ell(\alpha) = \begin{cases} \ulcorner 0, 0 \urcorner & \text{if } \alpha = 0 \\ \ulcorner 1, 0 \urcorner & \text{if } \alpha = \Omega \\ \ulcorner 2, \ell(\alpha_1), \cdots, \ell(\alpha_n) \urcorner & \text{if } \alpha = \omega^{\alpha_1} + \cdots + \omega^{\alpha_n} \\ \ulcorner 3, \ell(\beta), \ell(\Omega) \urcorner & \text{if } \alpha = \psi_\Omega(\beta), \end{cases}$$

where the distinction by cases refers to the unique representation of 4.5. With the aid of ℓ, the ordinal representation system of (53) can be defined by letting $\mathcal{OR}$ be the image of ℓ and setting $\lhd := \{(\ell(\gamma), \ell(\delta)) : \gamma < \delta \wedge \delta, \gamma \in \mathcal{T}(\Omega)\}$ etc. However, for a proof that this definition of $\langle \mathcal{OR}, \lhd, \hat{\mathfrak{R}}, \hat{\psi}, \dots \rangle$ in point of fact furnishes an elementary ordinal representation system, we have to refer to the literature (cf. [8, 12, 82]).

4.2.3 A reminder: Ordinal analysis of PA à la Schütte

It is well known that the axioms of Peano Arithmetic, **PA**, can be derived in a sequent calculus, $\mathbf{PA}_\omega$, augmented by an infinitary rule, the so–called *ω–rule*[9]

$$\frac{\Gamma, A(\bar{n}) \text{ for all } n}{\Gamma, \forall x A(x)}.$$

An ordinal analysis for **PA** is then attained as follows:

- Each **PA**–proof can be "unfolded" into a $\mathbf{PA}_\omega$–proof of the same sequent.
- Each such $\mathbf{PA}_\omega$–proof can be transformed into a cut–free $\mathbf{PA}_\omega$–proof of the same sequent of length $< \varepsilon_0$.

In order to obtain a similar result for set theories like **KP**, we have to work a bit harder. Guided by the ordinal analysis of **PA**, we would like to invent an infinitary rule which, when added to **KP**, enables us to eliminate cuts. As opposed to the natural numbers, it is not clear how to bestow a canonical name to each element of the set–theoretic universe. However, within the confines of the constructible universe, which is made from the ordinals, it is

[9] $\bar{n}$ stands for the nth numeral

pretty obvious how to "name" sets once we have names for ordinals at our disposal.

4.2.4 The language of RS_Ω

The problem of "naming" sets will be solved by erecting a formal constructible hierarchy using the ordinals from $\mathcal{T}(\Omega)$. Henceforth, we shall restrict ourselves to ordinals from $\mathcal{T}(\Omega)$.

Definition 4.6 We adopt a language of set theory, $\mathcal{L}$, which has only the predicate symbol $\in$. The *atomic formulae* of $\mathcal{L}$ are those of either form $(a\in b)$ or $\neg(a\in b)$. The *$\mathcal{L}$-formulae* are obtained from atomic ones by closing off under $\wedge, \vee, (\exists x\in a), (\forall x\in a), \exists x$, and $\forall x$.

Definition 4.7 The *RS_Ω-terms* and their *levels* are generated as follows.

1. For each $\alpha < \Omega$, $\mathbb{L}_\alpha$ is an RS_Ω–term of level α.
2. The formal expression $[x\in\mathbb{L}_\alpha : F(x, \vec{s})^{\mathbb{L}_\alpha}]$ is an RS_Ω–term of level α if $F(a, \vec{b})$ is an $\mathcal{L}$–formula (whose free variables are among the indicated) and $\vec{s} \equiv s_1, \cdots, s_n$ are RS_Ω–terms with levels $< \alpha$. $F(x, \vec{s})^{\mathbb{L}_\alpha}$ results from $F(x, \vec{s})$ by restricting all unbounded quantifiers to $\mathbb{L}_\alpha$.

We shall denote the level of an RS_Ω–term t by $|\,t\,|$; $t\in\mathcal{T}(\alpha)$ stands for $|\,t\,| < \alpha$ and $t\in\mathcal{T}$ for $t\in\mathcal{T}(\Omega)$.

The *RS_Ω-formulae* are the expressions of the form $F(\vec{s})$, where $F(\vec{a})$ is an $\mathcal{L}$–formula and $\vec{s} \equiv s_1, \ldots, s_n\in\mathcal{T}$.

For technical convenience, we let $\neg A$ be the formula which arises from A by (i) putting $\neg$ in front of each atomic formula, (ii) replacing $\wedge, \vee, (\forall x \in a), (\exists x \in a)$ by $\vee, \wedge, (\exists x \in a), (\forall x \in a)$, respectively, and (iii) dropping double negations.

Definition 4.8 We use the relation $\equiv$ to mean syntactical identity. For terms s, t with $|\,s\,| < |\,t\,|$ we set

$$s\mathring{\in}t \equiv \begin{cases} B(s) & \text{if } \ t \equiv [x\in\mathbb{L}_\beta : B(x)] \\ \mathsf{True}_s & \text{if } \ t \equiv \mathbb{L}_\beta \end{cases}$$

where True_s is a true formula, say $s\notin\mathbb{L}_0$.

Observe that $s\in t$ and $s\mathring{\in}t$ have the same truth value under the standard interpretation in the constructible hierarchy.

4.2.5 The rules of $\mathcal{L}_{RS}$

Having created names for a segment of the constructible universe, we can introduce infinitary rules analogous to the the ω-rule.

Let $A, B, C, \ldots, F(t), G(t), \ldots$ range over RS_Ω–formulae. We denote by upper case Greek letters $\Gamma, \Delta, \Lambda, \ldots$ finite sets of RS_Ω–formulae. The intended meaning of $\Gamma = \{A_1, \cdots, A_n\}$ is the disjunction $A_1 \vee \cdots \vee A_n$. Γ, A stands for $\Gamma \cup \{A\}$ etc.. We also use the shorthands $r \neq s := \neg(r = s)$ and $r \notin t := \neg(r \in t)$.

Definition 4.9 The *rules* of RS_Ω are:

$$(\wedge) \qquad \frac{\Gamma, A \quad \Gamma, A'}{\Gamma, A \wedge A'}$$

$$(\vee) \qquad \frac{\Gamma, A_i}{\Gamma, A_0 \vee A_1} \quad \text{if } i = 0 \text{ or } i = 1$$

$$(b\forall) \qquad \frac{\cdots \Gamma, s \mathring{\in} t \rightarrow F(s) \cdots (s \in \mathcal{T}(|t|))}{\Gamma, (\forall x \in t) F(x)}$$

$$(b\exists) \qquad \frac{\Gamma, s \mathring{\in} t \wedge F(s)}{\Gamma, (\exists x \in t) F(x)} \quad \text{if } s \in \mathcal{T}(|t|)$$

$$(\forall) \qquad \frac{\cdots \Gamma, F(s) \cdots (s \in \mathcal{T})}{\Gamma, \forall F(x)}$$

$$(\exists) \qquad \frac{\Gamma, F(s)}{\Gamma, \exists x\, F(x)} \quad \text{if } s \in \mathcal{T}$$

$$(\notin) \qquad \frac{\cdots \Gamma, s \mathring{\in} t \rightarrow r \neq s \cdots\cdots (s \in \mathcal{T}(|t|))}{\Gamma, r \notin t}$$

$$(\in) \qquad \frac{\Gamma, s \mathring{\in} t \ \wedge \ r = s}{\Gamma, r \in t} \quad \text{if } s \in \mathcal{T}(|t|)$$

$$(\text{Cut}) \qquad \frac{\Gamma, A \qquad \Gamma, \neg A}{\Gamma}$$

$$(\text{Ref}_{\Sigma(\Omega)}) \quad \frac{\Gamma, A}{\Gamma, (\exists z \in \mathbb{L}_\Omega) A^z} \quad \text{if } A \text{ is a } \Sigma\text{-formula,}$$

where a formula is said to be Σ if all unbounded quantifiers are existential. A^z results from A by restricting all unbounded quantifiers to z.

4.2.6 $\mathcal{H}$–controlled derivations

If we dropped the rule ($\mathrm{Ref}_{\Sigma(\Omega)}$) from RS_Ω, the remaining calculus would enjoy full cut elimination owing to the symmetry of the pairs of rules $\langle (\wedge),\ (\vee) \rangle$, $\langle (\forall),\ (\exists) \rangle$, $\langle (\notin),\ (\in) \rangle$. However, partial cut elimination for RS_Ω can be attained by delimiting a collection of derivations of a very uniform kind. Fortunately, Buchholz has provided us with a very elegant and flexible setting for describing uniformity in infinitary proofs, called *operator controlled derivations* (see [10]).

Definition 4.10 Let $P(ON) = \{X : X \text{ is a set of ordinals}\}$. A class function $\mathcal{H} : P(ON) \to P(ON)$ will be called *operator* if $\mathcal{H}$ is a *closure operator*, i.e monotone, inclusive and idempotent, and satisfies the following conditions for all $X \in P(ON)$: $0 \in \mathcal{H}(X)$, and, if α has Cantor normal form $\omega^{\alpha_1} + \cdots + \omega^{\alpha_n}$, then $\alpha \in \mathcal{H}(X) \iff \alpha_1, ..., \alpha_n \in \mathcal{H}(X)$. The latter ensures that $\mathcal{H}(X)$ will be closed under $+$ and $\sigma \mapsto \omega^\sigma$, and decomposition of its members into additive and multiplicative components. For $Z \in P(ON)$, the operator $\mathcal{H}[\mathcal{Z}]$ is defined by $\mathcal{H}[\mathcal{Z}](\mathcal{X}) := \mathcal{H}(\mathcal{Z} \cup \mathcal{X})$.

If $\mathfrak{X}$ consists of "syntactic material", i.e. terms, formulae, and possibly elements from $\{0,1\}$, then let $\mathcal{H}[\mathfrak{X}](X) := \mathcal{H}(k(\mathfrak{X}) \cup X)$, where $k(\mathfrak{X})$ is the set of ordinals needed to build this "material". Finally, if s is a term, then define $\mathcal{H}[\mathfrak{s}]$ by $\mathcal{H}[\{\mathfrak{s}\}]$.

To facilitate the definition of $\mathcal{H}$–controlled derivations, we assign to each RS_Ω–formula A, either a (possibly infinite) disjunction $\bigvee(A_\iota)_{\iota\in I}$ or a conjunction $\bigwedge(A_\iota)_{\iota\in I}$ of RS_Ω–formulae. This assignment will be indicated by $A \cong \bigvee(A_\iota)_{\iota\in I}$ and $A \cong \bigwedge(A_\iota)_{\iota\in I}$, respectively. Define: $r{\in}t \cong \bigvee(s\mathring{\in}t \wedge r = s)_{s\in\mathcal{T}_{|t|}}$; $(\exists x{\in}t)F(x) \cong \bigvee(s\mathring{\in}t \wedge F(s))_{s\in\mathcal{T}_{|t|}}$; $A_0 \vee A_1 \cong \bigvee(A_\iota)_{\iota\in\{0,1\}}$; $\neg A \cong \bigwedge(\neg A_\iota)_{\iota\in I}$, if $A \cong \bigvee(A_\iota)_{\iota\in I}$. Using this representation of formulae, we can define the *subformulae* of a formula as follows. When $A \cong \bigwedge(A_\iota)_{\iota\in I}$ or $A \cong \bigvee(A_\iota)_{\iota\in I}$, then B is a subformula of A if $B \equiv A$ or, for some $\iota{\in}I$, B is a subformula of A_ι.

Since one also wants to keep track of the complexity of cuts appearing in derivations, each formula F gets assigned an ordinal rank $rk(F)$ which is roughly the sup of the level of terms in F plus a finite number.

Using the formula representation, in spite of the many rules of RS_Ω, the notion of $\mathcal{H}$–controlled derivability can be defined concisely. We shall use $I{\restriction}\alpha$ to denote the set $\{\iota{\in}I : |\iota| < \alpha\}$.

Definition 4.11 Let $\mathcal{H}$ be an operator and let Γ be a finite set of RS_Ω–formulae. $\mathcal{H} \overset{\alpha}{\underset{\rho}{\vdash}} \Gamma$ is defined by recursion on α. It is always demanded that $\{\alpha\} \cup k(\Gamma) \subseteq \mathcal{H}(\emptyset)$. The inductive clauses are:

$$(\bigvee) \qquad \frac{\mathcal{H} \overset{\alpha_0}{\underset{\rho}{\vdash}} \Lambda, A_{\iota_0}}{\mathcal{H} \overset{\alpha}{\underset{\rho}{\vdash}} \Lambda, \bigvee(A_\iota)_{\iota \in I}} \qquad \begin{array}{r} \alpha_0 < \alpha \\ \iota_0 \in I \restriction \alpha \end{array}$$

$$(\bigwedge) \qquad \frac{\mathcal{H}[\iota] \overset{\alpha_\iota}{\underset{\rho}{\vdash}} \Lambda, A_\iota \textit{ for all } \iota \in I}{\mathcal{H} \overset{\alpha}{\underset{\rho}{\vdash}} \Lambda, \bigwedge(A_\iota)_{\iota \in I}} \qquad |\iota| \leq \alpha_\iota < \alpha$$

$$(Cut) \qquad \frac{\mathcal{H} \overset{\alpha_0}{\underset{\rho}{\vdash}} \Lambda, \mathcal{B} \qquad \mathcal{H} \overset{\alpha_0}{\underset{\rho}{\vdash}} \Lambda, \neg\mathcal{B}}{\mathcal{H} \overset{\alpha}{\underset{\rho}{\vdash}} \Lambda} \qquad \begin{array}{r} \alpha_0 < \alpha \\ rk(B) < \rho \end{array}$$

$$(Ref_{\Sigma(\Omega)}) \qquad \frac{\mathcal{H} \overset{\alpha_0}{\underset{\rho}{\vdash}} \Lambda, A}{\mathcal{H} \overset{\alpha}{\underset{\rho}{\vdash}} \Lambda, \exists z\, A^z} \qquad \begin{array}{r} \alpha_0 < \alpha \\ A \in \Sigma \end{array}$$

The specification of the operators needed for an ordinal analysis will, of course, hinge upon the particular theory and ordinal representation system.

To connect **KP** with the infinitary system RS_Ω one has to show that **KP** can be embedded into RS_Ω. Indeed, the finite **KP**-derivations give rise to very uniform infinitary derivations.

Theorem 4.12 *If* **KP** $\vdash B(a_1, \ldots, a_r)$, *then* $\mathcal{H} \overset{\Omega\cdot m}{\underset{\Omega+n}{\vdash}} \mathcal{B}(\mathcal{J}_\infty, \ldots, \mathcal{J}_\nabla)$ *holds for some* m, n *and all set terms* $s_1, \ldots, s_r$ *and operators* $\mathcal{H}$ *satisfying*

$$\{\xi : \xi \textit{ occurs in } B(\vec{s})\} \cup \{\Omega\} \subseteq \mathcal{H}(\emptyset).$$

m and n depend only on the **KP**-*derivation of* $B(\vec{a})$.

The usual cut elimination procedure works as long as the cut formulae have not been introduced by an inference $(Ref_{\Sigma(\Omega)})$. As the main formula of an inference $(Ref_{\Sigma(\Omega)})$ has rank Ω one gets the following result.

Theorem 4.13 (Cut elimination I)

$$\mathcal{H} \overset{\alpha}{\underset{\Omega+n+1}{\vdash}} \Gamma \;\Rightarrow\; \mathcal{H} \overset{\omega_n(\alpha)}{\underset{\Omega+1}{\vdash}} \Gamma$$

where $\omega_0(\beta) := \beta$ *and* $\omega_{k+1}(\beta) := \omega^{\omega_k(\beta)}$.

The reason why the usual cut-elimination method fails for cuts with rank Ω is that it is too limited to treat a cut in the following scenario:

$$\dfrac{\dfrac{\mathcal{H} \overset{\delta}{\underset{\Omega}{\vdash}} \Gamma, A}{\mathcal{H} \overset{\xi}{\underset{\Omega}{\vdash}} \Gamma, \exists z\, A^z}(\Sigma\text{-}Ref_\Omega) \qquad \dfrac{\cdots \mathcal{H}[s] \overset{\xi_s}{\underset{\Omega}{\vdash}} \Gamma, \neg A^s \cdots (|s| < \Omega)}{\mathcal{H} \overset{\xi}{\underset{\Omega}{\vdash}} \Gamma, \forall z \neg A^z}(\forall)}{\mathcal{H} \overset{\alpha}{\underset{\Omega+1}{\vdash}} \Gamma}(Cut)$$

Fortunately, it is possible to eliminate cuts in the above situation provided that the side formulae Γ are of complexity Σ. The technique is known as "collapsing" of derivations.

In the course of "collapsing" one makes use of a simple bounding principle.

Lemma 4.14 (Boundedness) *Let A be a Σ-formula, $\alpha \leq \beta < \Omega$, and $\beta \in \mathcal{H}(\emptyset)$. If $\mathcal{H} \,|\!\frac{\alpha}{\rho}\, \Gamma, A$, then $\mathcal{H} \,|\!\frac{\alpha}{\rho}\, \Gamma, A^{\mathbb{L}_\beta}$.*

If the length of the derivation is already $\geq \Omega$, then "collapsing" results in a shorter derivation, however, at the cost of a much more complicated controlling operator.

Theorem 4.15 (Collapsing Theorem) *Let Γ be a set of Σ-formulae. Then we have*

$$\mathcal{H}_\eta \,|\!\frac{\alpha}{\Omega+1}\, \Gamma \quad\Rightarrow\quad \mathcal{H}_{f(\eta,\alpha)} \,|\!\frac{\psi_\Omega(f(\eta,\alpha))}{\psi_\Omega(f(\eta,\alpha))}\, \Gamma,$$

where $\left(\mathcal{H}_\xi\right)_{\xi \in \mathcal{T}(\Omega)}$ is a uniform sequence of ever stronger operators.

From Lemma 4.14 it follows that all instances of $(Ref_{\Sigma(\Omega)})$ can be removed from derivations of length Ω. For the latter kind of derivations there is a well-known cut-elimination procedure, the so-called *predicative cut-elimination.* Below this is stated in precise terms. It should also be mentioned that the φ function can be defined in terms of the functions of $\mathcal{T}(\Omega)$ and that $\varphi\alpha\beta < \Omega$ holds whenever $\alpha, \beta < \Omega$.

Theorem 4.16 (Predicative cut elimination)

$$\mathcal{H} \,|\!\frac{\delta}{\rho}\, \Gamma \text{ and } \delta, \rho < \Omega \quad\Rightarrow\quad \mathcal{H} \,|\!\frac{\varphi\rho\delta}{0}\, \Gamma .$$

The ordinal $\psi_\Omega(\varepsilon_{\Omega+1})$ is known as the *Bachmann-Howard ordinal.* Combining the previous results of this section, one obtains:

Corollary 4.17 *If A is a Σ-formula and* **KP** $\vdash A$, *then $L_{\psi_\Omega(\varepsilon_{\Omega+1})} \models A$.*

The bound of Corollary 4.17 is sharp, that is, $\psi_\Omega(\varepsilon_{\Omega+1})$ is the first ordinal with that property. Below we list further results that follow from the ordinal analysis of **KP**.

Corollary 4.18 *(i)* $|\mathbf{KP}| = |\mathbf{KP}|_{\sup} = |\mathbf{KP}|_{\Pi_2} = |\mathbf{KP}|^E_{\Pi_2} = \psi_\Omega(\varepsilon_{\Omega+1})$.

(ii) $\mathbf{sp}_{\Sigma_1}(\mathbf{KP}) = \psi_\Omega(\varepsilon_{\Omega+1})$.

4.3 Ordinal analysis of KPM

In many respects, **KP** is a very special case. Several fascinating aspects of ordinal analysis do not yet exhibit themselves at the level of **KP**. An example for the latter is that, in general, $\mathbf{sp}_{\Sigma_1}(T)$ is not contained in the ordinal representation system; the connection between them only emerges at the level of stronger theories. Furthermore, up to now the approach of "using" large cardinals to devise strong ordinal representation systems is only exemplified in a very weak sense namely in the shape of an uncountable cardinal. For these reasons, I shall outline the ordinal analysis of the stronger theory **KPM**. **KPM** formalizes a recursively Mahlo universe of sets and is considerably stronger than $\Delta^1_2-\mathbf{CA}+\mathbf{BI}$. It is distinguished by the fact that it is essentially the 'strongest' classical theory for which a consistency proof in Martin-Löf type theory can be carried out. The particular formal system of Martin-Löf type theory that suffices for such a consistency proof is based on P. Dybjer's schema of simultaneous inductive-recursive definition (cf. [20]) or E. Palmgren's higher order universes (cf. [64]) and proceeds by showing the well-foundedness of the representation system $\mathcal{T}(\mathbf{M})$ that was used in the ordinal analysis of **KPM** (cf.[75]) in type theory. However, I should be a little cautious here as a full proof has not yet been written down, mainly because it taxes the limits of human tolerance. Though, for a strong fragment of **KPM** (wherein the foundation scheme is restricted to set-theoretic Π_2 formulas) there is a full proof, using techniques of [86, 77].

4.3.1 The theory KPM

KPM is an extension of **KP** by a schema stating that for every Σ_1-definable (class) function there exists an admissible set closed under this function. Its canonical models are the sets $\mathbf{L}_\mu$ with μ recursively Mahlo. To be more precise, the *language of* **KPM**, denoted by $\mathcal{L}_{Ad}$, is an extension of the language of **KP** by a unary predicate **Ad** which is used to express that a set is an admissible set. In addition to the axioms of **KP**, **KPM** has the following axioms:

Ad-*Limit:* $\forall x\,\exists y\,(x \in y \wedge \mathbf{Ad}(y))$.

Ad-*Linearity:* $\forall u \forall v\,[\mathbf{Ad}(u) \wedge \mathbf{Ad}(v) \rightarrow u \in v \vee u = v \vee v \in u]$.

(**Ad1**)*:* $\mathbf{Ad}(a) \rightarrow \omega \in a \wedge \forall x \in a \forall z \in x\, z \in a$.

(**Ad2**)*:* $\mathbf{Ad}(a) \rightarrow A^a$,
where the sentence A is a universal closure of one of the following axioms:

Pairing: $\exists x\ (x = \{a, b\})$.

Union: $\exists x\ (x = \bigcup a)$.

Δ_0–*Sep:*	$\exists x\,\big(x = \{y\in a : F(y)\}\big)$ for all Δ_0–formulae $F(b)$
Δ_0–*Coll:*	$(\forall x\in a)\exists y G(x,y) \rightarrow \exists z(\forall x\in a)(\exists y\in z)G(x,y)$ for all Δ_0–formulae $G(b)$.
$(\mathbf{M})$:	$\forall x\,\exists y\,G(x,y) \;\rightarrow\; \exists z\,[\mathbf{Ad}(z) \,\wedge\, (\forall x\in z)\,(\exists y\in z)G(x,y)]$ for all Δ_0–formulae $G(a,b)$.

4.3.2 Ordinal functions based on a weakly Mahlo cardinal

To develop a sufficiently strong ordinal representation system we first develop certain collapsing under the assumption that a weakly Mahlo cardinal exists (cf. [74]).

In a paper from 1911 Mahlo [56] investigated two hierarchies of regular cardinals. Mahlo called the cardinals considered in the first hierarchy π_α-*numbers.* In modern terminology they are spelled out as follows:

κ is 0-*weakly inaccessible*	iff	κ is regular;
κ is $(\alpha+1)$-*weakly inaccessible*	iff	κ is a regular limit of α-weakly inaccessibl
κ is λ-*weakly inaccessible*	iff	κ is α-weakly inaccessible for every $\alpha<\lambda$

for limit ordinals λ. This hierarchy could be extended through diagonalization, by taking next the cardinals κ such that κ is κ-weakly inaccessible and after that choosing regular limits of the previous kind etc.

Mahlo also discerned a second hierarchy which is generated by a principle superior to taking regular fixed-points. Its starting point is the class of ρ_0-numbers which later came to be called *weakly Mahlo cardinals.* Weakly Mahlo cardinals are larger than any of those that can be obtained by the above processes from below. Here we shall define an extension of Mahlo's π-hierarchy by using ordinals above a weakly Mahlo to keep track of diagonalization.

Definition 4.19 Let

$$\mathbf{M} \;:=\; \text{first weakly Mahlo cardinal} \tag{54}$$

and set

$$\Re^{\mathbf{M}} \;:=\; \{\pi<\mathbf{M} : \pi \text{ regular}, \pi>\omega\}. \tag{55}$$

Variables κ,π will range over $\Re^{\mathbf{M}}$.

An ordinal representation system for the analysis of **KPM** can be derived from the following functions and Skolem hulls of ordinals, defined by recursion on α:

$$C^{\mathbf{M}}(\alpha,\beta) = \begin{cases} \textit{closure of } \beta \cup \{0, \mathbf{M}\} \\ \textit{under:} \\ +, (\xi \longmapsto \omega^{\xi}) \\ (\xi\delta \mapsto \chi^{\xi}(\delta))_{\xi<\alpha} \\ (\xi\pi \longmapsto \psi^{\xi}(\pi))_{\xi<\alpha} \end{cases} \tag{56}$$

$$\chi^{\alpha}(\delta) \simeq \delta^{th} \text{ regular } \pi < \mathbf{M} \text{ s.t. } C^{\mathbf{M}}(\alpha,\pi) \cap \mathbf{M} = \pi \tag{57}$$

$$\psi^{\alpha}(\pi) \simeq \min\{\rho < \pi :\; C^{\mathbf{M}}(\alpha,\rho) \cap \pi = \rho \;\wedge\; \pi \in C^{\mathbf{M}}(\alpha,\rho)\}. \tag{58}$$

Lemma 4.20 *For all* α,

$$\chi^{\alpha} : \mathbf{M} \to \mathbf{M}$$

i.e. χ^{α} *is a total function on* $\mathbf{M}$.

Proof: Set

$$X_{\alpha} := \{\rho < \mathbf{M} : C^{\mathbf{M}}(\alpha,\rho) \cap \mathbf{M} = \rho\}.$$

We want to show that X_{α} is closed and unbounded in $\mathbf{M}$. As $\mathbf{M}$ is weakly Mahlo the latter will imply that X_{α} contains $\mathbf{M}$-many regular cardinals, ensuring that χ^{α} is total on $\mathbf{M}$.

Unboundedness: Given $\eta < \mathbf{M}$, define

$$\begin{aligned} \eta_0 &= \sup(C^{\mathbf{M}}(\alpha,\eta+1) \cap \mathbf{M}) \\ \eta_{n+1} &= \sup(C^{\mathbf{M}}(\alpha,\eta_n) \cap \mathbf{M}) \\ \eta^* &= \sup_n \eta_n. \end{aligned}$$

One easily verifies $C^{\mathbf{M}}(\alpha,\eta^*) \cap \mathbf{M} = \eta^*$. Hence, $\eta < \eta^*$ and $\eta^* \in X_{\alpha}$.

Closedness: If $X_{\alpha} \cap \lambda$ is unbounded in a limit $\lambda < \mathbf{M}$, then

$$C^{\mathbf{M}}(\alpha,\lambda) = \bigcup_{\xi \in X_{\alpha} \cap \lambda} C^{\mathbf{M}}(\alpha,\xi),$$

whence

$$C^{\mathbf{M}}(\alpha,\lambda) \cap \mathbf{M} = \sup\{\xi : \xi \in X_{\alpha} \cap \lambda\} = \lambda,$$

verifying $\lambda \in X_{\alpha}$. □

For a comparison with Mahlo's π_{α} numbers let $\mathbf{I}_{\alpha}$ be the function that enumerates, monotonically, the α-weakly inaccessibles. Neglecting finitely many exceptions, the function $\mathbf{I}_{\alpha}$ enumerates Mahlo's π_{α} numbers.

Proposition 4.21 *For* $\alpha < \mathbf{M}$ *let*

$$\Delta(\alpha) := \textit{the } \alpha^{th}\ \kappa < \mathbf{M} \textit{ such that } \kappa \textit{ is } \kappa\textit{-weakly inaccessible.}$$

(i) $\forall\alpha < \Delta(0)\,\forall\xi < \mathbf{M}\;\mathbf{I}_\alpha(\xi) = \chi^\alpha(\xi)$.

(ii) $\Delta(\alpha) \;=\; \chi^{\mathbf{M}}(\alpha)$.

(iii) *If* $\chi^{\mathbf{M}}(\alpha) \le \beta < \chi^{\mathbf{M}}(\alpha+1)$, *then* $\forall\xi \le \alpha\;\chi^\beta(\xi) = \chi^{\mathbf{M}}(\xi)$.

(iv) *If* $\beta = \chi^{\mathbf{M}}(\alpha)$, *then* $\forall\xi \le \mathbf{M}\;\chi^\beta(\alpha+\xi) \;=\; \mathbf{I}_\beta(\xi)$.

(v) *If* $\chi^{\mathbf{M}}(\alpha) < \beta < \chi^{\mathbf{M}}(\alpha+1)$, *then* $\forall\xi < \mathbf{M}\;\chi^\beta(\alpha+1+\xi) \;=\; \mathbf{I}_\beta(\xi)$.

Ever higher levels of diagonalizations are obtained by the functions χ^{M^M}, $\chi^{M^{M^M}}$, etc.

The preceding gives rise to an EORS $\mathcal{T}(\mathbf{M})$ (similarly as sketched for $\mathcal{T}(\Omega)$) which is essentially order isomorphic to $C^{\mathbf{M}}(\varepsilon_{\mathbf{M}+1}, 0)$. This EORS exactly captures the strength of **KPM**.

4.3.3 The rules of $RS_{\mathbf{M}}$

The next step consists in utilizing $\mathcal{T}(\mathbf{M})$ for an ordinal analysis of **KPM**. Here we restrict ourselves to ordinals from $\mathcal{T}(\mathbf{M})$. The $RS_{\mathbf{M}}$*-terms* and their *levels* are generated as the RS_Ω-terms, except that in the starting case, for each $\alpha < \mathbf{M}$, $\mathbb{L}_\alpha$ is an $RS_{\mathbf{M}}$–term of level α. We will use $s \in RS_{\mathbf{M}}$ to convey that s is an $RS_{\mathbf{M}}$-term. The *atomic formulae* of $RS_{\mathbf{M}}$ are those of either form $(s\in t)$, $\neg(s\in t)$, $\mathbf{Ad}(s)$, or $\neg\mathbf{Ad}(s)$.

Definition 4.22 The *rules* of $RS_{\mathbf{M}}$ comprise $(\wedge), (\vee), (b\forall), (b\exists), \forall, \exists, (\notin), (\in), (\mathrm{Cut})$ as for RS_Ω. The additional rules are:

$$(\neg\mathbf{Ad})\qquad \frac{\cdots\Gamma, \mathbb{L}_\kappa \neq t \cdots (\kappa \le |\,t\,|)}{\Gamma, \neg\mathbf{Ad}(t)}$$

$$(\mathbf{Ad})\qquad \frac{\Gamma, \mathbb{L}_\kappa = t}{\Gamma, \mathbf{Ad}(t)}\quad \text{if } \kappa \le |\,t\,|$$

$$(\mathrm{Ref}_{\Sigma(\pi)})\qquad \frac{\Gamma, A^{\mathbb{L}_\pi}}{\Gamma, (\exists z \in \mathbb{L}_\pi)\, A^z}\quad \text{if } A \text{ is a } \Sigma\text{-formula whose terms have levels} < \pi$$

$$(\mathbf{M})\qquad \frac{\cdots\Gamma,\ \exists y\, F(s,y) \cdots (s \in RS_{\mathbf{M}})}{\Gamma, \exists z\,[\mathbf{Ad}(z) \wedge (\forall x \in z)(\exists y \in z) F(x,y)]}\quad \text{if } F \text{ is } \Delta_0.$$

Extending Definition 4.10, we assign to the $RS_{\mathbf{M}}$–formula $\mathbf{Ad}(t)$ the disjunction $\mathbf{Ad}(t) \cong \bigvee(\mathbb{L}_\pi = t)_{\mathbb{L}_\pi \in I}$, where $I := \{\mathbb{L}_\kappa : \kappa \in \Re^{\mathbf{M}}; \kappa \le |\,t\,|\}$.

Definition 4.23 Let $\mathcal{H}$ be an operator and let Γ be a finite set of $RS_{\mathbf{M}}$–formulae. $\mathcal{H} \overset{\alpha}{\underset{\rho}{\vdash}} \Gamma$ is defined by recursion on α. It is always demanded that $\{\alpha\} \cup k(\Gamma) \subseteq \mathcal{H}(\emptyset)$. The inductive clauses are:

$$(\bigvee) \qquad \frac{\mathcal{H} \left|\frac{\alpha_0}{\rho}\right. \Lambda, A_{\iota_0}}{\mathcal{H} \left|\frac{\alpha}{\rho}\right. \Lambda, \bigvee(A_\iota)_{\iota\in I}} \qquad \begin{array}{r} \alpha_0 < \alpha \\ \iota_0 \in I \restriction \alpha \end{array}$$

$$(\bigwedge) \qquad \frac{\mathcal{H}[\iota] \left|\frac{\alpha_\iota}{\rho}\right. \Lambda, A_\iota \textit{ for all } \iota\in I}{\mathcal{H} \left|\frac{\alpha}{\rho}\right. \Lambda, \bigwedge(A_\iota)_{\iota\in I}} \qquad |\,\iota\,| \leq \alpha_\iota < \alpha$$

$$(Cut) \qquad \frac{\mathcal{H} \left|\frac{\alpha_0}{\rho}\right. \Lambda, \mathcal{B} \qquad \mathcal{H} \left|\frac{\alpha_0}{\rho}\right. \Lambda, \neg\mathcal{B}}{\mathcal{H} \left|\frac{\alpha}{\rho}\right. \Lambda} \qquad \begin{array}{r} \alpha_0 < \alpha \\ rk(B) < \rho \end{array}$$

$$(Ref_{\Sigma(\pi)}) \qquad \frac{\mathcal{H} \left|\frac{\alpha_0}{\rho}\right. \Lambda, A}{\mathcal{H} \left|\frac{\alpha}{\rho}\right. \Lambda, \exists z\, A^z} \qquad \begin{array}{r} \alpha_0 < \alpha \\ \pi \in \Re^{\mathbf{M}} \\ A \in \Sigma \end{array}$$

$$(\mathbf{M}) \qquad \frac{\mathcal{H}[s] \left|\frac{\alpha_s}{\rho}\right. \Lambda, \exists y\, F(s,y) \textit{ for all } s\in RS_{\mathbf{M}}}{\mathcal{H} \left|\frac{\alpha}{\rho}\right. \Lambda, \exists z\, [\mathbf{Ad}(z) \wedge (\forall x\in z)(\exists y\in z)\, F(x,y)]} \qquad \begin{array}{r} |\,s\,| \leq \alpha_s < \alpha \\ F \in \Delta_0 \end{array}$$

As in the case of **KP** and RS_Ω, the proof system $RS_{\mathbf{M}}$ is tailored for an embedding of **KPM**.

Theorem 4.24 *If* **KPM** $\vdash B(a_1, \ldots, a_r)$, *then* $\mathcal{H} \left|\frac{\Omega\cdot m}{\Omega+n}\right. \mathcal{B}(\int_\infty, \ldots, \int_\nabla)$ *holds for some* m, n *and all set terms* $s_1, \ldots, s_r$ *and operators* $\mathcal{H}$ *satisfying*

$$\{\xi : \xi \textit{ occurs in } B(\vec{s})\} \cup \{\mathbf{M}\} \subseteq \mathcal{H}(\emptyset).$$

m and n depend only on the **KPM**-*derivation of* $B(\vec{a})$.

The cut-elimination procedure for $RS_{\mathbf{M}}$ is rather intricate (cf. [75]) and involves many more steps than in the case of RS_Ω. Omitting further details, we just state the outcome of it and a well-ordering proof for all initial segments of $\mathcal{T}(\mathbf{M})$ in **KPM**.

Corollary 4.25 *Letting* $\Omega := \chi^0(0)$, *we have:*

$$|\mathbf{KPM}| = |\mathbf{KPM}|_{\sup} = |\mathbf{KPM}|_{\Pi_2(L(\omega_1^{CK}))} = |\mathbf{KPM}|^{E}_{\Pi_2} = \psi^{\varepsilon_{\mathbf{M}+1}}(\Omega).$$

4.3.4 Recursively large ordinals and ordinal representation systems

The large cardinal hypothesis that **M** is the first weakly Mahlo cardinal is outrageous when compared with the strength of **KPM**. However, it enters the definition procedure of the collapsing function χ, which is then employed in the shape of terms to "name" a countable set of ordinals. As one succeeds in establishing recursion relations for the ordering between those terms, the set of terms gives rise to an ordinal representation system. It has long been suggested (cf. [25], p. 436) that, instead, one should be able to interpret the collapsing functions as operating directly on the recursively large counterparts of those cardinals. For example, taking such an approach in Definition 4.19 would consist in letting

$$\mathbf{M} := \text{first recursively Mahlo ordinal}$$

and setting $\Re^{\mathbf{M}} := \{\pi < \mathbf{M} : \pi \text{ admissible}, \pi > \omega\}$. The difficulties with this approach arise with the proof of Lemma 4.20. One wants to show that, for all α, $\chi^{\alpha}(\beta) < \mathbf{M}$ whenever $\beta < \mathbf{M}$. However, the arguments of the cardinal setting no longer work here. To get a similar result for a recursively Mahlo ordinal μ one would have to work solely with μ-recursive operations. In addition, the functions ψ^{α} would have to operate on admissible ordinals π. Here one wants $\psi_{\pi}(\alpha) < \pi$. In the cardinal setting this comes down to a simple cardinality argument. To get a similar result for an admissible π one would have to work solely with π-recursive operations. How this can be accomplished is far from being clear as the definition of $C^{\mathbf{M}}(\alpha, \rho)$ for $\rho < \pi$ usually refers to higher admissibles than just π. Notwithstanding that, the admissible approach is workable as was shown in [78, 81, 92]. A key idea therein is that the higher admissibles which figure in the definition of $\psi_{\pi}(\alpha)$ can be mimicked via names within the structure $\mathbf{L}_{\pi}$ in a π-recursive manner.

The drawback of the admissible approach is that it involves quite horrendous definition procedures and computations, which when taken as the first approach are at the limit of human tolerance.

On the other hand, the admissible approach provides a natural semantics for the terms in the EORSs. Recalling the notion of *good* Σ_1*-definition* from Definition 3.32, it turns out that all the ordinals of $\mathcal{T}(\mathbf{M}) \cap \mathbf{M}$ possess a good Σ_1-definition in **KPM** (cf. [81]) under the interpretation which takes $\mathbf{M}$ to be the first recursively Mahlo ordinal and lets the functions ψ^{α} operate on admissible ordinals instead of regular cardinals.

Unlike in the case of **KP**, $\mathcal{T}(\mathbf{M}) \cap \mathbf{M}$ only forms a proper subset of the **KPM**-definable ordinals, having many 'holes'.[10] Therefore, to illuminate the nature of the ordinals in $\mathcal{T}(\mathbf{M})$, it would be desirable to find another property which singles them out from the **KPM**-definable ordinals.

[10] The ordinals of $\mathcal{T}(\mathbf{M}) \cap \mathbf{M}$ are cofinal in $\mathbf{sp}_{\Sigma_1}(\mathbf{KPM})$, though. Letting $\pi_0 := \chi^{\varepsilon_{\mathbf{M}+1}}(0)$, one has $\sup(\mathbf{sp}_{\Sigma_1}(\mathbf{KPM})) = \psi^{0}(\pi_0)$.

References

[1] P. Aczel: *The Type Theoretic Interpretation of Constructive Set Theory*, in: MacIntyre, A., Pacholski, L., and Paris, J. (eds.), *Logic Colloquium '77*, North–Holland, Amsterdam 1978.

[2] P. Aczel: *The Type Theoretic Interpretation of Constructive Set Theory: Choice Principles*, in: Troelstra, A. S., van Dalen, D. (eds), *The L.E.J. Brouwer Centenary Symposium*, North–Holland, Amsterdam 1982.

[3] P. Aczel: *The Type Theoretic Interpretation of Constructive Set Theory: Inductive Definitions*, in: Marcus, R. B. et al. (eds), *Logic, Methodology, and Philosopy of Science VII*, North–Holland, Amsterdam 1986.

[4] H. Bachmann: *Die Normalfunktionen und das Problem der ausgezeichneten Folgen von Ordinalzahlen*, Vierteljahresschrift Naturforsch. Ges. Zürich 95 (1950) 115–147.

[5] J. Barwise: *Admissible Sets and Structures* (Springer, Berlin 1975).

[6] P. Bernays: *Hilbert, David,* Encyclopedia of philosophy, Vol. 3 (Macmillan and Free Press, New York, 1967) 496–504.

[7] W. Buchholz: *Eine Erweiterung der Schnitteliminationsmethode*, Habilitationsschrift (München 1977).

[8] W. Buchholz: *A new system of proof–theoretic ordinal functions,* Ann. Pure Appl. Logic **32** (1986) 195–207.

[9] W. Buchholz: *An independence result for* $(\Pi^1_1\text{-}CA+BI)$, Ann. Pure Appl. Logic 33 (1987) 131–155.

[10] W. Buchholz: *A simplified version of local predicativity,* in: Aczel, Simmons, Wainer (eds.): *Leeds Proof Theory 90* (Cambridge University Press, Cambridge, 1993) 115–147.

[11] W. Buchholz, S. Wainer: *Provable computable functions and the fast growing hierarchy,* in: Contemporary Mathematics 65 (American Mathematical Society, Providence,1987) 179–198.

[12] W. Buchholz, K. Schütte: *Proof theory of impredicative subsystems of analysis* (Bibliopolis, Naples, 1988).

[13] W. Buchholz, A. Cichon, A. Weiermann: *A uniform approach to fundamental sequences and hierarchies.* Mathematical Logic Quarterly 40 (1994) 273–286.

[14] W. Buchholz, S. Feferman, W. Pohlers, W. Sieg: *Iterated inductive definitions and subsystems of analysis* (Springer, Berlin, 1981).

[15] W. Buchholz and K. Schütte: *Proof theory of impredicative subsystems of analysis* (Bibliopolis, Naples, 1988).

[16] T. Carlson: *Ordinal arithmetic and* Σ_1 *elementarity* (1997) 12 pages, to appear in: Archive for Mathematical Logic.

[17] R. Diestel: *Graph Theory* (Springer, New York-Berlin-Heidelberg, 1997).

[18] A.G. Dragalin: *New forms of realizability and Markov's rule* (Russian), Dokl. Acad. Nauk. SSSR 2551 (1980) 543–537; translated in: Sov. Math. Dokl. 10, 1417–1420.

[19] F.R. Drake: *How recent work in mathematical logic relates to the foundations of mathematics*, The Mathematical Intelligencer vol. 7, no. 4 (1985) 27–35.

[20] P. Dybjer: *A general formulation of simultaneous inductive-recursive definitions in type theory* (preprint 1994) 21 pages. To appear in: Journal of Symbolic Logic.

[21] S. Feferman: *Systems of predicative analysis*, Journal of Symbolic Logic 29 (1964) 1–30.

[22] S. Feferman: *Systems of predicative analysis II. Representations of ordinals*, Journal of Symbolic Logic 33 (1968) 193–220.

[23] S. Feferman: *A Language and Axioms for Explicit Mathematics*, Lecture Notes in Math. 450 (Springer, Berlin, 1975), 87–139.

[24] S. Feferman: *Constructive Theories of Functions and classes* in: Boffa, M., van Dalen, D., McAloon, K. (eds.), *Logic Colloquium '78* (North-Holland, Amsterdam, 1979) 159–224.

[25] S. Feferman: *Proof theory: a personal report*, in: G. Takeuti, Proof Theory, 2nd edition (North-Holland, Amsterdam, 1987) 445–485.

[26] S. Feferman: *Hilbert's program relativized: Proof-theoretical and foundational reductions*, J. Symbolic Logic 53 (1988) 364–384.

[27] S. Feferman: *Gödel's program for new axioms: Why, where, how and what?* to appear in: Gödel '96 conference, Brno. 23 pages.

[28] H. Friedman: *Classically and intuitionistically provably recursive functions.* In: G.H. Müller, D.S. Scott: *Higher set theory* (Springer, Berlin, 1978) 21–27.

[29] H. Friedman, K. McAloon, and S. Simpson: *A finite combinatorial principle which is equivalent to the 1-consistency of predicative analysis,* in: G. Metakides (ed.): *Patras Logic Symposium* (North-Holland, Amsterdam, 1982) 197–220.

[30] H. Friedman, N. Robertson, P. Seymour: *The metamathematics of the graph minor theorem*, Contemporary Mathematics 65 (1987) 229–261.

[31] H. Friedman and S. Ščedrov: *Large sets in intuitionistic set theory*, Annals of Pure and Applied Logic 27 (1984) 1–24.

[32] H. Friedman and S. Sheard: *Elementary descent recursion and proof theory*, Annals of Pure and Applied Logic 71 (1995) 1–45.

[33] H. Gaifman: *A generalization of Mahlo's method for obtaining large cardinal numbers,* Israel Journal of Mathematics 5 (1967) 188–200.

[34] G. Gentzen: *Die Widerspruchsfreiheit der reinen Zahlentheorie,* Mathematische Annalen 112 (1936) 493–565.

[35] J.-.Y. Girard: *A survey of Π^1_2-logic. Part I: Dilators*, Annals of Mathematical Logic 21 (1981) 75–219.

[36] E. Griffor and M. Rathjen: *The strength of some Martin–Löf type theories.* Archive for Mathematical Logic 33 (1994) 347–385.

[37] D. Hilbert: *Die Grundlegung der elementaren Zahlentheorie*, Mathematische Annalen 104 (1931).

[38] D. Hilbert and P. Bernays: *Grundlagen der Mathematik II* (Springer, Berlin, 1938).

[39] P.G. Hinman: *Recursion-theoretic hierarchies* (Springer, Berlin, 1978).

[40] D. Isles: *Regular ordinals and normal forms*, in: A. Kino, J. Myhill, R.E. Vesley (eds.): Intuitionism and proof theory (North-Holland, Amsterdam, 1968) 288–300.

[41] G. Jäger: *Beweistheorie von KPN*, Archiv f. Math. Logik 2 (1980) 53–64.

[42] G. Jäger: *Zur Beweistheorie der Kripke–Platek Mengenlehre über den natürlichen Zahlen*, Archiv f. Math. Logik 22 (1982) 121–139.

[43] G. Jäger: *A well-ordering proof for Feferman's theory* T_0, Archiv f. Math. Logik 23 (1983) 65–77.

[44] Jäger, G.: *Theories for admissible sets: a unifying approach to proof theory* (Bibliopolis, Naples, 1986).

[45] G. Jäger and W. Pohlers: *Eine beweistheoretische Untersuchung von* $\mathbf{\Delta^1_2}\text{-}\mathbf{CA}+\mathbf{BI}$ *und verwandter Systeme*, Sitzungsberichte der Bayerischen Akademie der Wissenschaften, Mathematisch–Naturwissenschaftliche Klasse (1982).

[46] A. Kanamori: *The higher infinite.* (Springer, Berlin, 1995).

[47] A. Kanamori, M. Magidor: *The evolution of large cardinal axioms in set theory.* In: G. H. Müller, D.S. Scott (eds.) *Higher Set Theory.* Lecture Notes in Mathematics 669 (Springer, Berlin, 1978) 99-275.

[48] G. Kreisel: *On the interpretation of non-finitist proofs II,* Journal of Symbolic Logic 17 (1952) 43–58.

[49] G. Kreisel: *Mathematical significance of consistency proofs.* Journal of Symbolic Logic 23 (1958) 155–182.

[50] G. Kreisel: *Generalized inductive definitions,* in: Stanford Report on the Foundations of Analysis (Mimeographed, Stanford, 1963) Section III.

[51] G. Kreisel: *A survey of proof theory,* Journal of Symbolic Logic 33 (1968) 321–388.

[52] G. Kreisel: *Notes concerning the elements of proof theory.* Course notes of a course on proof theory at U.C.L.A. 1967 - 1968.

[53] G. Kreisel, G. Mints, S. Simpson: *The use of abstract language in elementary metamathematics: Some pedagogic examples,* in: *Lecture Notes in Mathematics,* vol. 453 (Springer, Berlin, 1975) 38–131.

[54] E. Lopez-Escobar: *An extremely restricted ω-rule,* Fundamenta Mathematicae 90 (1976) 159–172.

[55] A. Macintyre, H. Simmons: *Algebraic properties of number theories*, Israel Journal of Mathematics, Vol. 22, No. 1 (1975) 7–27.

[56] P. Mahlo: *Über lineare transfinite Mengen*, Berichte über die Verhandlungen der Königlich Sächsischen Gesellschaft der Wissenschaften zu Leipzig, Mathematisch-Physische Klasse, 63 (1911) 187–225.

[57] P. Mahlo: *Zur Theorie und Anwendung der ρ_0-Zahlen*, ibid. 64 (1912) 108–112.

[58] P. Mahlo: *Zur Theorie und Anwendung der ρ_0-Zahlen*, ibid. 65 (1913) 268–282.

[59] P. Martin-Löf: *Intuitionistic Type Theory*, (Bibliopolis, Naples 1984).

[60] G.E. Mints: *Finite investigations of infinite derivations,* Journal of Soviet Mathematics 15 (1981) 45–62.

[61] Y.N. Moschovakis: *Recursion in the universe of sets*, mimeographed note, 1976.

[62] J. Myhill: *Constructive Set Theory*, J. Symbolic Logic 40 (1975) 347–382.

[63] D. Normann: *Set recursion*, in: Fenstad et al. (eds.): *Generalized recursion theory II* (North-Holland, Amsterdam, 1978) 303–320.

[64] E. Palmgren: *On universes in type theory.* To appear in: *Proceedings of "Twenty-five years of type theory"*, Venice, 1995 (Oxford University Press).

[65] J. Paris, L. Harrington: *A mathematical incompleteness in Peano arithmetic.* In: J. Barwise (ed.): *Handbook of Mathematical Logic* (North Holland, Amsterdam, 1977) 1133-1142.

[66] H. Pfeiffer: *Ausgezeichnete Folgen für gewisse Abschnitte der zweiten und weiterer Zahlklassen* (Dissertation, Hannover, 1964).

[67] W. Pohlers: *Cut elimination for impredicative infinitary systems, part I: Ordinal analysis of ID_1*, Arch. f. Math. Logik 21 (1981) 69–87.

[68] W. Pohlers: *Proof-theoretical analysis of ID_ν by the method of local predicativity*, in: W. Buchholz, S. Feferman, W. Pohlers, W. Sieg: *Iterated inductive definitions and subsystems of analysis* (Springer, Berlin, 1981) 261–357.

[69] W. Pohlers: *Cut elimination for impredicative infinitary systems, part II: Ordinal analysis for iterated inductive definitions*, Arch. f. Math. Logik 22 (1982) 113–129.

[70] W. Pohlers: *Contributions of the Schütte school in Munich to proof theory*, in: G. Takeuti, *Proof Theory,* 2nd edition (North-Holland, Amsterdam, 1987) 406–431.

[71] W. Pohlers: *A short course in ordinal analysis*, in: P. Aczel, H. Simmons, S. Wainer (eds.): *Proof Theory* (Cambridge University Press, Cambridge, 1992) 27–78 .

[72] W. Pohlers: *Subsystems of set theory and second order number theory,* in: S. Buss (ed.): *Handbook of proof theory* (Elsevier Science B.V.) to appear.

[73] M. Rathjen: *Untersuchungen zu Teilsystemen der Zahlentheorie zweiter Stufe und der Mengenlehre mit einer zwischen* $\Delta^1_2 - CA$ *und* $\Delta^1_2 - CA + BI$ *liegenden Beweisstärke* (Publication of the Institute for Mathematical Logic and Foundational Research of the University of Münster, 1989).

[74] M. Rathjen: *Ordinal notations based on a weakly Mahlo cardinal,* Archive for Mathematical Logic 29 (1990) 249–263.

[75] M. Rathjen: *Proof-Theoretic Analysis of KPM,* Arch. Math. Logic 30 (1991) 377–403.

[76] M. Rathjen: *The role of parameters in bar rule and bar induction,* Journal of Symbolic Logic 56 (1991) 715–730.

[77] M. Rathjen: *Fragments of Kripke-Platek set theory with infinity,* in: P. Aczel, H. Simmons, S. Wainer (eds.): *Proof Theory* (Cambridge University Press, Cambridge, 1992) 251–273.

[78] M. Rathjen: *How to develop proof–theoretic ordinal functions on the basis of admissible sets.* Mathematical Quarterly 39 (1993) 47–54.

[79] M. Rathjen and A. Weiermann: *Proof–theoretic investigations on Kruskal's theorem,* Annals of Pure and Applied Logic 60 (1993) 49–88.

[80] M. Rathjen: *Admissible proof theory and beyond,* in: Logic, Methodology and Philosophy of Science IX (D. Prawitz, B. Skyrms and D. Westerstahl, eds.), Elsevier Science B.V. (1994) 123–147.

[81] M. Rathjen: *Collapsing functions based on recursively large ordinals: A well–ordering proof for KPM.* Archive for Mathematical Logic 33 (1994) 35–55.

[82] M. Rathjen: *Proof theory of reflection.* Annals of Pure and Applied Logic 68 (1994) 181–224.

[83] M. Rathjen: *Recent advances in ordinal analysis:* Π^1_2*-CA and related systems.* Bulletin of Symbolic Logic 1 (1995) 468–485.

[84] M. Rathjen: *An ordinal analysis of* Π^1_2 *comprehension and related systems,* preprint.

[85] M. Rathjen: *The higher infinite in proof theory,* in: J.A. Makowsky and E.V. Ravve (eds.): *Logic Colloquium '95.* Lecture Notes in Logic, vol. 11 (Springer, New York, Berlin, 1998) 275–304.

[86] M. Rathjen, E. Griffor and E. Palmgren: *Inaccessibility in constructive set theory and type theory.* Annals of Pure and Applied Logic 94 (1998) 181–200.

[87] M. Rathjen: *Large set axioms in constructive set theory and their strengths,* in preparation.

[88] Richter, W. and Aczel, P.: *Inductive definitions and reflecting properties of admissible ordinals.* In: J.E. Fenstad, Hinman (eds.) Generalized Recursion Theory (North Holland, Amsterdam, 1973) 301-381.

[89] H.E. Rose: *Subrecursion: functions and hierarchies.* (Clarendon Press, Oxford, 1984).

[90] G.E. Sacks: *Higher recursion theory* (Springer, Berlin, 1990).

[91] A. Schlüter: *Zur Mengenexistenz in formalen Theorien der Mengenlehre.* (Thesis, University of Münster, 1993).

[92] A. Schlüter: *Provability in set theories with reflection*, preprint, 1995.

[93] D. Schmidt: *Well-partial orderings and their maximal order types*, Habilitationsschrift (Heidelberg, 1979) 77 pages.

[94] K. Schütte: *Beweistheoretische Erfassung der unendlichen Induktion in der Zahlentheorie*, Mathematische Annalen 122 (1951) 369–389.

[95] K. Schütte: *Beweistheorie* (Springer, Berlin, 1960).

[96] K. Schütte: *Eine Grenze für die Beweisbarkeit der transfiniten Induktion in der verzweigten Typenlogik*, Archiv für Mathematische Logik und Grundlagenforschung 67 (1964) 45–60.

[97] K. Schütte: *Predicative well-orderings*, in: Crossley, Dummett (eds.), Formal systems and recursive functions (North Holland, 1965) 176–184.

[98] K. Schütte: *Proof Theory* (Springer, Berlin, 1977).

[99] H. Schwichtenberg: *Proof theory: Some applications of cut-elimination.* In: J. Barwise (ed.): *Handbook of Mathematical Logic* (North Holland, Amsterdam, 1977) 867–895.

[100] S. Simpson: *Nichtbeweisbarkeit von gewissen kombinatorischen Eigenschaften endlicher Bäume*, Archiv f. Math. Logik 25 (1985) 45–65.

[101] A. Setzer: *A well-ordering proof for the proof theoretical strength of Martin-Löf type theory*, to appear in: Annals of Pure and Applied Logic.

[102] C. Smorynski: *The incompleteness theorems.* In: J. Barwise (ed.): *Handbook of Mathematical Logic* (North Holland, Amsterdam, 1977) 821–864.

[103] C. Smorynski: *Self-reference and modal logic* (Springer, New York - Berlin, 1985).

[104] R. Sommer: *Ordinal arithmetic in* $I\Delta_0$. In: P. Clote and J. Krajicek (eds.): *Arithmetic, proof theory, and computational complexity* (Clarendon Press, Oxford, 1993) 320–363.

[105] R. Sommer: *Ordinal functions in fragments of arithmetic*, Preprint (1992) 28 pages.

[106] G. Takeuti: *Consistency proofs of subsystems of classical analysis*, Ann. Math. 86, 299–348.

[107] G. Takeuti, M. Yasugi: *The ordinals of the systems of second order arithmetic with the provably* Δ^1_2*-comprehension and the* Δ^1_2*-comprehension axiom respectively*, Japan J. Math. 41 (1973) 1–67.

[108] G. Takeuti: *Proof theory*, second edition (North Holland, Amsterdam, 1987).

[109] A. S. Troelstra: *Metamathematical investigations of intuitionistic arithmetic and analysis*, (Springer, Berlin, 1973).

[110] A. S. Troelstra and D. van Dalen: *Constructivism in Mathematics: An Introduction*, volume I, II, North–Holland, Amsterdam 1988.

[111] O. Veblen: *Continous increasing functions of finite and transfinite ordinals,* Trans. Amer. Math. Soc. 9 (1908) 280–292.

[112] A. Weiermann: *How to characterize provably total functions by local predicativity,* Journal of Symbolic Logic 61 (1996) 52–69.

[113] W.H. Woodin: *Large cardinal axioms and independence: The continuum problem revisited*, The Mathematical Intelligencer vol. 16, No. 3 (1994) 31–35.

Covering Properties of Core Models

Ernest Schimmerling
Carnegie Mellon University
and University of Connecticut

Theorem (Jensen's covering lemma) *Assume* $0^\#$ *does not exist. Let* A *be any uncountable set of ordinals. Then there is a* $B \in L$ *such that* $B \supseteq A$ *and* $\mathrm{card}(B) = \mathrm{card}(A)$.

In this paper, we outline Jensen's proof from a modern perspective. We isolate certain key elements of the proof which have become important both within and outside of inner model theory. This leads into an intuitive discussion of what core models are and the difficulties involved in generalizing Jensen's theorem to higher core models. Our hope is to give the reader some insight into these generalizations by concentrating on the simplest core model, L.

Jensen's theorem has striking consequences for cardinal arithmetic. Its conclusion implies that if $\omega_2 \leq \beta$ and β is a successor cardinal of L, then $\mathrm{cf}(\beta) = \mathrm{card}(\beta)$. In particular, if $0^\#$ does not exist, then L computes successors of singular cardinals correctly. The covering lemma also implies that some of the combinatorial principles, which Jensen proved in L, really hold. (I.e., they hold in V.) For example, if $0^\#$ does not exist and κ is any singular cardinal, then $\Box_\kappa$ holds.

By an *inner model*, we mean a transitive proper class model of ZFC. If M is an inner model, then M has the *covering property* if for every uncountable set of ordinals A, there exists $B \in M$ such that $B \supseteq A$ and $\mathrm{card}(B) = \mathrm{card}(A)$. *Core models* are certain kinds of inner models which we do not define here, except to say that L is a core model. (See [MiSt] for the precise definition and our Section 2 for a general description.)

If $0^\#$ exists, then L does not have the covering property. Dodd and Jensen found a substitute core model, which they called K, and proved that K has the covering property if there is no core model with a measurable cardinal. In fact, if $0^\#$ does not exist, then $K = L$, but not in general.

Expectations must be limited for extensions of the covering lemma to core models with measurable cardinals, because of the example given by Prikry forcing. At this juncture, there are at least three possibilities:

Cardinal arithmetical consequences of the covering property are known as *weak covering properties.* For example, the property that $\mathrm{cf}(\beta) = \mathrm{card}(\beta)$ whenever β is a successor cardinal of M is a weak covering property of M. The correct computation of successors of singular cardinals is another weak covering property, as is the correct computation of successors of weakly compact cardinals. A first approach is to prove that certain core models have weak covering properties under more relaxed anti-large cardinal hypotheses. Mitchell defined a core model, which he also called K, and proved that K has these weak covering properties under the hypothesis that there is no core model with a measurable cardinal κ of order κ^{++}. (See [Mi1].) The Mitchell core model and the Dodd-Jensen core model are the same if there is no core model with a measurable cardinal, so there is no ambiguity in the meaning of K. Steel extended the definition of K further, by weakening the anti-large cardinal hypothesis to the non-existence of a core model with a Woodin cardinal. More recently, Mitchell, Steel and the author proved the corresponding weak covering theorems. (See [St2], [MiSchSt], [MiSch], and [SchSt2].)

In a second approach, instead of weakening the covering property, one skirts the problem presented by Prikry forcing by considering only core models without measurable cardinals. For example, the minimal inner model closed under the operation $X \mapsto X^{\dagger}$ has no measurable cardinals, but is "beyond" $L[U]$. The author and Woodin have proved that this and other core models satisfy the covering property if their "sharps" do not exist. (The precise statements can be found in the forthcoming [SchWo]. Related results were obtained in [Mi2].)

Here is a typical example of a core model without measurable cardinals which plays an important role in the theory of projective sets of reals. If X is a set of ordinals and $n < \omega$, then let $\mathcal{M}_n(X)$ be the minimal inner model with n Woodin cardinals which has X as an element. (In particular, $\mathcal{M}_0(X) = L(X)$.) Under an appropriate large cardinal hypothesis, by [St3], $\mathcal{M}_n = \mathcal{M}_n(\emptyset)$ exists and is a core model. Let W_n be the minimal inner model closed under the operation

$$X \mapsto \mathcal{M}_n(X) \cap \mathcal{P}(\sup(X)).$$

Then W_n is a core model which has no measurable cardinals and so, by [SchWo], W_n has the "full" covering property, like Jensen's theorem for L. (W_n is also characterized as the minimal inner model which is closed under $X \mapsto C_{n+1}(X)$ if n is even, and $X \mapsto Q_{n+1}(X)$ if n is odd. See [St3].)

A third kind of extension to the covering lemma has us deal directly with the problematic Prikry sequences. One may weaken the covering property for a core model W to just

$$A \subseteq f''(\rho \cup \vec{C})$$

for some function $f \in W$, ordinal $\rho < \mathrm{card}(A)^+$, and *system of indiscernibles* $\vec{C}$ for W. The first such result is the Dodd-Jensen theorem for $L[U]$. (See [DoJe3] and [Do].) Mitchell and Gitik's lower bounds on the consistency strength of failure of the Singular Cardinal Hypothesis require deep analysis in this direction. (See [Gi] and [GiMi].)

These advanced covering results only came about after the elements of Jensen's proof were compartmentalized. The modern perspective which proved suitable for generalizations is known to many researchers in the area, but has not appeared in introductory form. In Sections 1, 3, and 4, we outline Jensen's proof from this modern point of view. We focus on some weak covering properties in Sections 1 and 3, before tackling the full result in Section 4. Having seen an extender in Section 1, we are able to say more about what core models are and to make some comments regarding covering properties of higher core models in Section 2. Section 3 offers an introduction to a common kind of chain argument.

There are many known simplifications to the proof of covering in the case of L, which we do not incorporate here, some of which the reader may notice. Again, the proof we give reflects our underlying interest in generalizations to higher core models.

It must be emphasized that none of the ideas in this paper are due to the author. There are several alternative approaches to Jensen's theorem, each useful and important for different reasons. The reader certainly will want to compare the proof we sketch with the proofs presented in [DeJe], [De], and [Ma].

The author thanks Sy Friedman, Aki Kanamori, and the referee for their helpful comments.

1 First pass at Jensen's proof: weak covering at countably closed cardinals

We begin by sketching a proof of the following weak covering property. Assume that κ is a countably closed cardinal. This means that $\mu^{\aleph_0} < \kappa$ whenever $\mu < \kappa$. For example, if $2^{\aleph_0} < \aleph_\omega$, then $\aleph_\omega$ is countably closed, since

$$(\aleph_n)^{\aleph_0} = \max(\aleph_n, 2^{\aleph_0})$$

whenever $1 \leq n < \omega$. Let $\lambda = (\kappa^+)^L$, and assume that $\mathrm{cf}(\lambda) < \kappa$. In particular, $\lambda < \kappa^+$. We shall show that $0^\#$ exists.

Let $X \prec V_{\kappa^+}$ with

$$\sup(X \cap \lambda) = \lambda,$$

$$\mathrm{card}(X) < \kappa,$$

$$\kappa \in X,$$

and

$${}^{\omega}X \subseteq X$$

Such an X can be realized as the union of a continuous ω_1 length chain of elementary submodels of V_{κ^+} where the first submodel contains a witness to $\mathrm{cf}(\lambda) < \kappa$, and successive submodels contain all the ω-sequences from the earlier models. Of course, this approach uses the countable closure of κ.

Let $\pi : N \simeq X$ be elementary with N transitive. Say $\pi(\overline{\kappa}) = \kappa$, $\pi(\overline{\lambda}) = \lambda$, $\delta = \mathrm{crit}(\pi)$, and $\alpha = \mathrm{OR} \cap N$. Then $\delta \leq \overline{\kappa} < \overline{\lambda} < \alpha$ and π is continuous at $\overline{\lambda}$ in the sense that

$$\sup(\pi''\overline{\lambda}) = \lambda = \pi(\overline{\lambda}).$$

By the condensation principle for L, $L^N = L_\alpha$. (And, in fact, $L_\alpha = J_\alpha$. Our convention here will be to use the L-hierarchy at levels where it coincides with the J-hierarchy.)

Easy Case. $\mathcal{P}(\delta) \cap L \subseteq L_\alpha$.

Let U be the ultrafilter on $N \cap \mathcal{P}(\delta)$ derived from π, namely

$$U = \{A \subseteq \delta \mid A \in N \text{ and } \delta \in \pi(A)\}$$

Then, by the case hypothesis, $U \cap L$ is an ultrafilter on $L \cap \mathcal{P}(\delta)$. The ultrapower of L by U consists of equivalence classes of the form $[f]^L_U$ where $f \in L$ and f is a function from δ into L.

Claim 1. $\mathrm{ult}(L, U)$ *is wellfounded.*

It follows from Claim 1 that $\mathrm{ult}(L, U) \simeq L$, so the ultrapower map gives a non-trivial elementary embedding of L into L, hence $0^\#$.

Claim 1 is proved using the countable completeness of U. Namely, if the claim fails, then there is a sequence of functions $\langle f_n \mid n < \omega\rangle$ such that $f_n \in L$, and a sequence $\langle A_n \mid n < \omega\rangle$ such that $A_n \in U \cap L$, with the property that $f_{n+1}(\xi) \in f_n(\xi)$ for every $\xi \in A_n$. But X is closed under ω-sequences, so

$$\pi\left(\langle A_n \mid n < \omega\rangle\right) = \langle \pi(A_n) \mid n < \omega\rangle \in X.$$

Therefore, if we set $A = \bigcap\{A_n \mid n < \omega\}$, then $\delta \in \bigcap \pi(A)$, so $A \in U$. In particular, $A \neq \emptyset$. But $\xi \in A$ and $n < \omega$ implies that $f_{n+1}(\xi) \in f_n(\xi)$, which is impossible.

Hard Case. *Otherwise.*

In this case, we shall get a contradiction by finding an ordinal γ such that

$$J_{\gamma+1} \models \lambda < \kappa^+ .$$

This truly is a contradiction since $\lambda = (\kappa^+)^L$.

Our first step is to derive an extender from π. Consider an arbitrary finite subset a of λ. Say $\mathrm{card}(a) = n < \omega$. Let μ_a be the least ordinal μ such that $\pi(\mu) > \max(a)$. By analogy with how U was defined in the Easy Case, we define a countably complete ultrafilter E_a on $N \cap \mathcal{P}([\mu_a]^n)$ by

$$E_a = \{A \subseteq [\mu_a]^n \mid A \in N \text{ and } a \in \pi(Y)\} \ .$$

To orient the reader, we note that the two ultrafilters $E_{\{\delta\}}$ and U differ in a trivial way:

$$\mu_{\{\delta\}} = \delta$$

and for all $A \subseteq \delta$,

$$A \in U \Longleftrightarrow \{\{\xi\} \mid \xi \in A\} \in E_{\{\delta\}} \ .$$

The (δ, λ)-*extender derived from* π is defined to be the system of ultrafilters

$$E = \langle E_a \mid a \in [\lambda]^{<\omega} \rangle \ .$$

We digress to discuss ultrapowers by extenders. Suppose that M is a transitive set model of a reasonable fragment of set theory with

$$V_{\overline{\lambda}}^M \subset N \ .$$

Then, for each $a \in [\lambda]^{<\omega}$, $E_a \cap M$ is an ultrafilter over M. The ultrapower $\mathrm{ult}(M, E_a)$ consists of equivalence classes $[f]^M_{E_a}$ where, if $n = \mathrm{card}(a)$, then $f \in M$ and f is a function from $[\mu_a]^n$ into M. Moreover, the ultrapowers of M by these ultrafilters form a direct limit system as indicated by the following commutative diagram.

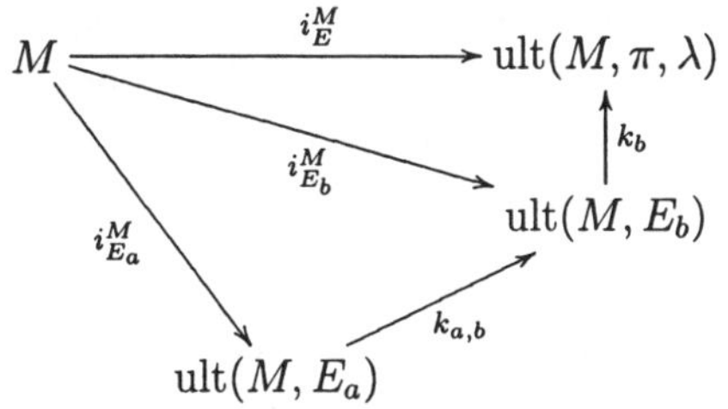

We have illustrated the case $a \subseteq b \in [\lambda]^{<\omega}$ and set $\mathrm{ult}(M, \pi, \lambda)$ equal to the direct limit of the structures $\mathrm{ult}(M, E_a)$ under the maps $k_{a,b}$. Then k_a is the limit of the embeddings $k_{a,b}$ and the elements of $\mathrm{ult}(M, \pi, \lambda)$ have the form

$$[a, f]^M_E = k_a([f]^M_{E_a}).$$

Admittedly, it takes some abstract nonsense to sort out the definition of $k_{a,b}$, which we leave to the reader. We have also set i^M_E equal to the limit of the ultrapower maps $i^M_{E_a}$.

We shall also be interested in restrictions of E, especially

$$E \restriction \kappa = \langle E_a \mid a \in [\kappa]^{<\omega} \rangle ,$$

and their associated ultrapowers. The ultrapower of M by $E \restriction \kappa$,

$$\text{ult}(M, \pi, \kappa),$$

is the direct limit of $\text{ult}(M, E_a)$ for $a \in [\kappa]^{<\omega}$. [It would be reasonable to write $\text{ult}(M, E \restriction \kappa)$ for $\text{ult}(M, \pi, \kappa)$, as is done elsewhere, although we shall not do so here.] There is an embedding of $\text{ult}(M, \pi, \kappa)$ into $\text{ult}(M, \pi, \lambda)$ with critical point at least κ. In many cases of interest, the two ultrapowers will be equal.

Here are a few remarks which are tie our notion of extender with some commonly used terminology and jargon. (These remarks could be skipped without loss of continuity.)

1. δ is the *critical point* of E. We write $\text{crit}(E) = \delta$.

2. λ is the *length* of E. We write $\text{lh}(E) = \lambda$.

3. Note that $\pi(\delta) < \lambda$. Equivalently, $\mu_a > \delta$ for some $a \in [\text{lh}(E)]^{<\omega}$. This property of E makes E a *long extender.*

4. The *superstrong extender derived from* π is $E \restriction \pi(\delta)$.

5. We shall see situations in which E is applied to M and $\text{OR} \cap M < \lambda$.

6. In certain contexts, the term "extender" is reserved for extenders which are total, that is, extenders which measure all subsets of their $[\mu_a]^n$'s for every $a \in [\lambda]^n$ and $n < \omega$. In cases such as ours, E would be called an *extender fragment,* since it only measures sets in N. In the terminology of [St2], (N, E) would be called a *background certificate.*

7. It is possible to express large cardinal axioms, such as the existence of measurable, strong, Woodin, superstrong, supercompact, and huge cardinals, by reference to long extenders. (See [St1].)

Returning to our outline, we next move towards applying the extender E to the longest initial segment of L possible. By the case hypothesis, $\alpha < (\delta^+)^L$. So we may define β to be the least ordinal such that for some $\mu < \overline{\lambda}$,

$$\mathcal{P}(\mu) \cap J_{\beta+1} \not\subseteq L_\alpha .$$

Because $\overline{\lambda}$ is the successor cardinal of $\overline{\kappa}$ in L_α , we have the some useful equivalent definitions of β. Namely, β is the least ordinal such that

$$\mathcal{P}(\overline{\kappa}) \cap J_{\beta+1} \not\subseteq L_\alpha .$$

And, also, β is the unique ordinal such that

$$J_\beta \models \overline{\lambda} = \overline{\kappa}^+$$

and

$$J_{\beta+1} \models \overline{\lambda} < \overline{\kappa}^+ .$$

The proofs of these equivalences use basic facts from [Je].

Clearly,

$$\delta < \alpha \leq \beta < (\delta^+)^L .$$

Now let $n < \omega$ be least such that for some $\mu < \overline{\lambda}$,

$$\mathcal{P}(\mu) \cap \Sigma_{n+1}^{\langle J_\beta , \in \rangle} \not\subseteq J_\beta .$$

Again, using the fact that $\overline{\lambda}$ is the successor cardinal of $\overline{\kappa}$ in J_β , n is least such that

$$\mathcal{P}(\overline{\kappa}) \cap \Sigma_{n+1}^{\langle J_\beta , \in \rangle} \not\subseteq J_\beta .$$

In terms of the projectum of J_β , this means that

$$\rho_n(J_\beta) \geq \overline{\lambda}$$

and

$$\rho_{n+1}(J_\beta) \leq \overline{\kappa} .$$

We now have that for every $\mu < \overline{\lambda}$,

$$\mathcal{P}(\mu) \cap L_{\overline{\lambda}} = \mathcal{P}(\mu) \cap L_\alpha = \mathcal{P}(\mu) \cap J_\beta = \mathcal{P}(\mu) \cap \Sigma_n^{\langle J_\beta , \in \rangle} \subsetneq \mathcal{P}(\mu) \cap \Sigma_{n+1}^{\langle J_\beta , \in \rangle} .$$

In particular, this holds for $\mu = \overline{\kappa}$.

At this point, we make another simplifying assumption, that $n = 0$. To complete the proof in the case $n \geq 1$, instead of working with J_β as is done below, one works with Jensen's Σ_n-coding structure $\langle J_\rho , \in , A \rangle$, where ρ is the Σ_n-projectum of J_β and A is the Σ_n-mastercode for J_β . (See [Je].)

For any structure $\mathfrak{A}$ and any set $S \subseteq |\mathfrak{A}|$, define

$$H_1^{\mathfrak{A}}(S) = \left\{ \tau^{\mathfrak{A}}[q] \mid q \in S \text{ and } \tau \text{ is a } \Sigma_1 \text{ Skolem term} \right\} .$$

Using the fact from [Je] that J_β is *sound*, we conclude that for some $x \in J_\beta$,

$$J_\beta = H_1^{\langle J_\beta , \in \rangle} (\overline{\kappa} \cup \{x\}) .$$

For example, the first standard parameter of J_β satisfies this equation. The specific choice of x is not relevant here however. (In this sense, less fine structure is used in the proof of the covering lemma than in the proof of Square.)

Claim 2. $\mathrm{ult}(J_\beta\ ,\pi,\lambda)$ *is wellfounded.*

The proof of Claim 2 uses countable closure as did that of Claim 1. (In Section 3, we shall sketch how to avoid countable closure altogether.) It follows from Claim 2 that $\mathrm{ult}(J_\beta\ ,\pi,\kappa)$ is wellfounded.

By Claim 2, we may identify $\mathrm{ult}(J_\beta\ ,\pi,\lambda)$ with J_γ for some ordinal γ. Let

$$\widetilde{\pi} : J_\beta \longrightarrow J_\gamma$$

be the ultrapower map. By Łoś's theorem adapted to long extenders, $\widetilde{\pi}$ is a Σ_0-elementary embedding of J_β into J_γ . Since the ultrapower is formed using only functions in J_β , $\widetilde{\pi}$ is a cofinal embedding in the sense that

$$\sup(\widetilde{\pi}''\beta) = \gamma\ .$$

So, in fact, $\widetilde{\pi}$ is Σ_1-elementary. [If β is a limit ordinal, $x \in J_\beta$, and φ is a Σ_1 formula, then $J_\beta \models \varphi[x]$ iff

$$\exists\beta_0 < \beta\ \ \forall\beta_1 > \beta_0\ \ J_{\beta_1} \models \varphi[x],$$

and cofinal Σ_0 embeddings preserve formulas of this form. If β is a successor ordinal, then one uses the S-hierarchy instead of the J-hierarchy.]

The relationship between $\pi : L_\alpha \longrightarrow L_{\kappa^+}$ and $\widetilde{\pi} : J_\beta \longrightarrow J_\gamma$ is obscured by the fact that $\alpha \leq \beta$ while $\kappa^+ > \gamma$. However, a little work shows that π and $\widetilde{\pi}$ agree below $\overline{\lambda}$. The main point is that for every $a \in \lambda$,

$$a = \pi(\mathrm{id})(a) = \Big[a, \mathrm{id}\Big]^{J_\beta}_{E\restriction\kappa},$$

where id is the identity map $u \mapsto u$.

We leave it as an exercise to verify that

$$\mathrm{ult}(J_\beta\ ,\pi,\lambda) = \mathrm{ult}(J_\beta\ ,\pi,\kappa)$$

and that $\widetilde{\pi}$ equals the ultrapower map from J_β into $\mathrm{ult}(J_\beta\ ,\pi,\kappa)$. The main point is that for every $\mu < \overline{l}ambda$, there is a well-order W of $\overline{\kappa}$ in $J_{\overline{\lambda}}$ and an ordinal $\iota < \kappa$ such that $\pi(\mu)$ is the rank in $\pi(W)$ of ι, and, so

$$\pi(\mu) = [\{\iota\}\,, \xi \mapsto \text{ the rank of } \xi \text{ in } W]^{J_\beta}_{E\restriction\kappa}\ .$$

The Σ_1 elementarity of $\widetilde{\pi}$ translates the soundness of J_β into the following properties of J_γ :

$$\widetilde{\pi}'' J_\beta = H_1^{\langle J_\gamma, \in\rangle}\Big((\widetilde{\pi}''\overline{\kappa}) \cup \{\widetilde{\pi}(x)\}\Big)$$

and

$$J_\gamma = H_1^{\langle J_\gamma, \in\rangle}\Big(\mathrm{lh}(E) \cup (\widetilde{\pi}'' J_\beta)\Big) = H_1^{\langle J_\gamma, \in\rangle}\Big(\lambda \cup \{\widetilde{\pi}(x)\}\Big) \ .$$

Using the fact that E and $E \restriction \kappa$ give the same ultrapower,

$$J_\gamma = H_1^{\langle J_\gamma, \in\rangle}\Big(\kappa \cup \{\widetilde{\pi}(x)\}\Big) \ .$$

Therefore

$$J_{\gamma+1} \models \widetilde{\pi}(\overline{\lambda}) < \kappa^+ \ .$$

Since $\overline{\lambda}$ is a regular in J_β and $\overline{\lambda} > \mu_a$ for all $a \in [\kappa]^{<\omega}$, it follows that $\widetilde{\pi}$ continuous at $\overline{\lambda}$. [Every function of the form

$$f : [\mu_a]^n \longrightarrow \overline{\lambda}$$

is bounded in $\overline{\lambda}$.] Recall that π is also continuous at $\overline{\lambda}$. Consequently,

$$\widetilde{\pi} \restriction J_{\overline{\lambda}+1} \ = \ \pi \restriction J_{\overline{\lambda}+1}$$

In particular, since $\lambda = \widetilde{\pi}(\overline{\lambda})$,

$$J_{\gamma+1} \models \lambda < \kappa^+ \ ,$$

which is the desired contradiction.

We have shown that if $0^\#$ does not exist, κ is a countably complete cardinal, and $\lambda = (\kappa^+)^L$, then either $\lambda = \kappa^+$ or $\mathrm{cf}(\lambda) = \kappa$. The proof easily adapts to show that if $0^\#$ does not exist, $\lambda = (\kappa^+)^L$, and $\mathrm{card}(\kappa)$ is countably closed, then either $\lambda = \mathrm{card}(\kappa)^+$ or $\mathrm{cf}(\lambda) = \mathrm{card}(\kappa)$.

2 Intermission: a few words about higher core models

We have seen how embeddings between transitive sets give rise to extenders and vice-versa. In this way, extenders can be seen as witnesses to large cardinal properties. For many reasons, it is desirable to construct inner models with large cardinals for which a Jensen style fine structural analysis is possible. That being the case, it is natural to consider inner models of the general form $L[\vec{E}]$, where $\vec{E}$ is a sequence of extenders which is intended to witness various large cardinal properties in $L[\vec{E}]$. Core models are certain inner models which have this form. We shall not give the precise definition, which can be found in [MiSt]. Rather, in order to highlight some of the key points, we shall tell a few small lies.

Initial segments of core models are structures of the form

$$\mathcal{J}_\alpha^{\vec{E}} = \langle J_\alpha^{\vec{E}}, \in, \vec{E} \restriction \alpha, E_\alpha \rangle$$

and are known as *premice.* [1] Part of the definition is that either $E_\alpha = \emptyset$, or there is an ordinal $\mu < \alpha$ such that E_α is a (μ, α)-extender over $J_\alpha^{\vec{E}}$. In other words, there is a premouse $J_\beta^{\vec{F}}$ and an embedding $\pi : J_\alpha^{\vec{E}} \longrightarrow J_\beta^{\vec{F}}$ such that $\text{crit}(\pi) = \mu$ and E_α is the extender of length α derived from π. There is also a *coherence* condition, which says that

$$\vec{F} \restriction \alpha = \vec{E} \restriction \alpha$$

and

$$F_\alpha = \emptyset.$$

The first part of the coherence condition says that E_α is "strong" in the sense of large cardinals. That is to say, $\mathcal{J}_\alpha^{\vec{E}}$ and its ultrapower by E_α agree below α. The second condition is used to compare premice: if two mice disagree at level α, then applying the extenders with index α improves the agreement to levels up to and including α for the ultrapowers. In the right context and with the technology of *iteration trees*, this naive approach to comparison can be made to work. Roughly speaking, premice which can be compared are known as *mice.*

We would like to say something about the difficulties in obtaining the covering results mentioned in the Introduction. So let us imagine that we are attempting to prove that a core model, $L[\vec{E}]$, has the weak covering property. Proceeding along the lines of Section 1, we have a cardinal κ, its $L[\vec{E}]$-successor cardinal λ,

$$N \underset{\pi}{\longrightarrow} X \prec V_{\kappa^+}$$

where N is transitive of cardinality $< \kappa$ and ordinal height α, and the critical point of π is δ. But there is an immediate difficulty with *condensation.* Namely, the pre-image of $L[\vec{E}]$ under π need not be an initial segment of $L[\vec{E}]$. In other words,

$$\pi^{-1}(L[\vec{E}]) = L_\alpha[\vec{F}]$$

[1]The actual definition of *core model* easily implies that

$$J_\alpha^{\vec{E}} = J_\alpha^{\vec{E}\restriction\alpha}$$

and that if $E_\alpha \neq \emptyset$, then

$$J_\alpha^{\vec{E}} = L_\alpha[\vec{E}].$$

for some extender sequence $\vec{F}$, but $\vec{E} \restriction \alpha$ and $\vec{F}$ may be different. This difficulty actually arises, and is just one reason for the comparison process using iteration trees mentioned above. (The next step would be to compare $L[\vec{E}]$ and $L[\vec{F}]$, but there is no reason to think that this comparison is trivial.)

Suppose that we manage to avoid the first difficulty altogether; there is still a second and more serious problem. Let us consider the case in which, like the Hard Case in Section 1, not every subset of δ from $L[\vec{E}]$ is in $L_\alpha[\vec{E}]$. We may proceed as before, letting $\mathcal{J}_\beta^{\vec{E}}$ be the first level of $L[\vec{E}]$ over which a subset of $\overline{\kappa} = \pi^{-1}(\kappa)$ which is missing from $L_\alpha[\vec{E}]$ is definable. Assume that Σ_1 is the minimal complexity for such a definition and proceed assuming that the fine structure of $L[\vec{E}]$ generalizes that of L in a straightforward way. Countable completeness allows us to conclude that

$$\text{ult}(\mathcal{J}_\beta^{\vec{E}}, \pi, \lambda) \text{ is wellfounded.}$$

Say

$$\text{ult}(\mathcal{J}_\beta^{\vec{E}}, \pi, \lambda) = \mathcal{J}_\gamma^{\vec{G}} .$$

As in the Hard Case of Section 1, $\mathcal{J}_\gamma^{\vec{G}}$ is a premouse over which λ is seen to have cardinality $\leq \kappa$. But, for this to be a contradiction, we would want that $\mathcal{J}_\gamma^{\vec{G}} \in L[\vec{E}]$. The difficulty is that $\vec{E} \restriction \gamma$ and $\vec{G}$ may be different. Again, the idea that leads to a solution (when there is a solution) is to compare $L[\vec{E}]$ and $L[\vec{G}]$ using iteration trees.

3 Second pass at Jensen's proof: weak covering without countable closure

We now describe how to replace the assumption of countable closure with $\aleph_2 \leq \kappa$ in Section 1. Once again, for simplicity of presentation, let κ be a cardinal, not just an L-cardinal. We assume that $\text{cf}(\lambda) < \kappa$ and show that $0^\#$ exists. In Section 1, countable closure was used to find an $X \prec V_{\kappa^+}$ with ${}^\omega X \subseteq X$, which, in turn, was used to prove the two claims of wellfoundedness. Even without countable closure, we can show that "many" X satisfy Claims 1 and 2 of Section 1. We shall find such X along an *internally approachable chain.*

Let ε be a regular cardinal with $\text{cf}(\lambda) < \varepsilon$ and $\aleph_2 \leq \varepsilon \leq \kappa$. Depending on whether or not λ has countable cofinality, either $\varepsilon = \aleph_2$ or $\varepsilon = (\text{cf}(\lambda))^+$ would do.

Let $\langle X_i \mid i < \varepsilon\rangle$ be a continuous chain of elementary substructures of V_{κ^+} such that for all $j < \varepsilon$,

$$\langle X_i \mid i \leq j\rangle \in X_{j+1} ,$$

$$X_j \cap \varepsilon \in \varepsilon,$$

and

$$\mathrm{card}(X_j) = \mathrm{card}(X_j \cap \varepsilon).$$

Assume also that $\kappa \in X_0$. For $i < \varepsilon$, let $\varepsilon_i = X_i \cap \varepsilon$. Note that $\langle \varepsilon_i \mid i < \varepsilon \rangle$ is a normal sequence converging to ε. For $i < \varepsilon$, let $\pi_i : N_i \longrightarrow V_{\kappa^+}$ be the inverse of the transitive collapse of X_i and α_i be the ordinal height of N_i . So $\mathrm{crit}(\pi_i) = \varepsilon_i$ and $\pi(\varepsilon_i) = \varepsilon$. Say $\pi_i(\kappa_i) = \kappa$ and $\pi_i(\lambda_i) = \lambda$. Let E_i be the extender of length λ derived from π_i .

Easy Case. *There is a stationary set*

$$S \subseteq \left\{ i < \varepsilon \mid \mathrm{cf}(i) \geq \aleph_1 \textit{ and } i = \varepsilon_i \right\}$$

such that $i \in S$ *implies*

$$\mathcal{P}(\varepsilon_i) \cap L \subset L_{\alpha_i} .$$

For $i \in S$, let U_i be the ultrafilter on N_i derived from π_i .

Claim 1. *There is an* $i \in S$ *such that* $\mathrm{ult}(L, U_i)$ *is wellfounded.*

Claim 1 implies that $0^{\#}$ exists. We leave the proof of Claim 1 as an exercise since its proof is similar to that of Claim 2 below.

Hard Case. *There is a stationary set*

$$S \subseteq \left\{ i < \varepsilon \mid \mathrm{cf}(i) \geq \aleph_1 \textit{ and } i = \varepsilon_i \right\}$$

such that $i \in S$ *implies*

$$\mathcal{P}(\varepsilon_i) \cap L \not\subset L_{\alpha_i} .$$

We call a partial function F on ε a *choice function* iff $F(i) \in X_i$ for all $i \in \mathrm{dom}(F)$. Fodor's lemma is used in the following general form in proofs of extensions of the covering lemma (cf. [MiSch]), although here we only need to consider choice functions into the integers.

Fodor's lemma. *Suppose that* F *is a choice function and that* $\mathrm{dom}(F)$ *is stationary in* ε. *Then there is a stationary* $S \subseteq \mathrm{dom}(F)$ *on which* F *is constant.*

Proof. Let $\langle G_i \mid i < \varepsilon \rangle$ be a sequence, which is strictly increasing and continuous with respect to inclusion, such that for all $i < \varepsilon$, G_i is a function from ε_i onto X_i. Let $C = \{i < \varepsilon \mid \varepsilon_i = i\}$. Then C is club in ε and if $i \in C$, then $\mathrm{crit}(\pi_i) = \varepsilon_i = i$. Define H on $\mathrm{dom}(F) \cap C$ by $H(i) = (G_i)^{-1}(F(i))$. Then $H(i) < \varepsilon_i = i$ for all $i \in \mathrm{dom}(H)$. By the usual Fodor's lemma, there is a stationary set $S \subseteq \mathrm{dom}(H)$ on which H is constant. Suppose that $i, j \in S$ and $i < j$. Then $F(i) = G_i(H(i)) = G_j(H(i)) = G_j(H(j)) = F(j)$. Therefore, F is constant on S.

Fix S as in the case hypothesis. For $i \in S$, let β_i be the least ordinal β such that

$$\mathcal{P}(\kappa_i) \cap J_{\beta+1} \not\subseteq L_{\alpha_i}$$

and n_i be the least $n < \omega$ such that there is a subset of κ_i which is $\mathbf{\Sigma}_{n+1}$ definable over $\langle J_{\beta_i}, \in \rangle$ but not an element of J_{β_i} . Let $n < \omega$ and $T \subseteq S$ be a stationary set such that $n_i = n$ for $i \in T$. For simplicity, let us assume that $n = 0$. (The other cases are handled using Σ_n fine structure for J_{β_i} as described earlier.)

Claim 2. *There is an $i \in T$ such that* $\mathrm{ult}(J_{\beta_i}, \pi_i, \kappa)$ *is wellfounded.* [2]

Claim 2 leads to a contradiction as did the corresponding claim in Section 1. Suppose, then, that $\mathrm{ult}(J_{\beta_i}, \pi_i, \kappa)$ is illfounded for every $i \in T$. One says that J_{β_i} *lifts badly* from i to ε. Let

$$j \in T \cap \lim(T).$$

Let Y be a countable elementary submodel of $V_{\kappa^{++}}$ with

$$\langle X_i \mid i < \varepsilon \rangle\ ,\ T\ ,\ j\ \in Y\ .$$

Fix an $i < j$ such that $i \in T$ and

$$Y \cap X_j \subset X_i\ .$$

[Recall that j has uncountable cofinality.] We shall need to consider the natural map $\pi_{i,j} : N_i \longrightarrow N_j$ and the extender $E_{i,j}$ of length λ_i derived from $\pi_{i,j}$. The next subclaim shows that if J_β lifts badly from i to ε, and J_β is definable (so that $\beta \in X_j$), then β lifts badly from i to j.

Subclaim A. *Suppose that $\beta \le \beta_i$, $\beta \in X_j$, and*

$$\mathrm{ult}(J_\beta, \pi_i, \kappa) \textit{ is illfounded.}$$

[2]In fact, this ultrapower is equal to $\mathrm{ult}(J_{\beta_i}, \pi_i, \lambda)$. In Section 1, we made do, to the extent possible, without assuming that λ is a successor cardinal in L. This generality is required for the full result which we sketch in Section 4. In this section, we shall use the fact that λ is a successor cardinal in L with little further comment.

Then

$$\text{ult}(J_\beta\ , \pi_{i,j}\ , \kappa_j) \text{ is illfounded.}$$

In particular,

$$\text{ult}(J_{\beta_i}\ , \pi_{i,j}\ , \kappa_j) \text{ is illfounded.}$$

Proof. Applying the elementarity of $\pi_j : N_j \longrightarrow V_{\kappa^+}$ we have that

$$N_j \models \text{“ult}\Big(\pi_j^{-1}(J_\beta), \pi_j^{-1}(\pi_i \restriction J_{\kappa_i}), \pi_j^{-1}(\kappa)\Big) \text{ is illfounded.”}$$

Hence,

$$N_j \models \text{“ult}(J_\beta\ , \pi_{i,j}\ , \kappa_j) \text{ is illfounded.”}$$

Subclaim A then follows by the absoluteness of illfoundedness.

Let β^{min} be the least $\beta \leq \beta_i$ such that

$$\text{ult}(J_\beta\ , \pi_i\ , \kappa) \text{ is illfounded.}$$

So $J_{\beta^{min}}$ is the least level of L which lifts badly from i to ε. This definition puts $\beta^{min} \in X_j$. Since β^{min} satisfies the hypothesis of Subclaim A,

$$\text{ult}(J_{\beta^{min}}\ , \pi_{i,j}\ , \kappa_j) \text{ is illfounded.}$$

In other words, $J_{\beta^{min}}$ lifts badly from i to j.

Subclaim B. *There is a $\beta^* \leq \beta_i$ such that*

$$\text{ult}(J_{\beta^*}\ , \pi_{i,j}\ , \kappa_j) \text{ is wellfounded,}$$

while

$$\text{ult}(J_{\beta^*}\ , \pi_i\ , \kappa) \text{ is illfounded.}$$

Suppose that β^* is as in Subclaim B. By the definition of β^{min} and the second clause of Subclaim B, $\beta^{min} \leq \beta^*$. Therefore, since $J_{\beta^{min}}$ lifts badly from i to j, J_{β^*} also lifts badly from i to j. But this is in direct contradiction with the first clause of Subclaim B.

It remains to find β^* as in Subclaim B. We shall realize J_{β^*} as a kind of "pullback" of J_{β_j} to i.

The soundness of J_{β_j} implies that there is an $x \in J_{\beta_j}$ such that

$$J_{\beta_j} = \bigcup\{Z_{\sigma,x} \mid \sigma < \beta_j\}\ ,$$

where $Z_{\sigma,x}$ be the Σ_1 hull in $\langle J_\sigma, \in \rangle$ of $\kappa_j \cup \{x\}$. [One may take x to be the standard parameter of J_{β_j}.] Moreover, for each $\sigma < \beta_j$, the transitive collapse of $Z_{\sigma,x}$ is J_γ for some $\gamma < \lambda_j$. Thus, there is a directed system $\mathcal{D} \subset J_{\lambda_j}$ whose direct limit is J_{β_j}. $\mathcal{D}$ consists of the transitive collapses of $Z_{\sigma,x}$ for $\sigma < \beta_j$ and $x \in J_\sigma$. Let $\mathcal{D}^* \subset J_{\lambda_i}$ be the direct limit system consisting of those J_η such that $\pi_{i,j}(J_\eta)$ is the transitive collapse of $Z_{\sigma,x}$ for some $\sigma < \beta_j$ and $x \in J_\sigma$. [Part of the point here is that $Z_{\sigma,x}$ might not be in the range of $\pi_{i,j}$ even if its transitive collapse is.]

There is a natural way in which $\pi_{i,j} \upharpoonright \kappa_i$ extends to an embedding from the direct limit of $\mathcal{D}^*$ into the direct limit J_{β_j} of $\mathcal{D}$. This has several consequences:

Facts.

1. *the direct limit of $\mathcal{D}^*$ is wellfounded,*

2. *there is an ordinal β^* such that J_{β^*} is the transitive collapse of the direct limit of $\mathcal{D}^*$,*

3. *$\beta^* \leq \beta_i$ and so $\mathrm{ult}(J_{\beta_i}, \pi_{i,j}, \kappa_j)$ is defined,*

4. *there is a commutative system of embeddings as in the diagram:*

$$
\begin{array}{ccccc}
J_{\beta^*} & \longrightarrow & J_{\beta_j} & \longrightarrow & \mathrm{ult}(J_{\beta_j}, \pi_j, \kappa) \\
 & \searrow & \uparrow & & \uparrow \\
 & \mathrm{ult}(J_{\beta^*}, \pi_{i,j}, \kappa_j) & \longrightarrow & \mathrm{ult}(J_{\beta^*}, \pi_i, \kappa)
\end{array}
$$

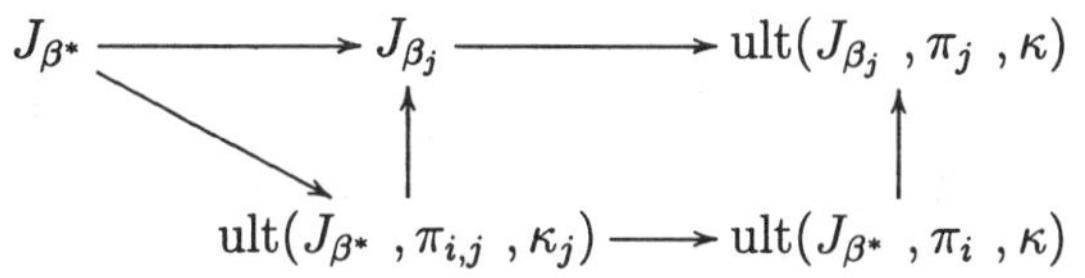

5. *$\mathrm{ult}(J_{\beta^*}, \pi_{i,j}, \kappa_j)$ is wellfounded since it embeds into J_{β_j} ,*

6. *$\mathrm{ult}(J_{\beta^*}, \pi_i, \kappa)$ is also defined, however*

7. *$\mathrm{ult}(J_{\beta^*}, \pi_i, \kappa)$ is illfounded.*

The reason for Fact 7 is that there is a witness to the illfoundedness of $\mathrm{ult}(J_{\beta_j}, \pi_j, \kappa)$ in the range of the embedding from $\mathrm{ult}(J_{\beta^*}, \pi_i, \kappa)$ to $\mathrm{ult}(J_{\beta_j}, \pi_j, \kappa)$. Here are a few hints why. Since $Y \prec V_{\kappa^{++}}$,

$$\langle Y, \in \rangle \models \text{``}\mathrm{ult}(J_{\beta_j}, \pi_j, \kappa) \text{ is illfounded.''}$$

So, there are functions

$$f_k \in J_{\beta_j} \cap Y ,$$

"coordinates"

$$a_k \in [\kappa]^{<\omega} ,$$

and "measure one" sets

$$A_k \in (E_j)_{a_k} \cap J_{\beta_j}$$

so that

$$\left\langle [a_k, f_k]_{E_j}^{J_{\beta_j}} \mid k < \omega \right\rangle$$

is an infinite descending chain of $\mathrm{ult}(J_{\beta_j}, \pi_j, \kappa)$ as witnessed by $\langle A_k \mid k < \omega \rangle$. Let $x_\ell = \langle f_k \mid k < \ell \rangle$ and choose sufficiently large $\sigma_\ell \in \beta_j \cap Y$ so that

$$J_{\sigma_\ell} \models \text{"} f_0 \ni \cdots \ni f_\ell \text{ on } A_\ell \text{"}$$

[More precisely, we may assume that $a_k \subset a_\ell$ whenever $k < \ell < \omega$. Whenever

$$a = \{a_1 < \cdots < a_m\} = \{b_{p_1} < \cdots < b_{p_m}\} \subseteq b = \{b_1 < \cdots < b_n\}$$

and

$$u = \{u_1 < \cdots < u_n\},$$

then we define

$$u^{a,b} = \{u_{p_1}, \ldots, u_{p_m}\}.$$

What we require above is that

$$J_{\sigma_\ell} \models \left\{ u \in [\mu_b]^n \mid f_k(u^{a,b}) \ni f_\ell(u) \right\} \subseteq A_\ell$$

whenever $k < \ell$, $a = a_k$, $b = a_\ell$, and $n = \mathrm{card}(b)$.] Then the transitive collapse of each Z_{σ_ℓ, x_ℓ} is in $J_{\lambda_j} \cap Y$. Some routine checking, which we leave as an exercise, shows that for every $k < \omega$,

$$\left[a_k, f_k \right]_{E_j}^{J_{\beta_j}}$$

is in the range of the natural embedding from $\mathrm{ult}(J_{\beta^*}, \pi_i, \kappa)$ to $\mathrm{ult}(J_{\beta_j}, \pi_i, \kappa)$. So Fact 7 holds.

We have shown that if $0^\#$ does not exist, κ is a cardinal, $\kappa \geq \aleph_2$, and $\lambda = (\kappa^+)^L$, then either $\lambda = \kappa^+$ or $\mathrm{cf}(\lambda) = \kappa$. The proof easily adapts to show that if $0^\#$ does not exist and λ is any successor cardinal of L such that $\aleph_2 \leq \lambda$, then $\mathrm{cf}(\lambda) = \mathrm{card}(\lambda)$.

4 Third pass at Jensen's proof: putting it all together

Let us assume that $0^\#$ does not exist. We say that Y *covers* X if $Y \supseteq X$ and $\mathrm{card}(Y) = \mathrm{card}(X)$. We prove by induction on ordinals λ that for every uncountable $X \subseteq \lambda$, there is a $Y \in L$ which covers X.

So fix λ and X. Clearly, by the induction hypothesis, we may assume that $\sup(X) = \lambda$. We may also assume that λ is an L-cardinal. [Otherwise, there is a $\kappa < \lambda$ and a constructible bijection $f : \kappa \longrightarrow \lambda$. By the induction hypothesis, there is a $Y \in L$ which covers $f^{-1}(X)$. Then $f''Y$ covers X.] And, also, we may assume that λ is not a cardinal, since otherwise λ itself covers X. For the same reason, we may assume that $\text{card}(X) < \text{card}(\lambda)$ and $\aleph_2 < \lambda$. Of course, $\text{cf}(\lambda) \leq \text{card}(X)$.

Let ε be a regular cardinal with $\text{card}(X) < \varepsilon < \lambda$ and $\aleph_2 \leq \varepsilon < \lambda$. For example, $\varepsilon = \text{card}(X)^+$ would do.

As in Section 3, select an internally approachable chain $\langle X_i \mid i < \varepsilon \rangle$ of substructures of V_{λ^+}, but make sure that $X_0 \supseteq X$. Let us use the same notation as in Section 3. Since $0^\#$ does not exist, we are in the Hard Case. We may not assume that λ is an L-successor cardinal. However, we may still define β_i to be the least ordinal β such that for some $\mu < \lambda$,

$$\mathcal{P}(\mu) \cap J_{\beta+1} \not\subseteq L_{\alpha_i}$$

and n_i to be the least $n < \omega$ such that there is a bounded subset of λ which is $\mathbf{\Sigma}_{n+1}$ definable over $\langle J_{\beta_i}, \in \rangle$ but not an element of J_{β_i}. As before, we restrict attention to the case $n_i = 0$.

As in Claim 2 of Section 3, we find $i < \varepsilon$ such that $\text{ult}(J_{\beta_i}, \pi_i, \lambda)$ is well-founded. (Again, some minor modifications must be made to allow for the possibility that λ is a limit cardinal of L.) Fix such an i and put $\pi = \pi_i$, $\delta = \varepsilon_i$, $\overline{\lambda} = \lambda_i$, $\alpha = \alpha_i$, and $\beta = \beta_i$. And, also, say

$$\widetilde{\pi} : J_\beta \longrightarrow J_\gamma = \text{ult}(J_{\beta_i}, \pi_i, \lambda).$$

Then, as in Section 1,

$$\text{ran}(\widetilde{\pi}) \cap (\lambda + 1) = \text{ran}(\pi) \cap (\lambda + 1) = X_i \cap (\lambda + 1) \supseteq X.$$

By soundness, there is a $\mu < \lambda$ and an $x \in J_\beta$ such that

$$J_\beta = H_1^{\langle J_\beta, \in \rangle}(\mu \cup \{x\}).$$

[Let μ be the Σ_1 projectum and x be the first standard parameter of J_β.] Then

$$X \subseteq \text{ran}(\widetilde{\pi}) = H_1^{\langle J_\gamma, \in \rangle}\Big((\widetilde{\pi}''\mu) \cup \{\widetilde{\pi}(x)\}\Big) = H_1^{\langle J_\gamma, \in \rangle}\Big((\pi''\mu) \cup \{\widetilde{\pi}(x)\}\Big).$$

Since $\pi(\mu) < \lambda$, by the induction hypothesis, there is a set $Y \in L$ which covers $\pi''\mu$. Therefore X is covered by

$$Z = \lambda \cap H_1^{\langle J_\gamma, \in \rangle}\Big(Y \cup \{\widetilde{\pi}(x)\}\Big)$$

and $Z \in L$, as desired.

References

[De] K.J. Devlin, *Constructibility*, Perspectives in Mathematical Logic, Springer-Verlag, Berlin-New York, 1984.

[DeJe] K.I. Devlin and R.B. Jensen, *Marginalia to a theorem of Silver*, in *Logic Conference, Kiel 1974*, Lecture Notes in Mathematics, 499, Springer-Verlag, 1975, 115–142.

[Do] A.J. Dodd, *The core model*, London Mathematical Society Lecture Note Series, 61, Cambridge University Press, Cambridge-New York, 1982.

[DoJe1] A.J. Dodd and R.B. Jensen, *The core model*, Ann. Math. Logic **20** (1981), no. 1, 43–75.

[DoJe2] A.J. Dodd and R.B. Jensen, *The covering lemma for K*, Ann. Math. Logic **22** (1982), no. 1, 1–30.

[DoJe3] A.J. Dodd and R.B. Jensen, *The covering lemma for $L[U]$*, Ann. Math. Logic **22** (1982), no. 2, 127–135.

[Je] R.B. Jensen, *The fine structure of the constructible hierarchy*, Ann. Math. Logic **4** (1972), no. 3, 229–308.

[Gi] M. Gitik, *The strength of the failure of the singular cardinal hypothesis*, Ann. Pure Appl. Logic **51** (1991), no. 3, 215–240.

[GiMi] M. Gitik and W.J. Mitchell, *Indiscernible sequences for extenders, and the singular cardinal hypothesis*, Ann. Pure Appl. Logic **82** (1996), no. 3, 273–316

[Ma] M. Magidor, *Representing sets of ordinals as countable unions of sets in the core model*, Trans. Amer. Math. Soc. **317** (1990), no. 1, 91–126.

[Mi1] W.J. Mitchell, *The core model for sequences of measures I*, Math. Proc. Cambridge Philos. Soc. **95** (1984), no. 2, 229–260.

[Mi2] W.J. Mitchell, *A hollow shell: covering lemmas without a core* in *Set theory (Curaçao, 1995; Barcelona, 1996)*, 183–198, Kluwer Acad. Publ., Dordrecht, 1998.

[MiSch] W.J. Mitchell and E. Schimmerling, *Weak covering without countable closure*, Math. Res. Lett. **2** (1995), no. 5, 595–609.

[MiSchSt] W.J. Mitchell, E. Schimmerling, and J.R. Steel, *The covering lemma up to a Woodin cardinal*, Ann. Pure Appl. Logic **84** (1997), no. 2, 219–255.

[MiSt] W.J. Mitchell and J.R. Steel, *Fine structure and iteration trees*, Lecture Notes in Logic, 3, Springer-Verlag, Berlin, 1994.

[Sch] E. Schimmerling, *Review of "The core model iterability problem" by John Steel*, J. Symbolic Logic **63** (1998) no. 1, 326–328.

[SchSt1] E. Schimmerling and J.R. Steel, *Fine structure for tame inner models*, J. Symbolic Logic **61** (1996), no. 2, 621–639.

[SchSt2] E. Schimmerling and J.R. Steel, *The maximality of the core model*, to appear in Trans. Amer. Math. Soc.

[SchWo] E. Schimmerling and W.H. Woodin, *The Jensen covering property*, to appear in J. Symbolic Logic.

[St1] J.R. Steel, *The well-foundedness of the Mitchell order*, J. Symbolic Logic **58** (1993), no. 3, 931–940.

[St2] J.R. Steel, *The core model iterability problem*, Lecture Notes in Logic, 8, Springer-Verlag, Berlin, 1996.

[St3] J.R. Steel, *Projectively well-ordered inner models*, Ann. Pure Appl. Logic **74** (1995), no. 1, 77–104.

[St4] J.R. Steel, *Inner models with many Woodin cardinals*, Ann. Pure Appl. Logic 65 (1993), no. 2, 185–209.

Ordinal Systems

Anton Setzer
Department of Mathematics, Uppsala University
P.O. Box 480, S-751 06 Uppsala, Sweden
email: setzer@math.uu.se

Abstract

Ordinal systems are structures for describing ordinal notation systems, which extend the more predicative approaches to ordinal notation systems, like the Cantor normal form, the Veblen function and the Schütte Klammer symbols, up to the Bachmann-Howard ordinal. σ-ordinal systems, which are natural extensions of this approach, reach without the use of cardinals the strength of the transfinitely iterated fixed theories ID_σ in an essentially predicative way. We explore the relationship with the traditional approach to ordinal notation systems via cardinals and determine, using "extended Schütte Klammer symbols", the exact strength of σ-ordinal systems.

1 Introduction

1.1 Motivation

The original problem, which motivated the research in this article, seemed to be a *pedagogical* one. Several times we have tried to teach ordinal notation systems above the Bachmann-Howard ordinal. The impression we got was that we were able to teach the technical development of these ordinal notation systems, but that some doubts in the audience always persisted. It remained unclear why one could get a well-ordered notation system by denoting small ordinals by big cardinals.

The situation was completely different with typical ordinal notation systems below the Bachmann-Howard ordinal. We had the impression that we always succeeded in teaching it once the audience had overcome some technical problems. And this included the Schütte Klammer symbols (an ordinal notation system extending the Veblen hierarchy — they will essentially be defined in this article): Although they are technically more complicated than the systems using one uncountable cardinal, they seem to be far more acceptable. Therefore the reason behind our pedagogical problems was not a technical one. The real problem was about *foundations.*

The original task of proof theory as understood by Hilbert was to show the consistency of systems in which mathematical reasoning can be formalized. After the proof of Gödel's incompleteness theorem, one had to modify this and demand the reduction of the consistency of a theory to some principles in whose correctness we have good reason to believe. One reason, why Gentzen's result was so much appreciated, when it was presented, was that he reduced the consistency of Peano Arithmetic to the principle of well-ordering up to ϵ_0, which we intuitively believe to be correct. The argument presented in Lemma 2.2 is an attempt to formalize what we believe is the reason for our confidence in this principle: the usual notation system for ϵ_0 is built from below and therefore has an intuitive well-ordering proof.

For the Veblen function and for the Schütte Klammer symbols the same holds. So, when analyzing a theory using these systems we have really gained more than just the reduction of the consistency of the theory to a primitive recursive well-ordering: the reduction is to the well-foundedness of an ordering, which we intuitively believe to be correct.

This leads to another aspect: the relationship to the notion of "natural well-ordering". There exists a trivial ordinal analysis for any consistent theory. Take as elements of a well-ordering essentially pairs consisting of a well-ordering proof in the theory and elements of the corresponding well-ordering and order them lexicographically by the Gödel-number of the proof and the ordering we are referring to. (A slight refinement is necessary in order to make it primitive recursive: replace the elements of the well-ordering by triples $<a, b, c>$ where a is an element of the well-ordering, b a calculation that determines that a is an element of this ordering and c is a calculation that determines for all $x < a$, where $<$ is the ordering on the natural numbers, whether x belongs to the well-ordering and, if yes, the order relation between x and a. Order triples $<a, b, c>$ by the ordering of a.) The corresponding ordering has as order type the proof theoretic ordinal of the theory.[1] In order to make clear that what one was doing is not trivial, one usually states that one determines the proof theoretic strength not in some arbitrary primitive recursive notation system, but in a natural one. However, nobody has succeeded up to now in defining precisely, what a natural well-ordering is. But there might be some systematic reason why we will never be able to formalize what a natural well-ordering is: if one has a precise notion, one will probably find a system diagonalizing over it, and this system can no longer be natural,

[1] We heard this example from Richard Sommer, but do not know of its origins.

although it will be in an intuitive sense.

Following on from the above considerations, we suggest that one should replace "natural well-ordering" by "ordering with an intuitive well-ordering proof". We believe that the reduction to such well-orderings is in fact the real motivation for designing stronger and stronger ordinal notation systems. We will see in the following that the usual systems as developed by the Schütte school (we have not studied ordinal diagrams sufficiently yet) fulfil essentially this requirement. With a little bit more structure it is easy to see intuitively, why the system is consistent.

In the following we are going to explore three types of (iterated) ordinal systems: (non-iterated) ordinal systems, n-ordinal systems and σ-ordinal systems. For each of these systems we will proceed as follows: First we motivate and introduce the structure. We will argue that each of the steps taken is very natural — it is not the only natural way of proceeding but one possible one. We will then present what we hope is an intuitive well-ordering proof, which will be formalized rigorously in a theory of the appropriate strength, and therefore also provides upper bounds for the order types of the structure. In the case of σ-ordinal systems the rigorous formalization is not completed yet and we will omit the argument. After this intuitive well-ordering proof we will give constructive well-ordering proofs, which will be far shorter than the intuitive argument. However, although they have the advantage that they can be formalized in constructive theories, which does not apply in the case of the intuitive well-ordering proofs, we personally believe in the well-ordering of the system, essentially because of the intuitive well-ordering proof, not the constructive ones. We do not understand yet why this is the case, and an analysis of this needs still to be done. It will become clear however that both proofs will be closely related.

Next we are going to introduce a sequence of ordinal notation systems, which exhausts the strength of the structure. It is no problem to develop it in a mere syntactical way. However, we also want to develop functions acting on ordinals and not just on notations, and, when developing first the functions, we get the notation system almost for free. So we will take the detour via ordinal functions, which might not be as convincing for those who like to cut out ideal concepts like ordinals, but might be quite satisfactory for those who want to compare the approach taken here with the traditional approach. We will succeed in recovering the functions used in the traditional approach in our setting. From the development adopted, it will be very clear that we can introduce the notation system without referring to ordinals, even for heuristic purposes, and it would be a boring task to rewrite these subsections so that no ordinals are used.

Technical advantages: This approach separates properties of an ordinal notation system, which do not have an influence on its whole strength, from structural properties, which are crucial for achieving that strength. The well-ordering proofs themselves will become easier (for instance it works in one step, one does not have first to verify that the accessible part is closed under $+$, then under $\omega^{\cdot}$, then under ψ etc.) and focuses on what is actually needed from the ordinal system, whereas the specific ordinal notation system plays only a role when verifying that it is an instance of a (non-iterated or n- or σ-) ordinal system. As a side-result we get simultaneously well-ordering proofs in the formal theories considered for a complete family of notation systems, not only for one specific system.

Background needed: We will work quite a lot with ordinals in this article. However, we believe that one can understand quite a lot of it, even without having a deep insight. Without the knowledge of ordinal notation systems up to ϵ_0 it will be probably difficult to understand anything. Knowledge of the Veblen function will be useful, but not necessary. We will consider Schütte Klammer symbols as examples. But those who do not know the Veblen function or the Klammer symbols can take the equation we state between those functions and the corresponding ordinal function generators as the definition. In this case we recommend skipping the versions with fixed points and just considering the fixed-point free versions of φ and the Klammer symbols, which are in our setting more natural than the usual ones. The comparison with the ϑ- and ψ-function is only relevant for those who know those systems. It will be useful — but not necessary — to read the first four sections of [Set98a], especially of the motivation given there.

Future developments: With σ-ordinal system we have not exhausted the power of combinations of ordinal systems, in the meantime we have developed ordinal systems which cover the strength of one recursive inaccessible and one recursive Mahlo ordinal, and this is certainly not the end of what can be done with our approach. For the author, this was quite satisfactory, since one can see, that even up to Mahlo we can work with extensions of Schütte's Klammer symbols in a way which is still from below. Before carrying out this research, we could work with and calculate with the strong ordinal notation systems, but always had some feeling that what we were doing was dubious. The only real argument for the well-foundedness of the systems was the well-ordering proof in a corresponding constructive theory and the main justification for carrying out ordinal analysis seemed to be to provide good tools for reducing the consistency of one theory to another where a direct way was impossible. But now we are really convinced that one gains something more: the reduction of the consistency to the well-ordering of a structure, for which we have an intuitive well-ordering proof, and which is — in an extended sense — built from below.

1.2 Notations

Definition 1.1 (a) As in [Set98a], A^* will be the set of finite sequences of elements of A, coded, if $A \subseteq \mathbb{N}$ as natural numbers, $(\vec{a})_i$ will be the ith element of the sequence $\vec{a}$ and (a slight change relative to [Set98a]) $\mathrm{seqlength}(\vec{a})$ will be the length of the sequence $\vec{a}$.

A class is an object $\{x \mid \varphi\}$, where φ is a formula. In this case, after some possibly necessary α-conversion, $a \in \{x \mid \varphi\} := \varphi[x := a]$. We identify unary predicates Q with the class $\{x \mid Q(x)\}$.
An *ordering* is a pair $(A, \prec)$ where A is a class and $\prec$ is a binary relation on A. It is *primitive recursive*, if A is a primitive recursive subset of $\mathbb{N}$ and $\prec$ is primitive recursive, and *linear*, if $\prec$ is a linear ordering on A.
If $A = (B, \prec)$, then $|A| := B$, $\prec_A := \prec$.
If $C \subseteq |A|$, we will write $(C, \prec)$ instead of $(C, \prec \cap (B \times B))$.
Transfinite induction over $(A, \prec)$ *with respect to the class* B, in short $\mathrm{TI}_{(A,\prec)}(B)$, is defined as $\forall x \in A(\forall y \in A(y \prec x \to y \in B) \to x \in B) \to \forall x \in A.x \in B$.
As in [Set98a], we define PRA^+ as the extension of PRA by additional predicates (called free predicates) without having induction over formulas containing these predicates. *Transfinite induction over* $(A, \prec)$ *is in* PRA *reducible to transfinite induction over* $(A_i, \prec_i)$ $(i = 1, \ldots, n)$, in short $\mathrm{TI}_{(A,\prec)}$ *is* PRA*-reducible to* $\mathrm{TI}_{(A_i,\prec_i)}$, if there exist $n_i \in \mathbb{N}$, variables $z_{i,j,k}$, classes $B_{i,j}$ with free variables $\subset \{z_{i,j,1}, \ldots, z_{i,j,m_{i,j}}\}$, such that $\mathrm{PRA}^+ \vdash (\bigwedge_{i=1}^{n} \bigwedge_{j=1}^{n_i} (\forall z_{ij1}, \ldots, z_{ijm_{i,j}}.\mathrm{TI}_{(A_i,\prec_i)}(B_{ij})) \to \mathrm{TI}_{(A,\prec)}(Q)$ for some free unary predicate Q.

(b) An ordering $(B, \prec)$ is an *elementary construction from orderings* $(A_1, \prec_1), \ldots, (A_m, \prec_m)$, if $(B, \prec)$ is primitive recursive, linear, linear provably in PRA respectively, provided this property holds for $(A_i, \prec_i)$, and, if, under the condition that $(A_i, \prec_i)$ are primitive recursive and linear provably in PRA, transfinite induction over $(B, \prec)$ reduces to transfinite induction over $(A_i, \prec_i)$.

(c) If $f : A \to B$, $M \subseteq A$, then $f[M] := \{f(x) \mid x \in M\}$.
$\mathcal{P}^{\mathrm{fin}}(A)$ is the set of finite sets of elements of A coded, if $A \subseteq \mathbb{N}$, as natural numbers, such that the usual properties, especially primitive recursiveness and decidable subset-relation hold.
In case $B \in \mathcal{P}^{\mathrm{fin}}(A)$, we write $t \in B$ for the statement expressing t is an element of B expressed in the language of PRA.
If $B \in \mathcal{P}^{\mathrm{fin}}(A)$, $a \in A$, $\prec$ is a binary relation of A, then $B \prec a :\Leftrightarrow \forall x \in B.x \prec a$, $a \prec B :\Leftrightarrow \exists x \in B.a \prec x$. These definitions extend to arbitrary classes B as well.
If $\prec$ is a binary relation on A, $a \preceq b :\Leftrightarrow a \prec b \lor a = b$, similarly we

define $\prec'$ from $\prec$, $\leq$ from $<$ etc.
If $k : A \to \mathcal{P}^{\mathrm{fin}}(B)$ and $C \subseteq B$, then $k^{-1}(C) := \{x \in A \mid k(x) \subseteq C\}$.
We write $f : A \to_\omega B$ for $f : A \to \mathcal{P}^{\mathrm{fin}}(B)$. If $f : A \to_\omega B$, $f' : A \to_\omega B$, $g : B \to_\omega C$, then $g \circ f$: $A \to_\omega C$, $(g \circ f)(a) := g[f(a)]$, and $f \subseteq f' :\Leftrightarrow \forall x \in A. f(x) \subseteq f'(x)$.

(d) If $A_1, \ldots, A_m$ are orderings, $f : (|A_1| \times \cdots \times |A_m|) \to M$ injective, then $f[A_1, \ldots, A_m]$ denotes the ordering $(f[|A_1| \times \cdots \times |A_m|], \prec)$ where $f(a_1, \ldots, a_m) \prec f(b_1, \ldots, b_m)$ if $(a_1, \ldots, a_m)$ lexicographically precedes $(b_1, \ldots, b_m)$ with respect to the orderings $A_1, \ldots, A_m$. We will use this definition only in case where $f(a_1, \ldots, a_n)$ is the result of substituting in a term t (such as $\varphi_{x_1} x_2$, $\psi(x_1)$, (x_1, x_2, x_3), $x_1 + x_2$, $\binom{x_1}{x_2}$ variables x_i by a_i, i. e. $f = \lambda x_1, \ldots, x_n.t$, and will write in this case the result of replacing in t the variable x_i by A_i instead of $f[A_1, \ldots, A_m]$ (i.e. $\varphi_{A_1} A_2$, $\psi(A_1)$, ... instead of $(\lambda x, y.\varphi_x y)[A_1, A_2]$, $(\lambda x.\psi(x))[A_1]$). The convention here is that the variables x_i are ordered from left to right and in case of $\binom{x}{y}$ from bottom to top (i.e. $\varphi_{A_1} A_2$ is ordered lexicographically on (A_1, A_2), $\binom{A_1}{A_2}$ lexicographically on (A_2, A_1)). Note that, after some standard Gödelization of terms, the new ordering in the examples with $f = \lambda \vec{x}.t$ is an elementary construction from the orderings A_i.

(e) If A is an ordering, A^*_{des} (A^*_{weakdes}) is the set/class of — possibly empty — strictly descending (weakly descending) sequences ordered lexicographically, which is an elementary construction from A. We omit double brackets for the elements of A^*_{des} and A^*_{weakdes}, writing for instance $(a_1, b_1, \ldots, a_m, b_m)$ instead of $((a_1, b_1), \ldots, (a_m, b_m))$ for an element of $(A, B)_{\mathrm{des}}$ and $\binom{a_1 \cdots a_m}{b_1 \cdots b_m}$ instead of $\left(\binom{a_1}{b_1} \cdots \binom{a_m}{b_m}\right)$ for an element of $\binom{A}{B}^*_{\mathrm{des}}$. Obviously A^*_{des} is an elementary construction from A.

(f) (Generalization of Schütte's Klammer symbols [Sch54]). If A, B are orderings, let
$\mathrm{Schütte}\binom{A}{B} := \binom{A}{B}_{\mathrm{des}} \cap \{\binom{a_1 \cdots a_m}{b_1 \cdots b_m} \mid m \in \omega, a_i \in A, b_i \in B, b_1 >_B \cdots >_B b_m\}$. Note that we have reversed the order relative to [Sch54]. Further we define
$\mathrm{Schütte}(A, B) := ((A, B))_{\mathrm{des}} \cap$
$\{(a_1, b_1, \ldots, a_m, b_m) \mid m \in \omega, a_i \in A, b_i \in B, a_1 >_A \cdots >_A a_m\}$.
Both are elementary constructions from A, B.

(g) If $A_1, \ldots, A_m$ are orderings and $|A_i|$ are disjoint, then

$$B := A_1 \otimes A_2 \otimes \cdots \otimes A_m$$

is the ordering with $|B|$ being the union of the A_i and $\prec_B$ being the union of the $\prec_{A_i}$ together with the pairs (a, b) for $a \in |A_i|$, $b \in |A_j|$,

$1 \leq i < j \leq n$. This is an elementary construction from the orderings A_i.

(h) We identify the one element set A with the ordering $(A, \emptyset)$.

(i) If A is an ordering, B a set, let $A \cap B := (|A| \cap B, \prec_A \cap((|A| \cap B) \times (|A| \cap B)))$ and $A \setminus B := A \cap (|A| \setminus B)$. Note that this is an elementary construction from A, if B is primitive recursive.

(j) If A is an ordering, let $\mathrm{Acc}(A)$ be the accessible part of $|A|$ with respect to $\prec_A$, i.e. the largest well-founded part of A, $\bigcap\{X \subseteq |A| \mid (X, \prec)$ well-ordered $\}$. $\mathrm{Acc}_\prec(B) := \mathrm{Acc}(B, \prec)$.

(k) If A is an ordering, $a, b \in |A|$, $B \subseteq |A|$, let $B \cap a := \{c \in B \mid c \prec_A a\}$, and $[a, b] := \{c \in |A| \mid a \preceq_A c \preceq_A b\}$. The half open and open intervals $[a, b[$, $]a, b]$ and $]a, b[$ are defined similarly. Again we obtain elementary constructions from A, if B is primitive recursive.

(l) If A is an ordering, $B, C \subseteq |A|$, then $B \sqsubseteq C$ (*B is an initial segment of C*) $:\Leftrightarrow B \subseteq C \wedge \forall x \in B.B \cap x = C \cap x$.

(m) Ord is the class of ordinals, $\mathbb{A}$ the class of additive principal numbers > 0.
We identify classes of ordinals A with the ordering $(A, <)$, where $<$ is the usual ordering on Ord.

2 Elementary Ordinal Systems

2.1 Definition of Ordinal Systems

We will in the following develop first the notation of ordinal systems from that of ordinal notation systems from below as defined in the first four chapters of [Set98a]. However, it is not necessary to read this article, since for the reader who doesn't know the previous approach we will then repeat the motivation given there, adapted to the new setting.

In [Set98a], the underlying structure of ordinal notation systems from below consisted of the set of notations T, a subset NF of T^*, linear orderings $\prec$ on T and $\prec'$ on NF and a function $f : \mathrm{NF} \to \mathrm{T}$. In order to deal with stronger systems, which extend the notion of ordinal notation system from below and allow to define ordinals beyond the Bachmann-Howard ordinal, one needs to deal with arguments that have more structure. Even in the case of ordinals below the Bachmann-Howard ordinal some more structure on the argument is needed. For instance in case of extended Schütte Klammer symbols (in English: "parenthesis symbols" or better "matrix symbols"), which will be defined in Example 2.7 (i) below, we need the information about the size of each of the sub-matrices and which ordinal notations belong to which sub-matrix. We could handle this by using some coding (see Subsection 2.2). However, the more elegant approach is to replace the set $\mathrm{NF} \subseteq \mathrm{T}^*$ by an arbitrary set Arg together with a function $\mathrm{k} : \mathrm{Arg} \to_\omega \mathrm{T}$. The intuition is, that $a \in \mathrm{Arg}$ is an argument for the function f which is built from ordinal notations $\mathrm{k}(a)$, but has some additional structural information. An ordinal notation system from below can be translated into the new structure by defining $\mathrm{Arg} := \mathrm{NF}$ and $\mathrm{k}(a_1, \ldots, a_m) := \{a_1, \ldots, a_m\}$.

Now f was always a bijection, and therefore we can identify Arg with T, define $f := \lambda x.x$ and omit f completely. We have therefore two orderings on T, $\prec$ and $\prec'$, $\prec$ being the ordering on the ordinal notation system and $\prec'$ being the ordering which determines the order, in which new ordinal notations are introduced.

In order to express that T is the closure under the above process, we add a function length : $\mathrm{T} \to \mathbb{N}$ and require $\mathrm{length}(a) < \mathrm{length}(b)$ for $a \in \mathrm{k}(b)$. We replace the long name "ordinal notation system from below" by "ordinal system".

The motivation for the resulting structure is now as follows: Ordinal notations t are built from a finite set of notations $\mathrm{k}(t)$. We want the system to be built from below: First, an ordinal should be denoted using smaller ones, i.e. $\mathrm{k}(t) \prec t$. Second, whenever we introduce a new notation t, we want to have constructed all smaller notations s before t. Either s could be below one of the components of t, i. e. $s \preceq \mathrm{k}(t)$, since, whenever we introduce an ordinal, we assume that we have constructed its components and therefore all ordinals below them as well. Or s must have been introduced before t with respect to the termination ordering $\prec'$, i. e. $s \prec' t$. In [Set98a] we showed that in case of simple systems like the standard system up to Γ_0 or the Schütte Klammer symbols, we can construct $\prec'$ from $\prec$ by using the lexicographic ordering on pairs and on strictly descending sequences together with some simple operations. For instance we ordered terms for the Cantor normal form by the lexicographic ordering on strictly descending sequences and terms $\varphi_a b$ by the lexicographic ordering on pairs (a, b). Orderings $\prec'$ constructed like this have the property that, whenever a set of notations is $\prec$-well-ordered, the set of notations built from it is $\prec'$-well-ordered, and in the definition of ordinal systems we will demand this condition. However, in the above examples it is possible to show this reduction in PRA, i.e. essentially in logic, and we define elementary ordinal systems by demanding additionally this stronger requirement. Elementary ordinal systems will have strength below the Bachmann-Howard ordinal, whereas with ordinal systems we will have no upper bound (choose for an arbitrary well-ordering $(\mathrm{T}, \prec)$, $\mathrm{k}(a) := \emptyset$, $\mathrm{length}(a) := 0$, $\prec' := \prec$).

Definition 2.1 (a) An *Ordinal System Structure*, in short *OS-structure*, is a quintuple $\mathcal{F} = (\mathrm{T}, \prec, \prec', \mathrm{k}, \mathrm{length})$ where T is a set, $\prec'$, $\prec$ are linear orderings on T, $\mathrm{k} : \mathrm{T} \to_\omega \mathrm{T}$ and $\mathrm{length} : \mathrm{T} \to \mathbb{N}$.
We will sometimes regard the first two or three elements of this quintuple as a unity, so when writing $\mathcal{F}$ is an OS-structure of the form $\mathcal{F} = (A, \prec', \mathrm{k}, \mathrm{length})$ or $(\mathcal{G}, \mathrm{k}, \mathrm{length})$ we mean that A is of the form $(\mathrm{T}, \prec)$ and $\mathcal{G}$ is of the form $(\mathrm{T}, \prec, \prec')$ with T, $\prec$ and $\prec'$ as above.

(b) In the following, if not mentioned otherwise, let $\mathcal{F}$ be as in (a), and for any index i $\mathcal{F}_i = (\mathrm{T}_i, \prec_i, \prec'_i, \mathrm{k}_i, \mathrm{length}_i)$.

(c) If $\mathcal{F}$ is an ordinal system structure as above, $A, B \subseteq \mathrm{T}$, then $B \restriction A := B \cap \mathrm{k}^{-1}(A)$, the *restriction of B to arguments built from A* or shorter *restriction of B to A*.

(d) An *Ordinal System*, in short *OS*, is an OS-structure $\mathcal{F}$, such that the following holds:
(OS 1) $\forall t \in \mathrm{Arg}.\mathrm{k}(t) \prec t$.
(OS 2) $\mathrm{length}[\mathrm{k}(t)] < \mathrm{length}(t)$.
(OS 3) $\forall t \in \mathrm{Arg}.\forall s \in \mathrm{T}(s \prec t \to (s \preceq \mathrm{k}(t) \vee s \prec' t))$.
(OS 4) If $A \subseteq \mathrm{T}$, A is $\prec$-well-ordered, then $\mathrm{T} \restriction A$ is $\prec'$-well-ordered.

(e) An OS $\mathcal{F}$ is *primitive recursively represented*, if T is a primitive recursive subset of $\mathbb{N}$, $\prec$, $\prec'$, k, length are primitive recursive, and all the properties of an OS, including linearity of $\prec$, $\prec'$, but except (OS 4), can be shown in PRA.

(f) A primitive recursively represented OS $\mathcal{F}$ is *elementary*, if additionally (OS 4) can be shown in PRA, i.e. for a free predicate R $\mathrm{TI}_{(\mathrm{T}\restriction R, \prec')}$ is PRA-reducible to $\mathrm{TI}_{(\mathrm{T}\cap R, \prec)}$.

(g) An OS-structure is *well-ordered*, if T is well-ordered. The *order type* of a well-ordered OS-structure $\mathcal{F}$ is the order type of T.

(h) Two OS-structures are isomorphic, if there exists a bijection between the underlying sets, which respects $\prec$, $\prec'$, k, length.

We illustrate an ordinal system by the following picture:

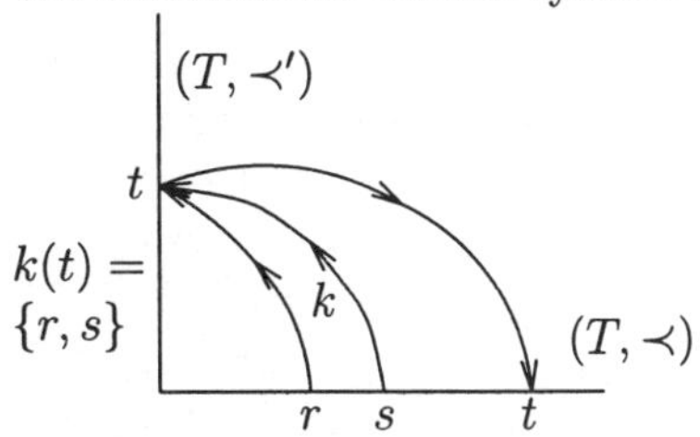

(The arrow from t in $(\mathrm{T}, \prec')$ to t in $(\mathrm{T}, \prec)$ represents the function f in the original definition, which is now the identity).

2.2 Equivalence of Elementary Ordinal Systems and Ordinal Notation Systems from below.

Let $\prec''$ be the ordering on NF in an ordinal notation system from below, f the function used there, define $\prec'$ on $f[\mathrm{NF}]$ by $f(\vec{a}) \prec' f(\vec{b}) :\Leftrightarrow \vec{a} \prec'' \vec{b}$, $\mathrm{k}(f(\vec{a})) := \{a_1, \ldots, a_n\}$ and length in the obvious way. Then, if we started with an ordinal notation system from below in which $\prec''$ is linear, provably in PRA, we obtain an elementary OS.

In the other direction, assume an elementary OS, together with an explicit enumeration of the natural numbers, i. e. there exists a primitive recursive function $g : \omega \to \mathrm{T}$, where we write $\underline{n}$ for $g(n)$, such that $\forall n.n \leq \underline{n}$ and $\forall n \in \mathbb{N}.\underline{n} = n$th element of T provable in PRA (more precisely the formula $\forall n \in \mathbb{N}(n \leq \underline{n} \wedge \forall x \in \mathrm{T}(x < \underline{n} \leftrightarrow \exists l < n.x = \underline{l}))$ is provable in PRA). Note that therefore $g[\omega]$ and $g^{-1} : g[\omega] \to \omega$ are primitive recursive. Define $\mathrm{T}' := \mathrm{T}$, $\mathrm{NF}' \subseteq \mathrm{T}'^{*}$,
$\mathrm{NF} := \{()\} \cup \{(\underline{0}, \underline{n}) \mid n \in \omega\} \cup$
$\quad \{(\underline{1}, \underline{t}, s_1, \ldots, s_m) \mid t \in \mathrm{T} \setminus g[\omega] \wedge m \in \omega \wedge s_1 \prec \cdots \prec s_m \wedge$
$\quad\quad \mathrm{k}(t) = \{s_1, \ldots, s_m\}\}$;
$f' : \mathrm{NF}' \to \mathrm{T}'$, $f'() := \underline{0}$, $f'(\underline{0}, \underline{n}) := \underline{n+1}$, $f'(\underline{1}, \underline{t}, s_1, \ldots, s_m) := t$;
$\prec''$ on NF by $() \prec'' (\underline{0}, \underline{0}) \prec'' (\underline{0}, \underline{1}) \prec'' (\underline{0}, \underline{2}) \prec'' \cdots \prec'' (\underline{1}, \underline{t}, s_1, \ldots, s_m)$ for all t and
$(\underline{1}, \underline{t}, s_1, \ldots, s_m) \prec'' (\underline{1}, \underline{t}', s_1', \ldots, s_l')$ iff $t \prec' t'$.
Then one sees easily that $(\mathrm{T}, \prec, \mathrm{NF}, \prec'', f')$ is an ordinal notation system from below, which has the same order type as the original ordinal system and has essentially "the same form", therefore under the above mentioned weak conditions, elementary OS can be seen as ordinal notation systems from below.

2.3 Intuitive Argument why OS are Well-ordered.

We will first illustrate what we regard as an intuitive argument for the well-ordering of ordinal system, by taking a standard example.

Take the notation system built from the Cantor Normal form and the Veblen-function, based on $\varphi_0\alpha = \epsilon_\alpha$. More precisely, it is defined as follows: Let $\widehat{\mathbb{N}}$ be the set of terms $\{0, \mathrm{S}(0), \mathrm{S}(\mathrm{S}(0)), \ldots\}$. $1 := \mathrm{S}(0)$. The set of ordinal notations T together with the ordering $\prec$ on T are simultaneously defined by: $\widehat{\mathbb{N}} \subseteq \mathrm{T}$, if $a_1, \ldots, a_m \in \mathrm{T}$, $n_i \in \widehat{\mathbb{N}} \setminus \{0\}$, $m > 0$, then $\varphi_{a_1} a_2 \in \mathrm{T}$ provided $m = 2$ and a_2 is not of the form $\varphi_{b_1} b_2$ with $a_1 \prec b_1$, and $\omega^{a_1} \cdot n_1 + \cdots + \omega^{a_m} \cdot n_m \in \mathrm{T}$ provided $a_m \prec \cdots \prec a_1$ and if $m = 1$, then $a_1 \neq 0$ and a_1 is not of the form

$\varphi_{b_1}b_2$. $\mathrm{S}^k(0) \prec \mathrm{S}^l(0) :\Leftrightarrow k < l$, $\mathrm{S}^k(0) \prec \omega^{a_1} \cdot n_1 + \cdots + \omega^{a_m} \cdot n_m$, $\mathrm{S}^k(0) \prec \varphi_a b$, $\omega^{a_1} \cdot n_1 + \cdots + \omega^{a_m} \cdot n_m \prec \omega^{b_1} \cdot l_1 + \cdots + \omega^{b_k} \cdot l_k :\Leftrightarrow (a_1, n_1, \ldots, a_m, n_m)$ is lexicographically smaller than $(b_1, l_1, \ldots, b_k, l_k)$, $c := \omega^{a_1} \cdot n_1 + \cdots + \omega^{a_m} \cdot n_m \prec \varphi_a b :\Leftrightarrow a_1, \ldots, a_m, n_1, \ldots, n_m \prec \varphi_a b$ and, if this condition does not hold, then $\varphi_a b \prec c$, and $\varphi_a b \prec \varphi_{a'} b' :\Leftrightarrow (a \prec a' \wedge b \prec \varphi_{a'} b') \vee (a = a' \wedge b \prec b') \vee (a' \prec a \wedge \varphi_a b \prec b')$. The termination ordering $\prec'$ on T is now defined by $\mathrm{S}^k(0) \prec' \omega^{a_1} \cdot n_1 + \cdots + \omega^{a_m} \cdot n_m \prec' \varphi_a b$, $\mathrm{S}^k(0) \prec' \mathrm{S}^l(0)$ iff $k < l$, $\prec'$ on terms $\omega^{a_1} \cdot n_1 + \cdots + \omega^{a_m} \cdot n_m$ is the lexicographic ordering, and $\prec'$ on terms $\varphi_a b$ is the lexicographic ordering on pairs (a, b) (the last two with respect to $\prec$). Further define $\mathrm{k}(0) = \emptyset$, $\mathrm{k}(\mathrm{S}(t)) := \{t\}$, $\mathrm{k}(\omega^{a_1} \cdot n_1 + \cdots + \omega^{a_m} \cdot n_m) := \{a_1, \ldots, a_m, n_1, \ldots, n_m\}$ and $\mathrm{k}(\varphi_a b) := \{a, b\}$.

It is easy to see that with the obvious definition of length we get an ordinal system $(\mathrm{T}, \prec, \prec', \mathrm{k}, \text{length})$, especially we have: If $A \subseteq \mathrm{T}$ is $\prec$-well-ordered, the set of notations "built from A" or "with components in A", namely $\mathrm{k}^{-1}(A)$, is $\prec'$-well-ordered. Now we proceed by selecting ordinal notations as follows: At the beginning we have no notation at all. The only notation which has components in the empty set is 0 and we select it. Now, if we take the ordinals with components in the set of notations selected, namely in $\{0\}$, we choose the $\prec'$-least one not selected before, namely $\mathrm{S}(0)$. Again, we take all notations with components in $\{0, \mathrm{S}(0)\}$, select the $\prec'$-least one not chosen before, $\mathrm{S}(\mathrm{S}(0))$, etc. Once we have selected all the natural numbers, the least one will be $\omega^1 \cdot 1$. The $\prec'$ least one in $\mathrm{T} \restriction \{0, \mathrm{S}(0), \mathrm{S}(\mathrm{S}(0)), \ldots\} \cup \{\omega^1 \cdot 1\}$ not selected before will be $\omega^1 \cdot 1 + \omega^0 \cdot 1$ and one sees easily that we will proceed selecting systematically all ordinals built by the Cantor normal form, i. e. all notations below $\varphi_0 0$.

The next ordinal selected will be $\varphi_0 0$. And here for the first time we will make a jump in the $\prec'$-ordering: there are notations $\prec' \varphi_0 0$ we have not chosen yet, but they are built from notations which we have not constructed before. After $\varphi_0 0$ we will choose $\omega^{\varphi_0 0} \cdot 1 + \omega^0 \cdot 1$ — we will therefore move downwards with respect to $\prec'$ — and then we will exhaust again the Cantor normal form and choose all ordinals below $\varphi_0 1$. Now $\varphi_0 1$ will be selected and we will then proceed selecting (with the steps for building the Cantor normal form in between) $\varphi_0 2$, $\varphi_0 3, \ldots, \varphi_0(\varphi_0 0), \ldots, \varphi_0(\varphi_0(\varphi_0 0)), \ldots$, then $\varphi_1 0$. etc.

In order to be able to select the $\prec'$-least element of the sets considered, we need the set of notations with components in the notations previously selected to be $\prec'$-well-ordered. Now by (OS 4) this will be the case, if the sequence of ordinals previously selected is known to be $\prec$-increasing and therefore $\prec$-well-ordered. Now we can easily see by induction over the process above that, whenever we select an ordinal, we select in fact the $\prec$-least one not chosen before and therefore the set of ordinals previously chosen will be a $\prec$-segment of T:

Assume A is the ordinals previously selected, which is by IH a segment of the ordinals and let a be the new notation chosen. a will be bigger than A, since A is a segment. We need to show, which we will do by induction over the length of b, that, if $b \prec a$, $b \in A$. If $b \prec a$, then $b \preceq \mathrm{k}(a) \subseteq A$, so $b \in A$ as A is a segment, or $b \prec' a$. b is built from smaller components, they must be, by the side-IH, all in A, therefore b has components in A, a was the $\prec'$-least such not selected before, so b must have been selected before, $b \in A$.

Now we assume that "we do not run out of ordinals". Therefore we must reach a point where we cannot select an ordinal any more. But then one sees easily that all notations must be selected, so the set of notations must be well-ordered (since we have selected them in increasing order) and we are done.

The argument "we do not run out of ordinals", can now be formalized in set theory for elementary ordinal systems by using ω_1^{ck}: if we could select as above ω_1^{ck} ordinals, we would have a primitive recursive well-ordering of order type ω_1^{ck}, a contradiction. For this argument we need the big ordinal ω_1^{ck}. However it is only needed to provide an ordinal "big enough". In fact, the process of selecting ordinals will terminate after α steps for some α which is below ω_1^{ck}. Something similar will be the case for all recursively large ordinals needed for stronger ordinal notation systems: they always play only the role of ordinals "big enough", in fact, all processes involved will terminate after α steps for some ordinal $< \omega_1^{\mathrm{ck}}$.

We are now going to formalize the well-ordering argument above rigorously, in order to get absolute precision. We will then show that for elementary ordinal systems, it can be formalized in Kripke-Platek set theory (with natural numbers as urelemente). Therefore we will get an upper bound for the strength of elementary ordinal systems.

Lemma 2.2 *(a) Let $\mathcal{F}$ be an ordinal system, and assume $\perp \notin \mathrm{T}$. Let $b \in \mathrm{T}$. By recursion on ordinals γ we define $a_\gamma \in \mathrm{T} \cup \{\perp\}$ as follows: Let $A_{<\gamma} := \{a_\alpha \mid \alpha < \gamma\}$. Define "$A_{<\gamma}$ is increasing" iff $A_{<\gamma} \subseteq \mathrm{T}$ and $\forall \alpha, \beta < \gamma(\alpha < \beta \rightarrow a_\alpha \prec' a_\beta)$.*

$$a_\gamma := \begin{cases} \min_{\prec'}((\mathrm{T} \restriction A_{<\gamma}) \setminus A_{<\gamma}) & \text{if } A_{<\gamma} \text{ is increasing and} \\ & \mathrm{T} \restriction A_{<\gamma} \not\subseteq A_{<\gamma}, \\ \perp & \text{otherwise.} \end{cases}$$

Then there exists a γ such that $A_{<\gamma}$ is increasing and $A_{<\gamma} = \mathrm{T}$. Therefore $(\mathrm{T}, \prec)$ is well-ordered.

(b) γ as in (a) is $< \omega^{\mathrm{ck}}$, if $\mathcal{F}$ is primitive recursively represented.

(c) If $\mathcal{F}$ is elementary, (a) can be formulated in KPω.

Proof: (a): a_γ is well-defined, since, if $A_{<\gamma}$ is increasing, $\mathrm{T} \restriction A_{<\gamma}$ is well-ordered with respect to $\prec'$, therefore as well $(T \restriction A_{<\gamma}) \setminus A_{<\gamma}$ and a_γ can be defined.

We show: If $A_{<\gamma}$ is increasing, then $A_{<\gamma}$ is an initial segment of T w.r.t. $\prec$, by recursion on γ: The cases $\gamma = 0$ or γ a limit ordinal follow trivially or by IH. Let now $\gamma = \delta + 1$. We need to show a_δ is the $\prec$-least element in $\mathrm{T} \setminus A_{<\delta}$: $a_\delta \notin A_{<\delta}$, which is an initial segment of T, therefore $A_{<\delta} \prec a_\delta$. We show by induction on length(b) $\forall b \prec a_\delta . b \in A_{<\delta}$, from which the assertion follows. $b \prec a_\delta$, therefore $b \preceq \mathrm{k}(a_\delta) \subseteq A_{<\delta} \sqsubseteq \mathrm{T}$ and therefore $b \in A_{<\delta}$ or $b \prec' a_\delta$. In the last case, by IH, since $\mathrm{k}(b) \prec b \prec a$, $\mathrm{k}(b) \subseteq A_{<\delta}$, $b \in \mathrm{T} \restriction A_{<\delta}$, by $\prec'$-minimality of a_δ $b \notin (\mathrm{T} \restriction A_{<\delta}) \setminus A_{<\delta}$, $b \in A_{<\delta}$.

Let κ be an admissible ordinal such that $\mathcal{F} \in \mathrm{L}_\kappa$ (if $\mathcal{F}$ is primitive recursively represented, κ can be chosen as ω_1^{ck}). If $A_{<\kappa}$ were now increasing, then $A_{<\kappa} = \{a \in \mathrm{T} \mid a \prec a_{\kappa+1}\}$ would be a well-ordering, which is an element of L_κ and has order type κ, a contradiction. Therefore there exists a least $\gamma < \kappa$ such that $A_{<\gamma}$ is not increasing. $\gamma = \gamma_0 + 1$, $A_{<\gamma_0}$ is increasing, therefore a well-ordered subset of T. $\mathrm{T} \restriction A_{<\gamma_0} \subseteq A_{<\gamma_0}$, therefore by induction on length(a) it follows that for all $a \in \mathrm{T}$ $a \in A_{<\gamma_0}$, $A_{<\gamma_0} = \mathrm{T}$ and the assertion.

(b): by the proof of (a).

(c): The formalization is straight-forward, the only difficulty is the argument referring to κ, which we replace by the following: Let $C := \{\gamma \in \mathrm{Ord} \mid A_{<\gamma} \text{ increasing }\}$, $C \sqsubseteq \mathrm{Ord}$, $f : C \to \mathrm{T}$, $f(\alpha) := a_\alpha$. $f[C]$ is a well-ordered initial segment of T. Assume $f[C] \neq \mathrm{T}$. Then $(\mathrm{T} \restriction f[C]) \not\subseteq f[C]$, let $a \in (\mathrm{T} \restriction f[C]) \setminus f[C]$ be $\prec'$ minimal. (Here we use reducibility of transfinite induction). As in the argument before it follows that a is the least element of T not in $f[C]$, therefore $f[C] = \{b \in \mathrm{T} \mid b \prec a\}$, a set, $f : \mathrm{Ord} \to f[C]$ bijective, $f^{-1} : f[C] \to \mathrm{Ord}$ bijective, therefore $\mathrm{Ord} = f^{-1}[f[C]]$ is a set, a contradiction. Therefore $f[C] = \mathrm{T}$ is well-ordered. □

Theorem 2.3 *The order type of elementary ordinal systems is less than the Bachmann-Howard ordinal.*

Proof: by Lemma 2.2 (c) and since $|\mathrm{KP}\omega|$ is the Bachmann-Howard ordinal. □

2.4 Constructive Well-ordering Proof

We regard the above well-ordering proof as rather intuitive. However, for treating intuitionistic theories we will need a constructive argument as well. This argument will as well be shorter and can be formulated for instance in intuitionistic ID_1 or type theory having the accessible part or one unnested W-type, which yields again the upper bound Bachmann-Howard for the order type of elementary OS.

Constructive well-ordering proof: Let $\mathrm{Acc} := \mathrm{Acc}_\prec(\mathrm{T})$ and $\mathrm{Acc}' := \mathrm{k}^{-1}(\mathrm{Acc})$, which is well-ordered w.r.t. $\prec'$. We show $\forall t \in \mathrm{Acc}' . t \in \mathrm{Acc}$ by induction on $t \in \mathrm{Acc}'$. It suffices to show $\forall s \prec t.s \in \mathrm{Acc}$ by side-induction on $\mathrm{length}(s)$. If $s \preceq \mathrm{k}(t)$, then $s \in \mathrm{Acc}$ since $\mathrm{k}(t) \subseteq \mathrm{Acc}$. Otherwise $s \prec' t$, $\mathrm{k}(s) \prec s \prec t$. By the side-IH $\mathrm{k}(s) \subseteq \mathrm{Acc}$ and by the main-IH therefore $s \in \mathrm{Acc}$ and we are done. Now it follows by induction on $\mathrm{length}(s)$ that $\forall s \in \mathrm{T}.s \in \mathrm{Acc}$, T is well-ordered.

2.5 Ordinal Function Generators.

We are going to define elementary ordinal systems, which exhaust the strength of this concept. As in the introduction, we want to get functions which act on ordinals as well. This does not require much extra work. As we said there, one can easily develop this without any reference to ordinals. The advantage of our approach is that we will get definitions for the usual ordinal functions like $+$, $\lambda\alpha.\omega^\alpha$, φ and the Schütte Klammer symbols and further new definitions for extensions of the Klammer symbols without much extra work.

Typically two kinds of ordinal functions are usually used: Versions having fixed points and fixed point free versions. One example is the Veblen function versus its fixed point free version (see for instance the function ψ in [Sch77], p. 84). When using versions with fixed points, we will usually not have the property that the arguments of an ordinal function are strictly smaller than its value. This is no harm, since in a next step one selects normal forms. The following definition allows to introduce both fixed point versions and fixed point free versions:

Definition 2.4 (a) An *ordinal function generator*, in short *OFG* is a quadruple $\mathcal{O} := (\mathrm{Arg}, \mathrm{k}, \mathrm{l}, <')$, where Arg is a class (in set theory), $\mathrm{k}, \mathrm{l} : \mathrm{Arg} \to_\omega \mathrm{Ord}$ such that $\forall a \in \mathrm{Arg}.\mathrm{l}(a) \subseteq \mathrm{k}(a)$ and $<'$ is a well-ordered relation on Arg. We define $<'_\mathcal{O} := <'$.

(b) If $\mathcal{O}$ is as above, we define simultaneously by recursion on $a \in \mathrm{Arg}$ sets $\mathrm{C}_\mathcal{O}(a) := \mathrm{C}(a) \subseteq \mathrm{Ord}$ and $\mathrm{eval}_\mathcal{O}(a) := \mathrm{eval}(a) \in \mathrm{Ord}$:
$\mathrm{C}(a) := \bigcup_{n<\omega} \mathrm{C}^n(a)$, where

$$\begin{aligned} \mathrm{C}^0(a) &:= (\bigcup \mathrm{k}(a)) \cup \mathrm{l}(a), \\ \mathrm{C}^{n+1}(a) &:= \mathrm{C}^n(a) \cup \{\mathrm{eval}(b) \mid b \in \mathrm{Arg}, b <' a, \mathrm{k}(b) \subseteq \mathrm{C}^n(a)\}. \end{aligned}$$

$\mathrm{eval}(a) := \min\{\alpha \in \mathrm{Ord} \mid \alpha \notin \mathrm{C}(a)\}$.

(c) Define for $\mathcal{O}$ as above, $\mathrm{NF} := \{a \in \mathrm{Arg} \mid \mathrm{k}(a) < \mathrm{eval}(a)\}$.

(d) Define for $\mathcal{O}$ as above $\mathcal{C}l \subseteq \mathrm{NF}$ inductively by: If $a \in \mathrm{NF}$, $\mathrm{k}(a) \subseteq \mathrm{eval}[\mathcal{C}l]$, then $a \in \mathcal{C}l$.
$\mathrm{Arg}[\mathcal{C}l] := \{a \in \mathrm{Arg} \mid \mathrm{k}(a) \subseteq \mathrm{eval}[\mathcal{C}l]\}$. Note that $\mathcal{C}l \subseteq \mathrm{Arg}[\mathcal{C}l]$.
Assuming $\mathrm{eval} \restriction \mathrm{NF}$ is injective, which will be shown later, we define

$k^0 : \mathrm{Arg}[\mathcal{Cl}] \to_\omega \mathcal{Cl}$ and $\mathrm{length} : \mathrm{Arg}[\mathcal{Cl}] \to \mathbb{N}$ by
$k^0(a) := \mathrm{eval}^{-1}[k(a)] \cap \mathcal{Cl}$, $\mathrm{length}(a) := \max(\mathrm{length}[k^0(a)] \cup \{-1\}) + 1$
for $a \in \mathrm{Arg}[\mathcal{Cl}]$ with $-1 < n$ for $n \in \mathbb{N}$, $-1 + 1 := 0$.
Further, for $a, b \in \mathrm{Arg}[\mathcal{Cl}]$ let $a \prec b :\Leftrightarrow \mathrm{eval}(a) < \mathrm{eval}(b)$.

Note that the definition of $\mathrm{eval}(a)$ expresses something similar as for OS: When defining $\mathrm{eval}(a)$, we know all $\alpha < \mathrm{eval}(a)$, since $\alpha < k(a)$, $\alpha \leq l(a)$, or α was introduced before a, i.e. $\alpha = \mathrm{eval}(b)$ for some $b <' a$. Further, if a is in normal form, then a is a notation for $\mathrm{eval}(a)$ referring to smaller ordinals only, namely $k(a)$.

Lemma and Definition 2.5 *Let $\mathcal{O}$ be as above.*

(a) $C(a)$ *is the least set such that* $(\bigcup k(a)) \cup l(a) \subseteq C(a)$ *and, if* $b \in \mathrm{NF}$, $b <' a$, $k(b) \subseteq C(a)$, *then* $\mathrm{eval}(b) \in C(a)$.

(b) $\forall a \in \mathrm{Arg}(k(a) \leq \mathrm{eval}(a) \wedge l(a) < \mathrm{eval}(a))$.

(c) $C(a)$ *is an initial segment of* Ord.

(d) eval *restricted to* NF *is injective.*

(e) $\mathrm{eval}[\mathcal{Cl}]$ *is an initial segment of* Ord.

(f) For $a, b \in \mathrm{Arg}[\mathcal{Cl}]$ *we have*

$$\begin{aligned}
\mathrm{eval}(a) < \mathrm{eval}(b) \quad &\Leftrightarrow \quad (a <' b \wedge k(a) < \mathrm{eval}(b)) \vee \\
&\qquad \mathrm{eval}(a) < k(b) \vee \mathrm{eval}(a) \leq l(b) \\
\mathrm{eval}(a) = \mathrm{eval}(b) \quad &\Leftrightarrow \quad (a <' b \wedge \mathrm{eval}(b) = \max(k(a)) \\
&\qquad \quad \wedge \mathrm{eval}(b) \notin l(a)) \vee \\
&\qquad (b <' a \wedge \mathrm{eval}(a) = \max(k(b)) \\
&\qquad \quad \wedge \mathrm{eval}(a) \notin l(b)) \vee \\
&\qquad a = b
\end{aligned}$$

(g) For $a, b \in \mathrm{NF}$ *we have* $\mathrm{eval}(a) < \mathrm{eval}(b) \Leftrightarrow (a <' b \wedge k(a) < \mathrm{eval}(b)) \vee \mathrm{eval}(a) \leq k(b)$.

(h) $\mathcal{F} := (\mathcal{Cl}, \prec, <', k^0, \mathrm{length})$ *is an ordinal system. We will call any OS-structure isomorphic to* $\mathcal{F}$ *an* ordinal system *based on* $\mathcal{O}$.

Proof: (a), (b): easy.

(c) Note that, if $C(a)$ is an initial segment, $\mathrm{eval}(a) = C(a)$. The proof of the assertion is by induction on a. We show $\forall \alpha \in C(a).\alpha \subseteq C(a)$ by induction on the definition of $C(a)$. If $\alpha \in \bigcup k(a)$, then $\alpha \subseteq C(a)$. If $\alpha \in l(a) \subseteq k(a)$,

$\alpha \subseteq \bigcup k(a) \subseteq C(a)$. If $\alpha = \mathrm{eval}(b)$, $b <' a$, $k(b) \subseteq C(a)$, by the side-IH $\bigcup k(b) \subseteq C(a)$, and by transitivity of $<'$ and $l(a) \subseteq k(a)$ it follows easily that $C(b) \subseteq C(a)$ and by the main IH $\alpha = C(b) \subseteq C(a)$.

(d) Assume $a \neq b$, $a, b \in \mathrm{NF}$, $\mathrm{eval}(a) = \mathrm{eval}(b)$. By linearity of $<'$, $a <' b$ or $b <' a$, so w.l.o.g. $a <' b$. $k(a) < \mathrm{eval}(a) = \mathrm{eval}(b)$, therefore $k(a) \subseteq C(b)$, $a <' b$, therefore $\mathrm{eval}(a) \in C(b) = \mathrm{eval}(b)$, a contradiction.

(e) We show $\forall a \in \mathcal{C}l.\mathrm{eval}(a) \subseteq \mathrm{eval}[\mathcal{C}l]$ by induction on $a \in \mathcal{C}l$: Using the IH, (a) and $\neg\mathrm{NF}(b) \to \mathrm{eval}(b) \in k(b)$ it follows that $C(a) \subseteq \mathrm{eval}[\mathcal{C}l]$, $\mathrm{eval}(a) \subseteq \mathrm{eval}[\mathcal{C}l]$.

(f) First formula "$\Leftarrow$": $\mathrm{eval}(a) \in C(b) = \mathrm{eval}(b)$.
Second formula "$\Leftarrow$" consider only the case $(a <' b \wedge \mathrm{eval}(b) = \max(k(a)) \wedge \mathrm{eval}(b) \notin l(a))$. $\mathrm{eval}(b) \in k(a) \leq \mathrm{eval}(a)$. Assume $\mathrm{eval}(b) < \mathrm{eval}(a)$. Then, since $\mathrm{eval}(b) \not< k(a)$, $\mathrm{eval}(b) \not\leq l(a)$, by $\mathrm{eval}(b) \in C(a)$ it follows $\mathrm{eval}(b) = \mathrm{eval}(b')$ for some $b' <' a$, $k(b') \subseteq C(a)$. By the definition of $C(a)$ and $\mathrm{eval}(b) \in C(a) \setminus C^0(a)$ there exist $b' <' a$ and n such that $k(b') \subseteq C^n(a)$, $\mathrm{eval}(b') \in C^{n+1}(a) \setminus C^n(a)$ and $\mathrm{eval}(b') = \mathrm{eval}(b)$. $b' <' b$, $\mathrm{eval}(b') = \mathrm{eval}(b)$, so $k(b') \not< \mathrm{eval}(b) = \mathrm{eval}(b')$, $\mathrm{eval}(b') \in k(b')$, contradicting the choice of b'. Therefore $\mathrm{eval}(b) = \mathrm{eval}(a)$.
"$\Rightarrow$" first formula: if the right hand side is false the right hand holds for $\mathrm{eval}(a) = \mathrm{eval}(b)$ or $\mathrm{eval}(b) < \mathrm{eval}(a)$, therefore the left hand side is false. For the second formula "$\Rightarrow$" follows in the same way.

(g) "$\Leftarrow$" is immediate, and "$\Rightarrow$" follows as before.

(h) (OS 1), (OS 2), (OS 4) are clear. (OS 3): If $s \prec t$, $s, t \in \mathrm{NF}$, then $\mathrm{eval}(s) \in C(t)$, $\mathrm{eval}(s) \leq k(t)$ or $s <' t$ and the assertion. □

Definition 2.6 If $\mathcal{O}_i = (\mathrm{Arg}_i, <'_i, k^i, l^i)$ are OFG $(i = 0, 1)$, $\mathrm{Arg}_0 \cap \mathrm{Arg}_1 = \emptyset$, Let $\mathcal{O}_0 \circledcirc \mathcal{O}_1 := (\mathrm{Arg}, <', k, l)$, be defined by $(\mathrm{Arg}, <') := (\mathrm{Arg}_0, <'_0) \circledcirc (\mathrm{Arg}_1, <'_1)$ and $k(r) := k^i(r)$, $l(r) := l^i(r)$ for $r \in \mathrm{Arg}_i$. $\mathcal{O}_0 \circledcirc \mathcal{O}_1$ is obviously an OFG and $\mathrm{eval}_{\mathcal{O}}(a) = \mathrm{eval}_{\mathcal{O}_0}(a)$ for $a \in \mathrm{Arg}_{\mathcal{O}_0}$.

Example 2.7 (a) Let $\alpha \in \mathrm{Ord}$ be fixed, $(\mathrm{Arg}_\alpha, <'_\alpha) := \alpha\underline{+}\mathrm{Ord}$ (i.e. expanding Definition 1.1 (d), $\mathrm{Arg}_\alpha = \{\alpha\underline{+}\beta \mid \beta \in \mathrm{Ord}\}$, $\alpha\underline{+}\beta <'_\alpha \alpha\underline{+}\gamma \Leftrightarrow \beta < \gamma$; here and in the following $\alpha\underline{+}\beta$, $\underline{\varphi}_\alpha\beta$ etc. are formal terms defined from ordinals α, β, etc. coded in set theory in some way). Let

$$k(\alpha\underline{+}\beta) := \{\alpha, \beta\},\ l(\alpha\underline{+}\beta) := \begin{cases} \emptyset & \text{if } \beta = 0, \\ \{\alpha\} & \text{otherwise,} \end{cases} \quad \mathcal{O}_\alpha := (\mathrm{Arg}_\alpha, <'_\alpha, k, l).$$

Then $\mathrm{eval}(\alpha\underline{+}\beta) = \alpha + \beta$. Note that $\mathcal{C}l = \emptyset$.
(We could change the definition of OFGs by omitting the condition "$<'$ linear". Then Lemma 2.5 (a) - (c) and (e) go through as well and one could define $(\mathrm{Arg}, <')$ as the union of $(\mathrm{Arg}_\alpha, <'_\alpha)$ with $\alpha\underline{+}\beta$ and $\gamma\underline{+}\rho$ incomparable for $\alpha \neq \gamma$. $\mathrm{eval}(\alpha\underline{+}\beta) = \alpha + \beta$, so this way we would get a definition of the full function $+$.)

(b) Let

$$\begin{aligned}
(\mathrm{Arg}, <) &:= \underline{\Sigma}(\mathbb{A}_{\mathrm{weakdes}}),\\
\mathrm{k}(\underline{\Sigma}(\alpha_1, \dots, \alpha_m)) &:= \{\alpha_1, \dots, \alpha_m\},\\
\mathrm{l}(\underline{\Sigma}(\alpha_1, \dots, \alpha_m)) &:= \begin{cases} \{\alpha_1\} & \text{if } m > 1,\\ \emptyset & \text{otherwise.}\end{cases}
\end{aligned}$$

Let $\mathcal{O}_{+,\mathrm{w}}$ be the resulting OFG. Then one easily verifies $\mathrm{eval}(\underline{\Sigma}()) = 0$ (similar cases occurring in future examples will not be mentioned below) $\mathrm{eval}(\underline{\Sigma}(\alpha_1, \dots, \alpha_m)) = \alpha_1 + \dots + \alpha_m$ for $m > 0$. The fixed point free version of it, i.e. $\mathrm{l}(t) := \mathrm{k}(t)$ yields the same result, except $\mathrm{eval}(\underline{\Sigma}(\alpha, \underbrace{1, \dots, 1}_{l})) = \alpha + l + 1$

(c) Let

$$\begin{aligned}
(\mathrm{Arg}, <) &:= \underline{\Sigma}'(\text{Schütte}(\mathbb{A}, \omega \setminus \{0\})),\\
\mathrm{k}(\underline{\Sigma}'(\alpha_1, n_1, \dots, \alpha_m, n_m)) &:= \{\alpha_1, \dots, \alpha_m, n_1, \dots, n_{m-1}, n_m - 1\},\\
\mathrm{l}(\underline{\Sigma}'(\alpha_1, n_1, \dots, \alpha_m, n_m)) &:= \begin{cases} \{\alpha_1, n_m - 1\} & \text{if } m > 1 \text{ or } n_m > 1,\\ \emptyset & \text{otherwise.}\end{cases}
\end{aligned}$$

Let $\mathcal{O}_+$ be the resulting OFG.
$\mathrm{eval}_{\mathcal{O}_+}(\underline{\Sigma}'(\alpha_1, n_1, \dots, \alpha_m, n_m)) = (\alpha_1 \cdot n_1) + \dots + (\alpha_m \cdot n_m)$.

(d) If we replace $\mathbb{A}$ in (b), (c) by Ord and $\underline{\Sigma}$ by CNF, we obtain ordinal function generators $\mathcal{O}_{\mathrm{CNF,w}}$, $\mathcal{O}_{\mathrm{CNF}}$ for the Cantor normal form, i.e. $\mathrm{eval}_{\mathcal{O}_{\mathrm{CNF,w}}}(\mathrm{CNF}(\alpha_1, \dots, \alpha_m)) = \omega^{\alpha_1} + \dots + \omega^{\alpha_m}$ and $\mathrm{eval}_{\mathcal{O}_{\mathrm{CNF}}}(\mathrm{CNF}'(\alpha_1, n_1, \dots, \alpha_m, n_m)) = \omega^{\alpha_1} \cdot n_1 + \dots + \omega^{\alpha_m} \cdot n_m$

(e) Let $\mathcal{O}_0$ be $\mathcal{O}_{+,\mathrm{w}}$ or $\mathcal{O}_+$, $\mathcal{O}_1 := (\underline{\omega}^{\mathrm{Ord}}, \mathrm{k}, \mathrm{l})$ with $\mathrm{k}(\underline{\omega}^{\alpha}) := \{\alpha\}$, $\mathrm{l}(\underline{\omega}^{\alpha}) := \emptyset$, $\mathcal{O} := \mathcal{O}_0 \oslash \mathcal{O}_1$, then $\mathrm{eval}(\underline{\omega}^{\alpha}) = \omega^{\alpha}$. The fixed point free version ($\mathrm{l}(\underline{\omega}^{\alpha}) := \{\alpha\}$) yields the same result, except $\mathrm{eval}(\underline{\omega}^{\alpha+n}) = \omega^{\alpha+n+1}$ for epsilon numbers α.

(f) Let $\mathcal{O}_0$ be $\mathcal{O}_{+,\mathrm{w}}$ or $\mathcal{O}_+$, $\mathcal{O}_1 := (\underline{\varphi}_{\mathrm{Ord}}\mathrm{Ord}, \mathrm{k}, \mathrm{l})$, $\mathrm{k}(\underline{\varphi}_{\alpha}\beta) := \{\alpha, \beta\}$,
$\mathrm{l}(\underline{\varphi}_{\alpha}\beta) := \begin{cases} \{\alpha\} & \text{if } \beta > 0,\\ \emptyset & \text{otherwise,}\end{cases}$ $\mathcal{O} := \mathcal{O}_0 \oslash \mathcal{O}_1$. Then $\mathrm{eval}(\underline{\varphi}_{\alpha}\beta) = \varphi_{\alpha}\beta$, where φ is the Veblen function with $\varphi_0\beta := \omega^{\beta}$. If we use the OFG from (d) or (e) instead for $\mathcal{O}_0$ we obtain the φ-function starting with $\varphi_0\alpha := \epsilon_{\alpha}$, and defining $\mathrm{l}(t) := \mathrm{k}(t)$ yields the fixed point free version of the Veblen function.

(g) Let $\mathcal{O}_0$ be $\mathcal{O}_{+,\mathrm{w}}$ or $\mathcal{O}_+$, $\mathcal{O}_1 := \underline{\varphi}(\text{Schütte}\binom{\mathrm{Ord}\setminus\{0\}}{\mathrm{Ord}})$. If $A = \binom{\beta_1 \cdots \beta_m}{\alpha_1 \cdots \alpha_m}$, define

$\mathrm{k}(\underline{\varphi}(A)) := \{\beta_1, \ldots, \beta_m, \alpha_1, \ldots, \alpha_m\}$ and

$\mathrm{l}(\underline{\varphi}(A)) := \{\beta_1, \ldots, \beta_{m-1}, \alpha_1, \ldots, \alpha_{m-1}\} \cup \begin{cases} \{\alpha_m\} & \text{if } \beta_m > 1, \\ \emptyset & \text{otherwise,} \end{cases}$

$\mathcal{O} := \mathcal{O}_0 \otimes \mathcal{O}_1$.

As in [Sch54] we allow addition of columns $\binom{0}{\alpha}$ and identify matrices which differ in such columns only. If $A = \binom{\beta_1 \cdots \beta_m}{\alpha_1 \cdots \alpha_m}$, $(A, \beta_{m+1}, \alpha_{m+1}) := \binom{\beta_1 \cdots \beta_m \beta_{m+1}}{\alpha_1 \cdots \alpha_m \alpha_{m+1}}$,

$(A, \beta_{m+1}, \alpha_{m+1}, \beta_{m+2}, \alpha_{m+2})$ etc. are defined similarly.

Then $\mathrm{eval}(\underline{\varphi}(A)) = \varphi(A)$, where $\varphi(A)$ is defined as in [Sch54], based on $\varphi(\alpha) := \omega^\alpha$, but with reversed order of the columns. We prove this by induction on $<'_{\mathcal{O}}$:

If $A = \underline{\varphi}(\alpha, 0)$, then we can easily show $\mathrm{eval}(A) = \omega^\alpha = \varphi(\alpha, 0)$.

If $A = \underline{\varphi}(B, \alpha, \beta, \gamma, 0)$, then it follows that $\mathrm{C}(\underline{\varphi}(A))$ is closed under $+$, $\lambda\delta.\mathrm{eval}(\underline{\varphi}(B, \alpha^*, \beta, \delta, \beta^*))$ for $\alpha^* < \alpha$, $\beta^* < \beta$ and contains, further, $\mathrm{eval}(\underline{\varphi}(B, \alpha, \beta, \gamma^*, 0))$ for $\gamma^* < \gamma$, therefore using the IH,

$$\mathrm{eval}(\underline{\varphi}(B, \alpha, \beta, \gamma, 0)) \geq \varphi(B, \alpha, \beta, \gamma, 0) \ .$$

On the other hand, $\varphi(A) \geq \mathrm{k}(\underline{\varphi}(A))$, $\varphi(A) > \mathrm{l}(\underline{\varphi}(A))$, and, if $B <' A$, $\mathrm{k}(\underline{\varphi}(B)) < \varphi(A)$, by the calculations in [Sch54] $\varphi(B) < \varphi(A)$, therefore using the IH $\forall \gamma \in \mathrm{C}(\underline{\varphi}(A))(\gamma < \varphi(A))$, $\varphi(A) \geq \mathrm{eval}(\underline{\varphi}(A))$.

(h) If we define, in (g), $\mathrm{l}(\underline{\varphi}(A)) := \mathrm{k}(\underline{\varphi}(A))$, we obtain a fixed point free version of the Klammer symbols or equivalently

$$\mathrm{eval}(\underline{\varphi}\begin{pmatrix} \beta_1 \cdots \beta_m \\ \alpha_1 \cdots \alpha_m \end{pmatrix}) = \vartheta(\Omega^{\alpha_1}\beta_1 + \cdots + \Omega^{\alpha_m}\beta_m) \ ,$$

where ϑ is defined as in [RW93], but without closing $\mathrm{C}(\alpha, \beta)$ under $\lambda\gamma.\omega^\gamma$. (We obtain the original ϑ function, if we take as $\mathcal{O}_0$ the OFG for CNF). See [Sei94] for details. If we restricting the definition to those α_i, β_i such that for $\gamma := \Omega^{\alpha_1}\beta_1 + \cdots + \Omega^{\alpha_m}\beta_m$ $\gamma \in \mathrm{C}(\gamma)$, where $\mathrm{C}(\gamma)$ is defined as for the ordinary ψ-function, then we will get the same result, but with ϑ replaced by the ψ-function ($\psi = \psi_0$ as in [Buc86] or $\psi = \psi_{\Omega_1}$ as in [Buc92], however with $\mathrm{C}(\alpha, \beta)$ not closed under φ). Note that this restriction has the effect that $\prec'$ and $\prec$ coincide in the corresponding OS for terms of the form $\underline{\varphi}(a)$.

(i) Let $\mathcal{S}_0 := \mathrm{Ord}$, $\mathcal{S}_{n+1} := \mathrm{Schütte}\binom{\mathrm{Ord}\setminus\{0\}}{\mathcal{S}_n}$,
$\mathrm{k}^i_0 : \mathcal{S}_i \to_\omega \mathrm{Ord}$,

$$\mathrm{k}^0_0(\alpha) \ := \ \{\alpha\},$$

$$\mathrm{k}_0^{i+1}\begin{pmatrix}\beta_1\cdots\beta_m\\ A_1\cdots A_m\end{pmatrix} \;:=\; \{\beta_1,\ldots,\beta_m\}\cup\bigcup_{j=1}^{m}\mathrm{k}_0^i(A_j).$$

Let $\mathrm{k}^i(\underline{\varphi}(A)) := \mathrm{k}_0^i(A)$,
$\mathcal{O}_1^i := (\underline{\varphi}(\mathcal{S}_i), \mathrm{k}^i, \mathrm{k}^i)$, $\mathcal{O}^i := \mathcal{O}_0 \otimes \mathcal{O}_1$ ($\mathcal{O}_0$ as before). We obtain, what we can call the "fixed point free version of extended Schütte Klammer symbols". Let $f_i : \mathcal{S}_i \to \mathrm{Ord}$, $f_0(\alpha) := \alpha$, $f_{i+1}\binom{\beta_1\cdots\beta_m}{A_1\cdots A_m} := \Omega^{f_i(A_1)}\beta_1 + \cdots + \Omega^{f_i(A_m)}\beta_m$, then one can easily see that $\mathrm{eval}_{\mathcal{O}^i}(\underline{\varphi}(A)) = \vartheta(f_i(A))$, ϑ as in (h). Applying a similar restriction as in (h) yields the ψ-function restricted to ordinals $< \underbrace{\Omega^{\Omega^{\cdots^{\Omega}}}}_{i\text{ times}}$.

(j) As in (g) and (h), we can define a version of the extended Schütte Klammer symbols with fixed points. The verifications takes more space than is available here. In some sense we believe however that in the context of OS, the fixed point free versions are at least as natural or even more natural than the versions with fixed points.

(k) Let $\mathcal{O}^i$ be as in (i). The union of $\mathcal{O}^i$ can be described as follows: Let

$$\begin{aligned}\mathcal{S}'_{-1} &:= \{0\},\\ \mathcal{S}'_0 &:= \mathrm{Ord},\\ \mathcal{S}'_{n+1} &:= \mathcal{S}'_n \otimes (\mathrm{Sch\ddot{u}tte}\begin{pmatrix}\mathrm{Ord}\setminus\{0\}\\ \mathcal{S}'_n\end{pmatrix} \setminus |\mathrm{Sch\ddot{u}tte}\begin{pmatrix}\mathrm{Ord}\setminus\{0\}\\ \mathcal{S}'_{n-1}\end{pmatrix}|),\end{aligned}$$

$\mathrm{k}_0^i : \mathcal{S}'_i \to_\omega \mathrm{Ord}$, $\mathrm{k}_0^0(\alpha) := \{\alpha\}$, $\mathrm{k}_0^{i+1}(A) = \mathrm{k}_0^i(A)$ for $A \in \mathcal{S}'_i$, $\mathrm{k}_0^{i+1}(A)$ is defined as in (i) for $A \in \mathcal{S}'_{i+1}\setminus\mathcal{S}'_i$. Let $|\mathcal{S}'| := \bigcup_{i<\omega}|\mathcal{S}'_i|$, $\prec_{\mathcal{S}'} := \bigcup_{i<\omega}\prec_{\mathcal{S}'_i}$, $\mathrm{k}^i(\underline{\varphi}(A)) := \mathrm{k}_0^i(A)$, $\mathrm{k} := \bigcup \mathrm{k}^i$, $\mathcal{O} := \mathcal{O}_0 \otimes (\underline{\varphi}(\mathcal{S}'), \mathrm{k}, \mathrm{k})$, which is an OFG.
Define $\iota_i : |\underline{\varphi}(\mathcal{S}_i)| \to |\underline{\varphi}(\mathcal{S}'_i)|$ by $\iota_0(\underline{\varphi}(\alpha)) := \alpha$, $\iota_1(\underline{\varphi}()) := \underline{\varphi}(0)$, $\iota_1(\underline{\varphi}\binom{\alpha}{0}) := \underline{\varphi}(\alpha)$, $\iota_1(a) := a$ otherwise, $\iota_{i+2}(a) := \iota_{i+1}(a)$ for $a \in \underline{\varphi}(\mathcal{S}_{i+1})$, $\iota_{i+2}(\underline{\varphi}\binom{\beta_1\cdots\beta_m}{A_1\cdots A_m}) := \underline{\varphi}\binom{\beta_1\cdots\beta_m}{\iota_{i+1}(A_1)\cdots\iota_{i+1}(A_m)}$ otherwise. Then ι_i is an isomorphism from $\underline{\varphi}(\mathcal{S}_i)$ to $\underline{\varphi}(\mathcal{S}'_i)$, $\underline{\varphi}(\mathcal{S}'_0) \sqsubseteq \underline{\varphi}(\mathcal{S}'_1) \sqsubseteq \cdots$, and the isomorphism can be extended to an isomorphism from $\mathrm{Arg}_{\mathcal{O}^i}$ to an initial segment of $\mathrm{Arg}_{\mathcal{O}}$ which preserves the component sets k_i. In this sense $\mathcal{O}$ is the union of the $\mathcal{O}_i$.

The following lemma, which will be used in the proof of 2.9, helps to derive from an OFG an OS based on it, which is primitive recursive: One introduces a set of terms T″, which have the syntactical form of being arguments, if we replace its components by ordinals. From these terms we now select a set T of terms and T′ of arguments based on T such that T represents all elements of $\mathcal{C}l$ and T′ all elements of $\mathrm{Arg}[\mathcal{C}l]$. Whether an element belongs to T′ or

T depends now on the ordering of its components, which need to be in T, and then one can show that under the assumptions formulated in the next lemma, T, T′, $\prec$ and $\prec'$ will be primitive recursive. Especially condition (e) will have some flavor of Π^1_2-logic, however, the ordering of the elements of T and whether and element of T″ belongs to T will depend not only on the order of its components, which allows some more generality.

Lemma 2.8 *Let $\mathcal{O}$ be an OFG as usual, and let* T'' *be a primitive recursive set,* $\mathrm{T} \subseteq \mathrm{T}' \subseteq \mathrm{T}''$, $f : \mathrm{T}' \to \mathrm{Arg}$, $\widehat{\mathrm{k}} : \mathrm{T}'' \to_\omega \mathrm{T}''$, $\widehat{\mathrm{l}} : \mathrm{T}'' \to_\omega \mathrm{T}''$, $\mathrm{length}' : \mathrm{T}'' \to \mathbb{N}$, *all primitive recursive. Let for* $a, b \in \mathrm{T}'$, $a \prec' b :\Leftrightarrow f(a) <' f(b)$, $a \prec b :\Leftrightarrow \mathrm{eval}(f(a)) < \mathrm{eval}(f(b))$. *Assume the following conditions:*

(a) $\forall a \in \mathrm{T}'.\widehat{\mathrm{k}}(a) \subseteq \mathrm{T}$.

(b) $\forall a \in \mathrm{T}'(f[\widehat{\mathrm{k}}(a)] = \mathrm{k}^0(f(a)) \wedge f[\widehat{\mathrm{l}}(a)] = \mathrm{l}'(f(a)))$.
$\forall a \in \mathrm{T}''.\mathrm{length}'[\widehat{\mathrm{k}}(a)] < \mathrm{length}'(a)$.

(c) For every subset A of T $f[\{t \in \mathrm{T}' \mid \widehat{\mathrm{k}}(t) \subseteq A\}] = \{a \in \mathrm{Arg} \mid \mathrm{k}(a) \subseteq \mathrm{eval}[f[A]]\}$.

(d) If $A \subseteq \mathrm{T}'$, $f \restriction A$ is injective, then $f \restriction \{t \in \mathrm{T}' \mid \widehat{\mathrm{k}}(t) \subseteq A\}$ is injective.

(e) For $a, b \in \mathrm{T}''$ such that $\widehat{\mathrm{k}}(a) \cup \widehat{\mathrm{k}}(b) \subseteq \mathrm{T}$ we can determine from $\prec$ restricted to $\widehat{\mathrm{k}}(a) \cup \widehat{\mathrm{k}}(b)$ (coded as a finite list which is coded as a natural number) in a primitive recursive way, whether $a, b \in \mathrm{T}'$ and, if this is the case, whether $a \prec' b$.

(f) $a \in \mathrm{T} \Leftrightarrow a \in \mathrm{T}' \wedge f(a) \in \mathrm{NF} \wedge \widehat{\mathrm{k}}(a) \subseteq \mathrm{T}$.

Then T, T′, $\prec'$, $\prec$, *are primitive recursive, we can define* $\widehat{\mathrm{length}} : \mathrm{T}' \to_\omega \mathrm{T}$ *such that* $\widehat{\mathrm{length}}(a) = \mathrm{length}(f(a))$ *for* $a \in \mathrm{T}'$, *and* $(\mathrm{T}, \prec, \prec', \widehat{\mathrm{k}}, \widehat{\mathrm{length}})$ *is an OS based on $\mathcal{O}$.*

Proof: We determine for $t \in \mathrm{T}''$ whether $t \in \mathrm{T}'$, $t \in \mathrm{T}$, and for $r, s \in \mathrm{T}'$ such that $\mathrm{length}'(r), \mathrm{length}'(s) \leq \mathrm{length}'(t)$ whether $r \prec' s$, and whether $r \prec s$ holds by recursion on $\mathrm{length}'(t)$ and side-recursion on $\mathrm{length}'(r) + \mathrm{length}'(s)$: If $\widehat{\mathrm{k}}(t) \not\subseteq \mathrm{T}$, $t \notin \mathrm{T}'$. Otherwise we can decide whether $t \in \mathrm{T}'$. Assume now r, s as above. Whether $r \prec' s$ follows from the induction hypothesis and then using Lemma 2.5 (f) we can determine whether $r \prec s$ holds. Now $t \in \mathrm{T} :\Leftrightarrow t \in \mathrm{T}' \wedge \widehat{\mathrm{k}}(t) \prec t$.

$f : \mathrm{T} \to \mathrm{Arg}$ is injective by condition (d). $f[\mathrm{T}] \subseteq \mathrm{NF}$, $f[\mathrm{T}] \subseteq \mathcal{C}l$. We show $f[\mathrm{T}] = \mathcal{C}l$. It suffices to show $\forall a \in \mathcal{C}l.\exists b \in \mathrm{T}.f(b) = a$. Induction on a. $\widehat{\mathrm{k}}(a) \subseteq f[\mathrm{T}]$. Therefore there exists a $t \in \mathrm{T}'$, $\widehat{\mathrm{k}}(t) \subseteq \mathrm{T}$, such that $f(t) = a$. Since $\mathrm{NF}(a)$, it follows that $t \in \mathrm{T}$.

$\widehat{\mathrm{length}}(t)$ is now defined by $\widehat{\mathrm{length}}(t) := \max(\widehat{\mathrm{length}}[\widehat{\mathrm{k}}(t)] \cup \{-1\}) + 1$. □

Lemma 2.9 *For all the OFG in Example 2.7 (a) - (i), except for those examples involving the lexicographic ordering on weakly descending sequences ((b) and the examples referring to it) there exists an elementary OS represented by them.*

Remark 2.10 *Note that the development of the OS can be done without having to refer to ordinals.*

Proof of Lemma 2.9: First, using Lemma 2.8 we can define a primitive recursively represented OS based on the OFG. We look only at example (g) in detail: Let $\underline{0} := \underline{\Sigma}'()$, $\underline{l+1} := \underline{\Sigma}'(\underline{0}, \underline{l}\underline{+}1)$, $\underline{\mathbb{N}} := \{\underline{l} \mid l \in \omega\}$. Define T'' together with $\widehat{\mathrm{k}} : \mathrm{T}'' \to_\omega \mathrm{T}''$, $\widehat{\mathrm{l}} : \mathrm{T}'' \to_\omega \mathrm{T}''$, $\mathrm{length}(\mathrm{T}'') \to \mathbb{N}$ recursively by:
If $m \in \omega$, $a_i \in \mathrm{T}''$, $b_i \in \underline{\mathbb{N}}$, $b_i \neq \underline{0}$ for $i < m$, then
$t_0 := \underline{\Sigma}'(a_1, b_1, \ldots, a_{m-1}, b_{m-1}, a_m, b_m\underline{+}1) \in \mathrm{T}''$,

$$\widehat{\mathrm{k}}(t_0) := \{a_1, b_1, \ldots, a_m, b_m\},\ \widehat{\mathrm{l}}(t_0) := \begin{cases} \{a_1, b_m\} & \text{if } m > 1 \text{ or } b_m \neq \underline{0}, \\ \emptyset & \text{otherwise,} \end{cases}$$

$\mathrm{length}(t_0) := \max\{\mathrm{length}(a_1), \ldots, \mathrm{length}(a_m), \mathrm{length}(b_1), \ldots, \mathrm{length}(b_m)\} + 1.$

If $a_1, \ldots, a_m, b_1, \ldots, b_m \in \mathrm{T}''$, $a_1, \ldots, a_m \neq \underline{0}$, $t_1 := \underline{\varphi}\binom{a_1 \cdots a_m}{b_1 \cdots b_m} \in \mathrm{T}''$, and we define $\widehat{\mathrm{k}}(t_1)$, $\widehat{\mathrm{l}}(t_1)$ by translating the conditions from Example 2.7 (g) and $\mathrm{length}(t_1)$ as before. T'', length, $\widehat{\mathrm{k}}$, $\widehat{\mathrm{l}}$ are primitive recursive.

Define $\mathrm{T} \subseteq \mathrm{T}''$, $\mathrm{T}' \subseteq \mathrm{T}''$, together with $f : \mathrm{T}' \to \mathrm{Arg}$ inductively by:
If t_0 is as above, $\widehat{\mathrm{k}}(t_0) \subseteq \mathrm{T}$, $\mathrm{eval}(f(a_1)) > \cdots > \mathrm{eval}(f(a_m))$, then $t_0 \in \mathrm{T}'$,
$f(t_0) := \underline{\Sigma}'(\mathrm{eval}(f(a_1)), \mathrm{eval}(f(b_1)), \ldots, \mathrm{eval}(f(a_{m-1})), \mathrm{eval}(f(b_{m-1})),$
$\qquad \mathrm{eval}(f(a_m)), \mathrm{eval}(f(b_m)) + 1)$.

If t_1 is as above, $\mathrm{eval}(f(b_1)) > \cdots > \mathrm{eval}(f(b_m))$, then $t_1 \in \mathrm{T}'$,
$f(t_1) := \underline{\varphi}\binom{\mathrm{eval}(f(a_1)) \cdots \mathrm{eval}(f(a_m))}{\mathrm{eval}(f(b_1)) \cdots \mathrm{eval}(f(b_m))}$.

Further $t \in \mathrm{T} :\Leftrightarrow t \in \mathrm{T}' \wedge \mathrm{NF}(f(t))$.

The conditions of Lemma 2.8 are now fulfilled and therefore most conditions of a primitive recursively represented OS are fulfilled, what is missing is verified easily.

The other OS are treated similarly. The constructions of new orderings considered in [Set98a] Lemma 3.5 (especially the lexicographic ordering on pairs and and strictly descending sequences) yield orderings such that transfinite induction over it reduces to transfinite induction over the underlying orderings in PRA and therefore we get actually get in all cases elementary OS. □

However, transfinite induction over weakly descending sequences reduces only in HA to the underlying ordering, so the examples where we used weakly descending sequences do not yield elementary OS.

Theorem 2.11 *(a) The bound in Theorem 2.3 is sharp: The supremum of the order types of elementary OS is exactly the Bachmann-Howard ordinal.*

(b) The limit of the order type of ordinal systems from below as introduced in [Set98a] is the Bachmann-Howard ordinal.

Proof: (a): Example 2.7 (i) yields OS of strength $\vartheta(\underbrace{\Omega^{\Omega^{\cdot^{\cdot^{\cdot^{\Omega}}}}}}_{i \text{ times}})$, which in the limit reaches the Bachmann-Howard ordinal.
(b): by subsection 2.2, (a) and the upper bound developed in the last two sections of [Set98a]. □

Note that the construction in the example used in (a) really exhausts the full strength of elementary OS: In the OS related to $\mathcal{O}_i$ $\prec'$ is built from $\prec$ using i-times nested lexicographic ordering on descending sequences, TI over which reduces to the underlying ordering by using formulas of increasing length. The union of the $\mathcal{O}_i$, $\mathcal{O}'$ yields an OFG, such that in the OS belonging to it TI over $\prec'$ no longer reduces to TI over $\prec$ in PRA, so is no longer elementary. What is required are twice iterated ordinal systems.

3 n-times Iterated Ordinal Systems

3.1 Introduction

The usual way of getting beyond the Bachmann-Howard ordinal is to violate condition (OS 1), which was the basis for our analysis in the last two sections of [Set98a]. However, the foundationally more interesting approach seems to keep the condition that the ordinals are built from below and instead weaken the requirement that TI over $\prec'$ reduces to TI over $\prec$ in an elementary way, namely in PRA.

If we look at the OS corresponding to Example 2.7 (k) we can see that we generated by meta recursion a sequence of matrices of increasing complexity. We can replace this meta recursion by using a second OS: Let $\mathcal{F}_0 := (\mathrm{T}_0, \prec, \prec', \mathrm{k}, \mathrm{length})$ be the primitive recursive (but non-elementary) OS based on $\mathcal{O}$, defined by the method used in Lemma 2.9. Let $\mathrm{T}_1 := \{A \mid \underline{\varphi}(A) \in \mathrm{T}_0\}$ with the ordering $A \prec_1 A' :\Leftrightarrow \underline{\varphi}(A) \prec' \underline{\varphi}(A')$. Define $\mathrm{k}_{0,j} : \mathrm{T}_0 \to_\omega \mathrm{T}_j$ for $i, j = 0, 1$ by: $\mathrm{k}_{0,0} := \mathrm{k}$, $\mathrm{k}_{0,1}(\underline{\Sigma}'(\vec{a})) := \emptyset$, $\mathrm{k}_{0,1}(\underline{\varphi}(A)) := \{A\}$ and for $D \subseteq \mathrm{T}_0$, $E \subseteq \mathrm{T}_i$, let $D \restriction_{0,j} E := \mathrm{k}_{0,j}^{-1}(E)$. Then if $B \subseteq \mathrm{T}_0$, $C \subseteq \mathrm{T}_1$ are well-ordered w.r.t. $\prec$, $\prec_1$, then $(\mathrm{T}_0 \restriction_{0,0} B) \restriction_{0,1} C$ is well-ordered. If we take as C the set of matrices built from notations in B only and, if we know that C is well-ordered, then $\mathrm{T}_0 \restriction_{0,0} B = (\mathrm{T}_0 \restriction_{0,0} B) \restriction_{0,1} C$ is well-ordered, and we have shown that $\mathcal{F}_0$ is an OS.

Now in order to show that C is well-ordered we use a second OS: Define $\mathrm{k}_{1,i} : \mathrm{T}_1 \to_\omega \mathrm{T}_i$ by: if $a \in \mathrm{T}_0$, $\mathrm{k}_{1,0}(a) := \{a\}$, $\mathrm{k}_{1,1}(a) := \emptyset$, and if $A = \binom{a_1 \cdots a_m}{A_1 \cdots A_n}$, then $\mathrm{k}_{1,0}(A) := \{a_1, \ldots, a_n\} \cup \bigcup_{j=1}^m \mathrm{k}_{1,0}(A_j)$, $\mathrm{k}_{1,1}(A) := \{A_1, \ldots, A_m\}$ and define $C \restriction_{1,j} D$ as before. Let $\mathcal{F}_1 := (\mathrm{T}_1, \mathrm{k}_{1,1}, \prec', \prec', \mathrm{length})$, and for $B \subseteq \mathrm{T}_0$ $\mathcal{F}_1[B] := (\mathrm{T}_1 \restriction_{1,0} B, \prec_1, \prec_1, \mathrm{k}_{1,1}, \mathrm{length})$. If B is $\prec$-well-ordered, then $\mathcal{F}_1[B]$ is now an ordinal system, and therefore $C := \mathrm{T}_1 \restriction_{1,0} B$ is well-ordered and with this C the above holds.

$\mathcal{F}_1$ is a system which internalizes what was before meta-induction. It is a relatively weak OS which is the relativized extension of the OS for the Cantor normal form. Expanding the OS we introduced in the last section we can get now more complex "matrices", which can be used by the first OS. As before, we will exhaust the strength of this concept (which will be called 2-OS) by using the hierarchy of extended Klammer symbols. In order to go beyond this, we can iterate the step from OS to 2-OS once more and yield 3-OS, 4-OS etc. We are going to formalize in the following n-OS for arbitrary natural numbers n.

3.2 n-Ordinal Systems

In the example before we had as basic structures two ordinal structures

$$\mathcal{F}_j := (\mathrm{T}_j, \prec_j, \prec_j', \mathrm{k}_{j,j}, \mathrm{length}_j) \quad (j = 0, 1)$$

together with functions $\mathrm{k}_{i,1-i} : \mathrm{T}_i \to_\omega \mathrm{T}_{1-i}$. It can be visualized as follows:

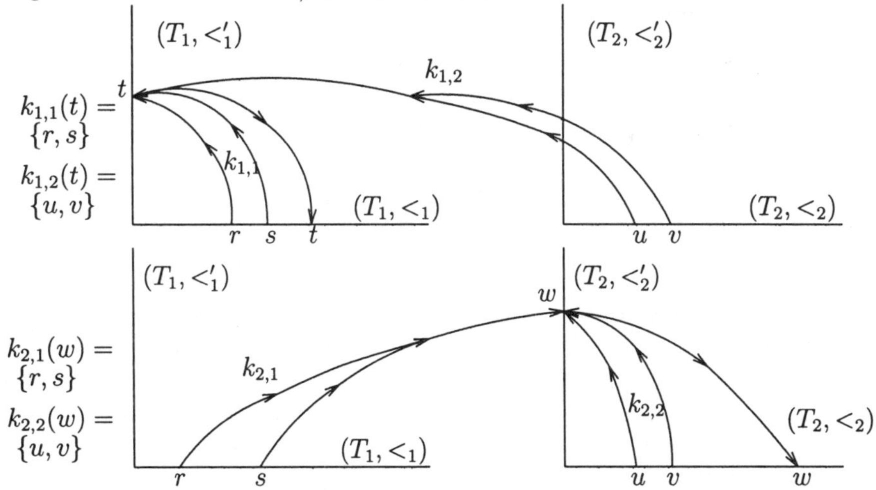

For simplicity assume $\mathrm{T}_0 \cap \mathrm{T}_1 = \emptyset$ and write $\prec$, $\prec'$, length, instead of $\prec_j$, $\prec_j'$, length_j, and write $\restriction_j$ instead of $\restriction_{i,j}$ as defined before.

In the example $\mathcal{F}_j$ fulfilled all conditions of an OS except (OS 4). We used that, if $A \subseteq \mathrm{T}_0$, $B \subseteq \mathrm{T}_1$ are well-ordered, then $(\mathrm{T}_0 \restriction_0 A) \restriction_1 B$ is $\prec'$-

well-ordered. In order for $\mathcal{F}_1[A]$ to be an OS, we need as well that under the same conditions for A, B $(\mathrm{T}_1 \restriction_0 A) \restriction_1 B$ is well-ordered. Further we needed that $(\mathrm{T}_0 \restriction_0 A) \restriction_1 (\mathrm{T}_1 \restriction_0 A) = \mathrm{T}_0 \restriction_0 A$ for all $A \subseteq \mathrm{T}_0$. Only $\supseteq$ needs to be fulfilled, which is equivalent to $\mathrm{k}_{1,0}[\mathrm{k}_{0,1}(a)] \subseteq \mathrm{k}_{0,0}(a)$. In order to get that $\mathcal{F}_1[A]$ is an OS-structure, we need $\mathrm{k}_{1,1} : \mathrm{T}_1 \restriction_0 A \to_\omega \mathrm{T}_1 \restriction_0 A$ for all $A \subseteq \mathrm{T}_0$, i. e. $\mathrm{k}_{1,0}[\mathrm{k}_{1,1}(a)] \subseteq \mathrm{k}_{1,0}(a)$. The generalization to n-ordinal systems is now straight-forward and we get the following definition:

Definition 3.1 (a) An *n-times iterated Ordinal-System-structure*, in short *n-OS-structure*, is a triple $\mathcal{F} = ((\mathcal{G}_i)_{i<n}, (\mathrm{k}_{i,j})_{i,j<n}, (\mathrm{length}_i)_{i<n})$ such that $\mathcal{F}_i := (\mathcal{G}_i, \mathrm{k}_{i,i}, \mathrm{length}_i)$ are OS-structures and $\mathrm{k}_{i,j} : \mathrm{T}_i \to_\omega \mathrm{T}_j$ (each $i, j < n$).
In the following, when introduced as an n-OS, $\mathcal{F}$ will be always as above (where n can be replaced by any natural number), $\mathcal{F}_i = (\mathcal{G}_i, \mathrm{k}_{i,i})$, $\mathcal{G}_i = (\mathrm{T}_i, \prec_i, \prec'_i)$ and and we will usually write $((\mathcal{G}_i), (\mathrm{k}_{i,j}), \mathrm{length}_i)$ instead of $((\mathcal{G}_i)_{i<n}, (\mathrm{k}_{i,j})_{i,j<n})$. Further we assume T_i are always disjoint and write $\prec$, $\prec'$, length, instead of $\prec_i$, $\prec'_i$, length_i.

(b) If $\mathcal{F}$ is an n-OS-structure as above, $A \subseteq \mathrm{T}_i$, $B \subseteq \mathrm{T}_j$, $A \restriction_j B := A \cap \mathrm{k}_{i,j}^{-1}[B]$. If $B_j \subseteq \mathrm{T}_j$, then $A \restriction_{j=l}^m B_j := A \cap \bigcap_{j=l}^m \mathrm{k}_{i,j}^{-1}[B_j]$. We write $\restriction_{i<j}$, $\restriction_{i\le j}$ for $\restriction_{i=0}^{j-1}$ and $\restriction_{i=0}^{j}$. If $\mathcal{F}$ is an $n+1$-OS-structure and $A \subseteq \mathrm{T}_0$, then $\mathcal{F}[A] := ((\mathrm{T}_{i+1} \restriction_0 A, \prec_{i+1}, \prec'_{i+1})_{i<n-1}, (\mathrm{k}_{i+1,j'+1})_{i,j'<n-1}, (\mathrm{length}_{i+1})_{i<n-1}))$. (More precisely we have to restrict $\prec_{i+1}$, $\prec'_{i+1}$, $\mathrm{k}_{i+1,j'+1}$ and length_{i+1} to $\mathrm{T}_{i+1} \restriction_0 A$.)

(c) An n-OS-structure $\mathcal{F}$ is an *n-times iterated Ordinal System*, in short *n-OS*, if for all $i, j, l < n$ the following holds
$(n - \mathrm{OS}\ 1)$ $\mathcal{F}_i$ fulfil (OS 1), (OS 2) and (OS 3);
$(n - \mathrm{OS}\ 2)$ if $l < j$, $i, j < n$, then $\mathrm{k}_{j,l} \circ \mathrm{k}_{i,j} \subseteq \mathrm{k}_{i,l}$;
$(n - \mathrm{OS}\ 3)$ if $A_i \subseteq \mathrm{T}_i$ are $\prec$-well-ordered $(i < n)$, $l < n$, then $\mathrm{T}_l \restriction_{i<n} A_i$ is $\prec'$-well-ordered.

(d) An n-OS $\mathcal{F}$ is *well-ordered*, if $(\mathrm{T}_0, \prec)$ is, and its *order type* is that of $(\mathrm{T}_0, \prec)$. It is *primitive recursive*, if the involved sets are primitive recursive subsets of the natural numbers, all functions, relations are primitive recursive and all properties except of the well-ordering condition can be shown in primitive recursive arithmetic. It is *elementary*, if additionally the well-ordering condition follows in PRA in the sense of reducibility of transfinite induction in PRA.

(e) Two n-OS are *isomorphic*, if there is are bijections between the underlying sets, which respect $\mathrm{k}_{i,j}$, length_i, $\prec_i$, $\prec'_i$ for all $i, j < n$.

We will need the following auxiliary definition:

Definition 3.2 (a) A *relativized n-OS* is a triple $(\mathcal{F}, \mathrm{T}_{-1}, \prec, (\mathrm{k}_{i,-1})_{0 \le i < n})$ such that $\mathcal{F}$ is an n-OS structure, $(\mathrm{T}_{-1}, \prec)$ is a linearly ordered set, $\mathrm{k}_{i,-1} : \mathrm{T}_i \to_\omega \mathrm{T}_{-1}$, for $j > -1$ $\mathrm{k}_{j,-1} \circ \mathrm{k}_{i,j} \subseteq \mathrm{k}_{i,-1}$ and such that $\mathcal{F}$ fulfils the conditions of an n-OS except for condition$(n -$ OS 3), which is replaced by: if $A_i \subseteq \mathrm{T}_i$ are $\prec$-well-ordered $(-1 \le i < n)$, then $\mathrm{T}_l \restriction_{i=-1}^{n-1} A_i$ is $\prec'$-well-ordered $(0 \le l < n)$.

(b) If $(\mathcal{F}, \mathrm{T}_{-1}, \mathrm{k}_{i,-1})$ is a relativized n-OS, $A \subseteq \mathrm{T}_{-1}$, then $\mathcal{F}[A]_{-1}$ is the relativization of $\mathcal{F}$ to A defined in a straightforward way.

Well-ordering of a $n+1$-OS reduces to the well-ordering of an n-OS as follows:

Lemma 3.3 *Assume $\mathcal{F}$ is an $n+1$-OS-structure.*

(a) If $A \subseteq \mathrm{T}_0$ and $\mathcal{F}$ fulfils $((n+1)-$OS 2), then $\mathcal{F}[A]$ is an n-OS-structure which fulfils $(n -$ OS 2).

(b) If $\mathcal{F}$ is an $n+1$-OS and $A \subseteq \mathrm{T}_0$ is $\prec$-well-ordered, then $\mathcal{F}[A]$ is an n-OS.

(c) If $\mathcal{F}$ is an 1-OS, then $\mathcal{F}$ is an OS.

(d) n-OS are well-ordered.

(e) For elementary n-OS, (d) can be shown in KPω extended by the existence of $n-1$ admissible sets $x_1, \dots, x_{n-1}$ such that $x_1 \in x_2 \in \dots \in x_{n-1}$.

Proof: (a) We need only to show that $\mathrm{k}_{i+1,j+1} : \mathrm{T}_{i+1} \restriction_0 A \to \mathrm{T}_{j+1} \restriction_0 A$, which follows by $\mathrm{k}_{j+1,0}(\mathrm{k}_{i+1,j+1}(a)) \subseteq \mathrm{k}_{i+1,0}(a)$.

(b): Easy.

(c): Trivial.

(d) We show by induction on $n \ge 1$: If $\mathcal{F}$ is an n-OS, then $(\mathrm{T}_i, \prec)$ are well-ordered for $i < n$. $n = 1$: (c) and Theorem 2.2 (a). $n \to n+1$: We show first that $\mathcal{F}_0$ is an OS. We have only to show (OS 4). Assume $A \subseteq \mathrm{T}$ is $\prec$-well-ordered. Then $\mathcal{F}[A]$ is an n-OS, and by the IH it follows that $\mathrm{T}_i \restriction_0 A$ are well-ordered $(i = 1, \dots, n+1)$. Therefore $(\mathrm{T}_0 \restriction_0 A) \restriction_{i=1}^{n} (\mathrm{T}_i \restriction_0 A)$ is $\prec'$-well-ordered. But $(\mathrm{T}_0 \restriction_0 A) \restriction_{i=1}^{n} (\mathrm{T}_i \restriction_0 A) = \mathrm{T}_0 \restriction_0 A$, since $\mathrm{k}_{i,0}(\mathrm{k}_{0,i}(a)) \subseteq \mathrm{k}_{0,0}(a)$.

(e) We show by Meta-induction on n in KPω: If $(\mathcal{F}, \mathrm{T}_{-1}, \prec, \mathrm{k}_{i,-1})$ is a relativized n-OS, $A \subseteq \mathrm{T}_{-1}$ is well-ordered, and there are $n-1$ admissibles above $\{A, \mathcal{F}, \mathrm{k}_{i,-1}\}$, then $\mathrm{T}_i \restriction_{-1} A$ are $\prec$-well-ordered.

The case $n = 1$ follows as in Theorem 2.2 (b). In the step $n \to n+1$ let κ be the least admissible ordinal such that $\{A, \mathcal{F}, \mathrm{k}_{i,-1}\} \in \mathrm{L}_\kappa$. Define a_α as in Lemma 2.2 (c) for $\alpha \le \kappa$. This can be done, by the same argument as in (d), using the fact that $\mathcal{F}[A]_{-1}[B]$ is an n-OS for $B \subseteq \mathrm{T}_0$ $\prec$-well-ordered and that therefore, if $B \in \mathrm{L}_{\kappa^+}$, $(\mathrm{T}_i \restriction_{-1} A) \restriction_0 B$ are well-ordered ($i = 1, \ldots, n$), where κ^+ is the next admissible ordinal above κ. If we now replace C in the proof of 2.2 (c) by $\{\alpha \in \kappa \mid A_{<\alpha} \text{ increasing}\}$, and the argument that there is no bijection between a set and Ord by that there is no (Δ_0-definable) bijection between an element of a and κ, we conclude that $\{a_\alpha \mid \alpha \in C\} = \mathrm{T}_0 \restriction_{-1} A$, $\mathrm{T}_0 \restriction_{-1} A$ is well-ordered and therefore $\mathrm{T}_i \restriction_{-1} A$ are also well-ordered. □

Theorem 3.4 *The strength of an n-OS ($n > 0$) is less than $|\mathrm{ID}_n| = \psi_{\Omega_1}(\epsilon_{\Omega_i+1})$ (where $\psi_\kappa\alpha$ is as in [Buc92]).*

Proof: By Lemma 3.3 (e) and, since the strength of KPω extended by the existence of $n-1$ admissibles is $|\mathrm{ID}_n|$. □

3.3 Constructive Well-ordering Proof.

We will give now a proof of Lemma 3.3 (e) which can be formalized even in constructive theories like Martin-Löf's type theory extended by an at most n times nested W-type or intuitionistic ID_n:

Define inductively $\mathrm{M}_i, \mathrm{Acc}_i \subseteq \mathrm{T}_i$: $\mathrm{M}_i := \mathrm{T}_i \restriction_{j<i} \mathrm{Acc}_j$, $\mathrm{Acc}_i := \mathrm{Acc}_\prec(\mathrm{M}_i)$. Let $\mathrm{Acc}'_i := \mathrm{T}_i \restriction_{j<n} \mathrm{Acc}_j$. Then $(\mathrm{Acc}_i, \prec)$, $(\mathrm{Acc}'_i, \prec')$ are well-ordered. We show by induction on $n-i$ that $\mathrm{Acc}_i = \mathrm{M}_i$. Assume i according to induction. We show $\forall s \in \mathrm{Acc}'_i . s \in \mathrm{Acc}_i$ by side-induction on $(\mathrm{Acc}'_i, \prec')$:

$s \in \mathrm{M}_i$. Further $\forall r \in \mathrm{M}_i (r \prec s \to r \in \mathrm{Acc}_i)$ by (side-side-)induction on length(r): Assume r according to induction, $r \in \mathrm{M}_i \wedge r \prec s$. If $r \preceq \mathrm{k}_{i,i}(s) \subseteq \mathrm{Acc}_i$, it follows that $r \in \mathrm{Acc}_i$. Otherwise $r \prec' s$. We show $r \in \mathrm{Acc}'_i$. If $j < i$, $\mathrm{k}_{i,j}(r) \subseteq \mathrm{Acc}_j$. $\mathrm{k}_{i,i}(r) \prec r \prec s$, if $j < i$, $\mathrm{k}_{i,j}[\mathrm{k}_{i,i}(r)] \subseteq \mathrm{k}_{i,j}(r) \subseteq \mathrm{Acc}_j$, therefore $\mathrm{k}_{i,i}(r) \subseteq \mathrm{M}_i$ and by side-side-IH $\mathrm{k}_{i,i}(r) \subseteq \mathrm{Acc}_i$. For $i < j < n$ we show by side[3]-induction on j $\mathrm{k}_{i,j}(r) \subseteq \mathrm{Acc}_j$. If $l < j$, $\mathrm{k}_{j,l}[\mathrm{k}_{i,j}(r)] \subseteq \mathrm{k}_{i,l}(r) \subseteq \mathrm{Acc}_l$ by the IH, $\mathrm{k}_{i,j}(r) \subseteq \mathrm{M}_j$, and by the main-IH $\mathrm{k}_{i,j}(r) \subseteq \mathrm{Acc}_j$. Therefore $r \in \mathrm{Acc}'_i$, $r \prec' s$, by the side-IH $r \in \mathrm{Acc}_i$, and the side-side-induction and therefore as well the side-induction are complete. Now it follows by induction on length(s) that $\forall s \in \mathrm{M}_i . s \in \mathrm{Acc}_i$, $\mathrm{Acc}_i = \mathrm{M}_i$ and the main induction is finished. Hence $\mathrm{T}_0 = \mathrm{M}_0 = \mathrm{Acc}_0$ and we are done.

3.4 n-Ordinal Function Generators

As before we are going to define the ordinal systems which exhaust the strength of elementary n-ordinal systems. As there, we want also as a side result to get functions defined on arbitrary ordinals. Again of course this detour via ordinals is not necessary, one could easily define the n-ordinal systems purely syntactically.

We need to represent the higher ordinal systems we used as ordinals and will do it in the usual way by taking ordinals of higher number classes: Let

$\Omega_0 := 0$, $\Omega_n := \aleph_n$. Ordinals in $[\Omega_n, \Omega_{n+1}[$ can be regarded as objects referring to ordinals in $[0, \Omega_n[$ and therefore be considered as representatives of the higher OS. In a refined approach one could replace Ω_n by the n-th admissible ordinal Ω_n^{rec} (starting with $\Omega_1^{\mathrm{rec}} := \omega_1^{\mathrm{ck}}$).

In the definition of (non-iterated) OFG we closed the sets $\mathrm{C}(a)$ under $(\mathrm{k}(b) \subseteq \mathrm{C}(a) \wedge b <' a) \to \mathrm{eval}(b) \in \mathrm{C}(a)$ with no restriction on b being in normal form. But one could easily show that having the restriction of b being in normal form yields the same result, since, if b is not in normal form, $\mathrm{eval}(b) \in \mathrm{k}(b)$. In the case of n-OFG, normal form will again mean that $\mathrm{k}(b) \subseteq \mathrm{C}_i(b)$. However, this condition can be violated by having $\mathrm{eval}_j(c) \in \mathrm{k}(b)$ such that $j > i$ and $\mathrm{k}(c) \cap [\Omega_i, \Omega_{i+1}[> \mathrm{eval}(a)$. So we no longer have $a \notin \mathrm{NF} \to \mathrm{eval}(a) \in \mathrm{k}(a)$, and the above argument no longer goes through. We will use the easiest way of dealing with these problems: we add only $\mathrm{eval}(c)$ to $\mathrm{C}(a)$, if $\mathrm{NF}(c)$ holds.

Definition 3.5 (a) An *n-ordinal function generator*, in short *n-OFG* is a quadruple $\mathcal{O} := (\mathrm{Arg}_i, \mathrm{k}_i, \mathrm{l}_i, <'_i)_{i<n}$ such that Arg_i are classes (in set theory), $\mathrm{k}_i, \mathrm{l}_i : \mathrm{Arg}_i \to_\omega \mathrm{Ord}$, $\forall a \in \mathrm{Arg}_i . \mathrm{l}_i(a) \subseteq \mathrm{k}_i(a)$, and $<'_i$ is a well-ordered relation on Arg_i ($i < n$). We assume in the following that Arg_i are disjoint and therefore omit the index i in k, l, $<'$. Let $\mathcal{O}$ in the following be as above.

(b) $\Omega_0 := 0$, $\Omega_{n+1} := \aleph_{n+1}$,

(c) An n-OFG is *cardinality based* iff for all countable $B \subseteq \mathrm{Ord}$ $\mathrm{k}_i^{-1}(B)$ is countable and for $a \in \mathrm{Arg}_i$, $\mathrm{l}_i(a) \subseteq \mathrm{k}_i(a) \cap [\Omega_i, \Omega_{i+1}[$. In this case we define $\mathrm{k}_{i,j}(a) := \mathrm{k}_i(a) \cap [\Omega_j, \Omega_{j+1}[$.

(d) If $\mathcal{O}$ as above is a cardinality based n-OFG, we define by main recursion on $n - i$ ($i = 1, \ldots, n$) by side-recursion on $<'$ ordinals for $a \in \mathrm{Arg}_i$ $\mathrm{eval}_i(a) \in \mathrm{Ord}$ and subsets $\mathrm{C}_i(a) \subseteq \mathrm{Ord}$ by: $\mathrm{C}_i(a) := \bigcup_{l=0}^{\omega} \mathrm{C}_i^l(a)$, where $\mathrm{C}_i^0(a) := [0, \Omega_i[\cup (\bigcup(\mathrm{k}_i(a) \cap \Omega_{i+1})) \cup \mathrm{l}(a)$, $\mathrm{C}_i^{l+1}(a) := \mathrm{C}_i^l(a) \cup$
$\bigcup_{i \le j < n} \{\mathrm{eval}_j(b) \mid (j = i \to b <' a), b \in \mathrm{Arg}_j, \mathrm{NF}_j(b), \mathrm{k}(b) \subseteq \mathrm{C}_i^l(a)\}$ and $\mathrm{NF}(b) :\Leftrightarrow \mathrm{NF}_j(b) :\Leftrightarrow \mathrm{eval}_j(b) \in \mathrm{C}_j(b)$.
$\mathrm{eval}_i(a) := \min\{\alpha \mid \alpha \notin \mathrm{C}_i(a)\}$, $\mathrm{eval}(a) := \mathrm{eval}_i(a)$ if $a \in \mathrm{Arg}_i$, and $\mathrm{NF}_i := \{a \in \mathrm{Arg}_i \mid \mathrm{NF}(a)\}$.

(e) For cardinality based n-OFG $\mathcal{O}$ we define $\mathcal{C}l_i \subseteq \mathrm{NF}_i$ simultaneously for all $i < n$ inductively defined by: if $b \in \mathrm{NF}_i$, $\mathrm{k}(b) \subseteq \bigcup_{i<n} \mathrm{eval}_i[\mathcal{C}l_i]$, then $b \in \mathcal{C}l_i$. $\mathcal{C}l := \bigcup_{i<n} \mathcal{C}l_i$, $\mathrm{Arg}[\mathcal{C}l]_i := \{a \in \mathrm{Arg}_i \mid \mathrm{k}(a) \subseteq \mathrm{eval}[\mathcal{C}l]\}$, $\mathrm{Arg}[\mathcal{C}l] := \bigcup_{i<n} \mathrm{Arg}[\mathcal{C}l]_i$. Note $\mathcal{C}l_i \subseteq \mathrm{Arg}[\mathcal{C}l]_i$.
Assuming that $\mathrm{eval}_i \restriction \mathrm{NF}_i$ is injective, which will be shown later, we define length : $\mathrm{Arg}[\mathcal{C}l] \to \mathbb{N}$, $\mathrm{k}_{i,j}^0, \mathrm{k}'_{i,j} : \mathrm{Arg}[\mathcal{C}l]_i \to_\omega \mathcal{C}l_j$ simultaneously inductively by:

$\mathrm{k}^0_{i,j}(b) := \mathrm{eval}_j^{-1}[\mathrm{k}_{i,j}(b)] \cap \mathcal{C}l_j$. $\mathrm{k}'_{i,j}(b) := \mathrm{k}^0_{i,j}(b) \cup \bigcup_{j<l<n} \mathrm{k}'_{l,j}[\mathrm{k}^0_{i,l}(b)]$.
$\mathrm{length}(b) := \max(\mathrm{length}[\bigcup \mathrm{k}'_{i,i}(b)] \cup \{-1\}) + 1$.

(f) Define for $a, b \in \mathrm{Arg}_i$ $a \prec b :\Leftrightarrow a \prec_i b :\Leftrightarrow \mathrm{eval}_i(a) < \mathrm{eval}_i(b)$.

Lemma 3.6 *Let $\mathcal{O}$ be a cardinality based n-OFG.*

(a) $\mathrm{C}_i(a)$ *is the least set M such that* $\mathrm{C}^0_i(a) \subseteq M$ *and such that, if* $b \in \mathrm{Arg}_j$, $j = i \wedge b <' a$ *or* $j > i$, *and, if* $\mathrm{NF}(b)$ *and* $\mathrm{k}(b) \subseteq M$, *then* $\mathrm{eval}_j(b) \in \mathrm{M}$.

(b) $\mathrm{C}_i(a) \cap \Omega_{i+1} \sqsubseteq \mathrm{Ord}$.
Therefore especially $\mathrm{eval}_i(a) = \mathrm{C}_i(a) \cap \Omega_{i+1}$

(c) If $a \in \mathrm{Arg}_i$, *then* $\mathrm{eval}_i(a) \in [\Omega_i, \Omega_{i+1}[$.

(d) If $a, b \in \mathrm{NF}_i$, *then* $\mathrm{eval}(a) < \mathrm{eval}(b) \Leftrightarrow (a <' b \wedge \mathrm{k}_{i,i}(a) < \mathrm{eval}(b)) \vee \mathrm{eval}(a) \leq \mathrm{k}_{i,i}(b)$.
In particular, $\mathrm{eval} \restriction \mathrm{NF}_i$ *is injective and therefore* length *and* $\mathrm{k}'_{i,j}$ *are well-defined.*

(e) If $a \in \mathrm{Arg}[\mathcal{C}l]_i$, $l < j$, *then* $\mathrm{k}'_{l,j}[\mathrm{k}'_{i,l}(a)] \subseteq \mathrm{k}'_{i,j}(a)$.

(f) If $a \in \mathrm{Arg}_i$, $b \in \mathcal{C}l_j \cap \mathrm{C}_i(a)$, $i, j, l < n$, $i \leq j$, *then* $\mathrm{eval}[\mathrm{k}'_{j,l}(b)] \subseteq \mathrm{C}_i(a)$.
If $a \in \mathcal{C}l_i$, *then* $\mathrm{eval}[\mathrm{k}'_{i,j}(a)] \subseteq \mathrm{C}_i(a)$.

(g) $\mathcal{F} := ((\mathcal{C}l_i, \prec_i)_{i<n}, (\mathrm{k}'_{i,j})_{i,j<n}, (\mathrm{length}_i)_{i<n})$ *is an n-ordinal system. We will call any n-OS structure isomorphic to $\mathcal{F}$ an* n-OS based on $\mathcal{O}$.

Proof: (a) is clear, since $\mathrm{k}(a)$ is finite.

(b): Main induction on $n-i$, Side Induction on $a \in \mathrm{Arg}_i$: Assume assertion for $b <' a$. We show $\forall \alpha \in \mathrm{C}^n_i(a) \cap \Omega_{i+1}.\alpha \subseteq \mathrm{C}_i(a)$ by side-side-induction on n: $n = 0$: clear, since $\mathrm{l}_i(a) \subseteq \mathrm{k}_i(a)$. $n \to n+1$: If $\alpha \in (\mathrm{C}^{n+1}_i(a) \setminus \mathrm{C}^n_i(a)) \cap \Omega_{i+1}$, then $\alpha = \mathrm{eval}_i(b)$ with $\mathrm{k}(b) \subseteq \mathrm{C}^n_i(a)$, by side-side-IH and, since $\mathrm{l}(b) \subseteq \mathrm{k}(b)$, $\bigcup(\mathrm{k}(b) \cap \Omega_{n+1}) \cup \mathrm{l}(b) \subseteq \mathrm{C}_i(a)$, and by an immediate induction $\mathrm{C}^l_i(b) \subseteq \mathrm{C}_i(a)$ for all $l \in \omega$, $\mathrm{C}_i(b) \subseteq \mathrm{C}_i(a)$, therefore $\alpha = \min\{\gamma \mid \gamma \notin \mathrm{C}_i(b)\} \subseteq \mathrm{C}_i(a)$.

(c) By an induction using the condition "cardinality based OFG" it follows that the cardinality of $\mathrm{C}^n_i(a)$ and therefore as well that of $\mathrm{C}_i(a)$ is $< \Omega_{i+1}$. We conclude that there exists an $\alpha < \Omega_{i+1}$ such that $\alpha \notin \mathrm{C}_i(a)$.

(d) "$\Leftarrow$" If $a <' b$, $\mathrm{k}_{i,i}(a) < \mathrm{eval}(b)$, $\mathrm{NF}(a)$, it follows easily that $\mathrm{k}(a) \subseteq \mathrm{C}(a) \subseteq \mathrm{C}(b)$, $\mathrm{eval}_i(a) \in \mathrm{C}(b) \cap \Omega_{i+1} = \mathrm{eval}_i(b)$. If $\mathrm{eval}(a) \leq \mathrm{k}_{i,i}(b)$, the assertion follows by $\mathrm{k}_{i,i}(b) < \mathrm{eval}(b)$. "$\Rightarrow$" If the right side is false, the right side holds for $\mathrm{eval}_i(b) < \mathrm{eval}_i(a)$ or $a = b$, so the left side is false.

(e): easy induction on $\mathrm{length}(a)$.

(f): First assertion by induction on $\mathrm{length}(a)$, side-induction on $\mathrm{length}(b)$. Second assertion: by $\mathrm{k}(a) \subseteq \mathrm{C}_i(a)$.

(g): by the above (in order to show $\mathrm{k}'_{i,i}(a) < \mathrm{eval}(a)$ use $\mathrm{k}'_{i,i}(a) \subseteq \mathrm{C}_i(a)$). □

Example 3.7 (a) Let

$\mathrm{Arg}'_0 := \mathrm{CNF}'_0([0,\Omega_1[,]0,\omega[)$,

$\mathrm{Arg}'_{i+1} := \mathrm{CNF}'_{i+1}([0,\Omega_{i+2}[,]0,\Omega_{i+1}[) \setminus \mathrm{CNF}'_{i+1}(\{0\},]0,\Omega_{i+1}[)$,

k'_0, l'_0 as k, l in Example 2.7 (d) (with CNF′ replaced by CNF'_0),

$\mathrm{k}'_{i+1}(\mathrm{CNF}'_{i+1}(\alpha_1,\beta_1,\dots,\alpha_m,\beta_m)) := \{\alpha_1,\dots,\alpha_m,\beta_1,\dots,\beta_m\}$,

$$\mathrm{l}'_{i+1}(\mathrm{CNF}'_{i+1}(\alpha_1,\beta_1,\dots,\alpha_m,\beta_m)) := \begin{cases}\{\alpha_1\} & \text{if } m>1 \text{ or } \beta_m>1,\\ \emptyset & \text{otherwise.}\end{cases}$$

$(|\mathrm{Arg}'_i|,\mathrm{k}'_i,\mathrm{l}'_i,<_{\mathrm{Arg}'_i})$ is a cardinality based n-OFG such that

$\mathrm{eval}_0(\mathrm{CNF}'_0(\alpha_1,n_1\dots,\alpha_m,n_m)) = \omega^{\alpha_1}n_1+\dots+\omega^{\alpha_m}n_m$,

$\mathrm{eval}_{i+1}(\mathrm{CNF}'_{i+1}(\alpha_1,\beta_1\dots,\alpha_m,\beta_m)) = \Omega_{i+1}^{\alpha_1}\beta_1+\dots+\Omega_{i+1}^{\alpha_m}\beta_m$.

(b) Let $\mathcal{S}_l$, k^l be defined as in Example 2.7 (i), but with the restriction always to ordinals $<\Omega_{n+1}$. Let Arg'_i, k'_i, l'_i be as before, $\mathcal{F}_i := \mathrm{Arg}'_i \ominus \underline{\psi}_i(\mathcal{S}_l)$, $\mathrm{k}_i \restriction \mathrm{Arg}'_i := \mathrm{k}'_i$, $\mathrm{l}_i \restriction \mathrm{Arg}'_i := \mathrm{l}'_i$, $\mathrm{k}_i(\underline{\psi}_i(a)) := \mathrm{k}^l(a)$, $\mathrm{l}_i(\underline{\psi}_i(a)) := \mathrm{k}^l(a)\cap[\Omega_i,\Omega_{i+1}[$ for $a\in|\mathcal{S}_l|$. The resulting cardinality based n-OFG can be seen as a generalization of the extended Schütte Klammer symbols.

(c) Let $f_l^n : \mathcal{S}_l \to \mathrm{Ord}$, $f_0^n(\alpha) := \alpha$, $f_{l+1}^n\binom{\beta_1\cdots\beta_m}{A_1\cdots A_m} := \Omega_n^{f_l^n(A_1)}\beta_1+\dots+\Omega_n^{f_l^n(A_m)}\beta_m$. If we restrict now $\underline{\psi}_i(\mathcal{S}^l)$ to $\underline{\psi}_i(A)$ such that $f_l^n(A)\in \mathrm{C}'_{\Omega_{i+1}}(f_l^n(A))$, where $\mathrm{C}'_{\Omega_{i+1}}(a)$ is the C-set for the function $\psi_{\Omega_{i+1}}$ as in [Buc92], but with 0, $+$, $\omega^{\cdot}$ as basic functions, we get $\mathrm{eval}_i(\underline{\psi}_i(A)) = \psi_{\Omega_{i+1}}(f_l^n(A))$. Again, with this last modification we get that in the resulting ordinal notation system $\prec_i$ and $\prec'_i$ coincide for terms of the form $\underline{\psi}_i(A)$.

Remark 3.8 *(a) The straightforward generalization of Lemma 2.8 holds as well for cardinality based n-OFG with the exception that we can conclude that $a \prec b$ is primitive recursive only for $a\in \mathrm{T}$, $b\in\mathrm{T}'$. Further under the assumptions of this generalized lemma the function corresponding to $\mathrm{k}'_{i,j}$, which takes the place of $\mathrm{k}_{i,j}$ in the resulting n-OS, is primitive recursive.*

(b) In order to replace Ω_i by Ω_i^{rec} (the i th admissible, starting with $\Omega_0 = 0$, $\Omega_1 = \omega_1^{\mathrm{ck}}$) we need some additional conditions, which essentially express that we have terms as in (a) for the higher number classes and arguments, but based on ordinals in $[0,\Omega_i^{\mathrm{rec}}[$. They can be developed in a similar way as the conditions in (a), but because of lack of space, we omit them here.

Proof of (a): As in Lemma 2.8. Only the primitive recursive determination of $r \prec s$ follows as follows: Note that $\widehat{\mathrm{k}}_{i,j}$ corresponds to $\mathrm{k}^0_{i,j}$. For $s \in \mathrm{T}'_i$ let $\mathrm{C}^j(s) := \{r \in \mathrm{T}_j \mid \mathrm{eval}(f(r)) \in \mathrm{C}_i(f(s))\}$. We determine primitive recursively, by recursion on $\mathrm{length}(r)$, for $s \in \mathrm{T}'_i$, $r \in \mathrm{T}_j$ such that $\mathrm{length}(r), \mathrm{length}(s) \leq \mathrm{length}(t)$, whether $r \in \mathrm{C}^j(s)$: If $j < i$, $r \in \mathrm{C}^j(s)$. If $j = i$, $r \in \mathrm{C}^j(s) \Leftrightarrow r \preceq \widehat{\mathrm{k}}_{i,i}(s) \vee (r \preceq' s \wedge \forall l < n(\widehat{\mathrm{k}}_{i,l}(r) \subseteq \mathrm{C}^l(s)))$ and, if $i < j$, $r \in \mathrm{C}^j(s) \Leftrightarrow \forall l < n.\widehat{\mathrm{k}}_{i,j}(r) \subseteq \mathrm{C}^l(s)$. Then in case $i = j$ it follows that $r \prec s \Leftrightarrow r \in \mathrm{C}^i(s)$ and $t \in \mathrm{T}'_i \Leftrightarrow \mathrm{NF}(f(t)) \Leftrightarrow \forall j < n.\widehat{\mathrm{k}}_{i,j}(t) \subseteq \mathrm{C}^j(t)$. □

Lemma 3.9 *(a) For the OFG in the Examples 3.7, there exists an elementary OS represented by them.*

(b) The supremum of the strength of n-OS is $|\mathrm{ID}_n|$.

Proof: (a) as for Lemma 2.9, (b) as for Theorem 2.11. □

4 Transfinitely Iterated Ordinal Systems

4.1 Definition of σ-OS

The naïve way of extending the approach used in the last section to the transfinite does not work. In the proof of well-ordering of n-OS we always reduced well-ordering of an n-OS $\mathcal{F}_n$ to $\mathcal{F}_n[A]$ for well-ordered sets $A \subseteq \mathrm{T}_0$. If we try this with n replaced by ω, we will reduce the well-ordering of an ω-OS to the well-ordering of an ω-OS, which does not work. In the proof using the accessibility predicate we had to use induction on $n - i$ instead of i, and this induction does not work if n is replaced by ω.

Proof theoretically, when moving to at least ω-iterated OS, we go beyond the strength of $\Pi^1_1 - \mathrm{CA}$, so Π^1_1-arguments, which we used in the intuitive well-ordering proofs there, no longer work.

Our proof proceeded in some sense by induction on the lexicographic ordering of $(\mathrm{T}_0, \mathrm{T}_1, \ldots, T_{n-1})$. Whereas the lexicographic ordering on tuples of fixed length is well-ordered, if the underlying orderings are, this no longer holds for tuples of arbitrary length. However, for sequences descending along some well-ordering it does. The solution for our problem is now the following: Introduce a function, level : $\mathrm{T}_\mu \to \mathrm{L}$, where $(\mathrm{L}, \prec_\mathrm{L})$ is an ordering. Let $\mathrm{T}^{\leq l}_\mu$ ($\mathrm{T}^{\prec l}_\mu$) be the restriction of T_μ to those a such that $\mathrm{level}_\mu(a) \leq l$ ($\mathrm{level}_\mu(a) \prec_\mathrm{L} l$). Now let $\mathrm{T}^{\leq l}_\mu$ refer only to $\mathrm{T}^{\prec l}_\nu$ (if $\mu < \nu$), i.e. $\mathrm{level}_\nu[\mathrm{k}_{\mu,\nu}(t)] \prec_\mathrm{L} \mathrm{level}_\mu(t)$ for $\mu < \nu$, and assume $\mathrm{T}^{\leq l}_\mu \sqsubseteq \mathrm{T}_\mu$, i.e. $r \prec_\mu s \to \mathrm{level}_\mu(r) \preceq_\mathrm{L} \mathrm{level}_\mu(s)$. Then we will proceed by working on sequences $\mathrm{T}^{\leq l_1}_{\mu_1}, \mathrm{T}^{\leq l_2}_{\mu_2}, \ldots$ such that $l_1 \succ_\mathrm{L} l_2 \succ_\mathrm{L} \cdots$.

We now need that $(\mathrm{L}, \prec_\mathrm{L})$ is a well-ordering. But to demand this directly would be too strong a requirement. What suffices is, to have functions $\widetilde{\mathrm{k}}_\mu$:

$L \to T_\mu$ such that, if we define $L \restriction_{\nu<\sigma} B_\nu$ as usual, but referring to $\widetilde{k}_\mu$, well-ordering of $L \restriction_{\nu<\sigma} B_\nu$ reduces to well-ordering of B_ν. We only need additionally to demand that, if we relativize the sets of terms and the levels to some set B, the levels of the terms in the relativized sets are in the relativized set of levels, i.e. $\forall a \in T_\nu . \widetilde{k}_\mu(\text{level}(a)) \subseteq k_{\nu,\mu}(a)$.

A last necessary modification is that, since we are no longer introducing first the nth OS completely, then the $(n-1)$th OS completely, etc., we need to demand that there is a descent in length when moving to higher components, i.e. $\nu > \mu \to \text{length}[k_{\mu,\nu}(r)] < \text{length}(r)$.

Apart from the conditions stated before, we need the (naïve) generalization of the conditions for n-OS, which will include that for every a, only finitely many ν satisfy $k_{\mu,\nu}(a) \neq \emptyset$. We can now (although for the concrete examples this is not necessary, but it might be useful in the future) weaken the condition (OS 2) in the sense that if $r \prec_\mu s$, then one new alternative is that $\text{level}_\mu(r) \prec_L \text{level}_\mu(s)$.

Assumption 4.1 *In the following we assume that some well-ordering $(\Sigma, <)$ of order type $\sigma > 0$, where σ is an ordinal below the Bachmann-Howard ordinal, is given. Let $0 = \min_< \Sigma$. We will identify $(\Sigma, <)$ with σ, writing $\mu < \sigma$ for $\mu \in \Sigma$, Ω_μ for $\Omega_{f(\mu)}$, where $f : \Sigma \to \sigma$ is the order isomorphism, etc.. In the following we assume $\mu, \nu, \xi < \sigma$. (Note that we use the same symbol $<$ as for the ordering on ordinals).*

Definition 4.2 (a) A *σ-times iterated Ordinal-System-structure relative to* $(\Sigma, <)$, in short *σ-OS-structure* is a quadruple

$$\begin{aligned}\mathcal{F} &= ((\mathcal{G}_\mu), (k_{\mu,\nu}), (L, \prec_L, (\text{level}_\mu), (\widetilde{k}_\mu)), \text{length}_\mu) \\ &= ((\mathcal{G}_\mu)_{\mu<\sigma}, (k_{\mu,\nu})_{\mu,\nu<\sigma}, (L, \prec_L, (\text{level}_\mu)_{\mu<\sigma}, (\widetilde{k}_\mu)_{\mu<\sigma}), (\text{length}_\mu)_{\mu<\sigma})\end{aligned}$$

such that, with $\mathcal{G}_\mu = (T_\mu, \prec_\mu, \prec'_\mu)$ and $\mathcal{F}_\mu := (\mathcal{G}_\mu, k_{\mu,\mu}, \text{length}_\mu)$, we have that $\mathcal{F}_\mu$ are OS-structures, $k_{\mu,\nu} : T_\mu \to_\omega T_\nu$, $(L, \prec_L)$ is an ordering, $\widetilde{k}_\mu : L \to_\omega T_\mu$ and $\text{level}_\mu : T_\mu \to L$ $(\mu, \nu \in \Sigma)$.
We will always assume that T_μ are disjoint and L is disjoint from T_μ and write therefore $\prec$, $\prec'$, length, level instead of $\prec_\mu$, $\prec'_\mu$, length_μ, level_μ and $\prec$ instead of $\prec_L$. In the following let $\mathcal{F}$ be as above.

(b) If $\mathcal{F}$ is a σ-OS-structure as above, $A \subseteq T_\mu$, $B \subseteq T_\nu$, $A \restriction_\nu B := A \cap k_{\mu,\nu}^{-1}[B]$. If $B_\nu \subseteq T_\nu$ for all $\nu < \xi$, then $A \restriction_{\nu<\xi} B_\nu := A \cap \bigcap_{\nu<\xi} k_{\mu,\nu}^{-1}[B_\nu]$. In the same way we define for $A \subseteq L$ $A \restriction_\nu B$, $A \restriction_{\nu<\xi} B_\nu$, referring to $\widetilde{k}_\nu$ instead of $k_{\mu,\nu}$.

(c) A σ-OS-structure as above is a *σ-times iterated Ordinal System* relative to Σ, in short *σ-OS*, if for all $\mu, \nu, \xi < \sigma$ and $r, s \in T_\mu$ the following hold

(σ – OS 1) $\mathrm{k}_{\mu,\mu}(r) \prec r$;

(σ – OS 2) $\mu \leq \nu \rightarrow \mathrm{length}[\mathrm{k}_{\mu,\nu}(r)] < \mathrm{length}(r)$;

(σ – OS 3) if $r \prec s$, then $r \preceq \mathrm{k}_{\mu,\mu}(s) \vee r \prec' s \vee \mathrm{level}(r) \prec \mathrm{level}(s)$;

(σ – OS 4) if $\xi < \nu$, then $\mathrm{k}_{\nu,\xi} \circ \mathrm{k}_{\mu,\nu} \subseteq \mathrm{k}_{\mu,\xi}$;

(σ – OS 5) $\mathrm{k}_{\mu,\xi'}(r) = \emptyset$ for almost all $\xi' < \sigma$;

(σ – OS 6) if $\mu < \nu$, then $\mathrm{level}[\mathrm{k}_{\mu,\nu}(r)] \prec \mathrm{level}(r)$;

(σ – OS 7) if $r \prec s$, then $\mathrm{level}(r) \preceq \mathrm{level}(s)$;

(σ – OS 8) $\widetilde{\mathrm{k}}_\nu(\mathrm{level}(r)) \subseteq \mathrm{k}_{\mu,\nu}(r)$;

(σ – OS 9) if $A_\xi \subseteq \mathrm{T}_\xi$ are $\prec$-well-ordered ($\xi < \sigma$), then

$(\mathrm{T}_\nu \restriction_{\mu<\sigma} A_\mu, \prec')$ and $(\mathrm{L} \restriction_{\mu<\sigma} A_\mu, \prec)$ are well-ordered, too.

(d) A σ-OS $\mathcal{F}$ is *well-ordered*, if $(\mathrm{T}_0, \prec)$ is (where $0 = \min_< \Sigma$), and its *order type* is that of $(\mathrm{T}_0, \prec)$. It is *primitive recursive*, if the involved sets (including Σ) are primitive recursive subsets of the natural numbers (parametrized in Σ, i. e. $t \in \mathrm{T}_\mu$ is primitive recursive in t and μ), all functions, relations (including $<$) are primitive recursive, the finitely many ν such that $\mathrm{k}_{\mu,\nu}(a) \neq \emptyset$ can be computed primitive recursively from μ, ν, a, and all properties (including linearity of $<$ and that the chosen ν such that $\mathrm{k}_{\mu,\nu}(a) \neq \emptyset$ are the only ones) except the well-ordering condition can be shown in primitive recursive arithmetic. It is *elementary*, if additionally the well-ordering condition follows in PRA in the sense of reducibility of transfinite induction in PRA to transfinite induction on $\{(\mu, a) \mid \mu < \sigma \wedge a \in \mathrm{T}_\mu\}$ with the lexicographic ordering $(\mu, a) < (\nu, b) \Leftrightarrow \mu < \nu \vee (\mu = \nu \wedge a <_\mu b)$.

(e) "Two σ-OS are isomorphic" is defined as for n-OS.

Comparison with n-OS. n-OS have now been defined twice, since σ can be finite. So more precisely we have to distinguish between "finite n-OS" and "transfinite n-OS". However, one can easily see that a finite n-OS $\mathcal{F}$ can be considered as a special cases of a transfinite n-OS. Let $\Sigma := \{0, \dots, n-1\}$, $<$ be the usual ordering on Σ, $\mathrm{L} := \{n-1, n-2, \dots, 0\}$, $\widetilde{\mathrm{k}}_i(j) := \emptyset$, $i \prec j :\Leftrightarrow j < i$, $\mathrm{level}(r) := i$ for $r \in \mathrm{T}_i$. Replace $\mathrm{length}_i(r)$ by $\mathrm{length}'_i(r) := \max\{\mathrm{length}_i(r)\} \cup \bigcup_{j \geq i} \mathrm{length}'_j[\mathrm{k}_{i,j}(r)]$ (defined by recursion on $n - i$ side-recursion on $\mathrm{length}(r)$). One can easily see, that, if we extend the structure by the above and replace length by length$'$, we get a transfinite n-OS. This illustrates again why a naïve generalization of n-OS to ω-OS does not work: We get the reverse ordering of the natural numbers, which is not well-ordered.

4.2 Constructive Well-ordering Proof.

Theorem 4.3 *(a) Every σ-OS is well-ordered.*

(b) Every elementary σ-OS has order type below $|\mathrm{ID}_\sigma| = \psi(\epsilon_{\Omega_\sigma'+1})$.

Proof: (a): Define, inductively by recursion on $\mu \in \Sigma$, $\mathrm{M}_\mu, \mathrm{Acc}_\mu \subseteq \mathrm{T}_\mu$:
$\mathrm{M}_\mu := \mathrm{T}_\mu \lceil_{\nu<\mu} \mathrm{Acc}_\nu$, $\mathrm{Acc}_\mu := \mathrm{Acc}_\prec(\mathrm{M}_\mu)$.
$\mathrm{Acc}'_\mu := \mathrm{T}_\mu \lceil_{\nu<\sigma} \mathrm{Acc}_\nu$, $\mathrm{Acc}'_\mathrm{L} := \mathrm{L} \lceil_{\nu<\sigma \mathrm{Acc}_\nu}$.
Further, if $l \in \mathrm{L}$, $A \subseteq \mathrm{T}_\mu$ $A^{\preceq l} := \{x \in A \mid \mathrm{level}(x) \preceq l\}$, we similarly define $A^{\prec l}$.

$(\mathrm{Acc}_\mu, \prec)$, $(\mathrm{Acc}'_\mu, \prec')$ and $(\mathrm{Acc}'_\mathrm{L}, \prec)$ are well-ordered.

We show by induction on $l \in \mathrm{Acc}'_\mathrm{L}$ that

$$\forall l \in \mathrm{Acc}'_\mathrm{L}.\forall \mu < \sigma.\mathrm{M}_\mu^{\preceq l} \subseteq \mathrm{Acc}_\mu \ ,$$

and assume l according to induction. We show

$$\forall s \in \mathrm{Acc}'^{\preceq l}_\mu .s \in \mathrm{Acc}_\mu$$

by (side-)induction on $(\mathrm{Acc}'_\mu, \prec')$ and assume s according to induction: $s \in \mathrm{M}_\mu$. We define $\mathrm{C}^\nu(s) \subseteq \mathrm{T}_\nu$ $(\nu < \sigma)$:
For $\nu < \mu$, $\mathrm{C}^\nu(s) := \mathrm{Acc}_\nu$.
$\mathrm{C}^\mu(s) := \{r \in \mathrm{M}_\mu \mid r \prec s\}$.
If $\mu < \nu$, $\mathrm{C}^\nu(s) := \{r \in \mathrm{T}_\nu \mid \mathrm{level}(r) \prec \mathrm{level}(s) \wedge \forall \xi < \nu.\mathrm{k}_{\nu,\xi}(r) \subseteq \mathrm{C}^\xi(s)\}$.
Note that by (σ-OS 4) for $\nu > \mu$
$\mathrm{C}^\nu(s) = \{r \in \mathrm{T}_\nu \mid \mathrm{level}(r) \prec \mathrm{level}(s) \wedge \forall \xi < \mu(\mathrm{k}_{\nu,\xi}(r) \subseteq \mathrm{Acc}_\xi) \wedge \mathrm{k}_{\nu,\mu}(r) \prec s \wedge$
$\forall \xi(\mu < \xi < \nu \rightarrow \mathrm{level}[\mathrm{k}_{\nu,\xi}(r)] \prec \mathrm{level}(s))\}$.
The last equation allows us to define $\mathrm{C}^\nu(s)$ in ID_σ as needed in (b).

We prove

$$\forall \nu(\mu \leq \nu \rightarrow \forall \xi < \sigma.\mathrm{k}_{\nu,\xi}[\mathrm{C}^\nu(s)] \subseteq \mathrm{C}^\xi(s)) \qquad (*)$$

by induction on ξ:
For $\xi < \nu$ this follows by the definition of $\mathrm{C}^\nu(s)$. Otherwise for $\xi' < \xi$, $\mathrm{k}_{\xi,\xi'}[\mathrm{k}_{\nu,\xi}[\mathrm{C}^\nu(s)]] \subseteq \mathrm{k}_{\nu,\xi'}[\mathrm{C}^\nu(s)] \subseteq \mathrm{C}^{\xi'}(s)$ by IH. Let $r \in \mathrm{C}^\nu(s)$. In case $\mu = \nu = \xi$ we have $\mathrm{k}_{\nu,\xi}(r) \prec r \prec s$ and, if $\mu < \xi$, it follows that $\mathrm{level}[\mathrm{k}_{\nu,\xi}(r)] \preceq \mathrm{level}(r) \preceq \mathrm{level}(s)$, and one of the $\preceq$ is actually $\prec$, so in both cases we get the assertion.

We show by (side-side-)induction on $\mathrm{length}(r)$ simultaneously for all ν that

$$\forall \nu.\forall r \in \mathrm{C}^\nu(s).r \in \mathrm{Acc}_\nu$$

and assume r according to induction, $r \in \mathrm{C}^\nu(s)$. If $\nu < \mu$, $r \in \mathrm{Acc}_\nu$. Assume $\mu \leq \nu$. By the side-side-IH and $(*)$ it follows that for all ξ, $\mathrm{k}_{\nu,\xi}(r) \subseteq \mathrm{Acc}_\xi$ and $\widetilde{\mathrm{k}}_\xi(\mathrm{level}(r)) \subseteq \mathrm{k}_{\nu,\xi}(r) \subseteq \mathrm{Acc}_\xi$, $r \in \mathrm{Acc}'_\nu$, $\mathrm{level}(r) \in \mathrm{Acc}'_\mathrm{L}$.

If $\mu = \nu$, $r \prec s$, $r \preceq \mathrm{k}_{\mu,\mu}(s) \subseteq \mathrm{Acc}_\mu$, $r \in \mathrm{M}_\mu$, $r \in \mathrm{Acc}_\mu$ or $r \prec' s$, $r \in \mathrm{Acc}'_\mu$, $\mathrm{level}(r) \preceq \mathrm{level}(s) \preceq l$, and by side-IH $r \in \mathrm{Acc}_\mu$, or $\mathrm{level}(r) \prec \mathrm{level}(s)$ and as in the next case "$\mu < \nu$" the assertion follows.

If $\mu < \nu$, $\mathrm{level}(r) \prec \mathrm{level}(s) \preceq l$, $\mathrm{level}(r) \in \mathrm{Acc}'_{\mathrm{L}}$, $r \in \mathrm{M}_\nu$, by main the IH $r \in \mathrm{Acc}_\nu$.

Now it follows that $\forall r \in \mathrm{M}_\mu(r \prec s \to r \in \mathrm{Acc}_\mu)$, and, since $s \in \mathrm{M}_\mu$, $s \in \mathrm{Acc}(\mathrm{M}_\mu) = \mathrm{Acc}_\mu$, and the side-induction is complete.

Now by induction on length(s) it follows that $\forall s \in \mathrm{M}^{\preceq l}_{\overline{\mu}}.s \in \mathrm{Acc}_\mu$: If $s \in \mathrm{M}^{\preceq l}_{\overline{\mu}}$ we first show $\forall \nu.\mathrm{k}_{\mu,\nu}(s) \subseteq \mathrm{Acc}_\mu$. For $\nu < \mu$ this follows by assumption. For $\mu \leq \nu$ we show this by induction on ν. With the usual argument we get using the IH $\mathrm{k}_{\mu,\nu}(s) \subseteq \mathrm{M}^{\preceq l}_{\overline{\nu}}$, and therefore by IH $\mathrm{k}_{\mu,\nu}(s) \subseteq \mathrm{Acc}_\nu$. Therefore $s \in \mathrm{Acc}'_\mu$ and by the proven statement of the side-induction it follows that $s \in \mathrm{Acc}_\mu$. Therefore the main-IH is completed.

It now follows by induction on length(r), that simultaneously for all μ, $\forall r \in \mathrm{T}_\mu(r \in \mathrm{Acc}_\mu \wedge \mathrm{level}(r) \in \mathrm{Acc}'_{\mathrm{L}})$: By IH $\mathrm{k}_{\mu,\nu}(r) \subseteq \mathrm{Acc}_\nu$, $\widetilde{\mathrm{k}}_\nu(\mathrm{level}(r)) \subseteq \mathrm{Acc}_\nu$, $\mathrm{level}(r) \in \mathrm{Acc}'_{\mathrm{L}}$, $r \in \mathrm{M}^{\preceq \mathrm{level}(r)}_{\overline{\mu}}$, $r \in \mathrm{Acc}_\mu$.

Therefore it follows T_μ is well-ordered and we are done.

(b) The proof of (a) can be carried out in ID_σ. (Note, that transfinite induction over σ is one of the axioms of ID_σ). □

4.3 σ-Ordinal Function Generators

The ordinal function generators referring to σ-OS are now defined similarly as for n-OFG. The only difference is that in the definition of $\mathrm{C}_\mu(a)$ we will refer only to $b \in \mathrm{Arg}_\nu$ $(\nu > \mu)$ such that $\mathrm{level}_\nu(b) < \mathrm{level}_\mu(a)$, which is the obvious adaptation of the principles for σ-OS to OFG.

Definition 4.4 (a) A *σ-ordinal function generator*, in short *σ-OFG* is a triple $\mathcal{O} := (\mathcal{A}, \mathcal{L}, \widetilde{\mathrm{k}})$, where $\mathcal{A} = (\mathrm{Arg}_\mu, \mathrm{k}_\mu, \mathrm{l}_\mu, <'_\mu, \mathrm{level}_\mu)_{\mu<\sigma}$, $\mathcal{L} = (\mathrm{L}, <_{\mathrm{L}})$, Arg_μ are classes (in set theory), $\mathrm{k}_\mu, \mathrm{l}_\mu : \mathrm{Arg}_\mu \to_\omega \mathrm{Ord}$, $\mathrm{level}_\mu : \mathrm{Arg}_\mu \to \mathrm{L}$, $\forall a \in \mathrm{Arg}_\mu.\mathrm{l}_\mu(a) \subseteq \mathrm{k}_\mu(a)$, $<'_\mu$ is a well-ordered relation on Arg_μ, $\forall a, b \in \mathrm{Arg}_\mu(a <'_\mu b \to \mathrm{level}_\mu(a) \leq'_{\mathrm{L}} \mathrm{level}_\mu(b))$, $\mathcal{L}$ is a well-ordering, $\widetilde{\mathrm{k}} : \mathrm{L} \to_\omega \mathrm{Ord}$, $\forall a \in \mathrm{Arg}_\mu.\widetilde{\mathrm{k}}[\mathrm{level}_\mu(a)] \subseteq \mathrm{k}_\mu(a)$.

We assume in the following that Arg_μ and L are disjoint and therefore omit the index μ in k, l, $<'$, level and the index L in $<'$. Let $\mathcal{O}$ always be as above.

(b) $\Omega_0 := 0$, $\Omega_\mu := \aleph_\mu$ otherwise.

(c) A σ-OFG is *cardinality based* iff for all $B \subseteq \mathrm{Ord}$ countable $\mathrm{k}^{-1}_{\overline{\mu}}(B)$ is countable and $\forall \mu < \sigma.\forall a \in \mathrm{Arg}_\mu.\mathrm{l}_\mu(a) \subseteq \mathrm{k}_\mu(a) \cap [\Omega_\mu, \Omega_{\mu+1}[$. In this case we define for $a \in \mathrm{Arg}_\mu$, $\mathrm{k}_{\mu,\nu}(a) := \mathrm{k}(a) \cap [\Omega_\nu, \Omega_{\nu+1}[$ and for $a \in \mathrm{L}$ $\widetilde{\mathrm{k}}_\nu(a) := \mathrm{k}(a) \cap [\Omega_\nu, \Omega_{\nu+1}[$.

(d) If $\mathcal{O}$ as above is a cardinality based σ-OFG, we define for $a \in \mathrm{Arg}_\mu$ simultaneously for all $\mu < \sigma$ by recursion on $\mathrm{level}_\mu(a) \in \mathrm{L}$, side-recursion on $(<', \mathrm{Arg}_\mu)$, $\mathrm{eval}_\mu(a) \in \mathrm{Ord}$ and subsets $\mathrm{C}_\mu(a) \subseteq \mathrm{Ord}$ by:

$$\begin{aligned}
\mathrm{C}_\mu(a) &:= \bigcup_{n=0}^{\omega} \mathrm{C}^n_\mu(a) \text{ where} \\
\mathrm{C}^0_\mu(a) &:= [0, \Omega_\mu[\cup (\bigcup (\mathrm{k}_\mu(a) \cap \Omega_{\mu+1})) \cup \mathrm{l}_\mu(a), \\
\mathrm{C}^{l+1}_\mu(a) &:= \mathrm{C}^l_\mu(a) \cup \\
&\quad \bigcup_{\nu \geq \mu} \{\mathrm{eval}_\mu(b) \mid b \in \mathrm{Arg}_\nu \wedge \\
&\qquad ((\nu = \mu \wedge b <' a \wedge \mathrm{level}(b) = \mathrm{level}(a)) \vee \\
&\qquad (\mathrm{level}(b) < \mathrm{level}(a))) \wedge \mathrm{k}(b) \subseteq \mathrm{C}^l_\mu(a) \wedge \mathrm{NF}_\nu(b)\},
\end{aligned}$$

where $\mathrm{NF}_\nu(b) :\Leftrightarrow \mathrm{k}(b) \subseteq \mathrm{C}_\nu(b)$.
$\mathrm{eval}_\mu(a) := \min\{\alpha \mid \alpha \notin \mathrm{C}_\mu(a)\}$.

(e) Let $\mathcal{O}$ be a cardinality based σ-OFG. Referring to the definition of $\mathrm{NF}(b)$ as above and assuming that $\mathrm{eval} \restriction \mathrm{NF}_\mu$ is injective, which will be shown later, NF_μ, Cl_μ, Cl, $\mathrm{Arg}[Cl]_\mu$ and $\mathrm{Arg}[Cl]$ are defined as in Definition 3.5.
Further we define $\mathrm{k}^0_{\mu,\nu}, \mathrm{k}'_{\mu,\nu} : \mathrm{Arg}[Cl]_\mu \rightarrow_\omega Cl_\nu$ by
$\mathrm{k}^0_{\mu,\nu}(b) := \mathrm{eval}_\nu^{-1}[\mathrm{k}_{\mu,\nu}(b)] \cap Cl_\nu$, $\mathrm{k}'_{\mu,\nu}(b) := \mathrm{k}^0_{\mu,\nu}(b) \cup \bigcup_{\nu \leq \xi < \sigma} \mathrm{k}'_{\xi,\nu}[\mathrm{k}^0_{\mu,\xi}(b)]$, length as in Definition 3.5 and $\mathrm{level}'_\mu : Cl_\mu \rightarrow \mathrm{L}$ by

$$\mathrm{level}'_\mu(r) := \max\{\mathrm{level}_\mu(r)\} \cup \mathrm{level}'_\mu[\mathrm{k}_{\mu,\mu}(r)] \ .$$

(f) For $r, s \in \mathrm{Arg}_\mu$ we define $r \prec s :\Leftrightarrow r \prec_\mu s :\Leftrightarrow \mathrm{eval}_\mu(r) < \mathrm{eval}_\mu(s)$.

Lemma 4.5 *Let $\mathcal{O}$ be a cardinality based σ-OFG.*

(a) Lemma 3.6 (a) - (f) hold mutatis mutandis.

(b)

$$\begin{aligned}
\mathcal{F} := {} & ((Cl_\mu, \prec_\mu)_{\mu<\sigma}, (\mathrm{k}'_{\mu,\nu})_{\mu,\nu<\sigma}, (\mathrm{length}_i)_{i<n}, (\mathrm{L}, <_{\mathrm{L}}), \\
& (\mathrm{level}'_\mu)_{\mu<\sigma}, (\mathrm{length}_\mu)_\mu)
\end{aligned}$$

is a σ-ordinal system. We will call any σ-OS-structure isomorphic to $\mathcal{F}$ a σ-ordinal system *based on $\mathcal{O}$.*

Proof: (a): We write (a).(x) for the assertion corresponding to Lemma 3.6(x). (a).(a): clear. (a).(b): Similarly as in Lemma 3.6 (b). In the argument, which showed there $\mathrm{C}^l_i(b) \subseteq \mathrm{C}_i(a)$ and in the new context now shows $\mathrm{C}^l_\mu(b) \subseteq \mathrm{C}_\mu(a)$, we use the fact that $b <' a$, therefore $\mathrm{level}(b) \leq \mathrm{level}(a)$. (a).(c): as before;

(a).(d): as before, using $a <' b \to \text{level}(a) \leq \text{level}(b)$; (a).(e): as before by an easy induction on length(a); (a).(f): as before.
(b): The only problems are (σ-OS 6), (σ-OS 7): We show for $r, s \in Cl$, that, if $\text{eval}_\nu(r) \in \text{C}_\mu(s)$, then, in case $\nu = \mu$, $\text{level}'_\nu(r) \leq \text{level}'_\mu(s)$, and, in case $\nu > \mu$, $\text{level}'_\nu(r) < \text{level}'_\mu(s)$, by main induction on length(s), side-induction on length(r), which is immediate. □

Example 4.6 (a) The straight forward generalization of Example 3.7 (a) together with $\text{L} := \{0\}$, $\widetilde{\text{k}}_\mu(0) := \emptyset$, $\text{level}_\mu(a) := 0$ yields a cardinality based σ-OFG for the Cantor normal form with basis ω ($\mu = 0$) and Ω_μ ($\mu > 0$).

(b) Example 3.7 (b) generalizes again to a cardinality based σ-OFG with $\text{L} := \mathcal{S}_l$ as in Example 2.7 (i), $\widetilde{\text{k}}(a) := \text{k}^l(a)$, $\text{level}(\Sigma'_\mu(\vec{t})) := 0$ (where $0 := () \in \text{L}$), $\text{level}(\underline{\psi}_\mu(A)) := (A)$ (only in case $l = 0$ we have to modify L in order to make it disjoint from T_μ). The resulting system can be seen as a further generalization of the extended Schütte Klammer symbols. Note that, whereas for n-OS for all terms $t = \underline{\psi}(A) \in \text{T}'$ $t \in \text{T}$ holds (in the fixed point free version), this is no longer the case here.

(c) If f^l_σ is defined similarly and we apply a similar restriction as in Example 3.7 (c), then we get $\text{eval}_\mu(\underline{\psi}_\mu(a)) = \psi_{\Omega_{\mu+1}}(f^l_\sigma(a))$ with $\psi_{\Omega_{\mu+1}}$ the usual ψ-function based on $0, +, \omega^{\cdot}$.

Remark 4.7 *Make the assumptions of the straight forward generalization of Lemma 2.8 to cardinality based σ-OFG. Additionally assume there are sets $\widehat{\text{L}} \subseteq \widehat{\text{L}}' \subseteq \widehat{\text{L}}''$ corresponding to* L *such that $\widehat{\text{L}}''$ is primitive recursive, a function $f : \widehat{\text{L}} \to \text{L}$ and primitive recursive functions $\widehat{\widetilde{\text{k}}}_\mu : \widehat{\text{L}}'' \to_\omega \text{T}''_\mu$, $\widehat{\text{level}}_\mu : \text{T}''_\mu \to \widehat{\text{L}}$ corresponding to $\widetilde{\text{k}}_\mu$, level_μ. Define for $l, l' \in \text{L}$, $l \prec_\text{L} l' :\Leftrightarrow f(l) <_\text{L} f(l')$. Assume the adaptation of condition (b) of Lemma 2.8 for the new structure and (a), (c) - (f) for $\widehat{\text{L}}$, $\widehat{\widetilde{\text{k}}}$ instead of* T, $\widehat{\text{k}}$, *where appropriate, and omitting* NF. *(E.g. the adaption of condition (d) to* L *reads: If $A_\mu \subseteq \text{T}'_\mu$, $f \restriction A_\mu$ is injective ($\mu < \sigma$), then $f \restriction \{t \in \text{L}' \mid \forall\mu < \sigma.\widehat{\widetilde{\text{k}}}_\mu(t) \subseteq A_\mu\}$ is injective.) Then the conclusion of this lemma generalized to our setting holds as well in the weakened version of Remark 3.8, and additionally $\widehat{\text{L}}$, $\widehat{\text{L}}'$, $\prec_\text{L}$ are primitive recursive.*

Proof: As before. □

Theorem 4.8 *(a) For every OFG in the Example 4.6 there exists an elementary OS based on it.*

(b) The supremum of the strength of σ-OS is $|\text{ID}_\sigma|$.

Proof: As for Theorem 3.9. □

References

[BS88] W. Buchholz and K. Schütte. *Proof Theory of Impredicative Subsystems of Analysis.* Bibliopolis, Naples, 1988.

[Buc86] W. Buchholz. A new system of proof-theoretic ordinal functions. *Ann. Pure and Appl. Logic*, 32:195 – 207, 1986.

[Buc92] W. Buchholz. A simplified version of local predicativity. In P. Aczel, H. Simmons, and S. S. Wainer, editors, *Proof Theory. A selection of papers from the Leeds Proof Theory Programme 1990*, pages 115 – 147, Cambridge, 1992. Cambridge University Press.

[Gir] J.-Y. Girard. Proof theory and logical complexity. Handwritten notes, 1135 pp.

[Gir80] J.-Y. Girard. Proof theoretic investigations of inductive definitions I. In E. Engeler, H. Läuchli, and V. Strassen, editors, *Logic & Algorithms. An international symposium held in honour of Ernst Specker*, pages 201 – 236, Zurich, Switzerland, 1980.

[Gir81] J.-Y. Girard. Π^1_2-Logic, part 1: Dilators. *Ann. Pure and Appl. Logic*, 21:75 – 219, 1981.

[Poh89] W. Pohlers. *Proof Theory. An introduction*, volume 1407 of *Springer Lecture Notes in Mathematics.* Springer, Berlin, Heidelberg, New York, 1989.

[RW93] Michael Rathjen and Andreas Weiermann. Proof–theoretic investigations on Kruskal's theorem. *Annals of Pure and Applied Logic*, 60:49–88, 1993.

[Sch54] K. Schütte. Kennzeichnung von Ordinalzahlen durch rekursiv definierte Funktionen. *Math. Ann.*, 127:16–32, 1954.

[Sch77] K. Schütte. *Proof Theory.* Springer, Berlin, Heidelberg, New York, 1977.

[Sei94] M. Seisenberger. Das Ordinalzahlbezeichnungssystem $\mathrm{OT}(\vartheta)$ und seine Verwendung im Beweis von Kruskals Satz. Master's thesis, Universität München, March 1994.

[Set93] A. Setzer. *Proof theoretical strength of Martin-Löf Type Theory with W-type and one universe.* PhD thesis, Universität München, 1993.

[Set98a] A. Setzer. An introduction to well-ordering proofs in Martin-Löf's type theory. To appear in: G. Sambin, J. Smith (Eds.), *Twenty-five years of constructive type theory*, Oxford University Press, 1998.

[Set98b] A. Setzer. Well-ordering proofs for Martin-Löf Type Theory. *Ann. Pure App. Logic*, 92:113 – 159, 1998.

POLISH GROUP TOPOLOGIES

SŁAWOMIR SOLECKI[1]

Department of Mathematics, Indiana University
Bloomington IN 47405
e-mail: ssolecki@indiana.edu

In what circumstances does a given group carry a Polish group topology? (A metric, separable, complete topology is called Polish, and a topology is a group if multiplication and inverse are continuous.) Two natural interpretations of this question can be considered.

1. We consider a group equipped only with its algebraic structure. This algebraic structure may or may not be compatible with the possibility of defining a Polish group topology on the group.

2. The group is endowed with a σ-algebra Σ of subsets, and the multiplication (regarded as function of two variables) and inverse operations are assumed to be measurable with respect to Σ, that is, preimages of sets from Σ are in the product σ-algebra $\Sigma \times \Sigma$ in case of multiplication and in Σ in case of inverse. Now, we would like to investigate the possibility of putting a Polish group topology on the group whose Borel sets coincide with Σ.

Problem 2 is more "canonical" than 1 in the following sense. A group can carry many Polish group topologies compatible with its algebraic structure. Take for example the reals, $\mathbf{R}$, with their natural topology and $\mathbf{R}^\omega$ with the product topology. Both these groups are topological groups with their respective Polish topologies. Now, $\mathbf{R}$ and $\mathbf{R}^\omega$ are isomorphic as groups (both of them are linear spaces over the rationals, $\mathbf{Q}$, with bases of cardinality continuum) but the topologies are clearly different ($\mathbf{R}$ is locally compact while $\mathbf{R}^\omega$ is not). On the other hand, if a group is equipped with a σ-algebra Σ as in 2, then the Polish group topology compatible with Σ (as in 2) is unique. (This follows from the fact that a Borel isomorphism between two groups with Polish group topologies is continuous [K1, 9.10].)

Section i is concerned with problem i for $i = 1, 2$. Section 2 contains also an application to the study of the equivalence relation of mutual absolute continuity between Borel probability measures on a Polish space.

---

[1]Partially supported by an NSF grant.

The material in the present paper overlaps only partially (the first half of Section 2) with my talk at Logic Colloquium 97.

1. The abstract problem.

1.1. In this section we consider the problem of finding algebraic obstacles for a given group to carry a Polish group topology. These considerations were inspired by a question of Kechris, still unresolved, of whether the free group with $2^{\aleph_0}$ generators carries a Polish group topology (we show that this is impossible for the free *abelian* group with $2^{\aleph_0}$ generators) and by a result of Woodin that the category algebra does not carry a Polish topology which makes the boolean operations continuous (we reprove this result below).

The results in this section are negative. The main theme is that it is very difficult for a subgroup of a direct sum of countable abelian groups to carry a Polish group topology. For example, as mentioned above, such a topology does not exists on any uncountable free abelian group. We give an application of this fact to model theory pointed out by Kechris. Corollaries to the main results in this section can be viewed as topological generalizations of algebraic theorems due to Baer, Łoś, and Fuchs (see Remark 1.10).

Another example of an interesting group that does not carry a Polish group topology was recently found by Hjorth [H1].

1.2. First we present a very simple lemma which can be applied in some interesting cases. A special case of Corollary 1.2(ii) was first observed by Louveau and Corollary 1.2(iii) was proved by Woodin. The argument given below is however different from the original proofs.

Lemma 1.1. *Let G be a Polish group, and let $H_n < G$, $n \in \omega$, be analytic subgroups of G with $G \setminus \bigcup_n H_n$ countable. Then for some n, $|G/H_n| \leq \aleph_0$.*

Proof. By the Baire category theorem, since $G \setminus \bigcup_n H_n$ is countable, there exists n such that H_n is not meager and hence, by Pettis' theorem (see [K, 9.9, 9.11]), open. But then $|G/H_n| \leq \aleph_0$.

Corollary 1.2. *(i) $\bigoplus_n (\mathbf{Z}_{2^n})^\omega$, where $\mathbf{Z}_{2^n}$ is the cyclic group of rank 2^n, does not carry a Polish group topology.*

(ii) An uncountable free (or free abelian) group does not carry a Polish group topology which makes some set freely generating the group analytic.

(iii) (Woodin) The category algebra of a Polish space without isolated points does not carry a Polish topology which makes the boolean operations continuous.

Proof. (i) Assume $G = \bigoplus_n (\mathbf{Z}_{2^n})^\omega$ can be equipped with a Polish group topology. Let $H_n = \{g \in G : 2^n g = 0\}$. Then each H_n is closed (since multiplication is continuous), $\bigcup_n H_n = G$, and clearly $|G/H_n| = 2^{\aleph_0}$, which contradicts Lemma 1.1.

(ii) Let G be an uncountable free (or free abelian) group equipped with a Polish group topology, and let A be an analytic subset of G freely generating G. Let A_n, $n \in \omega$, be uncountable, pairwise disjoint, relatively closed subsets of A, and let $H_n = \langle \bigcup_{i=0}^{n} A_i \rangle$. Then each H_n is an analytic subgroup of G, $\bigcup_n H_n = G$, and G/H_n is uncountable since two distinct elements of A_{n+1} lie in distinct left cosets of H_n. Thus we have a contradiction with Lemma 1.1.

(iii) Let X be a Polish space without isolated points and let CAT be its category boolean algebra, that is, the family of all Baire measurable subsets of X regarded modulo meager sets with the natural boolean operations. Assume that CAT admits a Polish topology which makes the boolean operations continuous. Note that CAT is a Polish group with symmetric difference as addition. Fix a basis $\{W_n : n \in \omega\}$ for the topology on X consisting of nonempty sets. Let $H_n = \{a \in \text{CAT} : a \cap [W_n] = [\emptyset]\}$. Here $[U]$ is the equivalence class of U in CAT. Note that all the H_n's are closed subgroups of CAT (since $\cap$ is continuous), $\bigcup_n H_n = \text{CAT} \setminus \{[X]\}$, and $|\text{CAT}/H_n| = 2^{\aleph_0}$, and we again have a contradiction with Lemma 1.1.

The idea of the proofs of Proposition 1.3 and Corollary 1.4 is related to the argument used by Łoś in [Łś] to obtain a certain algebraic result. We will need some new notation. For an abelian group G, let Div(G) denote the maximal divisible subgroup of G and let G_m, $m \in \omega$, denote the group of all elements annihilated by m, that is, $G_m = \{g \in G : mg = 0\}$. Note that if G is a topological group, then G_m is closed.

Proposition 1.3. *Let G be a metric complete abelian group. Then precisely one of the following holds:*

(i) G/G_m is discrete for some $m \in \omega$;

(ii) there exists a compact uncountable set $K \subseteq G$ and a countable subgroup $Q < G$ such that $K/Q \subseteq$ Div(G/Q).

Proof. If some G_m contains a neighborhood of 0, then G_m is open and G/G_m is discrete, that is, (i) holds. So assume that no G_m contains an open neighborhood of 0. We deduce (ii). Let d be an invariant, so complete, metric on G. We recursively construct $y_n \in G$ so that

$$(*) \qquad d(n!y_n, 0) > 2\sum_{k>n} k!d(y_k, 0).$$

Such a choice is possible since by our assumption for any $0 < m \in \omega$ and any $\epsilon > 0$, we can find $y \in G$ such that $d(y, 0) < \epsilon$ and $my \neq 0$. Define $f : \mathcal{P}(\omega) \to G$ by $f(x) = \sum_{n\in x} n!y_n$. ($\mathcal{P}(\omega)$ is the power set of ω, which we think of here as identified in the natural way with $\{0,1\}^\omega$, and which makes it a compact metric space.) Condition $(*)$ guarantees that the sum defining $f(x)$ converges and that f is continuous. Moreover, if $x, y \in \mathcal{P}(\omega)$

and $x \neq y$, let n be the smallest natural number in the symmetric difference of x and y. Then by $(*)$

$$d(f(x), f(y)) \geq d(n!y_n, 0) - \sum_{x \ni k > n} d(k!y_k, 0) - \sum_{y \ni k > n} d(k!y_k, 0)$$
$$\geq d(n!y_n, 0) - 2\sum_{k>n} k!d(y_k, 0) > 0.$$

It follows that f is 1-to-1. So $K = f[\mathcal{P}(\omega)]$ is compact and uncountable. Define $Q = \langle y_n : n \in \omega \rangle$. We claim that $K/Q \subseteq \mathrm{Div}(G/Q)$. Fix $x \in \mathcal{P}(\omega)$. Let

$$g_n = \sum_{x \ni k \geq n} \frac{k!}{n!} y_k.$$

Note that the sum on the right-hand side converges by $(*)$. Then $g_0 = f(x)$ and

$$(n+1)g_{n+1} = (n+1) \sum_{x \ni k \geq (n+1)} \frac{k!}{(n+1)!} y_k = \sum_{x \ni k \geq (n+1)} \frac{k!}{n!} y_k = g_n \text{ or } g_n - y_n.$$

Thus $(n+1)g_{n+1}/Q = g_n/Q$, so $f(x)/Q = g_0/Q \in \mathrm{Div}(G/Q)$.

An abelian group is called *reduced* if its maximal divisible subgroup consists of the zero element only.

Corollary 1.4. *Let G be a metric complete abelian group which is a subgroup of a direct sum of countable reduced groups.*

(i) For some $0 < m \in \omega$, G/G_m is discrete.

(ii) If G is additionally torsion free, then G is discrete.

Proof. Clearly (ii) follows from (i), so it is enough to show (i). We assume that G is a subgroup of $\bigoplus_{i \in I} R_i$ for some countable reduced abelian groups R_i, $i \in I$, and prove that condition (ii) of Proposition 1.3 fails. By $\mathrm{dom}(g)$, $g \in G$, let us denote the finite set of indices in I at which g is not 0. Let $A = \bigcup_{g \in Q} \mathrm{dom}(g)$. Since A and R_i, $i \in A$, are countable, we can find $g_1, g_2 \in K$ such that $g_1|A = g_2|A$ and $g_1 \neq g_2$. Let $i \in I \setminus A$ be a coordinate on which g_1 and g_2 differ, that is, $(g_1)_i \neq (g_2)_i$. Since $(g_1 - g_2)/Q \in \mathrm{Div}(G/Q)$ it follows that $(g_1 - g_2)_i$ is a nonzero element of R_i which is in $\mathrm{Div}(R_i)$ contradicting the fact that R_i is reduced.

Corollary 1.5. *Any complete metric group topology on a free abelian group is discrete.*

In particular, no uncountable free abelian group carries a Polish group topology.

Proof. This follows immediately from Corollary 1.4(ii).

Remark 1.6. The following application of Corollary 1.5 to model theory was pointed out by Kechris in answer to a question of S. Thomas. For M a countable model of some countable language, one considers $\mathrm{Aut}(M)$, the group of all automorphisms of M, that is, the group of all permutations of the universe of M preserving the structure of the model. A natural question here is whether a given group can be the automorphism group of some M. Corollary 1.5 implies that an uncountable free abelian group is not isomorphic to $\mathrm{Aut}(M)$ for a countable model M. The argument for this is as follows. Let S_∞ be the group of all permutations of ω with composition as group operation. S_∞ is a G_δ subset of ω^ω, so the topology on S_∞ inherited from this inclusion is Polish and, as is not difficult to see, it is also a group topology which makes S_∞ a Polish group. As is well-known $\mathrm{Aut}(M)$ is isomorphic to a closed subgroup of S_∞. (This can be derived from the fact that the universe of M can be identified with ω or a natural number.) Thus, $\mathrm{Aut}(M)$ being a closed subgroup of a Polish group is itself a Polish group and cannot be isomorphic to a free abelian uncountable group by Corollary 1.5.

As proved in [JST], the situation is very different for $\mathrm{Aut}(M)$ for uncountable M.

1.3. The next theorem enables us to tackle a more general problem of finding out which direct sums of countable abelian groups (which are not necessarily reduced) admit Polish group topologies. For example, a torsion free group which is a direct sum of countable abelian groups admits a Polish group topology iff it is algebraically isomorphic to A or $A \times \mathbf{R}$ for a countable group A.

Theorem 1.7. *Let G be a Polish abelian group. Then precisely one of the following possibilities holds:*
(i) for some $m \in \omega$, $\mathrm{Div}(G/G_m)$ is open in G/G_m.
(ii) there exists an uncountable compact set $K \subseteq G$ and a countable group $Q < G$ such that
(a) $K/Q \subseteq \mathrm{Div}(G/Q)$;
(b) if $g_1, g_2 \in K$ and $g_1 \neq g_2$, then $g_1 - g_2 \notin \mathrm{Div}(G)$.

Before proving the theorem we state its consequences. We need three additional definitions. An abelian group G is called *bounded* if $mG = \{mg : g \in G\} = \{0\}$ for some $m \in \omega$, $m > 0$. A subgroup H of an abelian group G is called *pure* if for any $h \in H$ and $n \in \omega$ if there exists $g \in G$ such that $h = ng$, then there exists $h' \in H$ with $h = nh'$. Note that if $H = G$, then H is a pure subgroup of G.

Corollary 1.8. *Let G be a Polish group. Assume G is torsion free and a pure subgroup of a direct sum of countable abelian groups. Then $\mathrm{Div}(G)$ is open in G.*

Proof. Assume G is a torsion free Polish group that is a pure subgroup of $\bigoplus_{i\in I} R_i$, where R_i are countable abelian. Assume towards contradiction that G fulfils the condition from Theorem 1.7(ii). Let $A = \bigcup\{\mathrm{dom}(g) : g \in Q\}$. Then A is countable, so there exist distinct $g_1, g_2 \in K$ such that $g_1|A = g_2|A$. By Theorem 1.7(ii)(a), $g_1 - g_2 \in \mathrm{Div}(\bigoplus_{i\in I} R_i)$. Since G is torsion free and a pure subgroup, $g_1 - g_2 \in \mathrm{Div}(G)$, which contradicts Theorem 1.7(ii)(b). Thus, since $G_m = \{0\}$, by Theorem 1.7, $\mathrm{Div}(G)$ is open in G.

Remark 1.9. If G is torsion free, $\mathrm{Div}(G)$ is a direct sum of some number of copies of the rationals, $\mathbf{Q}$, and by [F, vol. 1, Theorem 21.2], $G = A \times \mathrm{Div}(G)$ for some group A. Since a Polish group has cardinality $\aleph_0$ or $2^{\aleph_0}$, it follows that a group G is as in Corollary 1.8 iff it is countable or it is the product of a countable group and the direct sum of $2^{\aleph_0}$ copies of $\mathbf{Q}$. To state this more succinctly, note that the direct sum of $2^{\aleph_0}$ many copies of $\mathbf{Q}$ is isomorphic to the reals, $\mathbf{R}$. Thus a torsion free Polish group which is a direct sum of countable abelian groups (or even only a pure subgroup of such a direct sum) is isomorphic to A or $A \times \mathbf{R}$ where A is a countable group. Obviously, the reverse implication holds as well: both A and $A \times \mathbf{R}$ carry Polish group topologies.

Remark 1.10. Corollaries 1.4, 1.5, and 1.8 may be regarded as generalizations of certain algebraic results. Corollary 1.5 generalizes the classical theorem of Baer that $\mathbf{Z}^\omega$ is not a free group. Since $\mathbf{Z}^\omega$ carries the product topology (with each $\mathbf{Z}$ given the discrete topology) which is a Polish group topology, if it were free, it would have to be countable by Corollary 1.5. Corollaries 1.4 and 1.8 generalize the following theorems due to Łoś and Fuchs which in turn extend Baer's result.

Let R_n, $n \in \omega$, be abelian and torsion free. Then

(i) (Fuchs [F, vol. 2, pp. 161-163]) $\prod_n R_n$ embeds into a direct sum of countable, abelian, reduced groups iff $R_n = \{0\}$ for all but finitely many n.

(ii) (Łoś [Łś, Theorem 3], see also Fuchs [F, vol. 2, p. 163]) if each R_n is countable, then $\prod_n R_n$ is a direct sum of countable abelian groups iff R_n is divisible for all but finitely many n.

Again these follow from Corollaries 1.4(ii) and 1.8 respectively since $\prod_n R_n$ carries a group topology which is metric complete in case (i) and Polish in case (ii).

Lemma 1.11. *Let G be a Polish abelian group. Let $H_n < G$, $n \in \omega$, be closed and let $H < G$ be analytic. Assume that $H_{n+1} < H_n$ and that $H \cap H_n$ is meager in H_n for each n. Then there exists $K \subseteq G$ uncountable, compact such that*

(i) K/H_n is finite for each n and

(ii) if $x, y \in K$ and $x \neq y$, then $x/H \neq y/H$.

Proof. For $s \in 2^{<\omega}$, let $|s|$ be the unique $n \in \omega$ with $s \in 2^n$. We recursively construct $x_s \in G$, $O^s_n \subseteq G \times G$, and $V_s \subseteq G$, $s \in 2^{<\omega}$, $n \in \omega$, so that if we let $U_s = (x_s + H_m) \cap V_s$ for $m = |s|$, then

(1) $x_s - x_t \in H_{|s|}$ if $s \subseteq t$

(2) $x_s \in U_s$

(3) $O^s_n \subseteq U_{s0} \times U_{s1}$ is relatively open, $n \in \omega$

(4) if $s0 \subseteq t_0$, $s1 \subseteq t_1$, and $|t_0| = |t_1|$, then $O^s_n \cap (U_{t_0} \times U_{t_1})$ is dense in $U_{t_0} \times U_{t_1}$ for all $n \in \omega$

(5) if $s0 \subseteq t_0$, $s1 \subseteq t_1$, and $|t_0| = |t_1|$, then $U_{t_0} \times U_{t_1} \subseteq O^s_n$ for $n \leq |t_0|$

(6) $y_1 - y_2 \notin H$ for $y_1, y_2 \in \bigcap_n O^s_n$

(7) V_s is open with $\text{diam}(V_s) \leq 1/|s|$ where the diameter is taken with respect to some complete metric compatible with the topology on G

(8) $s \subseteq t$, $s \neq t$ implies $\overline{V_t} \subseteq V_s$

(9) $s \perp t$ implies $V_s \cap V_t = \emptyset$

Assume the construction has been carried out. Define $f : 2^\omega \to G$ by letting $f(\alpha)$ = the unique element of $\bigcap_n V_{\alpha|n}$ By (1), (2), (7), (8) and the definition of U_s, $\bigcap_n U_{\alpha|n} \neq \emptyset$, so $\{f(\alpha)\} = \bigcap_n U_{\alpha|n}$. By (7-9) f is well-defined, continuous, and 1-to-1. Let $K = f[2^\omega]$. To check (ii) let $\alpha, \beta \in 2^\omega$ be distinct. Fix the smallest m such that $\alpha(m) \neq \beta(m)$, say $\alpha(m) = 0$ and $\beta(m) = 1$. By (5), $(f(\alpha), f(\beta)) \in \bigcap_n O^s_n$ with $s = \alpha|m = \beta|m$. So by (6), $f(\alpha) - f(\beta) \notin H$. To see (i), fix $n \in \omega$. Note that by (2) and (7), $f(\alpha) = \lim_k x_{\alpha|k}$ whence by (1), $f(\alpha) - x_{\alpha|n} \in H_n$. Thus $K/H_n = \{x_s/H_n : s \in 2^n\}$ which is finite.

To perform the construction we need a slight strengthening of our assumption. Let $h : G \times G \to G$ be the operation of taking the difference, $h(g_1, g_2) = g_1 - g_2$.

Claim 1. For any $m \in \omega$ and any $y_1, y_2 \in G$, $h^{-1}(H)$ is meager in $(y_1 + H_m) \times (y_2 + H_m)$.

Proof of Claim 1. Assume that the claim fails for some m, y_1, and y_2. By Kuratowski-Ulam (see [K, 8.41]), we can find $z \in y_2 + H_m$ and a nonmeager in H_m set $A \subseteq H_m$ such that $A + y_1 - z \subseteq H$. Then for any $a \in A$, $A - a \subseteq H_m$ is nonmeager in H_m and $A - a = A + y_1 - z - (a + y_1 - z) \subseteq H$, so $H \cap H_m$ is nonmeager in H_m, contradiction.

We will also use the following general fact.

Claim 2. Let X, Y be Polish spaces and let $f : X \to Y$ be continuous and open. Let $W \subseteq X$ be open and $B \subseteq W$ comeager in W. Then

$$\{y \in f[W] : B \cap f^{-1}(y) \text{ is comeager in } W \cap f^{-1}(y)\}$$

is comeager in $f[W]$.

Proof of Claim 2. By restricting f to W, it is enough to prove the above claim for $W = X$ and $Y \doteq f[W]$. By the Baire category theorem applied to $f^{-1}(y)$, it suffices to show that if $U \subseteq X$ is dense open, then $Z = \{y \in$

$Y : U \cap f^{-1}(y)$ is dense in $f^{-1}(y)\}$ is comeager. Let $\{W_n : n \in \omega\}$ be a basis for X. Then

$$Z \supseteq \bigcap_n (Y \setminus f[W_n]) \cup f[U \cap W_n],$$

and, by openness and continuity of f, each set of the intersection is comeager.

A bit of notation: $\forall^* x \in A$ stands for "for comeagerly many x in A (in the relative topology on A)." Let $x_\emptyset = 0$ and $V_\emptyset = G$. Now assume we have constructed x_s, V_s, $s \in 2^m$, and O^t_n for $t \in 2^{<m}$ and $n \in \omega$. Let s_i, $i = 0, 1, \ldots, 2^m - 1$ list 2^m in such a way that if $i < j$, then s_i is below s_j in the lexicographic order. Let $W_i \subseteq G$, $i < 2^m$, be open and small enough so that

$\overline{W_i} \subseteq V_{s_i}$;

$\mathrm{diam}(W_i) < 1/(m+1)$;

$W_i \cap x_{s_i} + H_m \neq \emptyset$;

$(W_i \cap x_{s_i} + H_m) \times (W_j \cap x_{s_j} + H_m) \subseteq O^s_n$ for $i < j$ and $n \leq m+1$ where $s = s_i \cap s_j =$ the longest t with $t \subseteq s_i$ and $t \subseteq s_j$.

Now define $y_i \in W_i \cap (x_{s_i} + H_m)$ recursively and $D_i \subseteq y_i + H_{m+1}$ so that

(a) D_i is countable and dense in $W_i \cap (y_i + H_{m+1})$

(b) $\forall x \in D_i \forall^* y \in W_j \cap (x_{s_j} + H_m)\ (x, y) \in \bigcap_n O^s_n$ where $i < j$ and $s = s_i \cap s_j$

(c) $D_i \times D_j \subseteq \bigcap_n O^s_n$ where $i < j$ and $s = s_i \cap s_j$

Here is how to find D_j provided all D_i for $i < j$ have been constructed. Let $A \subseteq W_j \cap (x_{s_j} + H_m)$ be comeager and such that $D_i \times A \subseteq \bigcap_n O^s_n$ for all i with $i < j$ where $s = s_i \cap s_j$. This is possible by (b) since the D_i's are countable. Note also that from (3) and (4) by Kuratowski-Ulam

$$B = \{x \in W_j \cap (x_{s_j} + H_m) : \\ \forall j < k < 2^m \forall^* y \in W_k \cap (x_{s_k} + H_m)\ (x, y) \in \bigcap_n O^{s_j \cap s_k}_n\}$$

is comeager in $W_j \cap (x_{s_j} + H_m)$. Since the projection mapping $x_{s_j} + H_m \to (x_{s_j} + H_m)/H_{m+1}$ is continuous and open ($(x_{s_j} + H_m)/H_{m+1}$ is a Polish space with the natural topology), we get from Claim 2 that there exists $y_j \in A \cap B$ such that $(y_j + H_{m+1}) \cap A \cap B$ is comeager in $(y_j + H_{m+1}) \cap W_j$. Let D_j be a countable dense subset of $(y_j + H_{m+1}) \cap A \cap B$.

Conditions (a) and (c) guarantee that for $i < j$ and $s = s_i \cap s_j$, $\bigcap_n O^s_n \cap (y_i + H_{m+1} \times y_j + H_{m+1})$ is comeager in $(y_i + H_{m+1} \cap W_i) \times (y_j + H_{m+1} \cap W_j)$. Let $V_{s_i 0}$, $V_{s_i 1}$ be two disjoint open sets included in W_i with nonempty intersections with $y_i + H_{m+1}$. Let $x_{s_i 0}$ be such that $x_{s_i 0} \in V_{s_i 0} \cap (y_i + H_{m+1})$

and similarly for x_{s_i1}. By Claim 1,

$$C = \{(x, y) \in x_{s_i0} + H_{m+1} \times x_{s_i1} + H_{m+1} : \; x - y \in H\}$$

is meager in $x_{s_i0} + H_{m+1} \times x_{s_i1} + H_{m+1}$, so we can find $O_n^{s_i}$ relatively open and dense in $(V_{s_i0} \cap x_{s_i0} + H_{m+1}) \times (V_{s_i1} \cap x_{s_i1} + H_{m+1})$ so that $\bigcap_n O_n^{s_i}$ has empty intersection with C and $O_n^{s_i} = (V_{s_i0} \cap x_{s_i0} + H_{m+1}) \times (V_{s_i1} \cap x_{s_i1} + H_{m+1})$ for $n \leq |s_i| + 1$. We leave the verification of (1)–(9) to the reader.

Proof of Theorem 1.7. Let $\pi : \; G^\omega \to G$ be the projection on the 0'th coordinate. Define $H_m < G^\omega$, $m \in \omega$, by

$$(g_n) \in H_m \Leftrightarrow g_n = (n+1)g_{n+1} \text{ for } n < m.$$

Then H_m is a closed subgroup of G^ω and $\pi[H_m] = m!G$. Put $H = \pi^{-1}(\text{Div}(G))$. Note that

$$\text{Div}(G) = \{g \in G : \; \exists (g_n) \; (g_0 = g \text{ and } \forall n \; g_n = (n+1)g_{n+1})\}.$$

Thus Div(G) is analytic. (One can check that Div(G) is actually Borel and, moreover, the image via a continuous homomorphism of a Polish group.) It follows that H is analytic.

Case 1. There exists m such that $H \cap H_m$ is not meager in H_m.

Then, by Pettis' theorem (see [K, 9.9, 9.11]), $H \cap H_m$ is an open subgroup of H_m. Therefore

$$\{g \in G : \; (m!g, 2 \cdot 3 \cdot \dots \cdot mg, \dots, mg, g, 0, 0, \dots) \in H\}$$

is an open subgroup of G. This means that $G' = \{g \in G : \; m!g \in \text{Div}(G)\}$ is open whence so is $G'/G_{m!}$ (in $G/G_{m!}$) which is also included in Div$(G/G_{m!})$. Indeed, let $g \in G'$, so $m!g \in \text{Div}(G)$. Then there exist $h_k \in G$, $k \in \omega$, with $m!g = h_0$ and $h_{k+1} = (k+1)h_k$. Since obviously each h_k is in Div(G), we can find g_k with $h_k = m!g_k$. So $m!g = m!g_0$ and $m!g_{k+1} = (k+1)m!g_k$, whence $m!(g - g_0) = 0$ and $m!(g_{k+1} - (k+1)g_k) = 0$. Thus, $g/G_{m!} = g_0/G_{m!}$ and $g_{k+1}/G_{m!} = (k+1)g_k/G_{m!}$ for each k, which shows that $g \in \text{Div}(G/G_{m!})$. It follows therefore that Div$(G/G_{m!})$ is open in $G/G_{m!}$.

Case 2. For all m, $H \cap H_m$ is meager in H_m.

Let ρ be an invariant, so complete, metric on G bounded by 1, see [K3]. Define a metric on G^ω by

$$d((x_n), (y_n)) = \sum_n 2^{-n} \rho(x_n, y_n).$$

Then d is an invariant, complete metric on G^ω. Now apply Lemma 1.11 to H_m and H to obtain $K \subseteq G^\omega$. We can assume that K has no isolated points. Now it is easy to pick $x_s \in K$, $s \in 2^{<\omega}$, and a sequence $\epsilon_n > 0$ so that

(i) $x_s - x_t \in H_{|s|}$ if $s \subseteq t$,
(ii) $d(x_s, x_t) \leq \epsilon_n$ if $s \subseteq t$,
(iii) $\epsilon_n \leq (4^n n!)^{-1}$, and
(iv) $\min\{d(x_s, x_t) : s \neq t, s, t \in 2^n\} > 2\sum_{k>n} \epsilon_k$.

The construction is carried out recursively with respect to n and for $s \in 2^n$. It requires that, when finding x_s, one insure that $x_s + H_{|s|} \cap K$ has nonempty interior in K. But this can be easily accomplished since $K/H_{|s|}$ is finite.

By translating K we can arrange that

(v) $x_\emptyset = 0$.

Now if we define $f : 2^\omega \to G$ by

$$f(\alpha) = \lim_m \pi(x_{\alpha|m}),$$

then f is clearly continuous. We show that $f[2^\omega]$ has the required properties. First we show (a). Let Q be the subgroup of G generated by the coordinates of the x_s, that is, $Q = \langle (x_s)_n : s \in 2^{<\omega}, n \in \omega \rangle$. Fix $\alpha \in 2^\omega$. Let $y_m = (x_{\alpha|m+1})_m - (x_{\alpha|m})_m$. Then using (v) and (i) we get

$$\begin{aligned} f(\alpha) &= \sum_m \pi(x_{\alpha|m+1}) - \pi(x_{\alpha|m}) = \sum_m \pi(x_{\alpha|m+1} - x_{\alpha|m}) \\ &= \sum_m m!(x_{\alpha|m+1} - x_{\alpha|m})_m. \end{aligned}$$

Thus, $f(\alpha) = \sum_m m! y_m$. Moreover, by (ii) and (iii), $\rho(y_m, 0) < (2^m m!)^{-1}$, and clearly $y_m \in Q$ for each m. For $n \in \omega$, let

$$g_n = \sum_{m \geq n} \frac{m!}{n!} y_m.$$

The sum converges by the estimate on $\rho(y_m, 0)$. Note that $g_0 = f(\alpha)$ and

$$(n+1)g_{n+1} = (n+1) \sum_{m \geq (n+1)} \frac{m!}{(n+1)!} y_m = \sum_{m \geq (n+1)} \frac{m!}{n!} y_m = g_n - y_n.$$

Thus $(n+1)g_{n+1}/Q = g_n/Q$, so $f(\alpha)/Q = g_0/Q \in \mathrm{Div}(G/Q)$. To see (b), let $\alpha \neq \beta$ be in 2^ω. Then by (ii) and (iv), $x = \lim_n x_{\alpha|n}$ and $y = \lim_n x_{\beta|n}$ are distinct members of K, so $x - y \notin H$ hence $f(\alpha) - f(\beta) = \pi(x) - \pi(y) \notin \mathrm{Div}(G)$.

2. The Borel problem

2.1. In the second problem mentioned in the beginning of the paper a group H comes equipped with a σ-algebra Σ with respect to which the group operations are measurable and we are searching for a Polish group topology whose family of all Borel sets coincides with Σ. This problem was considered in [K3] and, for groups that are also linear spaces, in [SR]. If Σ is to come from a Polish group topology, we can certainly assume that it comes from some Polish topology (which is not necessarily a group topology) on H. Thus, without loss of generality we can assume that Σ is standard, that is, that Σ is the family of Borel subsets of H with respect to some Polish topology on H. Such groups are called *standard Borel*. We would like, however, to restrict our attention to a narrower class of groups. Namely, we will assume that H is a metric separable group (so multiplication and inverse are continuous) whose topology is such that H is a Borel subset in some (or, equivalently, any) Polish space into which it is embedded homeomorphically. Let us call such groups *metric Borel*. Since, by Kuratowski's theorem, each Borel subset of a Polish space can be given a Polish topology whose Borel sets coincide with Borel sets inherited from the inclusion in the ambient Polish space, we see that metric Borel groups are standard Borel. By a result of Hjorth [H2, Theorem 9.4], however, these two classes are not equal.

Most frequently metric Borel groups occur as Borel subgroups of Polish groups. Two points need to be made here. Firstly, each metric Borel group can be represented in this fashion. To see this one just needs to repeat the construction of the completion, which is itself a group, of a metric group. Let H be a metric Borel group. Recall that each metric topological group admits left invariant and right invariant metrics which are compatible with its topology (see [K3]). Fix d_l a left invariant metric on H which is compatible with the group topology on H. And similarly let d_r be the right invariant metric on H. Such metrics always exist, see [K3]. Define $d = d_l + d_r$ and let G be the completion of H with respect to d. One checks that G is a Polish group in which H is dense, see [K3, Theorem 1.7].

Secondly, the ambient Polish group G in which H is dense is uniquely determined by H. To be more precise, if G_1 and G_2 are two Polish groups containing H as a dense subgroup, then there exists a homeomorphic isomorphism between G_1 and G_2 which is equal to identity on H. To see this, let d_l^i, d_r^i be left and right invariant metrics on G_i, $i = 1, 2$, compatible with the group topologies. Now, a sequence (g_n) in G_i is d_l^i-Cauchy iff for each neighborhood V of 1, $g_n^{-1}g_m \in V$ for n, m large enough. It follows that a sequence (g_n) of elements of H is d_l^1-Cauchy iff it is d_l^2-Cauchy. Since the same holds for d_r^1 and d_r^2, we see that a sequence (g_n) in H is $d^1 = d_l^1 + d_r^2$-Cauchy iff it is $d^2 = d_l^2 + d_r^2$-Cauchy. Now let $\phi :\ H \to H$ be the identity map. If $g \in G_1$, let $g_n \in H$ be such that $g_n \to g$. Then (g_n) is d^1-Cauchy, so $(\phi(g_n))$ is d^2-Cauchy. But d^2 is complete (see [K3, Exercise 1.10]), so $(\phi(g_n))$ converges. Define $\tilde{\phi}(g) = \lim_n \phi(g_n)$. It is not difficult to check that $\tilde{\phi}$ is well-defined, onto, continuous with continuous inverse, and preserves group operations. Thus, $\tilde{\phi}$ is the required homeomorphic isomorphism between G_1 and G_2.

In the sequel, we will view metric Borel groups as subgroups of Polish groups. After [K3] we call a metric Borel group which admits a (unique) Polish group topology which preserves its Borel structure *Polishable.* The obvious question here is if for a given $\alpha < \omega_1$ there exist Borel Polishable subgroups of Polish groups which are not in Π^0_α. This was proved to be true by Saint-Raymond in [SR] who produced Polishable groups arbitrarily high in the Borel hierarchy which were additionally linear spaces. Another construction of arbitrarily complex Polishable groups is given in [HKL, Lemmas 5.4, 5.5, 5.6].

2.2. When one is presented with a Polishable subgroup H of a Polish group, one of the main issues is to find a way of recovering the unique Polish group topology on H which preserves H's Borel structure. As it turns out, this question was already addressed by Saint Raymond [SR] in case when H is a linear subspace of a locally convex metric linear space and it is a separable Frechet space with some topology which preserves its Borel structure. Recall that a linear space over the reals is separable Frechet if it carries a Polish topology which makes addition of vectors and multiplication by scalars continuous. The theorem below gives a similar analysis in the general case of Polishable group H. The main difference between the general case and the Frechet linear case is that here, unlike in [SR], we do not have the dual space of H (with the Frechet topology) at our disposal and additional complications result from H not being assumed abelian.

Theorem 2.1. *Assume H is a Polishable subgroup of a Polish group G. There exist $\alpha_1 < \omega_1$ and a sequence of groups H_α, $\alpha \leq \alpha_1$, all of which are Polishable with some Polish group topologies τ_α and such that*

(i) $H_0 = \overline{H}$ and $H_{\alpha_1} = H$;

(ii) $H_\alpha < H_\beta$ if $\alpha > \beta$ and $H_\lambda = \bigcap_{\alpha<\lambda} H_\alpha$ for limit λ;

(iii) H is τ_α-dense in H_α;

(iv) $H_{\alpha+1}$ is Π^0_3 in τ_α on H_α and such that, for any $A \subset H_\alpha$ Π^0_3 in τ_α with $H \subseteq A$, $A \cap H_{\alpha+1}$ is comeager in $\tau_{\alpha+1}$.

Remark 2.2. It is not difficult to check by transfinite induction that the sequence in Theorem 2.1 is uniquely determined by H, that is, if $\tilde{H}_\alpha$ for $\alpha < \alpha_2$ is another sequence fulfilling (i)-(iv) of the above theorem, then $\alpha_2 = \alpha_1$ and for each α, $\tilde{H}_\alpha = H_\alpha$.

The sequence of groups in Theorem 2.1 will be constructed by transfinite recursion. The following construction and Lemma 2.3 will enable us to go from α to $\alpha + 1$. Let H be a Polishable subgroup of a Polish group G. Let τ stand for the Polish group topology on H which has the same Borel structure as that inherited from the inclusion $H \subseteq G$. Define

$$\tilde{H} = \{g \in G : \forall 1 \in V \in \tau \exists h_1, h_2 \in H \; g \in \overline{h_1 V} \cap \overline{V h_2}\}.$$

Lemma 2.3. *$\tilde{H}$ is a Polishable Π^0_3 subgroup of G with $H \subseteq \tilde{H}$. Moreover, H is dense in the Polish group topology on $\tilde{H}$ and, for any Π^0_3 $A \subseteq G$ with $H \subseteq A$, $A \cap \tilde{H}$ is comeager in the Polish group topology on $\tilde{H}$.*

Proof. First, we check that $\tilde{H}$ is a subgroup of G. Let $g_1, g_2 \in \tilde{H}$ and let $1 \in V \in \tau$. We need to find an $h \in H$ with $g_1 g_2^{-1} \in \overline{hV}$. Then by symmetry we will also have an $h \in H$ with $g_1 g_2^{-1} \in \overline{Vh}$, whence $g_1 g_2^{-1} \in \tilde{H}$. Let $1 \in W \in \tau$ be such that $W^2 \subseteq V$. Let $h_2 \in H$ be such that $g_2 \in \overline{W^{-1} h_2}$ and let $h_1 \in H$ be such that $g_1 \in \overline{h_1 (h_2^{-1} W h_2)}$. All this can be arranged since $g_1, g_2 \in \tilde{H}$. Now

$$g_1 g_2^{-1} \in \overline{h_1 (h_2^{-1} W h_2) h_2^{-1} W} = \overline{h_1 h_2^{-1} W^2} \subseteq \overline{h_1 h_2^{-1} V}.$$

Let $V_n \in \tau$, $n \in \omega$, be a neighborhood basis at 1 in H such that $V_n^{-1} = V_n$ and $V_{n+1}^3 \subseteq V_n$. Let $F_n = \overline{V_n}$. Now we apply the standard technique of defining a left invariant metric on a group from a countable basis at 1 to the sequence (F_n). For $x, y \in G$, let

$$\delta_l(x, y) = \inf\{2^{-k} : \; x^{-1} y \in F_k\}$$

and

$$d_l(x, y) = \inf\{\sum_{i=0}^{n-1} \delta_l(x_i, x_{i+1}) : \; x_0 = x, x_n = y, x_i \in G\}.$$

The only non-trivial point in checking that d_l is a left invariant metric on G is to show that $d_l(x, y) = 0$ implies $x = y$. This follows from the following claim which will be used also in other situations.

Claim 1. $\delta_l \geq d_l \geq (1/2)\delta_l$.

The proof of this claim can be looked up in [K3].

Similarly we define the right invariant version of the above metric and we call it d_r. (The only difference is that in the definition of δ_r we now have $xy^{-1} \in F_k$.) Define $d = d_l + d_r$. We want to show that d restricted to $\tilde{H}$ is complete and separable and that multiplication viewed as a function from $\tilde{H} \times \tilde{H} \to \tilde{H}$ is continuous with respect to the topology induced by d on $\tilde{H}$.

We will also prove that H is a dense subgroup of $\tilde{H}$ in this Polish topology on $\tilde{H}$.

Let ρ stand for the restriction of d to $\tilde{H}$. We start by restating Claim 1 combined with its right-hand counterpart in terms of the F_n's.

Claim 2. Let $g_n, g \in \tilde{H}$, $n \in \omega$. Then $\rho(g_n, g) \to 0$ iff for each $k \in \omega$, $g_n \in gF_k$ and $g_n \in F_k g$ for n large enough.

The following claim will be crucial.

Claim 3. For each n and each $g \in \tilde{H}$ there exists k with $F_n \supseteq gF_k g^{-1}$.

Proof. Fix n and $g \in \tilde{H}$. Let $h \in H$ be such that $g \in F_{n+1}h^{-1}$. Then $h \in g^{-1}F_{n+1}$. Since $h \in H$, we can find a k such that $hV_{n+1}h^{-1} \supseteq V_k$ whence $hF_{n+1}h^{-1} \supseteq F_k$. Now it follows that

$$g^{-1}F_n g \supseteq g^{-1}F_{n+1}^3 g \supseteq hF_{n+1}h^{-1} \supseteq F_k.$$

The claim will allow us to show that H is ρ-dense in $\tilde{H}$. By Claim 2, it will be enough to show that if $g \in \tilde{H}$, then for any n there exists $h \in H$ with $h \in gF_n \cap F_n g$. By Claim 3, we can find $k \in \omega$ with $g^{-1}F_k g \subseteq F_n$. Without loss of generality we can suppose that $k \geq n$. Since $g \in \tilde{H}$ and so $g^{-1} \in \tilde{H}$, there exists $h \in H$ with $g^{-1} \in h^{-1}F_k$. This implies the following

$$h \in F_k g = g(g^{-1}F_k g) \subseteq gF_n.$$

Now, $h \in gF_n \cap F_k g \subseteq gF_n \cap F_n g$ as required.

We check now that the topology induced by ρ on $\tilde{H}$ is a group topology. Let $\rho(h_n, h) \to 0$ and $\rho(g_n, g) \to 0$, $g_n, g, h_n, h \in \tilde{H}$. We need to see that $\rho(g_n h_n^{-1}, gh^{-1}) \to 0$. By Claim 2, it suffices to show that given k, $g_n h_n^{-1} \in gh^{-1}F_k$ and $g_n h_n^{-1} \in F_k gh^{-1}$. We will only prove the first statement, the second following by a similar argument. Given k, by Claim 3, we can get m such that $gF_m g^{-1} \subseteq F_k$. By Claim 2, $g_n \in gF_{m+1}$ and $h_n \in hF_{m+1}$ for large enough n. It follows then that

$$h_n g_n^{-1} \in hF_{m+1}F_{m+1}^{-1}g^{-1} = hg^{-1}(gF_m g^{-1}) \subseteq hg^{-1}F_k$$

for large enough n. Since this happens for arbitrary k, we get the conclusion by invoking Claim 2 once again.

Now we check that ρ is a Polish metric on $\tilde{H}$. Let (g_n) be a ρ-Cauchy sequence of elements of $\tilde{H}$. By Claim 2, for any k, $g_n \in g_m F_k$ and $g_n \in F_k g_m$ for n, m large enough. Now, let b_l and b_r be left and right invariant metrics on G compatible with the topology on G (see [K3]). Then $b = b_l + b_r$ is a complete metric on G (see [K3, Exercise 1.10]). Moreover, since the b-diameters of the F_k's tend to 0 as k goes to infinity, we see that (g_n) is b-Cauchy, and therefore it converges in G to some $g \in G$. But now by letting n go to infinity in the above established relations $g_n \in g_m F_k$ and

$g_n \in F_k g_m$, n, m large, we get that for each k, $g \in g_m F_k \cap F_k g_m$ for m large enough, because F_k is closed in G. Thus, $g_m \in gF_k^{-1} = gF_k$ and $g_m \in F_k^{-1} g = F_k g$. By Claim 2, this will imply that (g_n) ρ-converges to g as soon as we show that $g \in \tilde{H}$. But for each m and each k, we can find $h_1, h_2 \in H$ with $g_m \in h_1 F_k \cap F_k h_2$. This combined with $g \in g_m F_k \cap F_k g_m$, for m large enough, gives for m large enough and any $k \geq 1$,

$$g \in h_1 F_k F_k \cap F_k F_k h_2 \subseteq h_1 F_{k-1} \cap F_{k-1} h_2$$

which indicates that $g \in \tilde{H}$.

Recall that τ is the Polish group topology on H with the Borel structure equal to that inherited from the inclusion $H \subseteq G$. It remains to see that if $H \subseteq A$ with $A \subseteq G$ Π^0_3, then $A \cap \tilde{H}$ is comeager in $\tilde{H}$ with the topology induced by ρ. Since each Π^0_3 set is the intersection of countably many Σ^0_2 sets it is enough to check the above statement for $A \in \Sigma^0_2$. Let $A = \bigcup_k L_k$ with $L_k \subseteq G$ closed, and let $U \subseteq H$ be a non-empty τ-open subset of H. Since τ is not weaker than the topology inherited from $H \subseteq G$, $L_k \cap H$ is τ-closed for each k. By the Baire Category Theorem, there exists a k_0 and a non-empty open set $V \subseteq H$ such that $V \subset U \cap L_{k_0}$. Thus for some n and $h \in V$, we have $hV_n \subseteq V$. (Recall that $\{V_n : n \in \omega\}$ was fixed at the beginning of the proof as an open basis at $1 \in H$.) Since each L_k is closed, it follows that $\overline{hV_n} \subseteq L_{k_0}$. Thus by Claim 2, h is in the interior of $L_{k_0} \cap \tilde{H}$ taken with respect to the topology generated by ρ. Since U was arbitrary, it follows that $\bigcup_k L_k \cap \tilde{H}$ contains a dense open subset of $\tilde{H}$.

Proof of Theorem 2.1. For $\alpha < \omega_1$ define Polishable groups H_α with Polish group topologies τ_α so that conditions (i)-(iv) from the formulation of the theorem hold. This can be done using Lemma 2.3 repeatedly and at limit stages taking $H_\lambda = \bigcap_{\alpha<\lambda} H_\alpha$, and letting the group topology be the supremum of the Polish group topologies on the H_α's, with $\alpha < \lambda$, restricted to H_λ. We now only need to check that the process will terminate at some stage $\alpha_1 < \omega_1$ and that at this last stage we get H with the Polish group topology on it.

Claim. There exists $\alpha_1 < \omega_1$ such that $\tau_\alpha | H = \tau_{\alpha_1} | H$ for $\alpha > \alpha_1$.

Proof of Claim. Recall that τ is the Polish group topology on H, and let $\tilde{\tau} = \bigcup_{\alpha<\omega_1} \tau_\alpha$. Since τ on H is not weaker than any of the topologies $\tau_\alpha | H = \{V \cap H : V \in \tau_\alpha\}$, we have $\tilde{\tau}|H \subseteq \tau$. Let (U_n) be a countable basis for τ. Put

$$A_n = H \setminus \bigcup \{V \in \tilde{\tau} : V \cap U_n = \emptyset\}.$$

Since $\tilde{\tau}|H \subseteq \tau$, there is a countable family $\mathcal{F}_n \subseteq \{V \in \tilde{\tau} : V \cap U_n = \emptyset\}$ with $A_n = H \setminus \bigcup \mathcal{F}_n$. Let $\alpha_0 < \omega_1$ be such that $\bigcup_n \mathcal{F}_n \subseteq \tau_{\alpha_0}$. Note that for all $\alpha \geq \alpha_0$, $A_n = \overline{U}^{\tau_\alpha | H}$. Let $\alpha_1 > \alpha_0$ be such that for any n, if $\mathrm{int}_{\tau_\alpha | H}(A_n) \neq \emptyset$ then $\mathrm{int}_{\tau_{\alpha_1} | H}(A_n) \neq \emptyset$.

Subclaim. For any $\alpha \geq \alpha_1$ and for any $V \in \tau_\alpha$ with $V \cap H \neq \emptyset$ there is $U \in \tau_{\alpha_1}$ with $\emptyset \neq U \cap H \subseteq V \cap H$.

Proof of Subclaim. Let $V \in \tau_\alpha$. We claim that $V \cap H = \bigcup_{n \in X} A_n$ for some $X \subseteq \omega$. Indeed, let $y \in V \cap H$. By regularity of τ_α, there is $U \in \tau_\alpha$ with $y \in U \subseteq \overline{U}^{\tau_\alpha} \subseteq V$. Since $H \cap U \in \tau$, there is n with $y \in U_n \subseteq U$, so

$$y \in A_n = \overline{U_n}^{\tau_\alpha | H} \subseteq \overline{U}^{\tau_\alpha} \subseteq V.$$

Note that $\bigcup_{n \in X} \overline{A_n}^{\tau_\alpha}$ is an F_σ, with respect to τ_α, and, since it covers $V \cap H$, countably many of its translates by elements in H cover H. So, by the last property listed in Lemma 2.3, for some $n \in X$, $\text{int}_{\tau_{\alpha+1}}(\overline{A_n}^{\tau_\alpha} \cap H_{\alpha+1}) \neq \emptyset$, and since $\overline{A_n}^{\tau_\alpha} \cap H = A_n$ and H is $\tau_{\alpha+1}$-dense in $H_{\alpha+1}$, $\text{int}_{\tau_{\alpha+1}|H}(A_n) \neq \emptyset$. By the choice of α_1, $\text{int}_{\tau_{\alpha_1}}(A_n) \neq \emptyset$. It follows that a $U \in \tau_{\alpha_1}$ as required exists.

Now, let $\alpha \geq \alpha_1$, and let $V \ni 1$, $V \in \tau_\alpha$. Let $1 \in V_1 \in \tau_\alpha$ be such that $V_1 V_1^{-1} \subseteq V$. By the Subclaim, there exists $U \in \tau_{\alpha_1}$ with $\emptyset \neq H \cap U \subseteq V_1$. Then

$$1 \in UU^{-1} \in \tau_{\alpha_1} \text{ and } UU^{-1} \cap H \subseteq V_1 V_1^{-1} \subseteq V.$$

So, each $(\tau_\alpha | H)$-neighborhood of 1 contains a $(\tau_{\alpha_1} | H)$-neighborhood of 1. Since both of them are group topologies and $\tau_{\alpha_1} \subseteq \tau_\alpha$, we get $\tau_{\alpha_1} | H = \tau_\alpha | H$ which finishes the proof of the claim.

Let α_1 be as in the Claim. If $H \neq H_{\alpha_1}$, then H is meager and dense in H_{α_1} with respect to τ_{α_1}. Let $H \subseteq F \subseteq H_{\alpha_1}$ be a meager F_σ (in τ_{α_1}). For some $\emptyset \neq U \in \tau_{\alpha_1+1}$, $\overline{U}^{\tau_{\alpha_1}} \subseteq F$. Since $\tau_{\alpha_1+1} | H = \tau_{\alpha_1} | H$, $U \cap H \in \tau_{\alpha_1} | H$, and since H is dense in τ_{α_1}, $\overline{U \cap H}^{\tau_{\alpha_1}}$ has a nonempty interior in τ_{α_1}. But $\overline{U \cap H}^{\tau_{\alpha_1}} \subseteq \overline{U}^{\tau_{\alpha_1}} \subseteq F$, so F has non-empty interior in τ_{α_1}, contradiction. It follows that $H = H_{\alpha_1}$. The topologies τ and τ_{α_1} coincide because there can only be one Polish group topology with a given Borel structure.

2.3. Another interesting question is to characterize Polishable subgroups of Polish groups. Define an equivalence relation E_1 on $[0,1]^\omega$ by letting $(x_n)E_1(y_n)$ iff $x_n = y_n$ for large enough n. If H is a Borel subgroup of a Polish group G, let $E_1 \leq G/H$ if there exists a Borel function $f : [0,1]^\omega \to G$ such that $(x_n)E_1(y_n)$ iff $f((x_n)) = hf((y_n))$ for some $h \in H$. This definition is saying that the equivalence relation on G induced by its division into left cosets of H is at least as complicated as E_1. The relationship between Polishability and E_1 was investigated by Kechris and Louveau in [KL]. They proved that if $E_1 \leq G/H$, then H is not Polishable and conjectured that the reverse implication holds as well, that is, that the condition $E_1 \not\leq G/H$ characterizes Polishability of H. This conjecture has been confirmed in some cases:

(1) (Solecki [S1, S2]) $G = \mathcal{P}(\omega)$ the power set of ω with symmetric difference as group operation; $H < G$ a Borel ideal. ($\mathcal{P}(\omega)$ is a compact

metric, so Polish, group, since by identifying a subset of ω with its characteristic function, we can think of $\mathcal{P}(\omega)$ as being equal to $\mathbf{Z}_2^\omega$ with the product topology. $H \subseteq \mathcal{P}(\omega)$ is an ideal if it is closed under taking subsets and finite unions. Each ideal is a subgroup of $\mathcal{P}(\omega)$.)

(2) (Solecki) G an arbitrary Polish group; H an abelian Π_3^0 subgroup of G.

(3) (Casevitz [C]) $G = \mathbf{R}^\omega$ with coordinatewise addition; $H < G$ a Borel linear subspace of G with the additional property that if $(x_n) \in H$ and $(y_n) \in \mathbf{R}^\omega$ is such that $|y_n| \leq |x_n|$ for each $n \in \omega$, then $(y_n) \in H$. ($\mathbf{R}^\omega$ is a Polish group with the product topology.)

(A proof of (2) will be published elsewhere.)

2.4. In this subsection, I give an interesting example of a Borel Polishable subgroup. (It is a subgroup of H^+, the Polish group of all increasing homeomorphisms of $[0, 1]$.) The example is interesting in the sense that Polishability of the group is not obvious. Moreover, the fact that the group is Polishable will turn out to be useful. It will be applied to study the equivalence relation of mutual absolute continuity between Borel probability measures on a Polish space (Theorem 2.8). This in turn will be used to give a new proof of a recent theorem of Kechris and Sofronidis [KS] that probability Borel measures on a given Polish space are difficult to classify up to mutual absolute continuity. (Details on this will be given in the appropriate place later on.) As already mentioned, the main point of our argument is that a certain subgroup G of H^+ is Polishable. This enables us to study continuous actions of G. It turns out that the equivalence relation of mutual absolute continuity between measures is induced, on a dense G_δ, by an action of G which is continuous, with respect to the Polish group topology on G, and turbulent. This will imply the Kechris-Sofronidis' theorem by Hjorth's theorem.

We view H^+, the set of all increasing homeomorphisms of $[0, 1]$, as equipped with the uniform topology. H^+ is a G_δ subset of the space $C[0, 1]$ of all continuous real valued functions on $[0, 1]$. Since $C[0, 1]$ is a Polish space, H^+ is Polish as well. Moreover, it is a topological group with composition as group operation (this fact rests on the observation that if a sequence of increasing homeomorphisms converges uniformly to an increasing homeomorphism h, then the sequence of inverses converges uniformly to h^{-1}); thus, H^+ is a Polish group. Recall that a function $f : [0, 1] \to \mathbf{R}$ is *absolutely continuous* if for any $\epsilon > 0$ there exists a $\delta > 0$ such that $\sum_{i=0}^n |f(b_i) - f(a_i)| < \epsilon$ for any $0 \leq a_0 < b_0 \leq a_1 < \cdots < b_n \leq 1$ with $\sum_{i=0}^n b_i - a_i < \delta$. Now define

$$H^+_{abs} = \{f \in H^+ : \text{ both } f \text{ and } f^{-1} \text{ are absolutely continuous}\}.$$

H^+_{abs} is a Π_3^0 subset of H^+. More importantly for us, since the composition of two absolutely continuous functions is absolutely continuous, H^+_{abs} is a subgroup of H^+.

Lemma 2.4. *H^+_{abs} is a Polishable subgroup of H^+.*

Proof. By λ we denote the Lebesgue measure on $[0,1]$. All the integrals below are taken with respect to this measure. Similarly, L_1 will always stand for $L_1(\lambda)$, the separable Banach space of all real valued λ-summable functions on $[0,1]$. L_1 is equipped with the norm $\|f\|_1 = \int_0^1 |f|$. Recall that for an absolutely continuous function $f : [0,1] \to [0,1]$, $f'(x)$ exists for λ-almost all x's, $f' \in L_1$ and $f(x) = f(0) + \int_0^x f'$ for each $x \in [0,1]$, see [Ł, Section 7.4]. Now for $f, g \in H^+_{abs}$ define

$$d(f,g) = \int_0^1 |f' - g'|.$$

One easily checks that d is symmetric and fulfils the triangle inequality, and from $f(x) = f(0) + \int_0^x f'$ for every $x \in [0,1]$ one deduces that $d(f,g) = 0$ iff $f = g$ and that the topology induced by d on H^+_{abs} is stronger than the uniform topology inherited from H^+. Thus, to prove the lemma it suffices to show that d induces a Polish group topology on H^+_{abs}.

Claim 1. d is separable.

Proof. This follows since, for H^+_{abs} equipped with d and L_1 with the L_1-norm, the mapping $H^+_{abs} \ni f \to f' \in L_1$ is an isometry and L_1 is separable.

For $f, g \in H^+_{abs}$ define

$$\rho(f,g) = \int_0^1 |f' - g'| + \int_0^1 |(f^{-1})' - (g^{-1})'|.$$

Just like d, ρ is a metric on H^+_{abs}. It is clearly not weaker than d but we actually have

Claim 2. ρ and d induce the same topology on H^+_{abs}.

Proof. It suffices to show that for $f_n, f \in H^+_{abs}$, $n \in \omega$, if $\int_0^1 |f'_n - f'| \to 0$, then $\int_0^1 |(f_n^{-1})' - (f^{-1})'| \to 0$. So assume $\int_0^1 |f'_n - f'| \to 0$. We use the following fact whose generalization can be found in [R, Chapter 3, Exercise (with a hint) 17(b)].

Fact: If $g_n, g \in L_1$ are non-negative, $\int_0^1 g_n = \int_0^1 g$, and $g_n \to g$ pointwise λ-a.e., then $g_n \to g$ in L_1.

Note that $g_n = (f_n^{-1})'$ and $g = (f^{-1})'$ easily fulfil the first two assumptions of the fact. Thus, it remains to show that there is a subsequence of (g_n) which converges pointwise λ-a.e. to g. (Then, obviously, by applying the argument below to subsequences of (g_n), we see that the whole sequence (g_n) must converge to g in L_1.) From the fact that $f' \in L_1$ and $f'_n \to f'$

in L_1 it follows easily that given $\epsilon > 0$ we can find $\delta > 0$ such that for any $E \subseteq [0,1]$ Borel with $\lambda(B) < \delta$ we have $\lambda(f_n(E)) = \int_E f'_n < \epsilon$ for all n. Using this we will prove that given $\epsilon > 0$ there is a $\delta > 0$ such that for any Borel E with $\lambda(E) < \delta$ and any subsequence (f_{n_l}) we can find a further subsequence $(f_{n_{l_k}})$ such that

$$\lambda(\bigcup_k f_{n_{l_k}}(E) \cup f(E)) < \epsilon.$$

Fix $\epsilon > 0$. Let δ be chosen so that $\lambda(f(E)) < \epsilon/3$ holds whenever $\lambda(E) < \delta$ for a Borel set E. We only show that the conclusion holds when we start with the full sequence (f_n). Find $\epsilon_k > 0$, $k \in \omega$, with $\sum_k \epsilon_k < \epsilon/3$, and furthermore pick for each ϵ_k a $\delta_k > 0$ with the property that $\lambda(f_k(E)) < \epsilon_k$ for each Borel set E with $\lambda(E) < \delta_k$ and each n. Now fix E Borel with measure less than δ. Pick an increasing sequence of compact sets $K_k \subseteq E$ with $\lambda(E \setminus K_k) < \delta_k$. Let $U \supseteq f(E)$ be an open set of measure less than $\epsilon/3$. Since (f_n) converges to f uniformly, given k, $f_n(K_k) \subseteq U$ for almost all n. By passing to a subsequence, we can assume that $f_n(K_n) \subseteq U$ for each n. Then for each n,

$$\lambda(f_n(E) \setminus U) \leq \lambda(f_n(E \setminus K_n)) < \epsilon_n.$$

Thus,

$$\lambda(f(E) \cup \bigcup_n f_n(E)) \leq \lambda(f(E)) + \lambda(U) + \sum_n \lambda(f_n(E) \setminus U) < 3\frac{\epsilon}{3} = \epsilon.$$

Now finally we are ready to show that a subsequence of (g_n) converges pointwise λ-a.e. to g. (Recall that $g_n = (f_n^{-1})'$ and $g = (f^{-1})'$.) We actually show that for any $\epsilon > 0$ there exist $A \subseteq [0,1]$ of measure less than ϵ and a sequence (n_k) such that $g_{n_k}(x) \to g(x)$ for $x \notin A$. Fix $\epsilon > 0$ and find $\delta > 0$ as above for this ϵ. Since $f'_n \to f'$ in L_1, by passing to a subsequence, we can assume that $f'_n \to f'$ uniformly outside a set E of measure less than δ (see [R, 3.12] and [Ł, 5.6.2]). Now we pass to a further subsequence, call it again (f_n), so that we can suppose that $\lambda(f(E) \cup \bigcup_n f_n(E)) < \epsilon$. Let $A = f(E) \cup \bigcup_n f_n(E)$. Note that for each x, $f_n^{-1}(x) \to f^{-1}(x)$ since uniform convergence of (f_n) to f implies uniform convergence of (f_n^{-1}) to f^{-1}. Now if additionally $x \notin A$, then $f_n^{-1}(x), f^{-1}(x) \notin E$ whence $f'_n(f_n^{-1}(x)) \to f'(f^{-1}(x))$. It follows that for $x \notin A$,

$$g_n(x) = \frac{1}{f'_n(f_n^{-1}(x))} \to \frac{1}{f'(f^{-1}(x))} = g(x)$$

which proves the claim.

Claim 2. The topology induced by d is group.

Proof. It suffices to show that multiplication is continuous in the first variable and that inverse is continuous. But multiplication in the first variable is a d-isometry while inverse is a ρ-isometry so the conclusion follows from Claim 1. The second assertion is obvious and the first one follows from the fact that for $f_1, f_2, g \in H^+_{abs}$ we have

$$\int_0^1 |(f_1 \circ g)' - (f_2 \circ g)'| = \int_0^1 (|f_1' - f_2'| \circ g) g' = \int_0^1 |f_1' - f_2'|.$$

Claim 3. The topology induced by d is Polish.

Proof. By Claim 1, it is enough to show that any ρ-Cauchy sequence d-converges. Let a sequence (f_n) of functions from H^+_{abs} be ρ-Cauchy. Then (f_n) converges uniformly to a continuous function f and (f_n') converges in $L_1(\lambda)$ to some function h. This easily implies that $f(x) = \int_0^x h$ whence f is absolutely continuous. Similarly we show that f^{-1} is absolutely continuous. Thus, $f \in H^+_{abs}$. Since $h = f'$, it follows that (f_n) d-converges to f. This finishes the proof of Claim 3 and the theorem.

Using Lemma 2.4 we will now study actions of H^+_{abs}. Two closely related actions of this group will be important to us. In both instances we think of H^+_{abs} as equipped with the Polish group topology from Lemma 2.4. First the left shift action on H^+ which is defined by $H^+_{abs} \times H^+ \ni (g, h) \to g \circ h \in H^+$. Since the topology on H^+_{abs} is stronger than that inherited from H^+ this action is clearly continuous. The second action of H^+_{abs} is defined on P the space of all probability Borel measures on $[0,1]$ equipped with the w^*-topology, that is, the weakest topology which makes the functions $P \ni \mu \to \int f d\mu \in \mathbf{R}$ continuous for each continuous $f: [0,1] \to \mathbf{R}$. P is a Polish space with this topology. For $\mu \in P$, let $h_\mu : [0,1] \to [0,1]$ be defined by $h_\mu(x) = \mu([0,x])$. It is well-known that for each nondecreasing, continuous from the right $h: [0,1] \to [0,1]$ with $h(1) = 1$ there exists a unique $\mu \in P$ with $h = h_\mu$ and

$$\mu_n \to \mu \text{ iff } \forall x\, (h_\mu \text{ continuous at } x \text{ implies } h_{\mu_n}(x) \to h_\mu(x)).$$

Define the action of H^+_{abs} on P as follows. For $h \in H^+_{abs}$ and $\mu \in P$, let

$$h\mu = \text{ the unique } \nu \in P \text{ with } h_\nu = h \circ h_\mu.$$

A measure $\mu \in P$ is *absolutely continuous* with respect to $\nu \in P$ if $\nu(A) = 0$ implies $\mu(A) = 0$ for any Borel $A \subseteq [0,1]$. We say that μ and ν, both in P, are *equivalent* if μ and ν are absolutely continuous with respect to each other. This is an equivalence relation between measures in P and we denote it by $\sim$.

The fact, included in the following lemma, that the measure equivalence on P is induced by a continuous action of a Polish group seems to be of independent interest.

Lemma 2.5. *The action of H^+_{abs} on P is continuous and the induced orbit equivalence relation coincides with the measure equivalence, that is, for $\mu, \nu \in P$, $\mu \sim \nu \Leftrightarrow \exists g \in H^+_{abs}\ \mu = g\nu$.*

Proof. Checking continuity is straightforward. The second part of the statement seems to be a standard fact formulated here in terms of group actions. We supply some details for completeness. Assume $h\mu = \nu$ for some $h \in H^+_{abs}$. Then $h \circ h_\mu = h_\nu$. We need to show that ν is absolutely continuous with respect to μ, that is, given $\epsilon > 0$, we need to find $\delta > 0$ such that $\mu(A) < \delta$ guarantees that $\nu(A) < \epsilon$ for any set $A \subseteq [0,1]$, which can be assumed to be the union of finitely many half-open intervals of the form $(a,b]$. Since $\mu((a,b]) = h_\mu(b) - h_\mu(a)$ and $\nu((a,b]) = h_\nu(b) - h_\nu(a) = h(h_\mu(b)) - h(h_\mu(a))$, a simple calculation using absolute continuity of h verifies that. Similarly, since $h_\mu = h^{-1} \circ h_\nu$, we see that μ is absolutely continuous with respect to ν. Thus, μ and ν are equivalent.

Now assume that μ and ν are equivalent and we want to find an $h \in H^+_{abs}$ with $h \circ h_\mu = h_\nu$. Note first the following general and easy to prove fact. Let $g_1, g_2 : [0,1] \to [0,1]$ be two continuous from the right, non-decreasing functions with $g_1(1) = g_2(1) = 1$ which additionally satisfy $g_1(x) = g_1(y)$ iff $g_2(x) = g_2(y)$ for any $x, y \in [0,1]$ and g_1 and g_2 are discontinuous at precisely the same points. Then there exists a unique increasing homeomorphism $h : [0,1] \to [0,1]$ which is linear on each interval $(x,y) \subseteq [0,1] \setminus g_1([0,1])$ and such that $h \circ g_1 = g_2$. Now, if μ and ν are equivalent, then h_μ and h_ν satisfy the conditions imposed on g_1 and g_2 above. So for some increasing homeomorphism h which is linear on all intervals included in $[0,1] \setminus h_\mu([0,1])$ we have $h \circ h_\mu = h_\nu$. But now using the fact that μ and ν are equivalent we easily deduce that both h and h^{-1} are absolutely continuous, so $h \in H^+_{abs}$.

Recall Hjorth's notion of turbulence, see [H2] or [K2]. (The precise definition below comes from [K2] but by a result of Hjorth, see [K2, 12.5], it is equivalent to that of [H2].) Let G be a Polish group and X a Polish space on which G acts continuously. The action is called *turbulent at* x if for any open sets $U \subseteq X$, $V \subseteq G$ with $x \in U$ and $1 \in V$ there exists $\emptyset \neq U_1 \subseteq U$ open such that for any $y \in U_1$ we can find $g_n \in V$, $n \in \omega$, with $g_n g_{n-1} \cdots g_0 x \in U$ for each n and

$$y \in \overline{\{g_n g_{n-1} \cdots g_0 x : n \in \omega\}}.$$

The action is called *turbulent* if each orbit is dense and meager and the action is turbulent at each point of X. It is called *generically turbulent* if its restriction of the action to an invariant dense G_δ subset of X is turbulent.

Lemma 2.6. *The left shift action of H^+_{abs} on H^+ is turbulent.*

Proof. Each piecewise linear function in H^+ is in H^+_{abs} which shows that H^+_{abs} is dense in H^+. Clearly H^+_{abs} is meager (otherwise being a dense subgroup of H^+ it would be equal to H^+). Thus, all orbits of the action of H^+_{abs} on H^+ are dense and meager. Since we are dealing with a shift action, to prove that it is turbulent at every point it is enough to check only that it is turbulent at id. Let now $U \subseteq H^+$ and $V \subseteq H^+_{abs}$ be open and such that $\mathrm{id} \in U$ and $\mathrm{id} \in V$. Let $U_1 \subseteq H^+$ be a ball centered at id with respect to $\|\cdot\|_\infty$ with $U_1 \subseteq U$. Here, $\|\cdot\|_\infty$ stands for the supremum norm. Let $h \in U_1$. Now let $\epsilon > 0$ be given. It suffices to show that there exist $g_i \in H^+_{abs}$, $i < n$ for some n, such that each g_i is in V, $g_i \circ g_{i-1} \circ \cdots \circ g_0 \circ \mathrm{id} \in U_1$, for $i < n$, and $g_{n-1} \circ \cdots \circ g_0 \circ \mathrm{id}$ is ϵ-close, with respect to $\|\cdot\|_\infty$, to h. We can easily find a piecewise linear mapping h_1 in H^+ which is ϵ-close to h and with the additional property that $\|h_1 - \mathrm{id}\|_\infty \leq \|h - \mathrm{id}\|_\infty$. Let $n \in \omega$ be large enough so that the d-ball in H^+_{abs} of radius $\leq 2/n$ centered at id is included in V, where d is the metric on H^+_{abs} defined in the proof of Lemma 2.4. For $i = 0, \ldots, n$ define

$$h_1^i = \frac{i}{n}h_1 + (1 - \frac{i}{n})\mathrm{id}.$$

Note that each h_1^i, being piecewise linear, is a member of H^+_{abs}. Let us also make note of the fact that

$$\|h_1^i - \mathrm{id}\|_\infty \leq \frac{i}{n}\|h_1 - \mathrm{id}\|_\infty \leq \|h - \mathrm{id}\|_\infty$$

from which it follows that $h_1^i \in U_1$. Let $g_i = h_1^{i+1} \circ (h_1^i)^{-1} \in H^+_{abs}$ for $i < n$. An easy calculation shows that

$$d(g_i, \mathrm{id}) = d(h_1^{i+1}, h_1^i) = \int_0^1 |\frac{i+1}{n}h_1' + (1 - \frac{i+1}{n}) - \frac{i}{n}h_1' - (1 - \frac{i}{n})| \leq 2/n.$$

Thus each g_i is in V. Moreover, $g_i \circ g_{i-1} \circ \cdots \circ g_0 \circ \mathrm{id} = h_1^{i+1} \in U_1$, for $i < n$, and $g_{n-1} \circ \cdots \circ g_0 \circ \mathrm{id} = h_1^n = h_1$ is ϵ-close to h.

Let $B \subseteq P$ consists of all measures which vanish on points and are positive on all nonempty open sets. It is easy to see that B is a dense G_δ subset of P and that it is $\sim$-invariant.

Lemma 2.7. *(i) The restriction of the action of H^+_{abs} on P to B is turbulent. The action is not turbulent at any other $\mu \in P$.*

(ii) The action of H^+_{abs} on P is generically turbulent.

Proof. (i) holds since the restriction of the action to B is essentially identical with the left shift action of H^+_{abs} on H^+. Indeed, for $f \in H^+$, let $\mu(f)$ be the unique $\mu \in P$ with $f = h_\mu$. The mapping $H^+ \ni f \to \mu(f) \in P$ is a homeomorphism onto B. (Continuity is obvious; continuity of the inverse map follows from the easy-to-prove fact that if a sequence of increasing homeomorphisms of $[0,1]$ converges pointwise to an increasing homeomorphism, then the convergence is uniform.) Moreover, $h\mu(f) = \mu(h \circ f)$ for $f \in H^+$. Now it follows by Lemma 2.6 that the action of H^+_{abs} on B is turbulent at each point of B. Note also that if $\mu \notin B$, then its orbit under the action of H^+_{abs} is nowhere dense in P. Thus, the action cannot be turbulent at such a μ.

(ii) follows from (i) since, as already noted, B is an invariant, dense G_δ.

We will now extend Lemma 2.7 to the space of all probability Borel measures defined on an arbitrary Polish space without isolated points. Let X be a Polish space. By $P(X)$ denote the space of all probability Borel measures on X with the weakest topology which makes the mappings $P(X) \ni \mu \to \int f d\mu$ continuous for $f:\ X \to \mathbf{R}$ continuous bounded. So, in particular, $P = P([0,1])$. $P(X)$ is a Polish space, see [K1, 17.23]. The topology on $P(X)$ can be induced by a smaller family of continuous functions. If we fix a metric compact space $\bar{X}$ which contains X as a dense subset, then the topology on $P(X)$ coincides with the weakest topology making the mappings $P(X) \ni \mu \to \int f d\mu$ continuous for all f which are restrictions to X of continuous real valued functions on $\bar{X}$, see [K1, 17.19]. It follows from this observation that if $Y \subseteq X$ is dense in X, then the mapping $\mu \to \mu|Y$ mapping $\{\mu \in P(X):\ \mu(Y) = 1\}$ onto $P(Y)$ is a homeomorphism. An argument for this goes as follows. If $\bar{X}$ is a compact metric space densely containing X, then Y is a dense subset of $\bar{X}$ as well. Now $\mu_n \to \mu$ in $P(X)$ precisely when for all f's which are restrictions of continuous functions on $\bar{X}$, we have $\int f d\mu_n \to \int f d\mu$. If, additionally, $\mu(Y) = \mu_n(Y) = 1$ this is equivalent to saying $\int (f|Y) d(\mu_n|Y) \to \int (f|Y) d(\mu|Y)$ which means $\mu_n|Y \to \mu|Y$ in $P(Y)$. Recall that $\{\mu \in P(X):\ \mu(F) < r\}$ and $\{\mu \in P(X):\ \mu(U) > r\}$, for F closed, U open, r a real number, are both open, see [K1]. Thus, $\{\mu \in P(X):\ \mu(F) = 0\}$ and $\{\mu \in P(X):\ \mu(G) = 1\}$, F closed, G G_δ, are G_δ. (To see it for the latter set, note that $\{\mu \in P(X):\ \mu(G) = 1\} = \bigcap_n \{\mu \in P(X):\ \mu(U_n) > 1 - \frac{1}{n}\}$ where $G = \bigcap_n U_n$, $U_{n+1} \subseteq U_n$, and each U_n is open.)

The notions of absolute continuity and equivalence between measures defined for elements of P before extend in the obvious manner to $P(X)$.

Theorem 2.8. *Let X be a Polish space without isolated points. There exists a $\sim$-invariant dense G_δ subset of $P(X)$ on which $\sim$ is induced by a continuous, turbulent action of a Polish group.*

Proof. X contains a dense G_δ set A which is homeomorphic to ω^ω. Let $\tilde{A} = \{\mu \in P(X):\ \mu(A) = 1\}$. Since A is G_δ, so is $\tilde{A}$. Obviously, $\tilde{A}$ is dense and $\sim$-invariant subset of $P(X)$. Let A' and $\tilde{A}'$ be analogously defined sets for $[0,1]$ and let $\phi : A \to A'$ be a homeomorphism. Then $\tilde{\phi} : \tilde{A} \to \tilde{A}'$ defined by specifying $\tilde{\phi}(\mu)(E) = \mu(\phi^{-1}(E))$ for any Borel set $E \subseteq X$ is a homeomorphism which preserves measure equivalence. Preservation of measure classes is obvious. To check that $\tilde{\phi}$ is a homeomorphism note that it can be obtained as follows. Start with the mapping $\tilde{A} \ni \mu \to \mu|A \in P(A)$, next identify $\mu|A$ with a $\nu \in P(A')$ using the homeomorphism ϕ between A and A', finally map ν to the unique $\rho \in \tilde{A}'$ with $\rho|A' = \nu$. Now the mapping $\mu|A \to \nu$ is obviously a homeomorphism while the mappings $\mu \to \mu|A$ and $\nu \to \rho$ are homeomorphisms by the remarks preceding the statement of Theorem 2.8. Since, by Lemma 2.5, $\sim$ on $\tilde{A}'$ is is induced by a Polish group action, we use $\tilde{\phi}$ to transfer this action on $\tilde{A}$. Moreover, by Lemma 2.7, the restriction of the action on $\tilde{A}'$ to some $\sim$-invariant dense G_δ is turbulent; thus, the same is true about the induced action on $\tilde{A}$ and the theorem follows.

As an application of the work done above (Theorem 2.8) I will prove, using a theorem of Hjorth [H, 3.2.10; K2, 12.5], an important non-classification result due to Kechris and Sofronidis [KS]. It is not difficult to see that one cannot classify in Baire measurable fashion measures in $P(X)$, X Polish without isolated points, up to equivalence using points of a Polish space as invariants. That is, it is not possible to assign in a Baire measurable fashion points of a Polish space Y to measures in $P(X)$, X Polish without isolated points, in such a way that two measures are equivalent iff they are mapped to the same element of Y. Kechris-Sofronidis' result goes much further and establishes that such a classification is impossible even if we allow points of Y to be identified by a Borel action of S_∞, the group of all permutations of ω, and only require that two measures in $P(X)$ are equivalent iff their images are identified by the S_∞ action. This is equivalent to the impossibility of assigning in a Baire measurable fashion countable models to equivalence classes of measures as complete invariants. Let us make the above discussion more precise. S_∞ is equipped with the topology inherited from the obvious inclusion $S_\infty \subseteq \omega^\omega$. Since S_∞ is a G_δ subset of ω^ω, this topology is Polish and as is not difficult to see it is also a group topology. Now, Kechris-Sofronidis' theorem can be stated as follows:

If Y is a Polish space which is acted on in a Borel fashion by S_∞ and if $f : P(X) \to Y$, X Polish, is Baire measurable and such that $\mu \sim \nu$ implies that $f(\mu)$ and $f(\nu)$ are in the same orbit of the S_∞ action, then there exists a dense G_δ subset of $P(X)$ all of whose members are mapped by f into one orbit in Y.

Remark 2.9. This theorem was proved in [KS] only for X Polish, compact and with no isolated points, but by an argument as in the proof following the statement of Theorem 2.8 and by the arguments below it is not difficult to obtain the general case from the particular one. Moreover, the above result can be considered a non-classification theorem only if there is no comeager equivalence class in $P(X)$, that is, when the isolated points in X are not dense in X.

The theorem of Hjorth that will be used states that if G is a Polish group acting continuously and turbulently on a Polish space Y, Z is a Polish space acted on in a Borel fashion by S_∞, and $f: Y \to Z$ is Baire measurable and such that for any $x, y \in Y$ lying in the same orbit of the action of G, $f(x)$ and $f(y)$ lie in the same orbit of the action of S_∞, then a dense G_δ subset of Y is transported by f into one orbit of Z.

Proof of Kechris-Sofronidis' theorem from Theorem 2.8. Let X be a Polish space. Let Y be a Polish space on which S_∞ acts in a Borel way. Assume $f: P(X) \to Y$ is Baire and such that $\mu \sim \nu$ implies $f(\mu)$ and $f(\nu)$ are isomorphic. If X does not have isolated points, then the result follows immediately from Theorem 2.8 and Hjorth's theorem. The argument below shows that the same holds true for a general Polish space. Let Z be the set of all isolated points of X.

Case 1. Z is dense in X.

Then $\{\mu \in P(X) : \ \mu(X \setminus Z) = 0 \text{ and } \forall x \in Z \ \mu(\{x\}) > 0\}$ is a dense G_δ $\sim$-equivalence class which, therefore, is a dense G_δ which is mapped by f to one orbit in Y.

Case 2. Z is not dense.

Let W be the interior of $X \setminus Z$. Then W is a (nonempty) Polish space without isolated points which is disjoint from Z and such that $W \cup Z$ is a dense G_δ in X. Moreover, W and Z are clopen subsets of $W \cup Z$. It follows immediately, that $Q = \{\mu \in P(X) : \ \mu(W \cup Z) = 1, \mu(U) > 0 \text{ and } \forall x \in Z \ \mu(\{x\}) > 0\}$ is a dense G_δ $\sim$-invariant subset of $P(X)$. It is G_δ since $W \cup Z$, W, and $\{x\}$ for $x \in Z$ are all open subsets of X. Let $P^+(Z) = \{\mu \in P(Z) : \ \forall x \in Z \ \mu(\{x\}) > 0\}$. Then $P^+(Z)$ is a dense G_δ subset of $P(Z)$ so it is a Polish space in its own right. Define $g : \ P(W) \times P^+(Z) \times (0,1) \to P(X)$ by specifying that for each Borel $A \subseteq X$,

$$g(\mu, \nu, a)(A) = (1-a)\mu(A \cap W) + a\nu(A \cap Z).$$

Using the remarks preceding Theorem 2.8, one easily checks that g is a homeomorphic embedding of $P(W) \times P^+(Z) \times (0,1)$ onto Q. Since Q is a dense G_δ, $f \circ g$ is a Baire measurable function. By the Kuratowski-Ulam theorem, we can find $\nu_0 \in P^+(Z)$ and $a_0 \in (0,1)$ such that the function $h : \ P(W) \to Y$ defined by $h(\mu) = f \circ g(\mu, \nu_0, a_0)$ is Baire measurable. Now it is easy to see that if $\mu, \nu \in P(W)$ and $\mu \sim \nu$, then $h(\mu)$ and $h(\nu)$ lie in the same orbit of the S_∞ action. Since W has no isolated points, by Theorem 2.8 and Hjorth's theorem, there exists a dense G_δ set $B \subseteq P(W)$ which is mapped by h to one orbit of the S_∞ action on Y, so each measure in $g[B \times \{\nu_0\} \times \{a_0\}]$ is mapped by f to the same orbit. But each measure in $g[B \times P^+(Z) \times (0,1)]$ is $\sim$-equivalent with some measure in $g[B \times \{\nu_0\} \times \{a_0\}]$, whence each measure in $g[B \times P^+(Z) \times (0,1)]$ is mapped to this orbit. Since g is a homeomorphism between $P(W) \times P^+(Z) \times (0,1)$ and Q, $g[B \times P^+(Z) \times (0,1)]$ is a dense G_δ subset of Q, so it is a dense G_δ in $P(X)$ and we are done.

References

[C] P. Casevitz, *Dichotomies pour les espaces de suites reelles*, to appear;

[F] L. Fuchs, *Abelian Groups*, Academic Press, New York-London, 1973;

[H1] G. Hjorth, *The group of categoricity preserving maps has no Polish topology*, to appear;

[H2] G. Hjorth, *Classification and Orbit Equivalence Relations*, to appear;

[HKL] G. Hjorth, A.S. Kechris, A. Louveau, *Borel equivalence relations induced by actions of the symmetric group*, Ann. Pure Appl. Logic, 92(1998), 63-112;

[JST] W. Just, S. Shelah, S. Thomas, *The automorphism tower problem revisited*, to appear;

[K1] A.S. Kechris, *Classical Descriptive Set Theory*, Springer-Verlag, 1995;

[K2] A.S. Kechris, *Actions of Polish Groups and Classification Problems*, to appear;

[K3] A.S. Kechris, *Lectures on Definable Group Actions and Equivalence Relations*, lecture notes;

[KL] A.S. Kechris, A. Louveau, *The structure of hypersmooth equivalence relations*, J. Amer. Math. Soc. 10(1997), 215-242;

[KS] A.S. Kechris, N.E. Sofronidis, *A strong generic ergodicity property of unitary conjugacy*, to appear;

[L] S. Łojasiewicz, *An Introduction to the Theory of Real Functions*, Wiley, 1988;

[Łś] J. Łoś, *On the complete direct sum of countable abelian groups*, Publ. Math. Debrecen 3(1954), 269-272;

[R] W. Rudin, *Real and Complex Analysis*, McGraw-Hill, 1987;

[SR] J. Saint-Raymond, *Espaces a modèle séparable*, Ann. Inst. Fourier, 26(1976), 211-256;

[S1] S. Solecki, *Analytic ideals*, Bull. Symb. Logic, 2(1996), 339-348;

[S2] S. Solecki, *Analytic ideals and their applications*, Ann. Pure Appl. Logic, to appear.

Forcing Closed Unbounded Subsets of $\aleph_{\omega+1}$

Maurice C. Stanley
San Jose State University

1. Introduction

This paper addresses a special case of the following problem: *Suppose that κ is an infinite cardinal. Characterize those $X \subseteq \kappa^+$ such that X contains a closed unbounded (club) subset of κ^+ in some κ and κ^+ preserving outer model.*

If V is a transitive standard model of ZFC, say that W is an ***outer model*** of V if $W \supseteq V$ is also a standard transitive model of ZFC and $V \cap \mathrm{OR} = W \cap \mathrm{OR}$.

Assume, as we shall everywhere, that the GCH holds in the inner model V.

If $\kappa = \omega$, then this problem has a well known solution. There exists a club subset of ω_1 contained in X in some ω_1-preserving outer model iff X is stationary in ω_1. The paper [**S**] to which this is a sequel shows that for regular $\kappa \geqslant \omega_1$ this characterization problem is generally unsolvable and never uniformly solvable in the way it is for $\kappa = \omega$. (See §6 for this statement in more detail.)

This paper addresses a special case of this problem when κ is singular, namely, the case of "bounded pattern width" subsets of $\aleph_{\omega+1}$. The ideas employed can be applied more generally, but further ideas are needed to settle the general problem. Consequently, here we shall be content to restrict ourselves to this case.

We need several definitions in order to state the theorem proved in this paper. For each ordinal α of uncountable cofinality, fix a monotonically

I would like to thank Uri Abraham for his valuable comments.

Research supported by N.S.F. Grant DMS 9505157.

Keywords: closed unbounded set, stationary set, forcing, strong covering lemma increasing continuous sequence

$$\langle\, \rho_i^\alpha : i < \mathrm{cf}(\alpha)\,\rangle$$

cofinal in α. If X is a set of ordinals, define the ***pattern*** of X at α by

$$\mathrm{ptn}(X,\alpha) = \Big\{\, i < \mathrm{cf}(\alpha) : \rho_i^\alpha \in X \,\Big\}.$$

Modulo the non-stationary ideal on $\mathrm{cf}(\alpha)$, the definition of $\mathrm{ptn}(X,\alpha)$ does not depend on the choice of the sequence $\langle\, \rho_i^\alpha : i < \mathrm{cf}(\alpha)\,\rangle$.

Suppose that κ is a regular uncountable cardinal. If X and Y are subsets of κ, then $X \subseteq_{\mathrm{NS}} Y$ indicates that $X \cap C \subseteq Y$, for some club $C \subseteq \kappa$. Similarly, $X =_{\mathrm{NS}} Y$ indicates that $X \subseteq_{\mathrm{NS}} Y$ and $Y \subseteq_{\mathrm{NS}} X$, and $X \in_{\mathrm{NS}} F$ indicates that $X =_{\mathrm{NS}} Y$, for some $Y \in F$. A defect in this notation is that it suppresses mention of κ. The intended κ will be evident in context.

The observation that the "pattern of" relation is transitive (modulo non-stationary ideals) will be useful.

Lemma. *Assume that $\lambda < \mu < \kappa$ are regular and uncountable, that X is a subset of κ, and that $\alpha \in \lim(X)$ has cofinality μ. Set $S = \mathrm{ptn}(X,\alpha)$. If $\beta \in \lim(S)$ has cofinality λ, then $\rho_\beta^\alpha \in \lim(X)$ and $\mathrm{ptn}(X,\rho_\beta^\alpha) =_{\mathrm{NS}} \mathrm{ptn}(S,\beta)$.*

Proof: Certainly $\rho_\beta^\alpha \in \lim(X)$. We must check the second claim. Let $f\colon \lambda \to \beta$ by $f(i) = \rho_i^\beta$; let $g\colon \beta \to \rho_\beta^\alpha$ by $g(j) = \rho_j^\alpha$; and let $h\colon \lambda \to \rho_\beta^\alpha$ by $h(i) = \rho_i^\alpha$. Each of these functions is increasing, continuous, and cofinal in its codomain. Thus there exist i club in λ such that $h(i) = (g \circ f)(i)$. The claim follows from the observations that

$$\mathrm{ptn}(X,\rho_\beta^\alpha) = h^{-1}(X)$$

and

$$\mathrm{ptn}(S,\beta) = \mathrm{ptn}\big(g^{-1}(X),\beta\big) = f^{-1}\big(g^{-1}(X)\big) = (g \circ f)^{-1}(X). \quad \square$$

We shall say that a subset $X \subseteq \kappa$ has "bounded pattern width" if fewer than the maximum possible number of patterns occur in X at limit points of X. Precisely, for each regular uncountable $\mu < \kappa$, let P_μ contain exactly one representative from each equivalence class, modulo the non-stationary ideal on μ, of

$$\Big\{\, S \subseteq \mu : S =_{\mathrm{NS}} \mathrm{ptn}(X,\alpha), \text{ for some } \alpha \in \lim(X) \text{ such that } \mathrm{cf}(\alpha) = \mu \,\Big\}.$$

Say that X has ***bounded pattern width*** iff

$$\sup_{\substack{\omega_1 \leqslant \mu < \kappa \\ \mu \text{ regular}}} \left|P_\mu\right| < \sup_{\substack{\omega_1 \leqslant \mu < \kappa \\ \mu \text{ regular}}} 2^\mu.$$

Say that ***covering*** holds between V and an outer model W if, given any set of ordinals $b \in W$, there exists a set of ordinals $a \in V$ such that $a \supseteq b$ and $|a|^W \leqslant \max(|b|^W, \aleph_1^W)$. Say that ***full covering*** holds between V and W if this inequality can be improved to $|a|^W = |b|^W$.

We shall use Shelah's version of strong covering. Suppose that W is an outer model of V. At move $i < \lambda$, Player I in the $(\lambda^*, \lambda, \kappa, \alpha)$-covering game, chooses a set of ordinals $a_i \in V$ such that $a_i \subseteq \alpha$ and $|a_i|^W < \lambda^*$ and $a_i \supseteq \bigcup_{j<i} b_j$. Then Player II chooses a set of ordinals $b_i \in W$ such that $b_i \subseteq \alpha$ and $|b_i|^W < \lambda^*$ and $b_i \supseteq \bigcup_{j \leqslant i} a_j$. Player I wins this game if in W there exists a club subset $C \subseteq \lambda$ such that $\bigcup_{j<i} a_j \in V$, whenever i is a cofinality κ limit point of C.

For current purposes, we say that ***strong covering*** holds between V and an outer model W if Player I has a winning strategy in the $(\lambda^*, \lambda, \kappa, \alpha)$-covering game, whenever α is an ordinal and $\kappa < \lambda < \lambda^*$ are uncountable regular cardinals in W.

It follows from Theorem 2.6 of [**Sh,** Ch.7] that W is a strongly covered outer model of V if GCH + $\square$ holds in the inner model V and (ordinary) covering holds between V and W. Thus if $V = L$, then any $\aleph_\omega$-preserving outer model is an $\aleph_\omega$ and $\aleph_{\omega+1}$ preserving strongly covered outer model.

The main result of this paper is the following

Theorem. *Assume GCH + $\square$. There is a definable operation from bounded pattern width subsets X of $\aleph_{\omega+1}$ to trees $\mathcal{T}_X$ such that the following three statements are equivalent.*

(1) *There is a club subset of $\aleph_{\omega+1}$ contained in X in some covered $\aleph_\omega$ and $\aleph_{\omega+1}$ preserving outer model.*
(2) *$\mathcal{T}_X$ is ill-founded.*
(3) *There is a club subset of $\aleph_{\omega+1}$ contained in X in a fully covered, GCH, $\aleph_\omega$, and $\aleph_{\omega+1}$ preserving set forcing extension.*

Of course, $\aleph_{\omega+1}$ is preserved in any covered outer model that preserves $\aleph_\omega$.

2. Motivation

As suggested at the outset, ideally we would like to characterize those subsets of $\aleph_{\omega+1}$ having club subsets in $\aleph_\omega$, and $\aleph_{\omega+1}$ preserving outer models.

$\aleph_\omega$*-preserving* requires only that V-cardinals unbounded in $\aleph_\omega$ be preserved in the outer model W. Why not ask which subsets of $\aleph_{\omega+1}$ have club subsets in cardinal preserving outer models? The reason is that an easy application of the results of [**S**] shows that this problem is in general unsolvable and never "uniformly solvable". Consequently, no satisfactory characterization is possible. Remark 1 in §6 addresses this.

Suppose, then, that $X \in V$ is a subset of $\aleph_{\omega+1}$ and that X has a club subset C in the $\aleph_\omega$ and $\aleph_{\omega+1}$ preserving outer model W. Anticipating later notation, let us write "ω_i" for $\aleph_i^V$ and "κ_n" for $\aleph_n^W$. When there is danger of confusion, we shall sometimes write explicitly "ω_i^V". We shall use $\aleph$'s to denote cardinals of the universe in which we are working at the moment. "$\aleph_\omega$" and "$\aleph_{\omega+1}$" are unambiguous because we are concerned with extensions that preserve these cardinals.

Working in W, for $n < \omega$ set

$$\mathcal{F}_n = \Big\{ \operatorname{ptn}^V(X,\alpha) : \alpha \in \lim(C) \text{ and } \operatorname{cf}^W(\alpha) = \kappa_{n+1} \Big\}.$$

Then $\mathcal{F}_n$ generates a filter on κ_{n+1}, each element of which has a club subset in W. A slightly stronger property is this: A family of subsets of a regular uncountable cardinal is ***principalized*** if there exists a club subset of that cardinal that is eventually contained in each member of the family. One strategy for adding a club subset to X by forcing over V would be to iteratively add club sets $C_n \subseteq \kappa_{n+1}$ principalizing $\mathcal{F}_n$. After this preparation, the set X would be "fat stationary" in the sense of [**AS**] and we could force to add a club subset to X. The difficulty with this plan, of course, is that in general the $\mathcal{F}_n$'s do not lie in V. To get around this, we need some counterpart of the $\mathcal{F}_n$'s that does lie in V.

The characterization we seek must work when $V = L$. Consequently, we may as well assume that covering holds between V and W (and restrict ourselves to this case). It follows from this assumption that $\operatorname{cf}^W(\omega_i) = \kappa_n$, when $\kappa_n \leqslant \omega_i < \kappa_{n+1}$ and $n \geqslant 2$. We may assume that this computation holds for all $n \geqslant 0$, since otherwise we could collapse κ_2 to secure it.

In V, given $X \subseteq \aleph_{\omega+1}$ that does have a club subset in an $\aleph_\omega$ and $\aleph_{\omega+1}$ preserving outer model, a first attempt at our task would find

$$\vec{\kappa} = \big\langle \kappa_n : 1 \leqslant n < \omega \big\rangle \quad \text{and} \quad \vec{F} = \big\langle F_i : \kappa_1 \leqslant \omega_{i+1} < \aleph_\omega \big\rangle$$

satisfying (a)–(d):

(a) $\kappa_1 < \kappa_2 < \cdots < \aleph_\omega$ are uncountable cardinals.

(b) $F_i \subseteq \{\operatorname{ptn}(X, \alpha) : \alpha \in \lim(X) \text{ and } \operatorname{cf}(\alpha) = \omega_{i+1}\} \cup \{\omega_{i+1}\}$ and $\omega_{i+1} \in F_i$.

(c) F_i is countable and $\bigcap F_i$ is stationary in ω_{i+1}.

Requirement (b) insists that $\omega_{i+1} \in F_i$ so that $F_i \neq \emptyset$, hence $\bigcap F_i$ is well defined in (c)—X might have no limit points of cofinality ω_{i+1}.

The requirement in (c) that F_i is countable may come as a surprise. It simplifies our work because principalizing F_i then amounts to adding a club subset to $\bigcap F_i$. Furthermore, this simplification comes at no cost in generality once we have restricted ourselves to X of bounded pattern width.

Our goal is to force to get an extension in which κ_n is the nth uncountable cardinal and the F_i's are principalized in preparation for adding a club subset to X, that is, for principalizing $\{X\}$. Since this last step will really be no different from the earlier ones, it is convenient to set

$$F_\omega = \{X\}.$$

Requirement (d) insures that principalizing the F_j's for $j < i$ helps with principalizing F_i when ω_{i+1} has cofinality greater than $\aleph_1$ in the generic extension.

(d) If $\kappa_2 \leqslant \kappa_n \leqslant \omega_{i+1} < \kappa_{n+1}$ or $i = \omega$ (so $F_i = \{X\}$), then

$$\Big\{ \alpha \in \lim(\textstyle\bigcap F_i) : \kappa_{n-1} \leqslant \operatorname{cf}(\alpha) < \kappa_n \text{ and }$$
$$\operatorname{ptn}(\textstyle\bigcap F_i, \alpha) \supseteq_{\mathrm{NS}} \bigcap F_j, \text{ where } \operatorname{cf}(\alpha) = \omega_{j+1} \Big\}$$

is stationary in ω_{i+1}.

It is not obvious that such sequences exist in V when $X \subseteq \aleph_{\omega+1}$ has a club subset in an $\aleph_\omega$ and $\aleph_{\omega+1}$ preserving outer model. It follows from the work in §5 that they do, at least when X has bounded pattern width and the outer model is strongly covered.

On the other hand, armed with such sequences, we can iteratively principalize the F_i's, in the process adding a club subset to X, as follows: At stage $i \leqslant \omega$ in the iteration assume that, for $j < i$, we have already

- principalized F_j;
- collapsed cardinals so that $|\omega_{j+1}^V| = \omega$, if $\omega_{j+1} < \kappa_1$ and $|\omega_{j+1}^V| = \kappa_n$, if $\kappa_n \leqslant \omega_{j+1} < \kappa_{n+1}$; and
- preserved the GCH, preserved cardinals greater than or equal to ω_{i+1}, and, in fact, preserved stationary subsets of cardinals greater than or equal to ω_{i+1}.

At stage i force with conditions c such that $c \subseteq \bigcap F_i$ is a closed bounded subset of ω_{i+1}^V having order type less than κ_n, if $\kappa_n \leqslant \omega_{i+1} < \kappa_{n+1}$. Let these conditions be ordered by reverse end-extension.

These conditions satisfy the $\leqslant\omega^V_{i+1}$-c.c. and are $<\kappa_n$-strategically closed, using hypotheses (c) and (d) and that F_j has been principalized for $j < i$. This forcing principalizes F_i, gives ω^V_{i+1} cofinality and cardinality κ_n, and preserves the induction hypotheses at stage $i + 1$.

However, a difficulty arises at stage ω of the iteration. We need now to see that the tail of the iteration beginning at stage i is $<\kappa_n$-distributive, when $\kappa_n \leqslant \omega_{i+1}$. For this, we need to be able to carry out a distributivity argument in all coordinates j such that $j \geqslant i$ simultaneously. For this we need a stronger version of (c) and (d) involving stationary subsets of $[\aleph_{\omega+1}]^{<\kappa_n}$ rather than simply stationary subsets of ω_{j+1}, for $j \geqslant i$ individually.

Fix a regular cardinal $\lambda \gg \aleph_{\omega+1}$. If $M \in [H_\lambda]^{<\kappa_n}$, for $i \leqslant \omega$ set

$$\sigma_M(i) = \sup(M \cap \omega_{i+1}).$$

What we need in place of (c) and (d) is that $R(n, \vec{\kappa}, \vec{F})$ is stationary in $[H_\lambda]^{<\kappa_n}$, for $1 \leqslant n < \omega$, where $M \in R(n, \vec{\kappa}, \vec{F})$ iff

- $M \in [H_\lambda]^{<\kappa_n}$ and $M^{<\kappa_{n-1}} \subseteq M$;
- if $\kappa_n \leqslant \omega_{i+1} \leqslant \aleph_{\omega+1}$, then $\sigma_M(i) \in \bigcap F_i$; and
- if $\kappa_2 \leqslant \kappa_n \leqslant \omega_{i+1} \leqslant \aleph_{\omega+1}$ and $S \in F_i$, then $\operatorname{ptn}\big(S, \sigma_M(i)\big) \supseteq_{\mathrm{NS}} \bigcap F_j$, where $\operatorname{cf}\big(\sigma_M(i)\big) = \omega_{j+1}$

The Set Forcing Lemma in §3 shows that under the additional hypothesis that $R(n, \vec{\kappa}, \vec{F})$ is stationary in $[H_\lambda]^{<\kappa_n}$, for $1 \leqslant n < \omega$, the iteration described above succeeds.

To prove the main theorem, given a set $X \subseteq \aleph_{\omega+1}$ of bounded pattern width, we shall construct a tree $\mathcal{T}_X$ with two properties. First, if X has a club subset in an $\aleph_\omega$ and $\aleph_{\omega+1}$ preserving strongly covered outer model, then $\mathcal{T}_X$ is ill-founded. Secondly, if $\mathcal{T}_X$ is ill-founded, then there exist sequences $\vec{\kappa}$ and $\vec{F}$ that, in virtue of the Set Forcing Lemma, allow us to add a club subset to X with GCH, $\aleph_\omega$, and $\aleph_{\omega+1}$ preserving set forcing.

Nodes in the tree $\mathcal{T}_X$ provide finite approximations to $\vec{\kappa}$ and $\vec{F}$. However, whether $R(n, \vec{\kappa}, \vec{F})$ is stationary in $[H_\lambda]^{<\kappa_n}$ is a question about the entire sequences $\vec{\kappa}$ and $\vec{F}$. How can $\mathcal{T}_X$ control a global property of its infinite paths? This is where bounded pattern width comes into the picture.

Suppose that X has bounded pattern width. Let r be such that $1 \leqslant r < \omega$ and X has fewer than ω_r many patterns, modulo the non-stationary ideals. Restrict consideration to sequences $\vec{\kappa}$ such that $\kappa_1 > \omega_r$. For each n, $\vec{\kappa}$, and $\vec{F}$ such that $R(n, \vec{\kappa}, \vec{F})$ is non-stationary, choose a club $K_{n,\vec{\kappa},\vec{F}} \subseteq [H_\lambda]^{<\kappa_n}$ disjoint from it. Let K_μ be the intersection of all $K_{n,\vec{\kappa},\vec{F}}$ such that $\mu = \kappa_n$. Then K_μ is club in $[H_\lambda]^{<\mu}$. The point is that, given $\vec{\kappa}$ and $\vec{F}$, a single M lying in $R(n, \vec{\kappa}, \vec{F}) \cap K_{\kappa_n}$ witnesses that $R(n, \vec{\kappa}, \vec{F})$ is stationary in $[H_\lambda]^{<\kappa_n}$—otherwise K_{κ_n} would be contained in the complement of $R(n, \vec{\kappa}, \vec{F})$.

The tree $\mathcal{T}_X$ consists of triples

$$\Big(\langle\kappa_1,\ldots,\kappa_\ell\rangle,\langle M_1,\ldots,M_\ell\rangle,\langle f_i:\kappa_1\leqslant\omega_{i+1}<\kappa_\ell\rangle\Big),$$

where $\langle\kappa_1,\ldots,\kappa_\ell\rangle$ is initial segment of the desired $\vec{\kappa}$; $\langle f_i:\kappa_1\leqslant\omega_{i+1}<\kappa_\ell\rangle$ is a finite approximation to an initial segment of the desired $\vec{F}$ (so f_i is a finite approximation to F_i); and M_n lies in K_{κ_n} and restrains the growth of the f_i's and κ_m's so that ultimately $M_n\in R(n,\vec{\kappa},\vec{F})$.

Strong covering is used to see that $\mathcal{T}_X$ has an infinite path when X has a club subset in an $\aleph_\omega$ and $\aleph_{\omega+1}$ preserving outer model. It is used to get suitable M_n's that lie in V, hence occur in nodes of $\mathcal{T}_X$.

3. The set forcing lemma

Suppose that

$$\vec{\kappa}=\big\langle\,\kappa_n:1\leqslant n<\omega\,\big\rangle$$

is a strictly increasing sequence of uncountable cardinals below $\aleph_\omega$ and that

$$\vec{F}=\big\langle F_i:\kappa_1\leqslant\omega_{i+1}\leqslant\aleph_{\omega+1}\big\rangle$$

is such that F_i is a non-empty countable family of subsets of ω_{i+1}. Set $\kappa_0=\omega$.

In this section we develop a hypothesis sufficient for producing a GCH and $\aleph_\omega$ preserving set generic extension in which each $\bigcap F_i$ has a club subset and in which $\kappa_n=\aleph_n$, for all $n<\omega$. When we use the Set Forcing Lemma to prove our main theorem, we shall set $F_\omega=\{X\}$.

Let us recall two definitions from the previous section. Fix a regular cardinal $\lambda\gg\aleph_\omega$. Let H_λ be the set of sets of hereditary cardinality less than λ. If $M\subseteq H_\lambda$ and $i\leqslant\omega$, set

$$\sigma_M(i)=\sup\big(M\cap\omega_{i+1}\big).$$

In Shelah's usual notation, $\sigma_M(i)=\chi_M(\omega_{i+1})$.

For $1\leqslant n<\omega$, define a set $R(n,\vec{\kappa},\vec{F})\subseteq[H_\lambda]^{<\kappa_n}$ by declaring that $M\in R(n,\vec{\kappa},\vec{F})$ iff

- $M\in[H_\lambda]^{<\kappa_n}$ and $M^{<\kappa_{n-1}}\subseteq M$;
- if $\kappa_n\leqslant\omega_{i+1}\leqslant\aleph_{\omega+1}$, then $\sigma_M(i)\in\bigcap F_i$; and
- if $\kappa_2\leqslant\kappa_n\leqslant\omega_{i+1}\leqslant\aleph_{\omega+1}$ and $S\in F_i$, then $\operatorname{ptn}\big(S,\sigma_M(i)\big)\supseteq_{\mathrm{NS}}\bigcap F_j$, where $\operatorname{cf}\big(\sigma_M(i)\big)=\omega_{j+1}$ (so $\kappa_{n-1}\leqslant\omega_{j+1}<\kappa_n$).

Note that if $R(n,\vec{\kappa},\vec{F})$ is unbounded in $[H_\lambda]^{<\kappa_n}$, then $\bigcap F_i$ is unbounded in ω_{i+1}, for $\kappa_n\leqslant\omega_{i+1}\leqslant\aleph_{\omega+1}$.

Set Forcing Lemma. *Assume the GCH and suppose that $R(n, \vec{\kappa}, \vec{F})$ is stationary in $[H_\lambda]^{<\kappa_n}$, for all $n \geqslant 1$. Then there exists a GCH preserving set generic extension $V[G]$ in which*

- $\aleph_n = \kappa_n$, *for $n < \omega$, and V-cardinals greater than or equal to $\aleph_\omega$ are preserved;*
- *full covering holds between V and $V[G]$; and*
- *there exists a club $C_i \subseteq \bigcap F_i$, for $\kappa_1 \leqslant \omega_{i+1}^V \leqslant \aleph_{\omega+1}$.*

PROOF: Recall that F_i is a non-empty countable family of subsets of ω_{i+1} for $\kappa_1 \leqslant \omega_{i+1} \leqslant \aleph_{\omega+1}$. For the sake of uniform notation, set $F_i = \{\omega_{i+1}\}$ when $\omega_{i+1} < \kappa_1$. Recall also that we have set $\kappa_0 = \{\omega\}$.

We define $\mathbb{P}$ iteratively to add a club subset C_i of order type κ_n to $\bigcap F_i$, if $\kappa_n \leqslant \omega_{i+1} < \kappa_{n+1}$, and to add a club subset C_ω of order type $\aleph_{\omega+1}$ to $\bigcap F_\omega$. Conditions in $\mathbb{P}$ are functions p with domain $\omega + 2$; $p(i)$ is a term for an initial segment of a subset of ω_i. Because we are concerned only to add subsets to ω_i for $1 \leqslant i < \omega$ and for $i = \omega + 1$, the terms $p(0)$ and $p(\omega)$ are trivial.

Formally, define $\mathbb{P}^{\omega_n}$ by recursion on $n \leqslant \omega+2$ by declaring that $p \in \mathbb{P}^{\omega_n}$ iff

- $\mathrm{dom}(p) = n$;
- $p \restriction i \in \mathbb{P}^{\omega_i}$, for $i < n$;
- if $n = i + 1$ and i is ω or is 0, then $p(i) = \emptyset$;
- if $n = i + 1$ and i is a successor, then $\mathbb{P}^{\omega_i}$ forces that $p(i)$ is a closed subset of $\bigcap F_{i-1}$ and, if i is finite, that $\mathrm{ot}(p(i)) < \kappa_m$, where $\kappa_m \leqslant \omega_i < \kappa_{m+1}$.

Order $\mathbb{P}^{\omega_n}$ by $\bar{p} \geqslant p$ iff $p \restriction i \Vdash$ "$p(i)$ end-extends $\bar{p}(i)$", for all $i < n$.

Set $\mathbb{P} = \mathbb{P}^{\aleph_{\omega+2}}$.

By induction on $n < \omega$, note that we may assume that $|\mathbb{P}^{\omega_n}| < \omega_n$. Then $|\mathbb{P}^{\aleph_\omega}| = |\mathbb{P}^{\aleph_{\omega+1}}| = \aleph_{\omega+1}$. So we may assume that $|\mathbb{P}| = \aleph_{\omega+1}$.

Note also that, when $n < \omega$, forcing with $\mathbb{P}^{\kappa_{n+1}}$ collapses all infinite cardinals below κ_{n+1} that are not among $\kappa_0, \ldots, \kappa_n$. This is because it adds a club subset to $\bigcap F_i$, which is unbounded in ω_{i+1}, having order type at most κ_m, where $\kappa_m \leqslant \omega_{i+1} < \kappa_{m+1} \leqslant \kappa_{n+1}$. Since $|\mathbb{P}^{\kappa_{n+1}}| < \kappa_{n+1}$, forcing with $\mathbb{P}^{\kappa_{n+1}}$ preserves all cardinals greater than or equal to κ_{n+1}.

Let $\mathbb{P}_{\omega_n} = \{p \restriction [n, \omega+1] : p \in \mathbb{P}\}$. Then $\mathbb{P}$ is equivalent to $\mathbb{P}^{\omega_n} * \mathbb{P}_{\omega_n}$.

Together with these observations, the following claim suffices to prove the lemma.

Claim. $\mathbb{P}^{\kappa_{n+1}} \Vdash$ "$\mathbb{P}_{\kappa_{n+1}}$ *is $\leqslant\kappa_n$-distributive", for each $n < \omega$.*

The claim is proved by induction on n. Note that the claim for $i < n$ implies that forcing with $\mathbb{P}$ preserves $\kappa_0, \ldots, \kappa_n$. Hence $\mathbb{P}^{\kappa_{n+1}} \Vdash \aleph_i = \check{\kappa}_i$, for $i \leqslant n + 1$.

Suppose now that $\bar{p} \in \mathbb{P}_{\kappa_{n+1}}$ and that $\vec{D} = \langle \mathring{D}_i : i < \kappa_n \rangle$ is a sequence of terms such that $\mathbb{P}^{\kappa_{n+1}} \Vdash$ "$\mathring{D}_i$ is dense in $\mathbb{P}_{\kappa_{n+1}}$". Assume that H_λ is equipped with a well ordering, so that it has canonical Skolem functions. Choose $M \prec H_\lambda$ such that $\kappa_n \cup \mathbb{P}^{\kappa_{n+1}} \cup \{\mathbb{P}^{\kappa_{n+1}}, \bar{p}, \vec{D}\} \subseteq M$ and $M \in R(n+1, \vec{\kappa}, \vec{F})$. Let G be $\mathbb{P}^{\kappa_{n+1}}$ generic over V. Then G is $\mathbb{P}^{\kappa_{n+1}}$ generic over M. Set $\widehat{M} = M[G]$. Then, working in $V[G]$, we have

- $\widehat{M} \prec H^V_\lambda[G] = H^{V[G]}_\lambda$, since $\mathbb{P}^{\kappa_{n+1}} \cup \{\mathbb{P}^{\kappa_{n+1}}\} \subseteq M$;
- $|\widehat{M}| = |\kappa_n| = \aleph_n$;
- $\langle \mathring{D}^G_i : i < \kappa_n \rangle \in \widehat{M}$;
- $\widehat{M}^{<\aleph_n} \subseteq \widehat{M}$, since $\left(M^{<\kappa_n}\right)^V \subseteq M$; and
- $\sigma_{\widehat{M}}(j) = \sigma_M(j) \in \bigcap F_j$, for j such that $\kappa_{n+1} \leqslant \omega^V_{j+1} \leqslant \aleph_{\omega+1}$.

As an aside, note that it follows that $\mathrm{cf}\big(\sigma_{\widehat{M}}(j)\big) = \aleph_n$, for all j such that $\kappa_n \leqslant \omega_{j+1} \leqslant \aleph_{\omega+1}$, so $\kappa_n \leqslant \mathrm{cf}^V\big(\sigma_{\widehat{M}}(j)\big) < \kappa_{n+1}$.

We must now break into two cases, depending on whether $n = 0$ (so $\aleph_n = \omega$) or $n > 0$. If $n = 0$, choose a descending sequence of conditions $\langle p_i : i \leqslant \omega \rangle$ such that

- $p_0 = \bar{p}$;
- p_{i+1} meets $\mathring{D}^G_i$;
- $p_i \in \widehat{M}$, if $i < \omega$; and
- $\mathbb{P}^{\omega_{j+1}} \Vdash \sup\big(p_\omega(j+1)\big) = \sup_{i<\omega}\big(\sup(p_i(j+1))\big) = \sigma_{\widehat{M}}(j)^\vee$, for j such that $\kappa_1 \leqslant \omega^V_{j+1} \leqslant \aleph_{\omega+1}$.

Then p_ω is as required.

In the other case, when $n > 0$, let $\langle x_i : i < \aleph_n \rangle$ enumerate $\widehat{M}$ and set

$$\widehat{M}_i = \text{the least } M \prec \widehat{M} \text{ such that } \{\mathbb{P}^{\kappa_{n+1}}, \bar{p}, \vec{D}, G\} \cup \\ \cup \{x_j : j < i\} \cup \{\widehat{M}_j : j < i\} \subseteq M.$$

Note by induction on i that $\widehat{M}_i \subseteq \widehat{M}$ and $|\widehat{M}_i| < \aleph_n$, so $\widehat{M}_i \in \widehat{M}$. Consequently, the recursion does not break down before stage $\aleph_n$.

Note by induction on i that $i \subseteq \widehat{M}_i$.

Suppose that $\kappa_{n+1} \leqslant \omega^V_{j+1} \leqslant \aleph_{\omega+1}$.

- If $i < i'$, then $\sigma_{\widehat{M}_i}(j) < \sigma_{\widehat{M}_{i'}}(j)$, since $\widehat{M}_{i'}$ satisfies that $|\widehat{M}_i| < \kappa_n$.
- If i is a limit ordinal, then $\sigma_{\widehat{M}_i}(j) = \sup_{i'<i} \sigma_{\widehat{M}_{i'}}(j)$, since $\widehat{M}_i = \bigcup_{i'<i} \widehat{M}_{i'}$.
- $\sup_{i<\aleph_n} \sigma_{\widehat{M}_i}(j) = \sigma_{\widehat{M}}(j)$, since $\widehat{M} = \bigcup_{i<\aleph_n} \widehat{M}_i$.

Thus $\langle \sigma_{\widehat{M}_i}(j) : i < \aleph_n \rangle$ is continuous, increasing, and cofinal in $\sigma_{\widehat{M}}(j)$, which lies in $\bigcap F_j$.

Now $n+1 \geqslant 2$, so if $S \in F_j$, then $\mathrm{ptn}\big(S, \sigma_{\widehat{M}}(j)\big) \supseteq_{\mathrm{NS}} \bigcap F_\ell$, for some ℓ such that $\kappa_n \leqslant \omega_{\ell+1}^V < \kappa_{n+1}$, namely, the ℓ such that $\mathrm{cf}^V\big(\sigma_{\widehat{M}}(j)\big) = \omega_{\ell+1}^V$. It follows that $\bigcap_{S \in F_j} \mathrm{ptn}\big(S, \sigma_{\widehat{M}}(j)\big)$ has a club subset in $V[G]$. By passing to a continuous subsequence of $\langle\, \widehat{M}_i : i < \aleph_n \,\rangle$ if necessary, we may assume that $\sigma_{\widehat{M}_i}(j) \in \bigcap F_j$, for all $i < \aleph_n$ and all j such that $\kappa_{n+1} \leqslant \omega_{j+1}^V \leqslant \aleph_{\omega+1}$.

Now $\langle\, \widehat{M}_j : j \leqslant i \rangle \in \widehat{M}$, for all $i < \aleph_n$, since $\widehat{M}^{<\aleph_n} \subseteq \widehat{M}$. So by passing to a further continuous subsequence, if necessary, we may assume that $\langle\, \widehat{M}_j : j \leqslant i \,\rangle \in \widehat{M}_{i+1}$, for all $i < \aleph_n$.

To finish the proof of the claim, choose a descending sequence of conditions p_i for $i \leqslant \aleph_n$ such that

- $p_0 \leqslant \bar{p}$;
- p_{i+1} meets $\mathring{D}_i^G$;
- $p_i \in \widehat{M}_{i+1}$; and
- $p_i \restriction (j{+}1) \Vdash \sup\big(p_i(j{+}1)\big) = \sigma_{\widehat{M}_i}(j)$, for all j such that $\kappa_{n+1} \leqslant \omega_{j+1}^V \leqslant \aleph_{\omega+1}$.

Then $p_{\aleph_n}$ is as required for the claim. □

4. The tree $\mathcal{T}_X$

Fix $X \subseteq \aleph_{\omega+1}$ having bounded pattern width. In this section, we define the tree $\mathcal{T}_X$ and its ordering $\lhd$.

For $n < \omega$, let P_n contain exactly one representative from each equivalence class, modulo the non-stationary ideal on ω_{n+1}, of

$$\Big\{\, S \subseteq \omega_{n+1} : S =_{\mathrm{NS}} \mathrm{ptn}(X, \alpha), \text{ for some } \alpha \in \lim(X) \text{ of cofinality } \omega_{n+1} \Big\}.$$

Then $\sup_{n \in \omega} |P_n| < \aleph_\omega$, since X has bounded pattern width and we are assuming the GCH.

Fix r such that $1 \leqslant r < \omega$ and $|P_n| < \omega_r$, for all n. Suppose we are given

- a natural number $n \geqslant 1$,
- an increasing sequence $\vec{\kappa} = \langle\, \kappa_n : 1 \leqslant n < \omega \,\rangle$ of uncountable cardinals such that $\omega_r < \kappa_1 < \kappa_2 < \cdots < \aleph_\omega$, and
- a sequence $\vec{F} = \langle\, F_i : \kappa_1 \leqslant \omega_{i+1} \leqslant \aleph_{\omega+1} \,\rangle$ such that $F_i \subseteq P_i \cup \{\omega_{i+1}\}$ is non-empty, for $i < \omega$, and $F_\omega = \{X\}$.

For each such n, each such $\vec{\kappa}$, and each such $\vec{F}$, choose a club $K_{n,\vec{\kappa},\vec{F}} \subseteq [H_\lambda]^{<\kappa_n}$ disjoint from $R(n, \vec{\kappa}, \vec{F})$, if possible. If $R(n, \vec{\kappa}, \vec{F})$ is stationary, let $K_{n,\vec{\kappa},\vec{F}} = [H_\lambda]^{<\kappa_n}$.

For each cardinal μ such that $\omega_r < \mu < \aleph_\omega$, set

$$K_\mu = \bigcap \Big\{\, K_{n,\vec{\kappa},\vec{F}} : \mu = \kappa_n \Big\}.$$

Using the GCH, note that there exist at most ω_r many triples $(n, \vec{\kappa}, \vec{F})$. And $K_{n,\vec{\kappa},\vec{F}}$ is club in $[H_\lambda]^{<\mu}$ when $\mu = \kappa_n$. Hence K_μ is club in $[H_\lambda]^{<\mu}$, since $\kappa_n > \omega_r$.

Define $\mathcal{T}_X$ to consist of all triples

$$\Big(\langle \kappa_1, \ldots, \kappa_\ell \rangle, \langle M_1, \ldots, M_\ell \rangle, \langle f_i : \kappa_1 \leqslant \omega_{i+1} < \kappa_\ell \rangle\Big),$$

where, letting $f_\omega = \{X\}$,

- $\ell < \omega$;
- $\omega_r < \kappa_1 < \cdots < \kappa_\ell < \aleph_\omega$ are cardinals;
- $f_i \subseteq P_i \cup \{\omega_{i+1}\}$ is finite and non-empty;
- $M_n \in K_{\kappa_n}$ and $M_n^{<\kappa_{n-1}} \subseteq M_n$, if $n > 1$;
- if $\kappa_n \leqslant \omega_{i+1} < \kappa_\ell$ or $i = \omega$, then $\sigma_{M_n}(i) \in \bigcap f_i$; and
- if $\kappa_2 \leqslant \kappa_n \leqslant \omega_{i+1} < \kappa_\ell$ or $i = \omega$, and if $S \in f_i$, then $\text{ptn}(S, \sigma_{M_n}(i)) \supseteq_{\text{NS}} \bigcap f_j$, where $\text{cf}(\sigma_{M_n}(i)) = \omega_{j+1}$ (so $\kappa_{n-1} \leqslant \omega_{j+1} < \kappa_n$).

Order $\mathcal{T}_X$ by $(\vec{\kappa}, \vec{M}, \vec{f}) \rhd (\vec{\kappa}', \vec{M}', \vec{f}')$ iff

- $\vec{\kappa}$ is a proper initial segment of $\vec{\kappa}'$;
- $\vec{M}$ is a proper initial segment of $\vec{M}'$; and
- $f_i \subseteq f_i'$, for all $i \in \text{dom}(\vec{f})$.

5. Proof of the main theorem

Assume the GCH. Fix $X \subseteq \aleph_{\omega+1}$ of bounded pattern width. We shall prove that the following three statements are equivalent.

(1) There is a club subset of $\aleph_{\omega+1}$ contained in X in an $\aleph_\omega$ and $\aleph_{\omega+1}$ preserving outer model in which Player II does not have a winning strategy in the $(\aleph_{n+1}, \aleph_n, \aleph_{n-1}, \alpha)$-covering game, for $2 \leqslant n < \omega$ and all α.

(2) $\mathcal{T}_X$ is ill-founded.

(3) There is a club subset of $\aleph_{\omega+1}$ contained in X in a fully covered GCH, $\aleph_\omega$, and $\aleph_{\omega+1}$ preserving set forcing extension. In this extension Player II does not have a winning strategy in the $(\aleph_{n+1}, \aleph_n, \aleph_{n-1}, \alpha)$-covering game, for $1 \leqslant n < \omega$ and all α.

Clearly (3) $\Rightarrow$ (1).

As recounted in §1, it is a theorem of Shelah's that, if we assume $\square$ as well as the GCH in the inner model containing X, then Player I has a winning strategy in the games mentioned in (1), provided that ordinary covering holds between that model and the outer model in which X has a club subset. Thus, assuming $\square$ in the inner model, (1) is implied by

(1′) There is a club subset of $\aleph_{\omega+1}$ contained in X in some covered $\aleph_\omega$ and $\aleph_{\omega+1}$ preserving outer model.

Clearly (3) implies its first sentence

(3′) There is a club subset of $\aleph_{\omega+1}$ contained in X in a fully covered, GCH, $\aleph_\omega$, and $\aleph_{\omega+1}$ preserving set generic extension.

Thus, once we have shown that (1)–(3) are equivalent, we shall have that, under the additional hypothesis that $\Box$ holds in the inner model, $(1') \Rightarrow (1) \Rightarrow (2) \Rightarrow (3) \Rightarrow (3')$. Certainly $(3') \Rightarrow (1')$. So we shall have that (1′), (2), and (3′) are equivalent under this extra hypothesis. This is the theorem stated in §1.

The following lemma proves that $(2) \Rightarrow (3)$.

Lemma. *If $(\mathcal{T}_X, \lhd)$ is ill founded, then X has a club subset in an $\aleph_\omega$, $\aleph_{\omega+1}$, and GCH preserving fully covered set generic extension $V[G]$ in which Player II does not have a winning strategy in the $(\aleph_{n+1}, \aleph_n, \aleph_{n-1}, \alpha)$-covering game, for $1 \leqslant n < \omega$ and all α.*

PROOF: Suppose that

$\langle \kappa_n : 1 \leqslant n < \omega \rangle$, $\langle M_n : 1 \leqslant n < \omega \rangle$, and $\langle f_i^\ell : \ell < \omega \ \& \ \kappa_1 \leqslant \omega_{i+1} < \kappa_\ell \rangle$

are provided by an infinite $\lhd$-descending sequence through $\mathcal{T}_X$. The ℓth element of this path is $\big(\langle \kappa_1, \dots, \kappa_\ell \rangle, \langle M_1, \dots, M_\ell \rangle, \langle f_i^\ell : \kappa_1 \leqslant \omega_{i+1} < \kappa_\ell \rangle\big)$. For the sake of uniform notation, set

$$\kappa_0 = \omega \quad \text{and set} \quad f_\omega^\ell = \{X\},$$

for all ℓ.

Set $F_i = \bigcup_{\ell<\omega} f_i^\ell$ and let $\vec{F} = \langle F_i : \kappa_1 \leqslant \omega_{i+1} \leqslant \aleph_{\omega+1} \rangle$. To apply the Set Forcing Lemma, we need this

Claim. *$R(n, \vec{\kappa}, \vec{F})$ is stationary, for $1 \leqslant n < \omega$.*

PROOF: Fix n. Then $M_n \in K_{\kappa_n}$ and $M_n^{<\kappa_{n-1}} \subseteq M_n$. Suppose that $\kappa_n \leqslant \omega_{i+1} \leqslant \aleph_{\omega+1}$. Then $\sigma_{M_n}(i) \in \bigcap f_i^\ell$, if $\kappa_\ell > \omega_{i+1}$. Hence $\sigma_{M_n}(i) \in \bigcap F_i$. If also $n \geqslant 2$ and $S \in F_i$, say $S \in f_i^\ell$, then $\operatorname{ptn}(S, \sigma_{M_n}(i)) \supseteq_{\text{NS}} \bigcap f_j^\ell$, where $\operatorname{cf}(\sigma_{M_n}(i)) = \omega_{j+1}$. So $\operatorname{ptn}(\bigcap F_i, \sigma_{M_n}(i)) = \bigcap_{S \in F_i} \operatorname{ptn}(S, \sigma_{M_n}(i)) \supseteq_{\text{NS}} \bigcap F_j$. Thus $M_n \in R(n, \vec{\kappa}, \vec{F}) \cap K_{\kappa_n}$. It follows that $R(n, \vec{\kappa}, \vec{F})$ is stationary in $[H_\lambda]^{<\kappa_n}$, since otherwise $K_{n,\vec{\kappa},\vec{F}}$ is disjoint from $R(n, \vec{\kappa}, \vec{F})$ and $K_{\kappa_n} \subseteq K_{n,\vec{\kappa},\vec{F}}$. This completes thc proof of the Claim.

Let $\mathbb{P}$ be the forcing property of the Set Forcing Lemma. If G is $\mathbb{P}$ generic over V, then in $V[G]$

- $\aleph_n = \kappa_n$, for $n < \omega$, and V-cardinals greater than or equal to $\aleph_\omega$ are preserved;
- full covering holds between V and $V[G]$; and
- there exists a club $C_i \subseteq \bigcap F_i$, for $\kappa_1 \leqslant \omega_{i+1}^V \leqslant \aleph_{\omega+1}$. In particular, C_ω is club in $\aleph_{\omega+1}$ and $C_\omega \subseteq X$, since $F_\omega = \{X\}$.

The one matter left to check is this

Claim. *For $1 \leqslant n < \omega$ and all α, Player II does not have a winning strategy in the $(\aleph_{n+1}, \aleph_n, \aleph_{n-1}, \alpha)$-covering game.*

PROOF: Fixing n, $\mathbb{P}$ factors as $\mathbb{P}^{\kappa_{n+1}} * \mathbb{P}_{\kappa_{n+1}}$, where $|\mathbb{P}^{\kappa_{n+1}}| < \kappa_{n+1}$ and $\mathbb{P}^{\kappa_{n+1}}$ forces that $\mathbb{P}_{\kappa_{n+1}}$ is $\leqslant\kappa_n$-distributive and that $\kappa_i = \aleph_i$, for $i \leqslant n+1$. So it suffices to see that in a $\mathbb{P}^{\kappa_{n+1}}$ generic extension, Player II does not have a winning strategy in the $(\kappa_{n+1}, \kappa_n, \kappa_{n-1}, \alpha)$-covering game, because subsequent forcing with $\mathbb{P}_{\kappa_{n+1}}$ adds no more sets of ordinals of size less than κ_{n+1}.

Fix α and a term $\mathring{s}$ for a strategy for Player II. Let λ be large and regular and let μ be the cardinal predecessor of κ_{n+1} in V. Supposing that H_λ is equipped with canonical Skolem functions, set

$$M_i = \text{the least } M \prec H_\lambda \text{ such that } \mu \cup \mathbb{P}^{\kappa_{n+1}} \cup \{\mathbb{P}^{\kappa_{n+1}}, \alpha, \mathring{s}\} \cup \{\, M_j : j < i \,\} \subseteq M,$$

for $i < \kappa_n$. Then $M_i = \bigcup_{j<i} M_j$, when $i < \kappa_n$ is a limit ordinal, and $\langle M_j : j \leqslant i \rangle \in M_{i+1}$. Suppose that G is $\mathbb{P}^{\kappa_{n+1}}$ generic over V. Then G is $\mathbb{P}^{\kappa_{n+1}}$ generic over M_i, so $M_i[G] \prec H^V_\lambda[G]$. Set $a_i = M_i \cap \alpha$. Then $a_i = M_i[G] \cap \alpha$. Set $b_i = \mathring{s}^G\big(\langle a_j, b_j : j < i \rangle^\frown \langle a_i \rangle\big)$. Note by induction that $b_i \in M_{i+1}[G]$. And $b_i \subseteq M_{i+1}[G]$, since $|b_i|^{V[G]} \leqslant \mu$. Hence $a_{i+1} \supseteq b_i$. And $\bigcup_{j<i} a_j = a_i \in V$, for all limit $i < \kappa_n$. Thus, playing the sequence $\langle a_i : i < \kappa_n \rangle$, Player I beats Player II playing according to the strategy $\mathring{s}^G$.□

Finally, we must see that (1) $\Rightarrow$ (2).

Lemma. *Assume the GCH in V. Suppose that there exists a club subset of $\aleph_{\omega+1}$ contained in X in some $\aleph_\omega$ and $\aleph_{\omega+1}$ preserving outer model W of V in which Player II does not have a winning strategy in the $(\aleph_{n+1}, \aleph_n, \aleph_{n-1}, \alpha)$-covering game, for $2 \leqslant n < \omega$ and all α. Then $(\mathcal{T}_X, \lhd)$ is ill-founded.*

PROOF: Work in W. By passing to a generic extension, if necessary, we may assume that $\aleph_1 > \omega^V_r$. (Recall that ω_r is chosen so that $1 \leqslant r < \omega$ and $|P_n| < \omega_r$, for all n, when $\mathcal{T}_X$ was defined in V.) Set $\kappa_n = \aleph_{2n+1}$, for $1 \leqslant n < \omega$, and set $\kappa_0 = \aleph_0$. Set

$$F_i = \Big\{\, S \in P_i : S \text{ has a club subset} \,\Big\} \cup \Big\{\, \omega^V_{i+1} \,\Big\},$$

for $\kappa_1 \leqslant \omega^V_{i+1} < \aleph_\omega$. Set $F_\omega = \{X\}$. Because $\aleph_1 > \omega^V_r > |P_i|^V$, we have that F_i is countable. Thus there exists a club $C_i \subseteq \bigcap F_i$, for $\kappa_1 \leqslant \omega^V_{i+1} \leqslant \aleph_{\omega+1}$.

Claim. *Fix $n \geqslant 1$. There exists $M \in V$ such that $M \in K_{\kappa_n}$ and*

- *$M^{<\kappa_{n-1}} \cap V \subseteq M$;*
- *if $\kappa_n \leqslant \omega^V_{i+1} \leqslant \aleph_{\omega+1}$, then $\sigma_M(i) \in \bigcap F_i$; and*
- *if $\kappa_2 \leqslant \kappa_n \leqslant \omega^V_{i+1} \leqslant \aleph_{\omega+1}$ and $S \in F_i$, then $\operatorname{ptn}(S, \sigma_M(i)) \supseteq_{\mathrm{NS}} \bigcap F_j$, where $\operatorname{cf}^V(\sigma_M(i)) = \omega^V_{j+1}$ (so $\kappa_{n-1} \leqslant \omega^V_{j+1} < \kappa_n$).*

PROOF: Note that $\kappa_n = \aleph_{2n+1}$ and that $\kappa_{n-1} = \aleph_{2n-1}$, if $n > 1$ and $\kappa_{n-1} = \aleph_0$, if $n = 1$.

By hypothesis, Player II does not have a winning strategy in the following $(\aleph_{2n+1}, \aleph_{2n}, \aleph_{2n-1}, |H^V_\lambda|)$-covering game:

Each player makes $\aleph_{2n}$ moves, in alternation, Player I playing the sequence $\langle M_i : i < \aleph_{2n} \rangle$, and Player II playing the sequence $\langle N_i : i < \aleph_{2n} \rangle$. At move i, Player I plays a set M_i such that

- $\bigcup_{j<i} N_j \subseteq M_i \subseteq H^V_\lambda$ and
- $|M_i|^W = \aleph_{2n}$.

At move i, Player II plays a set N_i such that

- $\bigcup_{j\leqslant i} M_j \subseteq N_i \subseteq H^V_\lambda$;
- $|N_i|^W = \aleph_{2n}$;
- $\sigma_{N_i}(k) > \sup_{j<i} \sigma_{N_j}(k)$, if $\aleph_{2n+1} \leqslant \omega^V_{k+1} \leqslant \aleph_{\omega+1}$;
- if $x \in \left[\bigcup_{j\leqslant i} M_j\right]^{<\aleph_{2n-1}} \cap V$, then $x \in N_i$; and
- $\left(N_i; K_{\kappa_n} \cap N_i, C_k \cap N_i\right)_{k\in I} \prec \left(H^V_\lambda; K_{\kappa_n}, C_k\right)_{k\in I}$, where $I = \{\, k \leqslant \omega : \aleph_{2n+1} \leqslant \omega^V_{k+1} \leqslant \aleph_{\omega+1} \,\}$.

Player I wins if there exists a club subset of $\aleph_{2n}$ such that if ℓ is a cofinality $\aleph_{2n-1}$ limit point of this set, then $\bigcup_{j<\ell} M_j \in V$.

Fix a play of this game in which Player I wins, then fix a cofinality $\aleph_{2n-1}$ limit point ℓ such that $\bigcup_{j<\ell} M_j \in V$. Set $M = \bigcup_{j<\ell} M_j$. On account of Player II's play, we have that

- $M^{<\kappa_{n-1}} \cap V \subseteq M^{<\aleph_{2n-1}} \cap V \subseteq M$;
- $M = \bigcup(K_{\kappa_n} \cap M)$, hence $M \in K_{\kappa_n}$; and
- $\sigma_M(i) \in \lim(C_i)$, hence $\sigma_M(i) \in \bigcap F_i$, if $\kappa_n \leqslant \omega^V_{i+1} \leqslant \aleph_{\omega+1}$.

Finally, suppose that $\kappa_2 \leqslant \kappa_n \leqslant \omega^V_{i+1} \leqslant \aleph_{\omega+1}$ and that $S \in F_i$. We may assume that $S \in P_i$ or, if $i = \omega$, that $S = X$. Say $\mathrm{cf}^V(\sigma_M(i)) = \omega^V_{j+1}$. Now $\mathrm{ptn}(S, \sigma_M(i)) \in_{\mathrm{NS}} P_j$, because $\mathrm{ptn}(S, \sigma_M(i))$ is a pattern of S, which is a pattern of X. Also $\mathrm{ptn}(S, \sigma_M(i))$ has a club subset in W because $\sigma_M(i) \in \lim(C_i)$. Hence $\mathrm{ptn}(S, \sigma_M(i)) \in_{\mathrm{NS}} F_j$, and so $\mathrm{ptn}(S, \sigma_M(i)) \supseteq_{\mathrm{NS}} \bigcap F_j$, completing the proof of the Claim. □

Let M_n be as in the Claim, for $1 \leqslant n < \omega$. Let $f^\ell_\omega = \{X\}$, for all ℓ, and let $\langle f^\ell_i : \kappa_1 \leqslant \omega^V_{i+1} < \kappa_\ell \text{ and } 1 \leqslant \ell < \omega \rangle$ be such that

- $f^\ell_i \subseteq F_i$ is finite and non-empty;
- if $\ell \leqslant \ell'$, then $f^\ell_i \subseteq f^{\ell'}_i$;
- $F_i = \bigcup_\ell f^\ell_i$; and
- if $\kappa_2 \leqslant \kappa_n \leqslant \omega^V_{i+1} < \kappa_\ell$ or $i = \omega$, and if $S \in f^\ell_i$, then $\mathrm{ptn}(S, \sigma_{M_n}(i)) \in_{\mathrm{NS}} f^\ell_j$, where $\mathrm{cf}^V(\sigma_{M_n}(i)) = \omega^V_{j+1}$.

Then

$$\Big(\langle\kappa_1,\ldots,\kappa_\ell\rangle, \langle M_1,\ldots,M_\ell\rangle, \langle\, f_i^\ell : \kappa_1 \leqslant \omega_{i+1}^V < \kappa_\ell\,\rangle\Big)_{\ell\in\omega}$$

is a $\lhd$-descending sequence through $\mathcal{T}_X$. □

6. Remarks

Set

$$\mathcal{S} = \Big\{\, S \subseteq \omega_2 : S \text{ has a club subset in an } \omega_1 \text{ and } \omega_2 \text{ preserving outer model}\Big\}.$$

This definition specifies $\mathcal{S}$ using a non-first-order condition over the universe V in which we are working. In part, **[S]** addresses the question whether $\mathcal{S}$ has a first-order definition, that is, allowing parameters in such definitions, whether $\mathcal{S}$ is an element of the universe. It is shown that

(1) If V is the minimum model of ZFC, then $\mathcal{S} \notin V$.
(2) If V is "sufficiently non-minimal", then $\mathcal{S} \in V$.
(3) However, there is no uniform definition of $\mathcal{S}^W$, for inner and outer models having the same ω_1 and ω_2.

It follows that there is no satisfactory characterization of the sets in $\mathcal{S}$ analogous to the characterization of subsets S of ω_1 that have club subsets in ω_1-preserving outer models, namely, "S is stationary".

Remark 1. In this paper's main theorem "$\aleph_\omega$ and $\aleph_{\omega+1}$ preserving" cannot be improved to "cardinal preserving." This is because the negative results of **[S]** apply to subsets of $\aleph_{\omega+1}$ if we insist on full cardinal preservation, even if we restrict our attention to sets of bounded pattern width.

For example, let L be the minimum model of ZFC. Then $\mathcal{S} \notin L$. Suppose that $S \subseteq \omega_2$. Using $\Box$ it is possible to define a sequence of functions $\langle\, f_\alpha : \alpha \text{ is a limit and } \mathrm{cf}(\alpha) \leqslant \omega_2\rangle$ such that $f_\alpha : \delta \to \alpha$, for some $\delta \leqslant \omega_2$, is continuous, monotonically increasing, and cofinal in α, and such that if $f_\alpha(\gamma) = f_\beta(\gamma')$, for some limit ordinal γ, then $\gamma = \gamma'$ and $f_\alpha \restriction \gamma = f_\beta \restriction \gamma$. Set

$$X = \bigcup_{\substack{\alpha<\aleph_{\omega+1}\\ \mathrm{cf}(\alpha)=\omega_2}} f_\alpha{}''S \cup \Big\{\, \alpha < \aleph_{\omega+1} : \mathrm{cf}(\alpha) \geqslant \omega_2 \Big\}.$$

Then X has bounded pattern width and S has a club subset in any cardinal preserving outer model in which X has a club subset. Furthermore, if S has a club subset C in a cardinal preserving outer model, then there exists a cardinal preserving set generic extension of $L[C]$ in which X has a club subset. Thus

$$\mathcal{X} = \Big\{\, X \subseteq \aleph_{\omega+1} : X \text{ has bounded pattern width and has a club subset in some cardinal preserving outer model}\Big\}$$

is not an element of the universe.

Remark 2. If λ is singular and uncountable, then there is no evident generalization of this paper's main theorem to characterize even bounded pattern width subsets of $\aleph_{\lambda+1}$ that have club subsets in $\aleph_\lambda$ and $\aleph_{\lambda+1}$ strongly covered outer models, at least if we insist that λ remains uncountable in the outer model. (Otherwise, we would be cheating.) This is because a tree analogous to $\mathcal{T}_X$ would not be a finite path tree.

This is not merely a shortcoming of our proof. Again, let us consider the case of the minimum model. Given a tree T of height and cardinality ω_1, there exists a bounded pattern width subset X of $\aleph_{\omega_1+1}$ (uniformly definable from T) such that T has a cofinal branch in any ω_1, $\aleph_{\omega_1}$ and $\aleph_{\omega_1+1}$ preserving outer model in which X has a club subset. A proof of this claim will appear elsewhere. The idea is to build the set X in such a way that its "pattern-of ordering" mirrors T. If $S \subseteq \mu$ and $S' \subseteq \mu'$ are patterns occurring in X, say that S is below S' in this ordering if $\mu < \mu'$ and $\{\, \alpha \in \lim(S') : \mathrm{cf}(\alpha) = \mu \text{ and } \mathrm{ptn}(S', \alpha) \supseteq_{\mathrm{NS}} S \,\}$ is stationary in μ'.

Another result of [S] is that in the minimum model it is not possible to characterize the trees of height and cardinality ω_1 that have cofinal branches in ω_1 preserving outer models. Thus it is not possible to characterize the subsets of $\aleph_{\omega_1+1}$ that have club subsets in ω_1, $\aleph_{\omega_1}$, and $\aleph_{\omega_1+1}$ preserving outer models.

Other negative results of [S] have analogs in this case, as well.

Remark 3. If κ is regular, then there does exist a uniform characterization of those bounded pattern width $X \subseteq \kappa^+$ such that in some κ and κ^+ preserving covered outer model there exists a club subset of κ^+ that is contained in X.

Indeed, "X is stationary" suffices when $\kappa^+ = \omega_1$. If κ is regular and uncountable and $X \subseteq \kappa^+$ has bounded pattern width, then the following three statements are equivalent for $X \subseteq \kappa^+$ of bounded pattern width:

(1) In some covered κ and κ^+ preserving outer model, there exists a club subset of κ^+ contained in X.

(2) $\left\{\, \alpha \in \lim(X) : \mathrm{cf}(\alpha) = \kappa \text{ and } \mathrm{ptn}(X, \alpha) =_{\mathrm{NS}} S \,\right\}$ is stationary in κ^+, for some stationary $S \subseteq \kappa$,

(3) In a fully covered GCH, κ, and κ^+ preserving set generic extension, there exists a club subset of κ^+ contained in X.

For (1) $\Rightarrow$ (2), suppose that $C \subseteq X$ is a club subset of κ^+ lying in an outer model of V as in (1). In this model κ remains regular and uncountable. For each $S \subseteq \kappa$ lying in V, set

$$Z_S = \left\{\, \alpha \in \lim(C) : \mathrm{cf}(\alpha) = \kappa \text{ and } V \models \text{“}\mathrm{ptn}(X, \alpha) =_{\mathrm{NS}} S\text{”} \,\right\}.$$

Then $\{\, Z_S : S \subseteq \kappa \text{ lies in } V \,\}$ splits $\lim(C)$ into fewer than κ^+ many pieces, since X has bounded pattern width and we are assuming the GCH in V. (2) follows.

Suppose now that $S \subseteq \kappa$ is as in (2). Then S remains stationary after κ is Lévy collapsed to become ω_1. Then X is a "fat stationary" subset of ω_2, in the sense of [**AS**], after forcing to add a club subset to S. It follows that X has a club subset in a further GCH, κ, and κ^+ preserving set generic extension. Thus (3) follows from (2).

Remark 4. Several questions are open.

(A) Is the hypothesis that Player II does not have a winning strategy in the $(\aleph_{n+1}, \aleph_n, \aleph_{n-1}, \alpha)$-covering game necessary for (1) $\Rightarrow$ (2) in this paper's main theorem?

(B) Is the restriction to sets of bounded pattern width necessary for such a theorem? That is, is there a uniform first-order characterization of $X \subseteq \aleph_{\omega+1}$ such that X has a club subset in some (strongly covered) $\aleph_\omega$ and $\aleph_{\omega+1}$ preserving outer model?

(C) (M. Foreman) Suppose that $X \subseteq \aleph_{\omega+1}$ has a club subset in some (strongly covered) $\aleph_\omega$ and $\aleph_{\omega+1}$ preserving outer model. Does there exist a $Y \subseteq X$ of bounded pattern width having a club subset in an $\aleph_\omega$ and $\aleph_{\omega+1}$ preserving (strongly covered) outer model?

(D) Suppose that $\alpha > \omega$ is a singular limit ordinal. Is there a uniform first-order characterization of (bounded pattern width) $X \subseteq \aleph_{\alpha+1}$ such that X has a club subset in some (strongly covered) $\aleph_\alpha$ and $\aleph_{\alpha+1}$ preserving outer model?

An interesting case of (D) occurs when $\alpha = \aleph_\alpha$. If $\alpha < \aleph_\alpha$, then we can collapse α to cardinality ω while preserving $\aleph_\alpha$ and $\aleph_{\alpha+1}$.

References

[AS] U. Abraham and S. Shelah, *Forcing closed unbounded sets*, Jour. Sym. Log. (48), 1983, 643–657.

[Sh] S. Shelah, *Cardinal Arithmetic*, Oxford Sci. Publ., 1994.

[S] M.C. Stanley, *Forcing closed unbounded subsets of* ω_2, 1998 (to appear).

First Steps into Metapredicativity in Explicit Mathematics

Thomas Strahm
Institut für Informatik und angewandte Mathematik
Universität Bern, Neubrückstrasse 10
CH-3012 Bern, Switzerland
strahm@iam.unibe.ch

Abstract

The system EMU of explicit mathematics incorporates the *uniform* construction of universes. In this paper we give a proof-theoretic treatment of EMU and show that it corresponds to transfinite hierarchies of fixed points of positive arithmetic operators, where the length of these fixed point hierarchies is bounded by ε_0.

1 Introduction

Metapredicativity is a new general term in proof theory which describes the analysis and study of formal systems whose proof-theoretic strength is beyond the Feferman-Schütte ordinal Γ_0 but which are nevertheless amenable to purely predicative methods. Typical examples of formal systems which are apt for scaling the initial part of metapredicativity are the transfinitely iterated fixed point theories $\widehat{\mathsf{ID}}_\alpha$ whose detailed proof-theoretic analysis is given by Jäger, Kahle, Setzer and Strahm in [18]. In this paper we assume familiarity with [18]. For natural extensions of Friedman's ATR that can be measured against transfinitely iterated fixed point theories the reader is referred to Jäger and Strahm [20].

In the mid seventies, Feferman [3, 4] introduced systems of *explicit mathematics* in order to provide an alternative foundation of constructive mathematics. More precisely, the origin of Feferman's program lay in giving a logical account of Bishop-style constructive mathematics. Right from the beginning, systems of explicit mathematics turned out to be of general interest for proof theory, mainly in connection with the proof-theoretic analysis of subsystems of first and second order arithmetic and set theory, cf. e.g. Jäger

[15] and Jäger and Pohlers [19]. More recently, systems of explicit mathematics have been used to develop a general logical framework for functional programming and type theory, where it is possible to derive correctness and termination properties of functional programs. Important references in this connection are Feferman [6, 7, 9] and Jäger [17].

Universes are a frequently studied concept in constructive mathematics at least since the work of Martin-Löf, cf. e.g. Martin-Löf [23] or Palmgren [27] for a survey. They can be considered as types of types (or names) which are closed under previously recognized type formation operations, i.e. a universe *reflects* these operations. Hence, universes are closely related to reflection principles in classical and admissible set theory. Universes were first discussed in the framework of explicit mathematics in Feferman [5] in connection with his proof of Hancock's conjecture. In Marzetta [25, 24] they are introduced via a so-called (non-uniform) limit axiom, thus providing a natural framework of explicit mathematics which has exactly the strength of predicative analysis, cf. also Marzetta and Strahm [26] and Kahle [22].

In this paper we discuss the system EMU of explicit mathematics which contains a *uniform* universe construction principle and includes full formula induction on the natural numbers. Our universes are closed under elementary comprehension and join (disjoint union), and there is an operation which uniformly takes a given type (name) and yields a universe containing that name. We show that EMU is proof-theoretically equivalent to the transfinitely iterated fixed point theory $\widehat{\mathsf{ID}}_{<\varepsilon_0}$ with proof-theoretic ordinal $\varphi 1 \varepsilon_0 0$ for φ a ternary Veblen function. Independently and very recently, similar results have been obtained in the context of Frege structures by Kahle [21] and in the framework of Martin-Löf type theory by Rathjen [29].

The plan of the paper is as follows. In Section 2 we give the formal definition of the system EMU. Section 3 is devoted to a wellordering proof for EMU, i.e. we establish $\varphi 1 \varepsilon_0 0 \leq |\mathsf{EMU}|$. In Section 4 we describe a proof-theoretic reduction of EMU to $\widehat{\mathsf{ID}}_{<\varepsilon_0}$. We conclude with some remarks concerning subsystems of EMU containing restricted induction principles on the natural numbers.

2 The theory EMU

In this section we introduce the theory EMU of explicit mathematics with a natural principle for the uniform construction of universes. We present EMU in the framework of types and names of Jäger [16] together with the finite axiomatization of elementary comprehension of Feferman and Jäger [10].

Let us first introduce the language $\mathcal{L}$ of EMU. It is a two-sorted language with countable lists of *individual variables* $a, b, c, f, g, h, x, y, z, \ldots$ and *type variables* $A, B, C, X, Y, Z, \ldots$ (both possibly with subscripts). $\mathcal{L}$ includes the following *individual constants*: k, s (combinators), $\mathsf{p}, \mathsf{p_0}, \mathsf{p_1}$ (pairing and projection), $\mathsf{0}$ (zero), $\mathsf{s_N}$ (successor), $\mathsf{p_N}$ (predecessor), $\mathsf{d_N}$ (definition by numerical cases), nat (natural numbers), id (identity), co (complement), int (intersection), dom (domain), inv (inverse image), j (join) and u (universe construction). There is only one binary function symbol $\cdot$ for (partial) application of individuals to individuals.

The relation symbols of $\mathcal{L}$ include equality $=$ for both individuals and types, the unary predicate symbols $\downarrow$ (defined) and N (natural numbers) on individual terms, U (universes) on types, and the binary relation symbols $\in$ (membership) and $\Re$ (naming, representation relation) between individuals and types.

The *individual terms* $(r, s, t, \ldots)$ of $\mathcal{L}$ are built up from individual variables and individual constants by means of $\cdot$, with the usual conventions for application in combinatory logic or λ calculus. We write (s, t) for $\mathsf{p}st$, s' for $\mathsf{s_N}s$, 1 instead of $\mathsf{0}'$ and so on. The *type terms* are just the type variables.

The *atoms* of $\mathcal{L}$ have one of the following forms: $s = t$, $A = B$, $s\downarrow$, $\mathsf{N}(s)$, $\mathsf{U}(A)$, $s \in A$, or $\Re(s, A)$. The *formulas* of $\mathcal{L}$ $(E, F, G, \ldots)$ are generated from the atoms by closing under the usual connectives as well as quantification of both sorts. The following table contains a useful list of abbreviations:

$$\begin{aligned}
s \simeq t &:= s\downarrow \vee t\downarrow \to s = t, \\
x \in \mathsf{N} &:= \mathsf{N}(x), \\
(\exists x \in A)F &:= (\exists x)(x \in A \wedge F), \\
(\forall x \in A)F &:= (\forall x)(x \in A \to F), \\
A \subset B &:= (\forall x \in A)(x \in B), \\
A \dot{\in} B &:= (\exists x)(\Re(x, A) \wedge x \in B), \\
s \dot{\in} t &:= (\exists X)(\Re(t, X) \wedge s \in X), \\
\Re(s) &:= (\exists X)\Re(s, X), \\
\mathsf{U}(s) &:= (\exists X)(\Re(s, X) \wedge \mathsf{U}(X)).
\end{aligned}$$

The logic of EMU is the classical logic of partial terms of Beeson [1] for the individuals, and classical logic with equality for the types[1]. The non-logical axioms of EMU are divided into the following groups.

I. Applicative axioms

Partial combinatory algebra

(1) $\mathsf{k}xy = x$,

(2) $\mathsf{s}xy\downarrow \wedge \mathsf{s}xyz \simeq xz(yz)$.

[1]All the results of this paper also hold in the presence of intuitionistic logic.

Pairing and projection

(3) $\mathsf{p_0}(x,y) = x \wedge \mathsf{p_1}(x,y) = y$.

Natural numbers

(4) $0 \in \mathsf{N} \wedge (\forall x \in \mathsf{N})(x' \in \mathsf{N})$,

(5) $(\forall x \in \mathsf{N})(x' \neq 0 \wedge \mathsf{p_N}(x') = x)$,

(6) $(\forall x \in \mathsf{N})(x \neq 0 \rightarrow \mathsf{p_N} x \in \mathsf{N} \wedge (\mathsf{p_N} x)' = x)$.

Definition by numerical cases

(7) $a \in \mathsf{N} \wedge b \in \mathsf{N} \wedge a = b \rightarrow \mathsf{d_N} xyab = x$,

(8) $a \in \mathsf{N} \wedge b \in \mathsf{N} \wedge a \neq b \rightarrow \mathsf{d_N} xyab = y$.

As usual one derives from the axioms of a partial combinatory algebra a theorem about λ abstraction as well as a form of the recursion theorem, cf. e.g. Beeson [1] or Feferman [3] for a proof of these standard facts. The axioms for types in general are given in the next block.

II. General axioms for types

Extensionality

(9) $(\forall x)(x \in A \leftrightarrow x \in B) \rightarrow A = B$.

Ontological axioms

(10) $\Re(a,B) \wedge \Re(a,C) \rightarrow B = C$,

(11) $(\exists x)\Re(x,A)$.

Axiom (10) tells us that there are no homonyms, i.e., different types have different names (representations), whereas axiom (11) states that every type has a name.

Natural numbers

(12) $(\exists X)[\Re(\mathsf{nat},X) \wedge (\forall x)(x \in X \leftrightarrow \mathsf{N}(x))]$.

Identity

(13) $(\exists X)[\Re(\mathsf{id},X) \wedge (\forall x)(x \in X \leftrightarrow (\exists y)(x = (y,y)))]$.

Complements

(14) $\Re(a,A) \rightarrow (\exists X)[\Re(\mathsf{co}\, a,X) \wedge (\forall x)(x \in X \leftrightarrow x \notin A)]$.

Intersections

(15) $\Re(a,A) \wedge \Re(b,B) \rightarrow$

$$(\exists X)[\Re(\mathsf{int}(a,b),X) \wedge (\forall x)(x \in X \leftrightarrow x \in A \wedge x \in B)].$$

Domains

(16) $\Re(a, A) \rightarrow (\exists X)[\Re(\mathsf{dom}\, a, X) \wedge (\forall x)(x \in X \leftrightarrow (\exists y)((x, y) \in A))]$.

Inverse images

(17) $\Re(a, A) \rightarrow (\exists X)[\Re(\mathsf{inv}(f, a), X) \wedge (\forall x)(x \in X \leftrightarrow fx \in A)]$.

An $\mathcal{L}$ formula is called *elementary*, if it contains no bound type variables nor the naming relation $\Re$. Axioms (12)-(17) provide a finite axiomatization of the scheme of uniform elementary comprehension, i.e. the usual scheme of elementary comprehension is derivable from (12)-(17), cf. Feferman and Jäger [10]. The final general type axiom is the principle of join. For its formulation, let us write $A = \Sigma(B, f)$ for the statement

$$(\forall x)(x \in A \leftrightarrow x = (\mathsf{p_0}x, \mathsf{p_1}x) \wedge \mathsf{p_0}x \in B \wedge \mathsf{p_1}x \in f(\mathsf{p_0}x)).$$

Join (disjoint sum)

(18) $\Re(a, A) \wedge (\forall x \in A)(\exists Y)\Re(fx, Y) \rightarrow (\exists Z)(\Re(\mathsf{j}(a, f), Z) \wedge Z = \Sigma(A, f))$.

Let us now turn to the axioms about universes, which are divided into three subsections.

III. Axioms for universes

Ontological axioms

(19) $\mathsf{U}(A) \wedge x \in A \rightarrow \Re(x)$,

(20) $\mathsf{U}(A) \wedge \mathsf{U}(B) \wedge A \in B \rightarrow A \subset B$.

The crucial axiom (19) claims that universes contain only names, and axiom (20) states a kind of transitivity condition.[2] Universes obey the following natural closure conditions.

Closure conditions

(21) $\mathsf{U}(A) \rightarrow \mathsf{nat} \in A$,

(22) $\mathsf{U}(A) \rightarrow \mathsf{id} \in A$,

(23) $\mathsf{U}(A) \wedge b \in A \rightarrow \mathsf{co}\, b \in A$,

(24) $\mathsf{U}(A) \wedge b \in A \wedge c \in A \rightarrow \mathsf{int}(b, c) \in A$,

[2] In [25, 24, 26, 22] a further ontological axiom for universes is present; it claims that $\in$ is total on universes. Totality is not an official axiom of EMU, but it can be added without raising its strength.

(25) $\mathsf{U}(A) \wedge b \in A \rightarrow \mathsf{dom}\, b \in A$,

(26) $\mathsf{U}(A) \wedge b \in A \rightarrow \mathsf{inv}(f, b) \in A$,

(27) $\mathsf{U}(A) \wedge b \in A \wedge (\forall x \in b)(fx \in A) \rightarrow \mathsf{j}(b, f) \in A$.

So far we have no axioms which guarantee the existence of universes at all. Therefore, we add the following principle of uniform universe construction, which uniformly for a given name yields a universe which contains that name.

Universe construction

(28) $\Re(a) \rightarrow \mathsf{U}(\mathsf{u}a) \wedge a \in \mathsf{u}a$.

In EMU we assume the induction schema, i.e. complete induction on the natural numbers is available for arbitrary statements of $\mathcal{L}$.

IV. Formula induction on N

For each $\mathcal{L}$ formula $F(x)$:

(29) $F(0) \wedge (\forall x \in \mathsf{N})(F(x) \rightarrow F(x')) \rightarrow (\forall x \in \mathsf{N})F(x)$.

This finishes the description of the systems EMU. In the next section we turn to the wellordering proof for EMU.

3 A wellordering proof for EMU

In this section we sketch the main lines of a wellordering proof for EMU. More precisely, we show that EMU proves transfinite induction for each initial segment of the ordinal $\varphi 1 \varepsilon_0 0$. This is also the proof-theoretic ordinal of the theory $\widehat{\mathsf{ID}}_{<\varepsilon_0}$ analyzed in Jäger, Kahle, Setzer and Strahm [18]; in the following we assume that the reader is familiar with the wellordering proofs for the theories $\widehat{\mathsf{ID}}_\alpha$ as they are presented in detail in Section 5 of [18].

In the sequel we presuppose the same ordinal-theoretic facts as given in Section 2 of [18]. Namely, we let Φ_0 denote the least ordinal greater than 0 which is closed under all n-ary φ functions, and we assume that a standard notation system of order type Φ_0 is given in a straightforward manner. We write $\prec$ for the corresponding primitive recursive wellordering with least element 0. When working in EMU in this section, we let $a, b, c, \ldots$ range over the field of $\prec$ and ℓ denote limit notations. There exist primitive recursive functions acting on the codes of this notation system which correspond to the usual operations on ordinals. In the sequel it is often convenient in order to simplify notation to use ordinals and ordinal operations instead of their codes and primitive recursive analogues. Then (for example) ω and $\omega + \omega$ stand for the natural numbers whose order type with respect to $\prec$ are ω and $\omega + \omega$. Finally, let us put as usual:

$$\begin{aligned} Prog(F) &:= (\forall a)[(\forall b \prec a)F(b) \to F(a)], \\ TI(F, a) &:= Prog(F) \to (\forall b \prec a)F(b). \end{aligned}$$

If we want to stress the relevant induction variable of a formula F, we sometimes write $Prog(\lambda a.F(a))$ instead of $Prog(F)$. If X is a type and x a name of a type, then $Prog(X)$ and $Prog(x)$ have their obvious meaning; $TI(X, a)$ and $TI(x, a)$ read analogously.

In the sequel it is our aim to derive $(\forall X)\,TI(X, \alpha)$ in EMU for each ordinal α less than $\varphi 1 \varepsilon_0 0$. A crucial step towards that aim is the following: given a type X with a name x, we can build a transfinite hierarchy of universes above a universe containing x along $\prec$, and indeed such a hierarchy can be shown to be well-defined up to each fixed α less than ε_0. The hierarchy h (depending on x) is given by the recursion theorem in order to satisfy the following recursion equations:

$$\begin{aligned} \mathsf{h}x0 &\simeq \mathsf{u}x, \\ \mathsf{h}x(a+1) &\simeq \mathsf{u}(\mathsf{h}xa), \\ \mathsf{h}x\ell &\simeq \mathsf{u}(\mathsf{j}(\{a : a \prec \ell\}, \mathsf{h}x)). \end{aligned}$$

In other words, the hierarchy starts with a universe containing x, at successor stages one puts a universe on top of the hierarchy defined so far, and at limit stages a universe above the disjoint union of the previously defined hierarchy is taken.

Lemma 1 *For each ordinal α less than ε_0, the following are theorems of* EMU*:*

1. $(\forall x)[\Re(x) \to (\forall a \prec \alpha)\mathsf{U}(\mathsf{h}xa)]$,
2. $(\forall x)[\Re(x) \to (\forall a \prec \alpha)(\forall b \prec a)(\mathsf{h}xb \in \mathsf{h}xa)]$.

Proof. For the proof of this lemma it is crucial to observe that we have transfinite induction up to each α less than ε_0 available in EMU with respect to *arbitrary* statements of $\mathcal{L}$. This is due to the fact that EMU includes the scheme of formula induction on the natural numbers. Hence, both claims can be proved by transfinite induction up to an $\alpha \prec \varepsilon_0$. For the first assertion this is immediate. For the second one makes use of the transitivity axiom (20). For example, assume that ℓ is a limit notation, and we want to establish that $\mathsf{h}xb \in \mathsf{h}x\ell$ for a specific $b \prec \ell$. Since ℓ is limit one also has $b+1 \prec \ell$, and of course we have $\mathsf{h}xb \in \mathsf{h}x(b+1)$. On the other hand, one easily sees that there is a name of the universe denoted by $\mathsf{h}x(b+1)$ which belongs to $\mathsf{h}x\ell$, since we have by definition $\mathsf{j}(\{c : c \prec \ell\}, \mathsf{h}x) \in \mathsf{h}x\ell$. But then $\mathsf{h}xb \in \mathsf{h}x\ell$ is immediate by transitivity. □

Crucial for carrying out the wellordering proof in EMU is the very natural notion $I^c_x(a)$ of *transfinite induction up to a for all types (respectively names) belonging to a universe* $\mathsf{h}xb$ *for* $b \prec c$, which is given as follows:

$$I^c_x(a) \ := \ (\forall b \prec c)(\forall u \in \mathsf{h}xb)\, TI(u, a).$$

The next lemma tells us that $I^\ell_x(a)$ can be represented by a type in $\mathsf{h}x\ell$.

Lemma 2 *For each ordinal α less than ε_0, the following is a theorem of* EMU*:*

$$(\forall x, \ell)[\Re(x) \wedge \ell \preceq \alpha \ \rightarrow \ (\exists y \in \mathsf{h}x\ell)(\forall a)(a \in y \leftrightarrow I^\ell_x(a))].$$

Proof. We sketch the proof of this claim by working informally in EMU. Assuming $\Re(x)$ and $\ell \preceq \alpha \prec \varepsilon_0$, we know by the definition of $\mathsf{h}x\ell$ that $\mathsf{j}(\{b : b \prec \ell\}, \mathsf{h}x) \in \mathsf{h}x\ell$. By closure of $\mathsf{h}x\ell$ under join this readily entails that also (a name of) the type

$$\{(b, u, v) : b \prec \ell \wedge u \in \mathsf{h}xb \wedge v \in u\}$$

belongs to $\mathsf{h}x\ell$. Therefore, by closure of $\mathsf{h}x\ell$ under elementary comprehension, there exists a y in $\mathsf{h}x\ell$ which satisfies the condition claimed by the lemma. □

The next lemma is used for the base case in Main Lemma I below. We do not give its proof here, since the relevant arguments can easily be extracted and adapted to the present context from Feferman [5, 8] or Schütte [30].

Lemma 3 *For each ordinal α less than ε_0 the following is a theorem of* EMU*:*

$$(\forall x, \ell, a)[\Re(x) \wedge \ell \preceq \alpha \wedge I^\ell_x(a) \ \rightarrow \ I^\ell_x(\varphi a \mathsf{0})].$$

The following corollary is an immediate consequence:

Corollary 4 *For each ordinal α less than ε_0 the following is a theorem of* EMU*:*

$$(\forall x, \ell)[\Re(x) \wedge \ell \preceq \alpha \ \rightarrow \ \mathit{Prog}(\lambda a. I^\ell_x(\Gamma_a))].$$

Main Lemma I below makes crucial use of the binary relation $\uparrow$, which reads as follows:

$$a \uparrow b \ := \ (\exists c, \ell)(b = c + a \cdot \ell).$$

We are now in a position to state Main Lemma I. It corresponds exactly to Main Lemma I in Jäger, Kahle, Setzer and Strahm [18], formulated in the framework of explicit mathematics with universes. Given the preparations outlined in this section, chiefly the last corollary and Lemma 2, its proof is very much the same as the proof given in [18] and, therefore, we omit it here.

Lemma 5 (Main Lemma I) *Let $Main_\alpha(a)$ be defined as follows:*

$$Main_\alpha(a) := (\forall x, b, c)[\Re(x) \land c \preceq \alpha \land \omega^{1+a} \uparrow c \land I_x^c(b) \rightarrow I_x^c(\varphi 1ab)].$$

Then EMU *proves $Prog(\lambda a.Main_\alpha(a))$ for each ordinal α less than ε_0.*

Using Main Lemma I, we are now in a position to derive the main theorem of this section.

Theorem 6 EMU *proves $(\forall X)\,TI(X, \alpha)$ for each ordinal α less than $\varphi 1\varepsilon_0 0$.*

Proof. It is enough to show that EMU proves $(\forall X)\,TI(X, \varphi 1\alpha 0)$ for each $\alpha < \varepsilon_0$. For that purpose, fix an arbitrary $\alpha < \varepsilon_0$. Then we also have $\omega^{1+\alpha} \cdot \omega < \varepsilon_0$ and, hence, we have $Prog(\lambda a.Main_{\omega^{1+\alpha}\cdot\omega}(a))$ as a theorem of EMU by Main Lemma I. Since transfinite induction below ε_0 is available in EMU with respect to arbitrary statements of $\mathcal{L}$, we obtain that EMU proves $Main_{\omega^{1+\alpha}\cdot\omega}(\alpha)$, i.e. the statement

$$(\forall x, b, c)[\Re(x) \land c \preceq \omega^{1+\alpha} \cdot \omega \land \omega^{1+\alpha} \uparrow c \land I_x^c(b) \rightarrow I_x^c(\varphi 1\alpha b)].$$

By choosing c as $\omega^{1+\alpha}\cdot\omega$ and b as 0 in this assertion, one derives the following as a theorem of EMU:

$$(\forall x)[\Re(x) \rightarrow I_x^{\omega^{1+\alpha}\cdot\omega}(\varphi 1\alpha 0)].$$

But now we can immediately derive EMU $\vdash (\forall X)\,TI(X, \varphi 1\alpha 0)$ as claimed. □

We finish this section by mentioning that it would be possible to obtain $\varphi 1\varepsilon_0 0$ as a lower bound for EMU even without assuming the transitivity axiom (20). However, the wellordering proof would require more "coding". Since transitivity of universes is a natural condition which holds in the standard structures of EMU discussed in the next section, we included (20) in the axioms of EMU.

4 Reduction of EMU to $\widehat{\mathsf{ID}}_{<\varepsilon_0}$

In this section we sketch a proof-theoretic reduction of EMU to the transfinitely iterated fixed point theory $\widehat{\mathsf{ID}}_{<\varepsilon_0}$; the latter theory is shown to possess proof-theoretic ordinal $\varphi 1\varepsilon_0 0$ in [18] and, hence, together with the results of the previous section, we obtain that $\varphi 1\varepsilon_0 0$ is also the proof-theoretic ordinal of EMU. Our reduction proceeds in two steps: first, we sketch a Tait-style reformulation of EMU which includes a form of the ω rule and, therefore, allows us to establish a partial cut elimination theorem for EMU, yielding quasi-normal derivations of length bounded by ε_0. In a second step we provide partial models for EMU which will subsequently be used in order to prove an asymmetric interpretation theorem for quasinormal derivations. It is argued that the whole procedure can be formalized in $\widehat{\mathsf{ID}}_{<\varepsilon_0}$; in particular, the partial models needed for an interpretation of EMU are available in $\widehat{\mathsf{ID}}_{<\varepsilon_0}$.

Let us start with an infinitary Tait-style reformulation of EMU. Since Tait formulations of systems of explicit mathematics are rather familiar from the literature, we confine ourselves to a sketchy description of the Tait calculus T_∞ of EMU. For more detailed expositions the reader is referred to Glaß and Strahm [13], or Marzetta and Strahm [26].

As usual, the language appropriate for setting up a Tait-style calculus for EMU presupposes complementary relation symbols for each relation of $\mathcal{L}$. Formulas are then generated from the positive and negative literals by closing under conjunction and disjunction as well as existential and universal quantification of both sorts. Negation is defined as usual by applying the law of double negation and De Morgan's laws. In the sequel we identify formulas of $\mathcal{L}$ and their translations in the Tait-style language corresponding to $\mathcal{L}$. Important classes of formulas are the so-called Σ^+ and Π^- formulas, cf. [13, 26]. A formula in the Tait-style language of $\mathcal{L}$ is called Σ^+, if it contains no negations of $\Re$ as well as no universal type quantifiers. Negations of Σ^+ formulas are called Π^- formulas. The rank $rn(F)$ of a formula F is defined in such a way that it is 0 if F is a Σ^+ or Π^- formula and it is computed as usual for more complex formulas, cf. [13, 26]. Axioms and rules of inference of T_∞ are formulated for finite sets of formulas, which have to be interpreted disjunctively. The capital greek letters $\Gamma, \Lambda, \ldots$ denote finite sets of formulas, and we write, e.g., Γ, Λ, F, G for the union of Γ, Λ and $\{F, G\}$.

The logical axioms and rules of inference of T_∞ are now as usual, cf. [13] for a detailed exposition. In particular, T_∞ includes the cut rule. As far as the non-logical axioms and rules are concerned, we notice that all axioms of EMU except axioms (18) and (29) can easily been written in a Tait style manner so that the relevant main formulas are always either in Σ^+ or in Π^-. For example, the universe construction axiom just reads as

$$\Gamma,\ \neg\Re(s),\ \mathsf{U}(\mathsf{u}s) \wedge s \in \mathsf{u}s.$$

Axiom (18) is replaced by the following two rules of inference, cf. [26].

$$\frac{\Gamma,\ t{\downarrow} \wedge \Re(s,A) \wedge (\forall x \in A)(\exists X)\Re(tx,X)}{\Gamma,\ (\exists Z)(\Re(\mathsf{j}(s,t),Z) \wedge Z{\subset}\Sigma(A,t))}\ (J_1)$$

$$\frac{\Gamma, t{\downarrow} \wedge \Re(s,A) \wedge (\forall x \in A)(\exists X)\Re(tx,X)}{\Gamma,\ (\exists Z)(\Re(\mathsf{j}(s,t),Z) \wedge Z{\supset}\Sigma(A,t))}\ (J_2)$$

where $Z{\subset}\Sigma(A,t)$ abbreviates

$$(\forall z)(z \in Z\ \rightarrow\ z = (\mathsf{p}_0 z, \mathsf{p}_1 z) \wedge \mathsf{p}_0 z \in A \wedge (\exists X)(\Re(t(\mathsf{p}_0 z), X) \wedge \mathsf{p}_1 z \in X)),$$

and $Z \supset \Sigma(A, t)$ is spelled out as

$$(\forall z)(z = (\mathsf{p}_0 z, \mathsf{p}_1 z) \wedge \mathsf{p}_0 z \in A \wedge (\forall X)(\Re(t(\mathsf{p}_0 z), X) \rightarrow \mathsf{p}_1 z \in X) \rightarrow z \in Z).$$

Finally, we replace the schema of formula induction (29) by the following version of the ω rule, cf. [13]. Here $\overline{n}$ denotes the nth numeral of $\mathcal{L}$.

$$\frac{\Gamma, \overline{n} \neq t \quad \text{for all } n < \omega}{\Gamma, \neg \mathsf{N}(t)}$$

$\mathsf{T}_\infty \vdash^{\alpha}_{k} \Gamma$ expresses that there is a derivation of the finite set Γ of $\mathcal{L}$ formulas such that α is an upper bound for the proof length and k is a strict upper bound for the ranks (in the sense of rn) of cut formulas occurring in the derivation.

We observe that EMU can be embedded into T_∞ in a straightforward manner; as usual, complete induction on the natural numbers is derivable by making use of the ω rule and at the price of infinite derivation lengths, cf. e.g. [13] for details.

Lemma 7 (Embedding of EMU into T_∞) *Assume that F is an $\mathcal{L}$ formula which is provable in* EMU*. Then there exist $\alpha < \omega + \omega$ and $k < \omega$ so that $\mathsf{T}_\infty \vdash^{\alpha}_{k} F$.*

Further, we observe that the axioms and rules of inference of T_∞ are tailored so that all main formulas are either Σ^+ or Π^-. Hence, usual cut elimination techniques from predicative proof theory (cf. e.g. [28, 30]) apply in order to show that all cuts of rank greater than 0 can be eliminated. The derivation lengths of the so-obtained quasinormal derivations can be measured as usual by the terms $\omega_k(\alpha)$, where we set $\omega_0(\alpha) = \alpha$ and $\omega_{k+1}(\alpha) = \omega^{\omega_k(\alpha)}$. We summarize our observations in the following partial cut elimination lemma.

Lemma 8 (Partial cut elimination for T_∞) *Assume that Γ is a finite set of $\mathcal{L}$ formulas so that $\mathsf{T}_\infty \vdash^{\alpha}_{1+k} \Gamma$. Then we have that $\mathsf{T}_\infty \vdash^{\omega_k(\alpha)}_{1} \Gamma$.*

A combination of the previous two lemmas yields the following corollary.

Corollary 9 *Assume that F is an $\mathcal{L}$ formula which is provable in* EMU*. Then there exists an $\alpha < \varepsilon_0$ so that $\mathsf{T}_\infty \vdash^{\alpha}_{1} F$.*

The second main step of our reduction of EMU to $\widehat{\mathsf{ID}}_{<\varepsilon_0}$ consists in setting up partial models $\mathfrak{M}(\alpha)$ for EMU, which will be used in order to prove an asymmetric interpretation theorem for quasinormal T_∞ derivations.

First, let us consider a fixed interpretation of the applicative (type-free) fragment of $\mathcal{L}$. We choose as universe for our operations the set of natural numbers $\mathbb{N}$ and interpret N by $\mathbb{N}$; term application $\cdot$ is interpreted as partial recursive function application, i.e. $a \cdot b$ just means $\{a\}(b)$. By ordinary recursion theory, it is now straightforward to find interpretations for $\mathsf{k}, \mathsf{s}, \mathsf{p}, \mathsf{p_0}, \mathsf{p_1}, 0, \mathsf{s_N}, \mathsf{p_N}, \mathsf{d_N}$ so that the applicative axioms of EMU are satisfied. In order to get an interpretation of the remaining individual constants of $\mathcal{L}$ we proceed as follows. Choose pairwise different natural numbers $\widehat{\mathsf{nat}}, \widehat{\mathsf{id}}, \widehat{\mathsf{co}}, \widehat{\mathsf{int}}, \widehat{\mathsf{dom}}, \widehat{\mathsf{inv}}, \widehat{\mathsf{j}}, \widehat{\mathsf{u}}$; interpret nat and id by $\langle\widehat{\mathsf{nat}}\rangle$ and $\langle\widehat{\mathsf{id}}\rangle$, respectively; interpret co by a natural number co so that $\{co\}(a) = \langle\widehat{\mathsf{co}}, a\rangle$; for int choose a natural number int so that $\{int\}(\langle a, b\rangle) = \langle\widehat{\mathsf{int}}, a, b\rangle$; the constants $\mathsf{dom}, \mathsf{inv}, \mathsf{j}, \mathsf{u}$ are interpreted analogously. Here we have used $\langle\ldots\rangle$ to denote standard sequence coding.

In a next step we now want to describe partial models $\mathfrak{M}(\alpha)$, $\mathfrak{N}(\alpha), \ldots$ of EMU. These are defined in such a way that they easily fit into the framework of iterated positive inductive definitions. Basically, one defines *codes* for types together with an *extension* and a *co-extension* for each such code. Essential use is made of fixed points of a positive arithmetic operator $\Phi_{X,\alpha}$ from the power set of $\mathbb{N}$ to the power set of $\mathbb{N}$, depending on a parameter set $X \subset \mathbb{N}$ and an ordinal α. We give the formal specification of $\Phi_{X,\alpha}$ first and afterwards comment on its informal meaning. For that purpose, fix naturals $r, \varepsilon, \bar{\varepsilon}$ which are different from all interpretations so far. Further, fix a parameter set $X \subset \mathbb{N}$ and an ordinal α. For $Y \subset \mathbb{N}$ we put $a \in \Phi_{X,\alpha}(Y)$, if there exist naturals b, c, d, f so that one of the following clauses (1)-(28) applies:

1. $a = \langle r, b\rangle \wedge a \in X$,

2. $a = \langle \varepsilon, b, c\rangle \wedge a \in X \wedge \langle r, c\rangle \in X$,

3. $a = \langle \bar{\varepsilon}, b, c\rangle \wedge a \in X \wedge \langle r, c\rangle \in X$,

4. $\alpha \in Suc^3 \wedge a = \langle r, \langle\widehat{\mathsf{u}}, b\rangle\rangle \wedge a \notin X \wedge \langle r, b\rangle \in X$,

5. $\alpha \in Suc \wedge a = \langle \varepsilon, b, \langle\widehat{\mathsf{u}}, c\rangle\rangle \wedge \langle r, \langle\widehat{\mathsf{u}}, c\rangle\rangle \notin X \wedge \langle r, c\rangle \in X \wedge \langle r, b\rangle \in X \wedge (\forall y)[\langle \varepsilon, y, b\rangle \in X \leftrightarrow \langle \bar{\varepsilon}, y, b\rangle \notin X]$,

6. $\alpha \in Suc \wedge a = \langle \bar{\varepsilon}, b, \langle\widehat{\mathsf{u}}, c\rangle\rangle \wedge \langle r, \langle\widehat{\mathsf{u}}, c\rangle\rangle \notin X \wedge \langle r, c\rangle \in X \wedge \langle r, b\rangle \notin X$,

7. $\alpha \in Suc \wedge a = \langle \bar{\varepsilon}, b, \langle\widehat{\mathsf{u}}, c\rangle\rangle \wedge \langle r, \langle\widehat{\mathsf{u}}, c\rangle\rangle \notin X \wedge \langle r, c\rangle \in X \wedge \langle r, b\rangle \in X \wedge \neg(\forall y)[\langle \varepsilon, y, b\rangle \in X \leftrightarrow \langle \bar{\varepsilon}, y, b\rangle \notin X]$,

8. $\alpha = 0 \wedge a = \langle r, \langle\widehat{\mathsf{nat}}\rangle\rangle$,

9. $\alpha = 0 \wedge a = \langle \varepsilon, b, \langle\widehat{\mathsf{nat}}\rangle\rangle$,

10. $\alpha = 0 \wedge a = \langle r, \langle\widehat{\mathsf{id}}\rangle\rangle$,

[3] *Suc* denotes the class of successor ordinals.

11. $\alpha = 0 \wedge a = \langle \varepsilon, b, \langle \widehat{\mathsf{id}} \rangle \rangle \wedge (\exists x)(b = \langle x, x \rangle)$,

12. $\alpha = 0 \wedge a = \langle \bar{\varepsilon}, b, \langle \widehat{\mathsf{id}} \rangle \rangle \wedge (\forall x)(b \neq \langle x, x \rangle)$,

13. $a = \langle r, \langle \widehat{\mathsf{co}}, b \rangle \rangle \wedge a \notin X \wedge \langle r, b \rangle \in Y$,

14. $a = \langle \varepsilon, b, \langle \widehat{\mathsf{co}}, c \rangle \rangle \wedge \langle \widehat{\mathsf{co}}, c \rangle \notin X \wedge \langle r, c \rangle \in Y \wedge \langle \bar{\varepsilon}, b, c \rangle \in Y$,

15. $a = \langle \bar{\varepsilon}, b, \langle \widehat{\mathsf{co}}, c \rangle \rangle \wedge \langle \widehat{\mathsf{co}}, c \rangle \notin X \wedge \langle r, c \rangle \in Y \wedge \langle \varepsilon, b, c \rangle \in Y$,

16. $a = \langle r, \langle \widehat{\mathsf{int}}, b, c \rangle \rangle \wedge a \notin X \wedge \langle r, b \rangle \in Y \wedge \langle r, c \rangle \in Y$,

17. $a = \langle \varepsilon, b, \langle \widehat{\mathsf{int}}, c, d \rangle \rangle \wedge \langle \widehat{\mathsf{int}}, c, d \rangle \notin X \wedge \langle r, c \rangle \in Y \wedge \langle r, d \rangle \in Y \wedge$
$\langle \varepsilon, b, c \rangle \in Y \wedge \langle \varepsilon, b, d \rangle \in Y$,

18. $a = \langle \bar{\varepsilon}, b, \langle \widehat{\mathsf{int}}, c, d \rangle \rangle \wedge \langle \widehat{\mathsf{int}}, c, d \rangle \notin X \wedge \langle r, c \rangle \in Y \wedge \langle r, d \rangle \in Y \wedge$
$[\langle \bar{\varepsilon}, b, c \rangle \in Y \vee \langle \bar{\varepsilon}, b, d \rangle \in Y]$,

19. $a = \langle r, \langle \widehat{\mathsf{dom}}, b \rangle \rangle \wedge a \notin X \wedge \langle r, b \rangle \in Y$,

20. $a = \langle \varepsilon, b, \langle \widehat{\mathsf{dom}}, c \rangle \rangle \wedge \langle \widehat{\mathsf{dom}}, c \rangle \notin X \wedge \langle r, c \rangle \in Y \wedge (\exists x)(\langle \varepsilon, \langle b, x \rangle, c \rangle \in Y)$,

21. $a = \langle \bar{\varepsilon}, b, \langle \widehat{\mathsf{dom}}, c \rangle \rangle \wedge \langle \widehat{\mathsf{dom}}, c \rangle \notin X \wedge \langle r, c \rangle \in Y \wedge (\forall x)(\langle \bar{\varepsilon}, \langle b, x \rangle, c \rangle \in Y)$,

22. $a = \langle r, \langle \widehat{\mathsf{inv}}, f, b \rangle \rangle \wedge a \notin X \wedge \langle r, b \rangle \in Y$,

23. $a = \langle \varepsilon, b, \langle \widehat{\mathsf{inv}}, f, c \rangle \rangle \wedge \langle \widehat{\mathsf{inv}}, f, c \rangle \notin X \wedge \langle r, c \rangle \in Y \wedge \langle \varepsilon, \{f\}(b), c \rangle \in Y$,

24. $a = \langle \bar{\varepsilon}, b, \langle \widehat{\mathsf{inv}}, f, c \rangle \rangle \wedge \langle \widehat{\mathsf{inv}}, f, c \rangle \notin X \wedge \langle r, c \rangle \in Y \wedge \{f\}(b)\uparrow$,

25. $a = \langle \bar{\varepsilon}, b, \langle \widehat{\mathsf{inv}}, f, c \rangle \rangle \wedge \langle \widehat{\mathsf{inv}}, f, c \rangle \notin X \wedge \langle r, c \rangle \in Y \wedge \langle \bar{\varepsilon}, \{f\}(b), c \rangle \in Y$,

26. $a = \langle r, \langle \widehat{\mathsf{j}}, b, f \rangle \rangle \wedge a \notin X \wedge \langle r, b \rangle \in Y \wedge$
$(\forall x)[\langle \bar{\varepsilon}, x, b \rangle \notin Y \to \langle r, \{f\}(x) \rangle \in Y]$,

27. $a = \langle \varepsilon, b, \langle \widehat{\mathsf{j}}, c, f \rangle \rangle \wedge \langle \widehat{\mathsf{j}}, c, f \rangle \notin X \wedge \langle r, c \rangle \in Y \wedge$
$(\forall x)[\langle \bar{\varepsilon}, x, c \rangle \notin Y \to \langle r, \{f\}(x) \rangle \in Y] \wedge$
$b = \langle (b)_0, (b)_1 \rangle \wedge \langle \varepsilon, (b)_0, c \rangle \in Y \wedge \langle \varepsilon, (b)_1, \{f\}((b)_0) \rangle \in Y$,

28. $a = \langle \bar{\varepsilon}, b, \langle \widehat{\mathsf{j}}, c, f \rangle \rangle \wedge \langle \widehat{\mathsf{j}}, c, f \rangle \notin X \wedge \langle r, c \rangle \in Y \wedge$
$(\forall x)[\langle \bar{\varepsilon}, x, c \rangle \notin Y \to \langle r, \{f\}(x) \rangle \in Y] \wedge$
$[b \neq \langle (b)_0, (b)_1 \rangle \vee \langle \bar{\varepsilon}, (b)_0, c \rangle \in Y \vee \langle \bar{\varepsilon}, (b)_1, \{f\}((b)_0) \rangle \in Y]$.

Natural numbers belonging to $\Phi_{X,\alpha}(Y)$ have one of the three forms $\langle r, a\rangle$, $\langle \varepsilon, b, a\rangle$ or $\langle \bar{\varepsilon}, b, a\rangle$ with the associated informal meaning, "a is a representation or name for a type", "b belongs to the type coded by a", and "b does not belong to the type coded by a", respectively. Clauses (1)-(3) inherit all type codes, ε relations and $\bar{\varepsilon}$ relations in X to $\Phi_{X,\alpha}(Y)$. In the case of α being a successor ordinal, clauses (4)-(7) associate to each type code in X a new type (universe), which contains exactly those type codes in X on which ε and $\bar{\varepsilon}$ are complementary. Clauses (8)-(28) state closure conditions for types in the sense of axioms (12)-(18) of EMU; in each case ε and $\bar{\varepsilon}$ are defined separately.

A sequence of sets of natural numbers $(X_\beta)_{\beta\leq\alpha}$ is called a Φ *sequence*, if it satisfies the following conditions for each $\beta \leq \alpha$:

(1) if $\beta = 0$, then X_β is a fixed point of $\Phi_{\emptyset,0}$;

(2) if β is a successor ordinal $\gamma + 1$, then X_β is a fixed point of $\Phi_{X_\gamma,\beta}$;

(3) if β is a limit ordinal, then X_β is a fixed point of $\Phi_{\bigcup_{\gamma<\beta} X_\gamma,\beta}$.

A Φ sequence $(X_\beta)_{\beta\leq\alpha}$ determines an interpretation $\mathfrak{M}(\alpha)$ of $\mathcal{L}$ as follows:

(i) the applicative fragment of $\mathcal{L}$ is interpreted as described above.

(ii) the types in $\mathfrak{M}(\alpha)$ range over the set T_α of natural numbers m so that $\langle r, m\rangle$ belongs to X_α and $\varepsilon, \bar{\varepsilon}$ are complementary with respect to m, i.e.

$$(\forall x)(\langle \varepsilon, x, m\rangle \in X_\alpha \leftrightarrow \langle \bar{\varepsilon}, x, m\rangle \notin X_\alpha).$$

(iii) the elementhood relation for T_α is $\in_\alpha$, and we have that $m \in_\alpha n$ if n belongs to T_α and $\langle \varepsilon, m, n\rangle$ is an element of X_α. Equality between types is just extensional equality.

(iv) the naming relation R_α of $\mathfrak{M}(\alpha)$ is given by pairs (m, n) so that m, n belong to T_α and are extensionally equal with respect to $\in_\alpha$.

(v) the collection of universes $U_\alpha \subset T_\alpha$ is obtained by taking those m for which there exists an $\langle \widehat{\mathsf{u}}, n\rangle$ in T_α that is extensionally equal to m.

This finishes the specification of $\mathfrak{M}(\alpha) = (T_\alpha, \in_\alpha, R_\alpha, U_\alpha)$. For each $\beta < \alpha$ we obtain an obvious restriction $\mathfrak{M}(\beta) = (T_\beta, \in_\beta, R_\beta, U_\beta)$ of $\mathfrak{M}(\alpha)$ by defining $T_\beta, \in_\beta, R_\beta, U_\beta$ from X_β analogously to (ii)-(v).

It is important to notice here that two structures $\mathfrak{M}(\alpha)$ and $\mathfrak{N}(\alpha)$ are in general different since they can be generated from two different Φ sequences. As we will see, however, our asymmetric interpretation theorem below is independent of a particular choice of a Φ sequence.

We are now ready to provide an asymmetrical interpretation of T_∞ into the structures $\mathfrak{M}(\alpha)$ for suitable α. In particular, we show that if a Σ^+ sentence A is provable in EMU, then there exists an ordinal α less than ε_0 so that A holds in each structure $\mathfrak{M}(\alpha)$. Asymmetrical interpretations are a well-known technique in proof theory, cf. e.g. [2, 14, 30]. They have previously been applied in the context of explicit mathematics e.g. in [11, 12, 13, 25, 24, 26].

Before we turn to the interpretation itself, let us state essential persistency properties of Σ^+ and Π^- formulas w.r.t. the structures $\mathfrak{M}(\alpha)$. The proof of the following lemma is immediate from the definition of the structures $\mathfrak{M}(\alpha)$.

Lemma 10 *Let $\mathfrak{M}(\alpha) = (T_\alpha, \ldots)$ be a structure for $\mathcal{L}$, and let $\gamma \le \beta \le \alpha$, $\vec{u} \in T_\gamma$ and $\vec{m} \in \mathbb{N}$. Then we have for all Σ^+ formulas $F[\vec{A}, \vec{a}]$[4] and all Π^- formulas $G[\vec{A}, \vec{a}]$:*

1. $\mathfrak{M}(\gamma) \models F[\vec{u}, \vec{m}] \quad \Longrightarrow \quad \mathfrak{M}(\beta) \models F[\vec{u}, \vec{m}]$.
2. $\mathfrak{M}(\beta) \models G[\vec{u}, \vec{m}] \quad \Longrightarrow \quad \mathfrak{M}(\gamma) \models G[\vec{u}, \vec{m}]$.

In the sequel let us assume that $\Gamma[\vec{A}, \vec{a}]$ is a set of Σ^+ and Π^- formulas. Further, let $\mathfrak{M}(\alpha)$ be a structure for $\mathcal{L}$ and let $\gamma \le \beta \le \alpha$. Then we write

$$\mathfrak{M}(\gamma, \beta) \models \Gamma[\vec{u}, \vec{m}] \qquad (\vec{u} \in T_\gamma, \vec{m} \in \mathbb{N}),$$

provided that one of the following conditions is satisfied:

(1) there is a Π^- formula $F[\vec{A}, \vec{a}]$ in Γ so that $\mathfrak{M}(\gamma) \models F[\vec{u}, \vec{m}]$;

(2) there is a Σ^+ formula $G[\vec{A}, \vec{a}]$ in Γ so that $\mathfrak{M}(\beta) \models G[\vec{u}, \vec{m}]$.

The asymmetric interpretation result mentioned above now reads as follows.

Lemma 11 (Main Lemma II) *Let γ be a fixed ordinal and $\mathfrak{M}(\omega^\gamma)$ an arbitrary $\mathcal{L}$ structure. Further assume that $\Gamma[\vec{A}, \vec{a}]$ is a finite set of Σ^+ and Π^- formulas so that $\mathsf{T}_\infty \overset{\alpha}{\underset{1}{\vdash}} \Gamma$ for an ordinal $\alpha < \gamma$. Then we have for all ordinals $\beta < \omega^\gamma$:*

$$\vec{u} \in T_\beta \text{ and } \vec{m} \in \mathbb{N} \quad \Longrightarrow \quad \mathfrak{M}(\beta, \beta + 2^\alpha) \models \Gamma[\vec{u}, \vec{m}].$$

Proof. The assertion is proved by induction on $\alpha < \gamma$. As an example we discuss the axiom about universe construction as well as the cut rule. In all other cases the claim follows from the construction of $\mathfrak{M}(\omega^\gamma)$, the induction hypothesis and the persistency lemma. In particular, observe that the complement property of the element relation is preserved by all type constructors.

[4]We write $F[\vec{A}, \vec{a}]$ in order to indicate that *all* parameters of F come from the list $\vec{A}, \vec{a}$.

Let us first assume that $\Gamma[\vec{A},\vec{a}]$ is an axiom about universe construction (28). Then $\Gamma[\vec{A},\vec{a}]$ has the form

$$\Lambda[\vec{A},\vec{a}],\ \neg(\exists X)\Re(s[\vec{a}],X),\ (\exists X)[\Re(\mathsf{us}[\vec{a}],X)\wedge\mathsf{U}(X)\wedge s[\vec{a}]\in X]. \quad (1)$$

Now fix an ordinal $\beta<\omega^\gamma$, $\vec{u}\in T_\beta$ and $\vec{m}\in\mathbb{N}$. Further suppose that $\mathfrak{M}(\beta)$ models $(\exists X)\Re(s[\vec{m}],X)$, i.e. we have that $s[\vec{m}]$ belongs to T_β. If $\mathsf{us}[\vec{m}]$ already belongs to T_β, then it is easily seen that $s[\vec{m}]\in_\beta \mathsf{us}[\vec{m}]$. Hence, assume that $\mathsf{us}[\vec{m}]$ is not in T_β. But then we have by construction of $\mathfrak{M}(\beta+1)$ that $\mathsf{us}[\vec{m}]$ belongs to $T_{\beta+1}$, and also $s[\vec{m}]\in_{\beta+1}\mathsf{us}[\vec{m}]$. All together we obtain by persistency for all ordinals $\alpha<\gamma$:

$$\mathfrak{M}(\beta,\beta+2^\alpha)\models\Gamma[\vec{u},\vec{m}]. \quad (2)$$

As a second illustrative example let us consider the case where $\Gamma[\vec{A},\vec{a}]$ is the conclusion of a cut rule. Then the cut formula has rank 0, i.e. there is a Σ^+ formula $F[\vec{A},\vec{a}]$ and $\alpha_0,\alpha_1<\alpha<\gamma$ so that

$$\mathsf{T}_\infty \overset{\alpha_0}{\underset{1}{\vdash}} \Gamma[\vec{A},\vec{a}],\ F[\vec{A},\vec{a}] \quad\text{and}\quad \mathsf{T}_\infty \overset{\alpha_1}{\underset{1}{\vdash}} \Gamma[\vec{A},\vec{a}],\ \neg F[\vec{A},\vec{a}]. \quad (3)$$

Choose $\beta<\omega^\gamma$, $\vec{u}\in T_\beta$ and $\vec{m}\in\mathbb{N}$. We have to show $\mathfrak{M}(\beta,\beta+2^\alpha)\models\Gamma[\vec{u},\vec{m}]$. If we apply the induction hypothesis to (3) with β and $\beta+2^{\alpha_0}$, respectively, then we get

$$\mathfrak{M}(\beta,\beta+2^{\alpha_0}) \models \Gamma[\vec{u},\vec{m}],\ F[\vec{U},\vec{m}], \quad (4)$$

$$\mathfrak{M}(\beta+2^{\alpha_0},\beta+2^{\alpha_0}+2^{\alpha_1}) \models \Gamma[\vec{u},\vec{m}],\ \neg F[\vec{U},\vec{m}]. \quad (5)$$

Observe that $\beta+2^{\alpha_0}+2^{\alpha_1}\le\beta+2^\alpha$. Hence, if it is

(i) $\mathfrak{M}(\beta,\beta+2^{\alpha_0})\models\Gamma[\vec{u},\vec{m}]$ or (ii) $\mathfrak{M}(\beta+2^{\alpha_0},\beta+2^{\alpha_0}+2^{\alpha_1})\models\Gamma[\vec{u},\vec{m}]$,

then our assertion immediately follows by persistency. But one of (i) and (ii) applies, since otherwise (4) and (5) imply

$$\mathfrak{M}(\beta+2^{\alpha_0})\models F[\vec{u},\vec{m}] \quad\text{and}\quad \mathfrak{M}(\beta+2^{\alpha_0})\models\neg F[\vec{u},\vec{m}]. \quad (6)$$

This, however, is not possible, and hence our claim is proved. □

Together with Corollary 9 we have thus established the following result.

Corollary 12 *Assume that the Σ^+ sentence F is provable in* EMU. *Then there exists an ordinal $\alpha<\varepsilon_0$ so that $\mathfrak{M}(\alpha)\models F$ for arbitrary $\mathcal{L}$ structures $\mathfrak{M}(\alpha)$.*

This finishes the treatment of quasinormal T_∞ derivations by means of asymmetric interpretation into partial models of EMU. We finish this section by briefly addressing how the reduction procedure for EMU described so far can be formalized in the transfinitely iterated fixed point theory $\widehat{\mathsf{ID}}_{<\varepsilon_0}$ of [18] in order to yield conservativity of EMU over $\widehat{\mathsf{ID}}_{<\varepsilon_0}$ with respect to arithmetic statements. Together with the results of the previous section and the fact that $|\widehat{\mathsf{ID}}_{<\varepsilon_0}| = \varphi 1 \varepsilon_0 0$ (cf. [18]) this shows the proof-theoretic equivalence of EMU and $\widehat{\mathsf{ID}}_{<\varepsilon_0}$ as desired.

The *first step* in reducing EMU to $\widehat{\mathsf{ID}}_{<\varepsilon_0}$ is provided by Corollary 9. Here we observe that a straightforward formalization of infinitary derivations and cut elimination procedures is required within $\widehat{\mathsf{ID}}_{<\varepsilon_0}$, cf. e.g. Schwichtenberg [31] for similar arguments. The *second step* of our reduction consists in formalizing Main Lemma II in $\widehat{\mathsf{ID}}_{<\varepsilon_0}$. Recall that this lemma holds for structures $\mathfrak{M}(\omega^\gamma)$ which are given by an arbitrary fixed point hierarchy of a (parameterized) positive arithmetic operator, and exactly such arbitrary fixed point hierarchies of length bounded below ε_0 are available in $\widehat{\mathsf{ID}}_{<\varepsilon_0}$; observe that we can do with structures of a fixed level less than ε_0 in Main Lemma II, since we are always working with a fixed EMU derivation. Of course, some straightforward formal truth definitions have to be described in $\widehat{\mathsf{ID}}_{<\varepsilon_0}$ for a proper formalization of Main Lemma II. Summing up, we have established the following result.

Theorem 13 EMU *can be embedded into $\widehat{\mathsf{ID}}_{<\varepsilon_0}$; moreover, arithmetic sentences are preserved under this embedding.*

Together with Theorem 6 we can thus state the following main corollary.

Corollary 14 EMU *is proof-theoretically equivalent to $\widehat{\mathsf{ID}}_{<\varepsilon_0}$ and has proof-theoretic ordinal $\varphi 1 \varepsilon_0 0$.*

5 Final remarks

In this paper we have given a proof-theoretic analysis of EMU, a system of explicit mathematics with a principle for *uniform* universe construction and including the schema of formula induction. Let us now briefly look at subsystems of EMU with restricted forms of complete induction on the natural numbers. Let $\mathsf{EMU}\!\restriction$ denote EMU with complete induction restricted to types, and $\mathsf{EMU}\!\restriction + (\Sigma^+\text{-}\mathsf{I_N})$ be EMU with complete induction restricted to formulas in the class Σ^+, cf. the previous section. Then the methods of the last section can be applied in order to get a reduction of $\mathsf{EMU}\!\restriction$ and $\mathsf{EMU}\!\restriction + (\Sigma^+\text{-}\mathsf{I_N})$ to $\widehat{\mathsf{ID}}_{<\omega}$ and $\widehat{\mathsf{ID}}_{<\omega^\omega}$, respectively, and indeed it can be shown that these bounds are sharp. The equivalence $\mathsf{EMU}\!\restriction \equiv \widehat{\mathsf{ID}}_{<\omega}$ has previously been obtained in

Kahle [22], who relied heavily on the treatment of a non-uniform formulation of the limit axiom in Marzetta [25, 24] and Marzetta and Strahm [26]. Let us summarize all these results in the following theorem.

Theorem 15 *We have the following proof-theoretic equivalences:*

1. $\mathsf{EMU}\!\upharpoonright\ \equiv \widehat{\mathsf{ID}}_{<\omega}$,

2. $\mathsf{EMU}\!\upharpoonright + (\Sigma^{+}\text{-}\mathsf{I_N}) \equiv \widehat{\mathsf{ID}}_{<\omega^\omega}$,

3. $\mathsf{EMU} \equiv \widehat{\mathsf{ID}}_{<\varepsilon_0}$.

The corresponding proof-theoretic ordinals are $\Gamma_0, \varphi 1\omega 0$, *and* $\varphi 1 \varepsilon_0 0$, *respectively.*

References

[1] BEESON, M. J. *Foundations of Constructive Mathematics: Metamathematical Studies.* Springer, Berlin, 1985.

[2] CANTINI, A. On the relationship between choice and comprehension principles in second order arithmetic. *Journal of Symbolic Logic 51* (1986), 360–373.

[3] FEFERMAN, S. A language and axioms for explicit mathematics. In *Algebra and Logic*, J. Crossley, Ed., vol. 450 of *Lecture Notes in Mathematics.* Springer, Berlin, 1975, pp. 87–139.

[4] FEFERMAN, S. Constructive theories of functions and classes. In *Logic Colloquium '78*, M. Boffa, D. van Dalen, and K. McAloon, Eds. North Holland, Amsterdam, 1979, pp. 159–224.

[5] FEFERMAN, S. Iterated inductive fixed-point theories: application to Hancock's conjecture. In *The Patras Symposion*, G. Metakides, Ed. North Holland, Amsterdam, 1982, pp. 171–196.

[6] FEFERMAN, S. Polymorphic typed lambda-calculi in a type-free axiomatic framework. In *Logic and Computation*, W. Sieg, Ed., vol. 106 of *Contemporary Mathematics.* American Mathematical Society, Providence, Rhode Island, 1990, pp. 101–136.

[7] FEFERMAN, S. Logics for termination and correctness of functional programs. In *Logic from Computer Science*, Y. N. Moschovakis, Ed., vol. 21 of *MSRI Publications.* Springer, Berlin, 1991, pp. 95–127.

[8] FEFERMAN, S. Reflecting on incompleteness. *Journal of Symbolic Logic 56*, 1 (1991), 1–49.

[9] FEFERMAN, S. Logics for termination and correctness of functional programs II: Logics of strength PRA. In *Proof Theory*, P. Aczel, H. Simmons, and S. S. Wainer, Eds. Cambridge University Press, Cambridge, 1992, pp. 195–225.

[10] FEFERMAN, S., AND JÄGER, G. Systems of explicit mathematics with non-constructive μ-operator. Part II. *Annals of Pure and Applied Logic 79*, 37–52 (1996).

[11] GLASS, T. *Standardstrukturen für Systeme Expliziter Mathematik.* PhD thesis, Westfälische Wilhelms-Universität Münster, 1993.

[12] GLASS, T. Understanding uniformity in Feferman's explicit mathematics. *Annals of Pure and Applied Logic 75*, 1–2 (1995), 89–106.

[13] GLASS, T., AND STRAHM, T. Systems of explicit mathematics with non-constructive μ-operator and join. *Annals of Pure and Applied Logic 82* (1996), 193–219.

[14] JÄGER, G. Beweistheorie von *KPN*. *Archiv für mathematische Logik und Grundlagenforschung 20* (1980), 53–64.

[15] JÄGER, G. A well-ordering proof for Feferman's theory T_0. *Archiv für mathematische Logik und Grundlagenforschung 23* (1983), 65–77.

[16] JÄGER, G. Induction in the elementary theory of types and names. In *Computer Science Logic '87*, E. Börger, H. Kleine Büning, and M.M. Richter, Eds., vol. 329 of *Lecture Notes in Computer Science*. Springer, Berlin, 1988, pp. 118–128.

[17] JÄGER, G. Type theory and explicit mathematics. In *Logic Colloquium '87*, H.-D. Ebbinghaus, J. Fernandez-Prida, M. Garrido, M. Lascar, and M. R. Artalejo, Eds. North Holland, Amsterdam, 1989, pp. 117–135.

[18] JÄGER, G., KAHLE, R., SETZER, A., AND STRAHM, T. The proof-theoretic analysis of transfinitely iterated fixed point theories. *Journal of Symbolic Logic*. To appear.

[19] JÄGER, G., AND POHLERS, W. Eine beweistheoretische Untersuchung von $(\Delta^1_2\text{-CA}) + (\text{BI})$ und verwandter Systeme. In *Sitzungsberichte der Bayerischen Akademie der Wissenschaften, Mathematisch-naturwissenschaftliche Klasse.* 1982, pp. 1–28.

[20] JÄGER, G., AND STRAHM, T. Fixed point theories and dependent choice. *Archive for Mathematical Logic*. To appear.

[21] KAHLE, R. *Applicative Theories and Frege Structures.* PhD thesis, Institut für Informatik und angewandte Mathematik, Universität Bern, 1997.

[22] KAHLE, R. Uniform limit in explicit mathematics with universes. Tech. Rep. IAM-97-002, Institut für Informatik und angewandte Mathematik, Universität Bern, 1997.

[23] MARTIN-LÖF, P. *Intutionistic Type Theory*, vol. 1 of *Studies in Proof Theory*. Bibliopolis, 1984.

[24] MARZETTA, M. *Predicative Theories of Types and Names*. PhD thesis, Institut für Informatik und angewandte Mathematik, Universität Bern, 1993.

[25] MARZETTA, M. Universes in the theory of types and names. In *Computer Science Logic '92*, E. Börger et al., Ed., vol. 702 of *Lecture Notes in Computer Science*. Springer, Berlin, 1993, pp. 340–351.

[26] MARZETTA, M., AND STRAHM, T. The μ quantification operator in explicit mathematics with universes and iterated fixed point theories with ordinals. *Archive for Mathematical Logic*. To appear.

[27] PALMGREN, E. On universes in type theory. In *Twentyfive Years of Type Theory*, G. Sambin and J. Smith, Eds. Oxford University Press. To appear.

[28] POHLERS, W. *Proof Theory: An Introduction*, vol. 1407 of *Lecture Notes in Mathematics*. Springer, Berlin, 1988.

[29] RATHJEN, M. The strength of Martin-Löf type theory with a superuniverse. Part I, 1997. Preprint.

[30] SCHÜTTE, K. *Proof Theory*. Springer, Berlin, 1977.

[31] SCHWICHTENBERG, H. Proof theory: Some applications of cut-elimination. In *Handbook of Mathematical Logic*, J. Barwise, Ed. North Holland, Amsterdam, 1977, pp. 867–895.

What Makes A (Pointwise) Subrecursive Hierarchy Slow Growing?

Andreas Weiermann
Institut für Mathematische Logik und Grundlagenforschung
der Westfälischen Wilhelms-Universität Münster
Einsteinstr. 62, D-48149 Münster, Germany

Abstract

A subrecursive hierarchy $(P_\alpha)_{\alpha<\varepsilon_0}$ of unary number-theoretic functions is called the slow growing if each P_α is dominated by a Kalmar elementary recursive function. A subrecursive hierarchy $(P_\alpha)_{\alpha<\varepsilon_0}$ is called pointwise if the value of $P_\alpha(x)$ can be computed recursively in terms of $P_\beta(x)$ with $\beta < \alpha$ but with the same argument x. The pointwise hierarchy $(G_\alpha)_{\alpha<\varepsilon_0}$ is defined recursively as follows: $G_0(x) := 0; G_{\alpha+1}(x) := 1 + G_\alpha(x); G_\lambda(x) := G_{\lambda[x]}(x)$ where λ is a limit and $(\lambda[x])_{x<\omega}$ is a distinguished fundamental sequence for λ. If $(G_\alpha)_{\alpha<\varepsilon_0}$ is defined with respect to the standard system of fundamental sequences then $(G_\alpha)_{\alpha<\varepsilon_0}$ is slow growing.

In the first part of the article we try to classify as well as seems possible those (natural) assignments of fundamental sequences for which the induced pointwise hierarchy remains slow growing or becomes fast growing in the sense that the resulting hierarchy matches up with the Schwichtenberg-Wainer hierarchy $(F_\alpha)_{\alpha<\varepsilon_0}$. To exclude pathological cases we only consider elementary recursive assignments which are natural at least in the sense that their associated fast growing hierarchies classify exactly the provably recursive functions of PA. It turns out that the growth rate of the pointwise hierarchy is intrinsically connected with strong continuity properties of the assignment of fundamental sequences with respect to ordinal addition. In fact the pointwise hierarchy becomes fast growing when its underlying system of fundamental sequences is defined as follows: $(\omega^\alpha + \lambda)[x] := \omega^\alpha + \lambda[x+1]; \omega^{\beta+1}[x] := \omega^\beta \cdot (x+1); \omega^\lambda[x] := \omega^{\lambda[x]}$ where $\omega^\alpha + \lambda > \lambda \in Lim$. On the other hand the pointwise hierarchy remains slow growing when the underlying system of fundamental sequences is defined as follows: $(\omega^\alpha + \lambda)[x] := \omega^\alpha + \lambda[x]; \omega^{\beta+1}[x] := \omega^\beta \cdot 2^x; \omega^\lambda[x] := \omega^{\lambda[2^x]}$ where $\omega^\alpha + \lambda > \lambda \in Lim$.

In the second part of the paper we investigate the growth rate of functions which are defined pointwise with respect to a norm function. Let $N0 := 0$ and $N\alpha := N\alpha_1 + \cdots + N\alpha_n$ if $\varepsilon_0 > \alpha = \omega^{\alpha_1} + \cdots + \omega^{\alpha_n} > \alpha_1 \geq \ldots \geq \alpha_n$. Let $A_\alpha(x) := \max(\{0\} \cup \{1 + A_\beta(x) : \beta < \alpha \ \&\ N\beta \leq N\alpha + x\})$ and $B_\alpha(x) := \max(\{0\} \cup \{1 + B_{\max\{\beta<\alpha : N\beta \leq N\alpha+x\}}(x)\})$. Then $(B_\alpha)_{\alpha<\varepsilon_0}$ is slow growing but $(A_\alpha)_{\alpha<\varepsilon_0}$ is fast growing. If $(A_\alpha)_{\alpha<\varepsilon_0}$ and $(B_\alpha)_{\alpha<\varepsilon_0}$ are defined with respect to the Cichon norm then according to results of Arai and Weiermann both hierarchies become fast growing.

Finally we reformulate some of the obtained results in terms of unprovability results for PA. Although true the following assertion is not provable in first order Peano arithmetic: $(\forall k)(\exists i)(\forall \alpha_0, \ldots, \alpha_i < \varepsilon_0)[N\alpha_0 \leq k \ \&\ (\forall l < i)[N(\alpha_{l+1}) \leq N(\alpha_l) + 1] \implies (\exists m, n)[m < n \leq i \ \&\ \alpha_m \leq \alpha_n]]$. As a corollary we re-obtain Friedman's result stating that PA does not prove that ε_0 is slowly well-ordered.

1 Introduction and Motivation

Subrecursive majorization hierarchies play an important role in proof theory and the theory of recursive functions. They provide convenient and well-established scales for classifying provably recursive functions and for characterizing other interesting subclasses of the general recursive functions. For example, if the proof-theoretic ordinal $\| T \|$ of a formal system T (which contains PRA) has been computed in a profound way then the (fast growing) Hardy hierarchy $(H_\alpha)_{\alpha<\|T\|}$ classifies the T-provably recursive functions (See, for example, [6] for a proof). In particular, function classes like the primitive recursive, multiple recursive and $< \varepsilon_0$-recursive functions can be characterized in terms of $(H_\alpha)_{\alpha<\omega^\omega}$, $(H_\alpha)_{\alpha<\omega^{\omega^\omega}}$ and $(H_\alpha)_{\alpha<\varepsilon_0}$. (Recall that the Hardy hierarchy of number-theoretic functions is defined recursively as follows: $H_0(x) := x$, $H_{\alpha+1}(x) := H_\alpha(x+1)$ and $H_\lambda(x) := H_{\lambda[x]}(x)$ where $(\lambda[x])_{x<\omega}$ denotes a canonically chosen fundamental sequence for the limit $\lambda < \varepsilon_0$.)

Some motivations for investigations on the pointwise hierarchy – which sometimes has also been called slow growing hierarchy – have been given, for example, by Girard in [5] from where we select some quotations [comments from our side are in square brackets]:

"The interest of the hierarchy γ lies in the fact that, in some sense, << e is the number of steps needed to compute γ_e >>. Certainly, when $e = 2e'$ we need just one additional step to compute $\gamma_{2e}(n)$ from $\gamma_e(n)$: add 1; but when $e = \langle e', e'' \rangle + 1$, this is not so clear. However, γ is a hierarchy which reflects more faithfully the computation process than others, like λ." [Quotation from p.330 l.31-35. Here γ denotes the pointwise hierarchy whereas λ denotes the fast growing hierarchy. Furthermore e, e' and e'' range over elements of Kleene's ordinal notation system $\mathcal{O}$, $2e$ denotes the successor of e and $\langle e', e''\rangle + 1$ denotes the limit ordinal coded by e''.] "In practice, one will often prefer λ, which can easily be composed. (Similarly, in practice, one will prefer proofs with cuts which are easier to handle.) However, γ is more interesting to study (as well as cut free proofs are more interesting to study)." [Quotation from p.331 l.5-8.]

Girard explained the difference between the pointwise and the fast growing hierarchy as follows. The pointwise hierarchy is "styled pointwise because, at each stage, we compute the value of $\gamma_e(n)$ by means of other values $\gamma_f(n)$ with the same n." [Quotation from p.330 l.21-22.] On the other hand he pointed out that the hierarchy "λ is clearly a non pointwise hierarchy: the computation of λ_e makes use of $\vartheta(\lambda_e)$ and $\vartheta(\lambda_e)(n)$ is determined by means of the << full >> function λ_e, not at all by its value $\lambda_e(n)$!

(ii) The rate of growth of the λ-hierarchy is tremendous whereas for usual indices, γ_e, remained a rather slow function (polynomial; exponential), λ_e will be the right hierarchy to use for crude estimates, but it cannot match the subtler hierarchy γ_e, which is closer to the computation process, and which can measure small differences which are << eaten >> by the coarse hierarchy λ." [Quotation from p.334 l.29-37. Here ϑ denotes the functional $\vartheta(f)(n) = f(n) + 1 + f(n+1)$.]

For classifying provably recursive functions of formal systems via the pointwise hierarchy it is convenient to apply Girard's hierarchy comparison theorem which states that the pointwise hierarchy when extended up to the Howard-Bachmann ordinal matches up with the fast growing hierarchy at level ε_0.

"For the first time, it was possible to give a reasonable definition of << ordinal of a theory >> which yielded values different from the usual ones: the ordinal $\| PA \|_e^\gamma$ for a suitable (<< *natural* >>) choice of e, is the Howard ordinal η_0; similarly if one computes the ordinal of ID_1 in this way, one gets the ordinal η_1 which is traditionally attached to $ID_2 \ldots$" [Quotation from p.438 l.28-32.]

Another interesting application of the pointwise hierarchy has been proposed implicitly by Cichon in [4]. Roughly speaking, the derivation lengths of a finite rewrite system R over a finite signature can be related to the order type of an R-termination ordering $\succ$ (which satisfies $\rightarrow_R \subseteq \succ$) via the pointwise hierarchy. In the meantime this principle has been verified for a large class of termination orderings [11]. These investigations also led naturally to further investigations on the nature of the pointwise hierarchy.

Of course, some very basic questions encounter more or less immediately when one looks at the definition of the pointwise hierarchy: What happens with the pointwise hierarchy when it is defined with respect to an assignment of fundamental sequences which is (slightly) different from the standard assignment? Are there convenient criteria on assignments of fundamental sequences to decide whether the induced pointwise hierarchy is slow growing? In a more general context one may ask: Are there convenient criteria to decide whether a given hierarchy of number-theoretic functions, which is defined pointwise, is slow growing? What makes a subrecursive hierarchy slow growing? In this article we shall try to give contributions to these questions.

It is known, for example from [2], that the fast growing Hardy-hierarchy does not change its growth rate essentially when in its definition the standard assignment of fundamental sequences is replaced by a slight modification of it or by a norm based assignment in the style of Cichon. In addition from results of Zemke 1977 [13] it is known that primitive recursive regulatedness is a condition on assignments of fundamental sequences which guarantees that the associated fast growing Schwichtenberg-Wainer hierarchies coincide at each ordinal level.

Therefore, reasoning via analogy, it seems very plausible, and perhaps it would also be desirable, to have a similar independence result for the pointwise hierarchy.

In this article we show more or less that such an independence result in terms of primitive recursive regulatedness or related conditions does not hold for the pointwise hierarchy. Indeed, as we will point out, small innocent looking changes in the definition of the assignment of fundamental sequences may yield drastic changes in the growth rate of the resulting pointwise hierarchy.

In the author's opinion these investigations may shed some light on the concept of *standard assignment of fundamental sequences* as we will try to explain in the following. For the standard assignment of fundamental sequences the pointwise hierarchy is slow growing. Among all reasonable assignments of fundamental sequences one may consider (at most) those as standard which yield slow growing pointwise hierarchies. This criterion would distinguish several norm based assignments of fundamental sequences as well as several slight modifications of the standard assignment from the standard assignment.

From the investigations in this paper it turns out that with respect to this criterion a form of strong Lipschitz continuity for ordinal addition, i.e. $(\omega^\alpha + \lambda)[x] = \omega^\alpha + \lambda[x]$ is perhaps a crucial property of the assignment for being standard.

In the second part of the paper we investigate the interdependencies between pointwise recursion and slow growingness from a more general point of view. Our examples indicate that pointwise recursions do by no means necessarily lead to slow growing hierarchies. These investigations lead naturally to combinatorial principles which are true but unprovable in PA.

The paper is essentially self-contained. Throughout the paper we denote ordinals less than ε_0 by small Greek letters. *Lim* denotes the set of limit ordinals less than ε_0. Small Latin numbers range over natural numbers.

2 What makes a pointwise hierarchy fast growing?

In this section we are going to show that the pointwise hierarchy becomes fast growing if the underlying system of fundamental sequences does not satisfy a strong Lipschitz continuity condition for ordinal addition or does not satisfy a pointwise continuity condition for the limit exponentiation case. To start let us recall some basic facts about ε_0 and the pointwise hierarchy.

Lemma 1 *For any nonzero ordinal $\alpha < \varepsilon_0$ there exist a unique natural number n and uniquely determined ordinals $\alpha_1, \ldots, \alpha_n < \varepsilon_0$ such that $\alpha = \omega^{\alpha_1} + \cdots + \omega^{\alpha_n}$ and $\alpha > \alpha_1 \geq \ldots \geq \alpha_n$.*

In the sequel we will write $\alpha =_{NF} \omega^{\alpha_1} + \cdots + \omega^{\alpha_n}$ if $\alpha = \omega^{\alpha_1} + \cdots + \omega^{\alpha_n}$ and $\alpha > \alpha_1 \geq \ldots \geq \alpha_n$. We will write $NF(\alpha, \beta)$ if $\alpha =_{NF} \omega^{\alpha_1} + \cdots + \omega^{\alpha_m}$, $\beta =_{NF} \omega^{\beta_1} + \cdots + \omega^{\beta_n}$ and $\alpha_m \geq \beta_1$.

Definition 1 (An assignment of fundamental sequences) *A mapping* $\cdot[\cdot]$: $(\varepsilon_0 \cap Lim) \times \omega \to \varepsilon_0$ *is called an* assignment of fundamental sequences *for the limits less than* ε_0 *if*

1. $\lambda[x] < \lambda[x+1]$ *for every* $x < \omega$.
2. $\sup\{\lambda[x] : x \in \omega\} = \lambda$.

For convenience we also put $0[x] := 0$ *and* $(\alpha+1)[x] := \alpha$ *for a given assignment.*

Definition 2 *The standard assignment of fundamental sequences is defined as follows.*

1. $(\omega^\alpha + \lambda)[\![x]\!] := \omega^\alpha + \lambda[\![x]\!]$ *if* $\omega^\alpha + \lambda > \lambda$ *and* $\lambda \in Lim$.
2. $\omega^\lambda[\![x]\!] := \omega^{\lambda[\![x]\!]}$ *if* $\lambda \in Lim$.
3. $\omega^{\alpha+1}[\![x]\!] := \omega^\alpha \cdot (x+1)$.

Obviously this definition meets the conditions for an assignment of fundamental sequences. Moreover it has the so called Bachmann property (cf., for example, Lemma 5). As shown, for example, in [2, 7] this property yields a decent theory of the induced fast growing hierarchy. In the case of ordinal addition and limit exponentiation the fundamental sequences are defined by a form of strong Lipschitz continuity. This notion can be understood as follows. Let $f : \varepsilon_0 \to \varepsilon_0$ be a continuous function. For any limit $\lambda < \varepsilon_0$ then $f(\lambda[x])$ converges to $f(\lambda)$ if $\lambda[x]$ converges to λ. If we take $f(\lambda[x])$ as the x-th member of the fundamental sequence for $f(\lambda)$ then the modulus of continuity is the simplest one, so that we may speak in such a situation about strong Lipschitz continuity.

Definition 3 (The pointwise hierarchy)
Let $\cdot[\cdot]$ be an assignment of fundamental sequences for the limits less than ε_0. With respect to $\cdot[\cdot]$ the pointwise hierarchy $(G_\alpha)_{\alpha<\varepsilon_0}$ is defined as follows.

1. $G_0(x) := 0$.
2. $G_{\alpha+1}(x) := 1 + G_\alpha(x)$.
3. $G_\lambda(x) := G_{\lambda[x]}(x)$ *if* $\lambda \in Lim$.

Lemma 2
Let the pointwise hierarchy be defined with respect to the standard assignment of fundamental sequences. Then we have the following:

1. $G_0(x) = 0$.
2. $G_\alpha(x) = (x+1)^{G_{\alpha_1}(x)} + \cdots + (x+1)^{G_{\alpha_n}(x)}$ *if* $\alpha =_{NF} \omega^{\alpha_1} + \cdots + \omega^{\alpha_n}$.
3. *For any $\alpha < \varepsilon_0$ the function G_α is bounded by an elementary recursive function.*

Proof. Assertion 2) is proved by induction on α. Assertion 3) follows from assertion 2). □

For investigations on more general pointwise hierarchies we introduce certain auxiliary concepts.

Definition 4 (The norm function) *1.* $N(0) := 0$.

2. $N(\alpha) := n + N(\alpha_1) + \cdots + N(\alpha_n)$ *if* $\alpha =_{NF} \omega^{\alpha_1} + \cdots + \omega^{\alpha_n}$.

In the sequel we will write $N\alpha$ in short for $N(\alpha)$.

Definition 5 *An assignment $\cdot[\cdot]$ of fundamental sequences for the limits less than ε_0 satisfies the* norm property *if $N(\lambda) \le N(\lambda[x])$ holds for every $\lambda < \varepsilon_0$ and $x \ge 2$.*

The standard assignment satisfies the norm property and every assignment considered in this paper also satisfies this requirement.

Definition 6
Let $\cdot[\cdot]$ be an assignment of fundamental sequences for the limits less than ε_0. With respect to this assignment $\cdot[\cdot]$ we define certain relations on the set of ordinals less than ε_0 as follows:

1. $\alpha \prec_x \beta$
$:\iff (\exists n > 0)(\exists \gamma_0, \ldots, \gamma_n)[\beta = \gamma_0 \;\&\; \alpha = \gamma_n \;\&\; (\forall i < n)[\gamma_{i+1} = \gamma_i[x]]]$.
2. $\alpha \preceq_x \beta :\iff \alpha \prec_x \beta \vee \alpha = \beta$.
3. $m \sqsubseteq_x \beta :\iff (\exists \alpha)[\alpha \preceq_x \beta \;\&\; m \leq N\alpha]$.

The following lemma provides some useful properties for investigating the growth rate of pointwise hierarchies.

Lemma 3 *Let $\cdot[\cdot]$ be an assignment of fundamental sequences for the limits less than ε_0 which satisfies the norm property. Let the pointwise hierarchy $(G_\alpha)_{\alpha<\varepsilon_0}$ be defined with respect to $\cdot[\cdot]$. Then we have:*

1. $\alpha \preceq_x \beta \implies G_\alpha(x) \leq G_\beta(x)$.
2. $G_\beta(x) \geq N\beta$ *for any* $x \geq 2$.
3. $m \sqsubseteq_x \beta \implies m \leq G_\beta(x)$ *for any* $x \geq 2$.

Proof. Straightforward. □

Definition 7 (Strongly elementary regulated assignments) *Let p be a ternary and q, r be binary elementary recursive functions such that p, q, r are weakly monotonic increasing in each argument and strictly monotonic increasing in their last argument. An assignment $\cdot[\cdot]$ is called* strongly elementary regulated *(via p, q, r) if the following properties are satisfied:*

1. $(\omega^\alpha + \lambda)[x] = \omega^\alpha + \lambda[p(N\alpha, N\lambda, x)]$ *if* $\omega^\alpha + \lambda > \lambda$ *and* $\lambda \in Lim$.
2. $\omega^\lambda[x] = \omega^{\lambda[q(N\lambda, x)]}$ *if* $\lambda \in Lim$.
3. $\omega^{\alpha+1}[x] = \omega^\alpha \cdot (1 + r(N\alpha, x))$.

In the sequel we give a complete classification of the pointwise hierarchies which result from strongly elementary regulated assignments of fundamental sequences.

Lemma 4 *Assume that $\cdot[\cdot]$ is strongly elementary regulated. Then $\cdot[\cdot]$ satisfies the norm property.*

Proof. This follows by definition. □

Lemma 5 *Assume that* $\cdot[\cdot]$ *is strongly elementary regulated. Let* $\lambda \in Lim$ *and assume that* $\lambda[x] < \alpha < \lambda$. *Then* $\lambda[x] \leq \alpha[y]$ *for all* y.

Proof. By induction on $N\lambda$. □

Corollary 1 *Assume that* $\cdot[\cdot]$ *is strongly elementary regulated. Let* $\prec_y$ *be defined with respect to* $\cdot[\cdot]$.

1. *Let* $\lambda \in Lim$. *Then* $\lambda[x] \prec_y \lambda[x+1]$.

2. *Assume that* $\alpha \prec_x \beta < \omega^\gamma + \beta$. *Then* $\omega^\gamma + \alpha \prec_x \omega^\gamma + \beta$.

3. *Assume that* $\alpha \prec_x \beta$. *Then* $\omega^\alpha \prec_x \omega^\beta$.

Proof. Assertion 1) follows from Lemma 5. The assertions 2) and 3) are proved by induction on β with the use of 1). □

The following lemma shows that monotonicity for the regulating functions yields the expected monotonicity for the induced assignments and pointwise hierarchies.

Lemma 6 *Let* $\cdot[\cdot]$ *be strongly elementary regulated via* p, q, r *and let* $\cdot[\cdot]'$ *be strongly elementary regulated via* p', q', r'. *Assume that* $p(x,y,z) \leq p'(x,y,z)$ $q(x,y) \leq q'(x,y)$ *and* $r(x,y) \leq r'(x,y)$ *holds for all* $x, y, z < \omega$. *Let* $(G_\alpha)_{\alpha<\varepsilon_0}$ *be defined with respect to* $\cdot[\cdot]$ *and let* $(G'_\alpha)_{\alpha<\varepsilon_0}$ *be defined with respect to* $\cdot[\cdot]'$. *Let* $\preceq_y$ *be defined with respect to* $\cdot[\cdot]$. *Then the following holds:*

1. *If* $\lambda \in Lim$ *then* $\lambda[x] \preceq_y \lambda[x]'$.

2. $G_\alpha(x) \leq G'_\alpha(x)$ *for any* $\alpha < \varepsilon_0$ *and* $x < \omega$.

Proof. Straightforward. □

We are now going to show that certain pointwise hierarchies consist in fact of fast growing functions. For this purpose we recall some basic facts from hierarchy theory.

Definition 8 (The Schwichtenberg-Wainer-Hierarchy) *With regard to the standard assignment of fundamental sequences we define recursively number-theoretic functions* F_α *as follows.*

1. $F_0(x) := 2^x$.

2. $F_{\alpha+1}(x) := F_\alpha^{x+1}(x)$.

3. $F_\lambda(x) := F_{\lambda[x]}(x)$ *if* λ *is a limit.*

The growth rate of this hierarchy does not depend on whether we define it in terms of $\cdot[\![\cdot]\!]$, or its variants considered in this paper. In any case we get a classification of the provably recursive functions of PA in terms of $(F_\alpha)_{\alpha<\varepsilon_0}$. More information on this hierarchy can be found, for example, in [3, 7, 9, 10].

Lemma 7 *Let $\preceq_x$ be defined with respect to the standard assignment $\cdot[\![\cdot]\!]$. Then $\alpha \preceq_x \beta$ yields $F_\alpha(y) \leq F_\beta(y)$ for all $y \geq x$. Furthermore each function F_α is strictly monotonic increasing.*

Definition 9 *We define explicitly two strongly elementary regulated variants $\cdot[\cdot]^1$ and $\cdot[\cdot]^2$ of the standard assignment of fundamental sequences as follows:*

1. *(a)* $(\omega^\alpha + \lambda)[x]^1 := \omega^\alpha + \lambda[x+1]^1$ *if* $\omega^\alpha + \lambda > \lambda$ *and* $\lambda \in Lim$.
   (b) $\omega^\lambda[x]^1 := \omega^{\lambda[x]^1}$ *if* $\lambda \in Lim$.
   (c) $\omega^{\alpha+1}[x]^1 := \omega^\alpha \cdot (1+x)$.
2. *(a)* $(\omega^\alpha + \lambda)[x]^2 := \omega^\alpha + \lambda[x]^2$ *if* $\omega^\alpha + \lambda > \lambda$ *and* $\lambda \in Lim$.
   (b) $\omega^\lambda[x]^2 := \omega^{\lambda[N\lambda+x]^2}$ *if* $\lambda \in Lim$.
   (c) $\omega^{\alpha+1}[x]^2 := \omega^\alpha \cdot (1+x)$.

As usual we put $2_0(l) := l$ and $2_{k+1}(l) := 2^{2_k(l)}$.

Theorem 1 *Let $\cdot[\cdot]^1$ and $\cdot[\cdot]^2$ be as in Definition 9. According to Definition 6 let $\sqsupseteq_3^1$ be defined with respect to $\cdot[\cdot]^1$ and let $\sqsupseteq_3^2$ be defined with respect to $\cdot[\cdot]^2$. Assume $\varepsilon_0 > \alpha_1 \geq \ldots \geq \alpha_n$ where $n \geq 0$. Let $\rho := \omega^{2+\alpha_1} + \cdots + \omega^{2+\alpha_n}$, $\alpha := \omega^{\delta+\omega^{\rho+\omega\cdot k+l}}$ and assume $NF(\delta, \omega^{\rho+\omega\cdot k+l})$. Then*

$$\alpha \sqsupseteq_3^i F_{\alpha_1}(\ldots F_{\alpha_n}(2_k(l+1))\ldots)$$

for $i = 1, 2$.

Proof. By induction on α. The proof is a modification of the proof of Theorem 1 from [12]. Assume that $i \in \{1, 2\}$ is fixed. Let $\cdot[\cdot] := \cdot[\cdot]^i$ and let $\sqsupseteq_3$ and $\succeq_3$ be defined with respect to $\cdot[\cdot]$. Let $F_{\hat\rho}(x) := F_{\alpha_1}(\ldots F_{\alpha_n}(x)\ldots)$ where $F_{\hat\rho}(x) := x$ for $n = 0$. We have to show $\alpha \sqsupseteq_3 F_{\hat\rho}(2_k(l+1))$. In the following calculations we frequently make use of assertions 1), 2) and 3) of Corollary 1, of assertion 1) of Lemma 6, and of Lemma 7. By (IH) we indicate an application of the induction hypothesis.

Case 1. $l = 0$.

Case 1.1. $k > 0$. Then

$$\begin{aligned} \alpha &= \omega^{\delta+\omega^{\rho+\omega\cdot k}} \\ &\succeq_3 \omega^{\delta+\omega^{\rho+\omega\cdot(k-1)+1}} && \text{by IH} \\ &\sqsupseteq_3 F_{\hat\rho}(2_{k-1}(2)) = F_{\hat\rho}(2_k(1)). \end{aligned}$$

Case 1.2. $k = 0$.
Case 1.2.1. $n > 0$.
Let $\sigma := \omega^{2+\alpha_1} + \cdots + \omega^{2+\alpha_{n-1}} \geq 0$ and $F_{\hat{\sigma}}(x) := F_{\alpha_1}(\ldots F_{\alpha_{n-1}}(x)\ldots)$ where $F_{\hat{\sigma}}(x) := x$ for $n = 1$.
Case 1.2.1.1. $\alpha_n = 0$. Then

$$\begin{aligned} \alpha &= \omega^{\delta+\omega^{\sigma}+\omega^{2}} \\ &\succeq_3 \omega^{\delta+\omega^{\sigma}+\omega\cdot 1} && \text{by IH} \\ &\sqsupseteq_3 F_{\hat{\sigma}}(2_1(1)) = F_{\hat{\sigma}}(F_0(1)) = F_{\hat{\rho}}(2_0(1)). \end{aligned}$$

Case 1.2.1.2. $\alpha_n = \alpha_n{}' + 1$. Then

$$\begin{aligned} \alpha &= \omega^{\delta+\omega^{\sigma}+\omega^{2+\alpha_n{}'+1}} \\ &\succeq_3 \omega^{\delta+\omega^{\sigma}+\omega^{2+\alpha_n{}'}\cdot 2} && \text{by IH} \\ &\sqsupseteq_3 F_{\hat{\sigma}}(F^2_{\alpha_n{}'}(1)) = F_{\hat{\sigma}}(F_{\alpha_n{}'+1}(1)) = F_{\hat{\rho}}(2_0(1)). \end{aligned}$$

Case 1.2.1.3. $\alpha_n \in Lim$ and $\alpha_n \neq \omega$.
In this case we have $\alpha_n[x] \geq \omega$ hence $2 + \alpha_n[x] = \alpha_n[x]$ for all $x \in \omega$. Thus

$$\begin{aligned} \alpha &= \omega^{\delta+\omega^{\sigma}+\omega^{2+\alpha_n}} \\ &= \omega^{\delta+\omega^{\sigma}+\omega^{\alpha_n}} \\ &\succeq_3 \omega^{\delta+\omega^{\sigma}+\omega^{\alpha_n[3]}} \\ &\succeq_3 \omega^{\delta+\omega^{\sigma}+\omega^{\alpha_n[1]}} \\ &= \omega^{\delta+\omega^{\sigma}+\omega^{2+\alpha_n[1]}} && \text{by IH} \\ &\sqsupseteq_3 F_{\hat{\sigma}}(F_{\alpha_n[1]}(1)) && \text{by Lemma 6 and Lemma 7} \\ &\geq F_{\hat{\sigma}}(F_{\alpha_n[1]}(1)) = F_{\hat{\sigma}}(F_{\alpha_n}(1)) = F_{\hat{\rho}}(2_0(1)). \end{aligned}$$

Case 1.2.1.4. $\alpha_n = \omega$.
Then $F_\omega(1) = F_2(1)$ and

$$\begin{aligned} \alpha &= \omega^{\delta+\omega^{\sigma}+\omega^{2+\omega}} \\ &= \omega^{\delta+\omega^{\sigma}+\omega^{\omega}} \\ &\succeq_3 \omega^{\delta+\omega^{\sigma}+\omega^{4}} \\ &= \omega^{\delta+\omega^{\sigma}+\omega^{2+2}} && \text{by IH} \\ &\sqsupseteq_3 F_{\hat{\sigma}}(F_2(1)) = F_{\hat{\sigma}}(F_\omega(1)) = F_{\hat{\rho}}(2_0(1)). \end{aligned}$$

Case 1.2.2. $n = 0$. Then $\alpha \sqsupseteq_3 N\alpha \geq 1 = F_{\hat{\rho}}(2_0(1))$.
Case 2. $l > 0$.

Case 2.1. $k > 0$. Then

$$\begin{aligned}
\alpha &= \omega^{\delta+\omega^{\rho+\omega\cdot k+l}} \\
&\succeq_3 \omega^{\delta+\omega^{\rho+\omega\cdot k+l-1}.(2^1+1)} \\
&\succeq_3 \omega^{\delta+\omega^{\rho+\omega\cdot k+l-1}.2+\omega^{\rho+\omega\cdot k+l-2}.(2^2+1)} \\
&\succeq_3 \cdots \\
&\succeq_3 \omega^{\delta+\omega^{\rho+\omega\cdot k+l-1}.2+\cdots+\omega^{\rho+\omega\cdot k}.(2^l+1)} \\
&\succeq_3 \omega^{\delta+\omega^{\rho+\omega\cdot k+l-1}.2+\cdots+\omega^{\rho+\omega\cdot k}.2^l+\omega^{\rho+\omega\cdot(k-1)+2^{l+1}}} && \text{by IH} \\
&\sqsupseteq_3 F_{\hat{\rho}}(2_{k-1}(2^{l+1}+1)) \geq F_{\hat{\rho}}(2_k(l+1)).
\end{aligned}$$

Case 2.2. $k = 0$.
Case 2.2.1. $n > 0$.
Case 2.2.1.1. $\alpha_n = 0$. Then

$$\begin{aligned}
\alpha &= \omega^{\delta+\omega^{\sigma+\omega^2+l}} \\
&\succeq_3 \omega^{\delta+\omega^{\sigma+\omega^2+l-1}.(2^1+1)} \\
&\succeq_3 \omega^{\delta+\omega^{\sigma+\omega^2+l-1}.2+\omega^{\sigma+\omega^2+l-2}.(2^2+1)} \\
&\succeq_3 \cdots \\
&\succeq_3 \omega^{\delta+\omega^{\sigma+\omega^2+l-1}.2+\cdots+\omega^{\sigma+\omega^2}.(2^l+1)} \\
&\succeq_3 \omega^{\delta+\omega^{\sigma+\omega^2+l-1}.2+\cdots+\omega^{\sigma+\omega^2}.2^l+\omega^{\sigma+\omega\cdot 2^{l+1}}} && \text{by IH} \\
&\sqsupseteq_3 F_{\hat{\sigma}}(2_{2^{l+1}}(1)) \geq F_{\hat{\sigma}}(2^{l+1}) = F_{\hat{\sigma}}(F_0(2_0(l+1))) = F_{\hat{\rho}}(2_0(l+1)).
\end{aligned}$$

Case 2.2.1.2. $\alpha_n = {\alpha_n}' + 1$. Then

$$\begin{aligned}
\alpha &= \omega^{\delta+\omega^{\sigma+\omega^{2+{\alpha_n}'+1}+l}} \\
&\succeq_3 \omega^{\delta+\omega^{\sigma+\omega^{2+{\alpha_n}'+1}+l-1}.(2^1+1)} \\
&\succeq_3 \omega^{\delta+\omega^{\sigma+\omega^{2+{\alpha_n}'+1}+l-1}.2+\omega^{\sigma+\omega^{2+{\alpha_n}'+1}+l-2}.(2^2+1)} \\
&\succeq_3 \cdots \\
&\succeq_3 \omega^{\delta+\omega^{\sigma+\omega^{2+{\alpha_n}'+1}+l-1}.2+\cdots+\omega^{\sigma+\omega^{2+{\alpha_n}'+1}}.(2^l+1)} && \text{since } \textstyle\sum_{i=1}^{l} 2^l = 2^{l+1}-2 \\
&\succeq_3 \omega^{\delta+\omega^{\sigma+\omega^{2+{\alpha_n}'+1}+l-1}.2+\cdots+\omega^{\sigma+\omega^{2+{\alpha_n}'+1}}.2^l+\omega^{\sigma+\omega^{2+{\alpha_n}'}.(2^{l+1}+1)}} && \text{by IH} \\
&\sqsupseteq_3 F_{\hat{\sigma}}(F_{{\alpha_n}'}^{2^{l+1}+1}(1)) && \text{since } 2^{l+1}+1 \geq l+2+l+1 \\
&\geq F_{\hat{\sigma}}(F_{{\alpha_n}'}^{l+2}(l+1)) = F_{\hat{\sigma}}(F_{\alpha_n}(l+1))) = F_{\hat{\rho}}(2_0(l+1)).
\end{aligned}$$

Case 2.2.1.3. $\alpha_n \in Lim$ and $\alpha_n > \omega$.
Then $\alpha[x] = 2 + \alpha[x]$ for all $x < \omega$. Hence

$$\alpha = \omega^{\delta+\omega^{\sigma+\omega^{2+\alpha_n}+l}}$$

$$
\begin{array}{ll}
= & \omega^{\delta+\omega^{\sigma}+\omega^{\alpha_n}+l} \\
\succeq_3 & \omega^{\delta+\omega^{\sigma}+\omega^{\alpha_n}+l-1}\cdot 2+\cdots+\omega^{\sigma}+\omega^{\alpha_n}\cdot(2^l+1) \\
\succeq_3 & \omega^{\delta+\omega^{\sigma}+\omega^{\alpha_n}+l-1}\cdot 2+\cdots+\omega^{\sigma}+\omega^{\alpha_n}\cdot 2^l+\omega^{\sigma}+\omega^{\alpha_n[2^{l+1}]} \quad \text{by IH} \\
\sqsupseteq_3 & F_{\hat{\sigma}}(F_{\alpha_n[2^l+1]}(1)) \quad \text{by Lemma 6 and Lemma 7} \\
\geq & F_{\hat{\sigma}}(F_{\alpha_n[2^l+1]}(1)) \geq F_{\hat{\sigma}}(F_{\alpha_n[l+1]}(l+1)) \\
= & F_{\hat{\sigma}}(F_{\alpha_n}(l+1))) = F_{\hat{\rho}}(2_0(l+1)).
\end{array}
$$

Case 2.2.1.4. $\alpha_n = \omega$. Then

$$
\begin{array}{lll}
\alpha & = & \omega^{\delta+\omega^{\sigma}+\omega^{\omega}+l} \\
& \succeq_3 & \omega^{\delta+\omega^{\sigma}+\omega^{\omega}+l-1}\cdot 2+\cdots+\omega^{\sigma}+\omega^{\omega}\cdot(2^l+1) \\
& \succeq_3 & \omega^{\delta+\omega^{\sigma}+\omega^{\omega}+l-1}\cdot 2+\cdots+\omega^{\sigma}+\omega^{\omega}\cdot 2^l+\omega^{\sigma}+\omega^{2^{l+1}+3} \quad \text{by IH} \\
& \sqsupseteq_3 & F_{\hat{\sigma}}(F_{2^{l+1}+1}(1)) \geq F_{\hat{\sigma}}(F_{l+2}(l+1)) \geq F_{\hat{\sigma}}(F_{\omega}(l+1)) = F_{\hat{\rho}}(2_0(l+1)).
\end{array}
$$

Case 2.2.2. $n = 0$. Then $\alpha \sqsupseteq_3 N\alpha \geq l = F_{\hat{\rho}}(2_0(l))$. □

Definition 10 *Let $(G^i_\alpha)_{\alpha<\varepsilon_0}$ be the pointwise hierarchy when it is defined with respect to $\cdot[\cdot]^i$ for $i = 1, 2$. Thus for $i = 1, 2$ we have:*

1. *$G^i_0(x) := 0$.*
2. *$G^i_{\alpha+1}(x) := 1 + G^i_\alpha(x)$.*
3. *$G^i_\lambda(x) := G^i_{\lambda[x]^i}(x)$ if $\lambda \in Lim$.*

Corollary 2 *The hierarchies $(G^1_\alpha)_{\alpha<\varepsilon_0}$ and $(G^2_\alpha)_{\alpha<\varepsilon_0}$ are fast growing. In fact we have for any $\alpha_n \leq \ldots \leq \alpha_1$ and δ with $NF(\delta, \omega^{\omega^{2+\alpha_1}+\cdots+\omega^{2+\alpha_n}+\omega\cdot k+\omega})$ and*

$$\alpha := \omega^{\delta+\omega^{\omega^{2+\alpha_1}+\cdots+\omega^{2+\alpha_n}+\omega\cdot k+\omega}}$$

that

$$G^i_\alpha(l) \geq F_{\alpha_1}(\ldots F_{\alpha_n}(2_k(l+1))\ldots)$$

for $i = 1, 2$, $k < \omega$ and $3 \leq l < \omega$.

Proof. Let $\succeq^i_l$ be defined with respect to $\cdot[\cdot]^i$. Then $\omega^{\delta+\omega^{\omega^{2+\alpha_1}+\cdots+\omega^{2+\alpha_n}+\omega\cdot k+\omega}} \succeq^i_l \omega^{\delta+\omega^{\omega^{2+\alpha_1}+\cdots+\omega^{2+\alpha_n}+\omega\cdot k+l}}$ holds for $i = 1, 2$ and $l \geq 3$. The assertion follows from Lemma 3 and Theorem 1. □

Corollary 3 *1. Assume that $\cdot[\cdot]$ is strongly elementary regulated via p, q, r. Let $(G_\alpha)_{\alpha<\varepsilon_0}$ be defined with respect to $\cdot[\cdot]$. Assume that p, q, r satisfy the following monotonicity requirements:*

   (a) $p(x, y, z) \geq z + 1$.
   (b) $q(x, y) \geq y$.
   (c) $r(x, y) \geq y$.

   Then the hierarchy $(G_\alpha)_{\alpha<\varepsilon_0}$ *is fast growing.*

2. *Assume that* $\cdot[\cdot]$ *is strongly elementary regulated via* p, q, r. *Let* $(G_\alpha)_{\alpha<\varepsilon_0}$ *be defined with respect to* $\cdot[\cdot]$. *Assume that* p, q, r *satisfy the following monotonicity requirements:*

   (a) $p(x, y, z) \geq z$.
   (b) $q(x, y) \geq x + y$.
   (c) $r(x, y) \geq y$.

   Then the hierarchy $(G_\alpha)_{\alpha<\varepsilon_0}$ *is fast growing.*

Proof. This follows from Corollary 2 and Lemma 6. □

3 What makes a pointwise hierarchy slow growing?

In this section we try to single out some general conditions on assignments of fundamental sequences under which the resulting pointwise hierarchies remain slow growing. Throughout this section let q be a unary elementary recursive strictly monotonic increasing function and let r be a binary elementary recursive function which is strictly monotonic increasing in each argument. Let N be the norm function as defined in Definition 4.

Definition 11 *Let* q, r *as above. By recursion on* $N\alpha$ *we define a natural number* $C_x(\alpha)$ *as follows.*

1. $C_x(0) := 0$.
2. $C_x(\alpha) := C_x(\omega^{\alpha_1}) + \cdots + C_x(\omega^{\alpha_n})$ *if* $\alpha =_{NF} \omega^{\alpha_1} + \cdots + \omega^{\alpha_n}$.
3. $C_x(\omega^\alpha) := (r(C_{q(x)}(\alpha), x) + 1)^{C_{q(x)}(\alpha)+1}$

Lemma 8 $N(\alpha) \leq C_x(\alpha)$.

Proof. By a straightforward induction on $N(\alpha)$. □

Definition 12 *Let q' be a unary elementary recursive function and r' be a binary elementary recursive function. An assignment* $\cdot[\cdot]\colon (\varepsilon_0 \cap Lim) \times \omega \to \omega$ *of fundamental sequences is* slowly regulated *by q' and r' if*

1. $\alpha[x] = \omega^{\alpha_1} + \cdots + \omega^{\alpha_n}[x]$ *if* $\alpha =_{NF} \omega^{\alpha_1} + \cdots + \omega^{\alpha_n}$ *[Strong Lipschitz continuity].*

2. $\omega^{\alpha}[x] = \omega^{\alpha[q'(x)]} \cdot r'(N\alpha, x) + \alpha_x$ *for some* $\alpha_x < \omega^{\alpha[q'(x)]}$ *with* $N(\alpha_x) < r(0,x)$ *[elementary recursive convergence].*

Lemma 9 *Let* $\cdot[\cdot]\colon (\varepsilon_0 \cap Lim) \times \omega \to \omega$ *be slowly regulated by the elementary recursive functions q' and r'. Assume that $q'(x) \leq q(x)$ and $r'(x,y) \leq r(x,y)$ for all $x, y < \omega$. Then we have the following.*

1. $C_x(\alpha) + 1 \leq C_x(\alpha + 1)$.

2. $\alpha < \beta \;\&\; N(\alpha) < r(0,x) \implies C_x(\alpha) < C_x(\beta)$.

3. $\alpha \in Lim \;\&\; x \geq y \implies C_x(\alpha[y]) < C_x(\alpha)$.

Proof. Assertion 1 is obvious. Assertion 2 is proved by induction on $N(\beta)$. We may assume that $\alpha > 0$. Assume first that $\beta =_{NF} \omega^{\beta_1} + \cdots + \omega^{\beta_n}$ where $n \geq 2$. Assume that $\alpha =_{NF} \omega^{\alpha_1} + \cdots + \omega^{\alpha_m}$. If $m < n$ and $\alpha_j = \beta_j$ for $1 \leq j \leq m$ then $C_x(\alpha) < C_x(\beta)$. Assume now that there is an $i \leq n$ such that $\alpha_i < \beta_i$ and $\alpha_j = \beta_j$ for $1 \leq j < i$. Then the induction hypothesis yields $C_x(\omega^{\alpha_i} + \cdots + \omega^{\alpha_m}) < C_x(\omega^{\beta_i})$, hence $C_x(\alpha) < C_x(\beta)$.

For the critical case assume that $\beta = \omega^{\beta_1}$. Assume that $\alpha =_{NF} \omega^{\alpha_1} + \cdots + \omega^{\alpha_m}$. Then $\alpha_j < \beta_1$ for $1 \leq j \leq m$ and $m \leq N\alpha < r(0,x)$. The induction hypothesis yields $C_{q(x)}(\alpha_j) < C_{q(x)}(\beta_1)$ for $1 \leq j \leq m$ hence

$$\begin{aligned} C_x(\alpha) &= (r(C_{q(x)}(\alpha_1), x) + 1)^{C_{q(x)}(\alpha_1)+1} + \ldots + (r(C_{q(x)}(\alpha_m), x) + 1)^{C_{q(x)}(\alpha_m)+1} \\ &< (r(C_{q(x)}(\beta_1), x) + 1)^{C_{q(x)}(\beta_1)} \cdot r(0,x) \\ &\leq (r(C_{q(x)}(\beta_1), x) + 1)^{C_{q(x)}(\beta_1)+1}. \end{aligned}$$

Assertion 3 is proved by induction on $N\alpha$. Assume that $\alpha =_{NF} \omega^{\alpha_1} + \cdots + \omega^{\alpha_n}$ where $n > 1$. Then $\alpha[y] = \omega^{\alpha_1} + \cdots + \omega^{\alpha_n}[y]$ and the induction hypothesis yields $C_x(\alpha[y]) = C_x(\omega^{\alpha_1}) + \cdots + C_x(\omega^{\alpha_n}[y]) < C_x(\omega^{\alpha_1}) + \cdots + C_x(\omega^{\alpha_n}) = C_x(\alpha)$.

For the critical case let $\alpha = \omega^{\beta}$. For $\beta \in Lim$ the induction hypothesis yields

$$C_{q(x)}(\beta[q'(y)]) < C_{q(x)}(\beta)$$

since $q'(y) \leq q(y) \leq q(x)$. The last inequality also holds for $\beta \notin Lim$ by assertion 1. Assertion 2 yields $C_x(\beta_y) < C_x(\beta)$. By Lemma 8 we therefore obtain

$$\begin{aligned} C_x(\alpha[y]) &= r'(N(\beta), y) \cdot (r(C_{q(x)}(\beta[q'(y)]), x) + 1)^{C_{q(x)}(\beta[q'(y)])+1} + C_x(\beta_y) \\ &< (r(C_{q(x)}(\beta), x) + 1)^{C_{q(x)}(\beta)+1} \\ &= C_x(\alpha). \end{aligned}$$

□

Theorem 2 *If $\cdot[\cdot]$ is slowly regulated then the resulting pointwise hierarchy is slow growing.*

Proof. Lemma 9 together with a straightforward induction on α yields $G_\alpha(x) \leq C_x(\alpha)$. By induction on $N\alpha$ one easily verifies that the mapping $x \mapsto C_x(\alpha)$ is bounded by an elementary recursive function. □

Summing up, the pointwise growing hierarchy is slow growing if the underlying system of fundamental sequences satisfies a strong Lipschitz continuity condition for ordinal addition and a pointwise continuity condition for the limit exponentiation case. It can be seen that this result extends to much further reaching ordinal notation systems than the one for ε_0. It turns out from these extensions that the binary ordinal addition function plays a distinguished role for the pointwise hierarchy.

4 What happens with norm based fundamental sequences?

The theory of norm based assignments of fundamental sequences and their associated fast growing hierarchies has been developed in [2]. Here we investigate the growth rate behavior of the pointwise hierarchy $(G_\alpha)_{\alpha<\varepsilon_0}$ when it is defined with respect to a norm based assignment.

Definition 13 (The Cichon norm, cf. [4]) *1.* $N_C(0) := 0$.

2. $N_C(\alpha) := \max\{n, 1+N_C(\alpha_1), \ldots, 1+N_C(\alpha_n)\}$ *if* $\alpha =_{NF} \omega^{\alpha_1} + \cdots + \omega^{\alpha_n}$.

Definition 14 (The Cichon assignment, cf. [4]) *For* $\alpha > 0$ *we define* $\alpha[x]_C := \max\{\beta < \alpha : N_C(\beta) \leq N_C(\alpha) + x\}$.

Theorem 3 *If* $(G_\alpha)_{\alpha<\varepsilon_0}$ *is defined with respect to* $\cdot[\cdot]_C$ *then the resulting pointwise hierarchy is fast growing.*

Proof. See, for example, [12] for a proof. □

Theorem 3 is in accordance with Theorem 1 since $\cdot[\cdot]_C$ does not satisfy the strong Lipschitz continuity. What happens if we replace in the definition of $\cdot[\cdot]_C$ the Cichon norm N_C by the "length"-norm N from Definition 4? The answer is provided by the following theorem.

Theorem 4 *For* $\alpha > 0$ *let* $\alpha[x]_L := \max\{\beta < \alpha : N\beta \leq N\alpha + x\}$. *Then* $\cdot[\cdot]_L$ *is slowly regulated. Hence the induced pointwise hierarchy* $(G_\alpha)_{\alpha<\varepsilon_0}$ *is slow growing.*

Proof. In this proof we abbreviate $\cdot[\cdot]_L$ by $\cdot[\cdot]$. For any λ in *Lim* one has $N(\lambda[x]) = N\lambda + x$. Of course by definition one has $N(\lambda[x]) \leq N\lambda + x$. If $N(\lambda[x]) < N(\lambda) + x$ then $N(\lambda[x] + 1) \leq N(\lambda) + x$ and $\lambda[x] + 1$ would be a better candidate for $\lambda[x]$ (cf. [2]).

Assume now that $\alpha = \omega^{\alpha_1} + \lambda > \lambda$. Assume further that $\alpha[x] = \omega^{\beta_1} + \beta_2 > \beta_2$. Then $\alpha_1 = \beta_1$. For otherwise $\beta_1 < \alpha_1$ but then $\omega^{\beta_1} + \beta_2 < \omega^{\alpha_1}$ and $N(\omega^{\alpha_1}) \leq N\alpha + x$ thus $\alpha[x]$ would not be chosen maximally. Thus $\alpha[x] = \omega^{\alpha_1} + \beta_2$ where $\beta_2 < \lambda$. $N(\alpha[x]) = N\alpha + x$ yields $N\beta_2 = N\lambda + x$. Therefore $\beta_2 \leq \lambda[x]$. We claim that $\beta_2 = \lambda[x]$. For otherwise $\beta_2 < \lambda[x]$. Then $\alpha[x] = \omega^{\alpha_1} + \beta_2 < \omega^{\alpha_1} + \lambda[x]$. From $N(\lambda[x]) = N\lambda + x$ we obtain $N(\omega^{\alpha_1} + \lambda[x]) = 1 + N\alpha_1 + N\lambda + x = N\alpha + x$, thus $\alpha[x]$ would not have been chosen maximally. Summing up we have proved $\alpha[x] = \omega^{\alpha_1} + \lambda[x]$ and $\cdot[\cdot]$ satisfies the strong Lipschitz continuity.

Assume now that $\alpha = \omega^\lambda$ with $\lambda \in$ *Lim*. We claim that $\alpha[x] = \omega^{\lambda[x]}$. Indeed, let $\alpha[x] = \omega^{\beta_1} + \beta_2 > \beta_2$. If $\beta_1 < \lambda[x]$ then $\alpha[x] = \omega^{\beta_1} + \beta_2 < \omega^{\lambda[x]}$. But then $\alpha[x]$ would not have been chosen maximally since $N(\omega^{\lambda[x]}) = 1 + N(\lambda[x]) = 1 + N\lambda + x = N\alpha + x$. If $\lambda[x] < \beta_1 < \lambda$ then $N\beta_1 < 1 + N\beta_1 + N\beta_2 = N(\alpha[x]) = 1 + N(\lambda[x])$ hence $N\beta_1 \leq N(\lambda[x])$ and $\lambda[x]$ would not have been chosen maximally. Thus $\lambda[x] = \beta_1$. Since $N(\lambda[x]) + 1 = N(\alpha[x]) = N(\omega^{\lambda[x]} + \beta_2)$ we have $N\beta_2 = 0$ thus $\beta_2 = 0$.

Finally assume that $\alpha = \omega^{\alpha_1+1}$. Assume that $\alpha[x] = \omega^{\beta_1} \cdot m + \beta_2$ with $m > 0$ and $\beta_2 < \omega^{\beta_1}$. Then $\beta_1 = \alpha_1$ for maximality reasons. Further

$$\begin{aligned} N(\alpha[x]) &= N\alpha + x \\ &= N(\alpha_1) + 2 + x \\ &= m \cdot (N\alpha_1 + 1) + N\beta_2 \\ &\geq N\alpha_1 + 1 + N\beta_2 \end{aligned}$$

Thus $N\beta_2 \leq 1 + x$. Let $n := \max\{y \leq x + 2 : y \cdot (N\alpha_1 + 1) \leq N\alpha_1 + 2 + x\}$. (Then n depends elementary recursively on $N\alpha_1$ and x.) From $N\alpha_1 + 2 + x = m \cdot (N\alpha_1 + 1) + N\beta_2$ we obtain $m \leq x + 2$ hence $n = m$. For if $m < n$ then $\alpha[x]$ would not have been chosen maximally. □

Remarks. Theorem 4 can be extended to larger segments of ordinals for example to Γ_0. Thus norm based fundamental sequences when defined with respect to a "length"-norm yield slow growing pointwise hierarchies. It is not clear to the author if every elementary recursive function can be majorized by a function G_α for some $\alpha < \varepsilon_0$ when G_α is defined with respect to $\cdot[\cdot]_L$.

For a given real number $c \geq 0$ let $\alpha[x]_c := \max\{\beta < \alpha : N\beta \leq c \cdot N\alpha + x\}$ for $\alpha > 0$. For $c \leq 1$ the resulting hierarchy $(G_\alpha)_{\alpha<\varepsilon_0}$ is slow growing. For $c \geq 2$ the resulting hierarchy $(G_\alpha)_{\alpha<\varepsilon_0}$ is fast growing as can be seen by an inspection of the proof of Theorem 1. For $1 < c < 2$ the resulting hierarchy $(G_\alpha)_{\alpha<\varepsilon_0}$ is fast growing but a proof of this is rather tedious and hence is omitted. Note that the strong Lipschitz continuity of $\cdot[\cdot]_c$ is violated for $c > 1$. One may consider $c = 1$ as a point of strong singularity for $(G_\alpha)_{\alpha<\varepsilon_0}$ when it is defined with respect to $\cdot[\cdot]_c$.

5 Looking at other hierarchies which are defined pointwise

In the final section we investigate hierarchies which are defined by pointwise transfinite recursion with respect to a norm function. The origin of these hierarchies goes back to investigations on the computational complexity of rewrite systems (cf. [4]).

Definition 15 (cf. [1]) *1.* $A'_0(x) := 0$.

2. $A'_\alpha(x) := \max\{1 + A'_\beta(x) \colon \beta < \alpha \;\&\; N_C(\beta) \leq N_C(\alpha) + x\}$ *if* $\alpha > 0$.

Arai proved via proof-theoretic means in [1] that the hierarchy $(A'_\alpha)_{\alpha<\varepsilon_0}$ is fast growing.

Definition 16 *1.* $A_0(x) := 0$.

2. $A_\alpha(x) := \max\{1 + A_\beta(x) \colon \beta < \alpha \;\&\; N\beta \leq N\alpha + x\}$ *if* $\alpha > 0$.

We are going to show that $(A_\alpha)_{\alpha<\varepsilon_0}$ is fast growing. Since $A_\alpha(x) \leq A'_{\omega^\alpha \cdot N\alpha}(x+1)$ we obtain an alternative proof of Arai's result concerning $(A'_\alpha)_{\alpha<\varepsilon_0}$.

Definition 17 *1.* $\alpha \lhd_x \beta :\iff (\exists n > 0)(\exists \gamma_0, \ldots, \gamma_n)[\beta = \gamma_0 \;\&\; \alpha = \gamma_n \;\&\; (\forall l < n)[\gamma_{l+1} < \gamma_l \;\&\; N(\gamma_{l+1}) \leq x + N(\gamma_l)]]$.

2. $\alpha \unlhd_x \beta :\iff \alpha \lhd_x \beta \vee \alpha = \beta$.

3. $m \underline{\ll}_x \beta :\iff (\exists \alpha)[\alpha \unlhd_x \beta \;\&\; m \leq N\alpha]$.

The natural sum is denoted by $\oplus$.

Lemma 10 *1.* $\alpha \lhd_x \beta \implies \gamma \oplus \alpha \lhd_x \gamma \oplus \beta$.

2. $\omega^\alpha \lhd_x \omega^{\alpha+1}$.

3. $\alpha \lhd_x \beta \implies \omega^\alpha \lhd_x \omega^\beta$.

Lemma 11 *1.* $N(\alpha+1) = N\alpha + 1$.

2. $\alpha \unlhd_x \beta \implies A_\alpha(x) \le A_\beta(x)$.

3. $N\alpha \unlhd_x \alpha$.

4. $A_\beta(x) \ge N\beta$.

5. $m \lll_x \beta \implies m \le A_\beta(x)$.

Proof. 4) by induction on β using the assertions 1) and 2).
5) follows from 3) and 4). □

Lemma 12 $\lambda \in Lim \implies N(\lambda[\![x]\!]) \le N\lambda \cdot (x+1)$.

We are going to show that $(A_\alpha)_{\alpha<\varepsilon_0}$ is fast growing. By the result of the previous section this cannot be established by a "maximum strategy" since always stepping down consecutively from α to $\max\{\beta < \alpha : N\beta \le N\alpha + x\}$ would lead to a slow growing process. The proof strategy consists in choosing consecutively a $\beta < \alpha$ which is typically smaller than $\max\{\beta < \alpha : N\beta \le N\alpha + x\}$ but which has a more complex number-theoretic content.

Theorem 5 *Assume* $\varepsilon_0 > \alpha_1 \ge \ldots \ge \alpha_n$, $n \ge 0$. *Then*

$$\omega^{\omega^{\alpha_1+3}\cdot 2} + \cdots + \omega^{\omega^{\alpha_n+3}\cdot 2} + \omega^{\omega^2+k} + \omega^{\omega+l} \gtrsim_1 F_{\alpha_1}(\ldots F_{\alpha_n}(2^k \cdot (l+1))\ldots).$$

Proof. By induction on $\alpha := \omega^{\omega^{\alpha_1+3}\cdot 2} + \cdots + \omega^{\omega^{\alpha_n+3}\cdot 2} + \omega^{\omega^2+k} + \omega^{\omega+l}$.
Case 1: $k > 0$.
Let $\beta := \omega^{\omega^{\alpha_1+3}\cdot 2} + \cdots + \omega^{\omega^{\alpha_n+3}\cdot 2} \ge 0$, $F_{\hat\beta}(x) := F_{\alpha_1}(\ldots F_{\alpha_n}(x)\ldots)$ and $\gamma := \beta + \omega^{\omega^2+k}$. Then

$$\begin{array}{rll} \alpha & \unrhd_1 & \gamma + \omega^{\omega+l-1} + 2 \\ & \unrhd_1 & \gamma + \omega^{\omega+l-2} + 2 + 2 \\ & \unrhd_1 & \gamma + 2 + 2\cdot(l+1) \\ & \unrhd_1 & \beta + \omega^{\omega^2+k-1} + \omega^{\omega+2\cdot(l+1)} \qquad \text{by IH} \\ & \gtrsim_1 & F_{\hat\beta}(2^{k-1}\cdot(2\cdot(l+1))) = F_{\hat\beta}(2^k\cdot(l+1)). \end{array}$$

Case 2: $k = 0$.
Case 2.1: $n > 0$.

Let $\gamma = \omega^{\omega^{\alpha_1+3}\cdot 2} + \cdots + \omega^{\omega^{\alpha_{n-1}+3}\cdot 2} \geq 0$ and $F_{\hat{\gamma}}(x) := F_{\alpha_1}(\ldots F_{\alpha_{n-1}}(x)\ldots)$.
Case 2.1.1: $\alpha_n = 0$. Then

$$\begin{array}{rll} \alpha & = & \gamma + \omega^{\omega^3\cdot 2} + \omega^{\omega^2} + \omega^{\omega+l} \\ & \trianglerighteq_1 & \gamma + \omega^{\omega^2+l+1} + \omega^{\omega} \qquad \text{by IH} \\ & \gtrsim_1 & F_{\hat{\gamma}}(2^{l+1}\cdot 1) \geq F_{\hat{\gamma}}(F_0(2^0\cdot(l+1))). \end{array}$$

Case 2.1.2: $\alpha_n = \delta + 1$.
Let $\beta = \gamma + \omega^{\omega^{\delta+1+3}}$. Then

$$\begin{array}{rll} \alpha & = & \gamma + \omega^{\omega^{\delta+1+3}\cdot 2} + \omega^{\omega^2} + \omega^{\omega+l} \\ & \trianglerighteq_1 & \beta + \omega^{\omega^3} + \omega^{\omega+l} + N\delta + 4 \\ & \trianglerighteq_1 & \beta + \omega^{\omega^2+N\delta+4} + \omega^{\omega+l} \\ & \trianglerighteq_1 & \beta + \omega^{\omega^2+N\delta+4} + 2\cdot(l+1) \\ & \trianglerighteq_1 & \beta + \omega^{\omega^2+N\delta+3} + \omega^{\omega+2\cdot(l+1)} \\ & \trianglerighteq_1 & \beta + \omega^{\omega^2+N\delta+3} + 2\cdot 2\cdot(l+1) \\ & \trianglerighteq_1 & \ldots \\ & \trianglerighteq_1 & \beta + 2^{N\delta+4}\cdot(l+1) \\ & \trianglerighteq_1 & \gamma + \omega^{\omega^{\delta+3}\cdot 2}\cdot(l+1) + \omega^{\omega^2} + \omega^{\omega+l} \qquad \text{by IH} \\ & \gtrsim_1 & F_{\hat{\gamma}}(F_{\delta}^{(l+1)}(2^0\cdot(l+1))) = F_{\hat{\gamma}}(F_{\alpha_n}(2^0\cdot(l+1))). \end{array}$$

Case 2.1.3: $\alpha_n \in Lim$.
Let $\beta = \gamma + \omega^{\omega^{\alpha_n+3}}$. Then

$$\begin{array}{rll} \alpha & = & \gamma + \omega^{\omega^{\alpha_n+3}\cdot 2} + \omega^{\omega^2} + \omega^{\omega+l} \\ & \trianglerighteq_1 & \beta + \omega^{\omega^3} + \omega^{\omega+l} + N\alpha_n + 4 \\ & \trianglerighteq_1 & \beta + \omega^{\omega^2+N\alpha_n+4} + \omega^{\omega+l} \\ & \trianglerighteq_1 & \beta + 2^{N\alpha_n+4}\cdot(l+1) \\ & \trianglerighteq_1 & \gamma + \omega^{\omega^{\alpha_n[l+1]+3}\cdot 2} + \omega^{\omega^2} + \omega^{\omega+l} \qquad \text{by IH} \\ & \gtrsim_1 & F_{\hat{\gamma}}(F_{\alpha_n[l+1]}(l+1)) = F_{\hat{\gamma}}(F_{\alpha_n}(l+1)). \end{array}$$

Case 2.2: $n = 0$. Then $\omega^{\omega^2} + \omega^{\omega+l} \gtrsim_1 l + 1$. □

Let $\omega_0(0) = 0$ and $\omega_{m+1}(0) := \omega^{\omega_m(0)}$.

Corollary 4 $A_{\omega_{m+17+k}(0)}(1) \geq A_{\omega^{\omega^{\omega_m(0)+3}}+\omega^{\omega+k}}(1) \geq F_{\omega_m(0)}(k+1)$.
Thus $k \mapsto A_{\omega_k(0)}(1)$ is not provably recursive in PA.

Application. A well known theorem of Friedman states that PA does not prove that ε_0 is slowly well-ordered, i.e. although true the following assertion is not provable in PA: $(\forall k)(\exists i)(\forall \alpha_0, \ldots, \alpha_i < \varepsilon_0)[(\forall l \leq i)[N(\alpha_l) \leq k+l \implies (\exists m, n)[m < n \leq i \ \& \ \alpha_m \leq \alpha_n]]$. See, for example, [8] for a reference. Let $A_{\varepsilon_0}(x) := \max\{1 + A_\beta(1) : \beta < \varepsilon_0 \ \& \ N\beta \leq x\}$. Then for any $k < \omega$ the value $A_{\varepsilon_0}(k)$ is the least natural number i such that for all $\alpha_0, \ldots, \alpha_i < \varepsilon_0$: If $N\alpha_0 \leq k$ and if $N(\alpha_{l+1}) \leq N(\alpha_l) + 1$ for all $l < i$ then there exist natural numbers m, n such that $m < n \leq i$ and $\alpha_m \leq \alpha_n$. Since as shown above A_{ε_0} is not provably total in PA we obtain the following refinement of Friedman's result. Although true the following assertion is not provable in PA: $(\forall k)(\exists i)(\forall \alpha_0, \ldots, \alpha_i < \varepsilon_0)[N\alpha_0 \leq k \ \& \ (\forall l < i)[N(\alpha_{l+1}) \leq N(\alpha_l) + 1 \implies (\exists m, n)[m < n \leq i \ \& \ \alpha_m \leq \alpha_n]]$. A similar refinement holds for Friedman's result on the slow well-quasi-orderedness of the set of finite binary trees with respect to the homeomorphic embeddability relation (cf. [8]).

Open problem. It would be interesting to have a natural interpretation of the results obtained in this paper from the viewpoint of Π^1_2-logic.

Acknowledgments. The author would like to thank T. Arai, W. Buchholz, S. Feferman, E. A. Cichon and S. S. Wainer for discussions on the subject.

References

[1] T. Arai: *Variations on a theme by Weiermann.* The Journal of Symbolic Logic (to appear).

[2] W. Buchholz, A. Cichon and A. Weiermann: *A uniform approach to fundamental sequences and hierarchies.* Mathematical Logic Quarterly 40 (1994), 273-286.

[3] W. Buchholz und S.S. Wainer: Provably computable functions and the fast growing hierarchy. Logic and Combinatorics. Contemporary Mathematics 65, American Mathematical Society (1987), 179-198.

[4] E.A. Cichon. *Termination proofs and complexity characterisations.* Proof Theory, P. Aczel et al. (eds.), Cambridge Univ. Press (1992), 173-193.

[5] J.Y. Girard: *Proof Theory and Logical Complexity.* Vol. 1. Studies in Proof Theory. Bibliopolis, Naples 1986.

[6] W. Pohlers: *A short course in ordinal analysis.* Proof Theory, Leeds 1990. P. Aczel et al. (eds.), Cambridge Univ. Press (1992), 27-78.

[7] H.E. Rose. *Subrecursion: functions and hierarchies.* Clarendon Press, Oxford, 1984.

[8] R. Smith: *The consistency strength of some finite forms of the Higman and Kruskal theorems.* Harvey Friedman's Research on the Foundations of Mathematics, L. A. Harrington et al. (eds.), North Holland (1985), 119-136.

[9] H. Schwichtenberg: *Eine Klassifikation der ε_0-rekursiven Funktionen.* Zeitschrift für mathematische Logik und Grundlagen der Mathematik 17 (1971), 61-74.

[10] S. S. Wainer: *Ordinal recursion and a refinement of the ordinal recursive functions.* The Journal of Symbolic Logic 37 (1972), 281-292.

[11] A. Weiermann: *Bounding derivation lengths with functions from the slow growing hierarchy.* Archive for Mathematical Logic (to appear).

[12] A. Weiermann: *Sometimes slow growing is fast growing.* Annals of Pure and Applied Logic 90 (1997), 91-99.

[13] F. Zemke: *P. R.-regulated systems of notation and the subrecursive hierarchy equivalence property.* Transactions of the American Mathematical Society 234 (1977), 89-118.

Minimality Arguments for Infinite Time Turing Degrees

Philip D. Welch
Graduate School of Science & Technology,
Kobe University,
Rokko-dai, Nada-ku
Kobe 657, Japan.
welch@pascal.seg.kobe-u.ac.jp

Abstract

We show that the length of the naturally occurring jump hierarchy of the infinite time Turing degrees is precisely ω, and construct continuum many incomparable such degrees which are minimal over **0**. We show that we can apply an argument going back to that of H. Friedman to prove that the set ∞-degrees of certain Σ^1_2-correct KP-models of the form L_σ $(\sigma < \omega_1^L)$ have minimal upper bounds.

1 Introduction

Obtaining minimality results in degree theory has a long history: the methods go back to those of Spector when he constructed minimal Turing degrees, and to Gandy-Sacks, [2], for minimal hyperdegrees. The perfect set construction is the common thread to these proofs. A further feature, which is shared, either directly or indirectly, by such arguments, is the use of a selection principle in order to typically, directly shrink a perfect set $T \subseteq {}^\omega\omega$ to a T' so that a particular function is either continuous one-to-one, or constant on the branches of T'. For example in Sack's construction [8], using uniformly hyperarithmetically pointed perfect trees, Kreisel Selection is appealed to in order to directly shrink hyperarithmetically pointed trees to uniformly pointed ones, and ultimately, so that relevant functions fall into this dichotomy on some tree. In Kechris' results ([6],[5]) category theoretic arguments are used to find a suitable T' on which, for example in [6], a Δ^1_{2n} function $F : {}^\omega\omega \to {}^\omega\omega$ can be seen to be (1-1) and continuous, or just constant. For a set $D = \{\boldsymbol{d_n}\}_{n<\omega}$ of Δ^1_2 degrees H. Friedman used a slightly different tack: he wished to construct a shrinking sequence of perfect trees with certain pointedness properties, whose final intersection would not necessarily have to be so rigourously pointed (the source of difficulties for Sacks in [8]), but such that branches f "remote" from $\{\boldsymbol{d_n}\}_{n<\omega}$, that is $f \notin \cup_n L[d_n]$, would be Δ^1_2-minimal over D. The trick was to use a notion of pointedness that was Π^1_2 and appeal to Schoenfield's Σ^1_2 absoluteness theorem, and, *via* Uniformisation, to the Basis Theorem for Σ^1_2. This was done in order to indirectly shrink perfect T's which might have a branch f lying in a particular $\Sigma^1_2(T)$ set, with f itself not in $L[T]$, by using Mansfield's Perfect Set argument (or its hyperarithmetic analogue.) We lifted this trick to obtain minimal upper bounds for Δ^1_3-degrees, assuming sharps for reals, and $\neg O^\dagger$ ([10]) or the failure of some $\Delta^1_2(z)$-Determinacy ([11]). Something of the Friedman argument remains also in [6] where Friedman's key lemma has as analogue a shrinking construction on trees to see that if a branch in a certain Σ^1_{2n} set is remote, (now this means not in the smallest $\Sigma^1_{2n}(T)$ set $C_{2n}(T)$), then by Π^1_{2n-1}-Uniformization (we are assuming Δ^1_{2n}-Determinacy here) we may shrink T to T' so that now all branches fall within the set and have the same ordinal value arising from the Π^1_{2n-1} norm.

We exploit this trick one more time in constructing minimal upper bounds for degrees of computability abstracted from the Infinite Time Turing machines of [3]. The notion of relative computability here is more akin to hyperdegrees (even Δ^1_2-degrees) than Turing degrees. We use "∞"-pointed trees (a Π^1_2-notion) to construct recursively a splitting system of trees all of whose paths dominate the degrees of a countable Σ^1_2-correct model. As in [1] the branches f of the final fusion tree, if remote from M, are seen to be minimal upper bounds for the degrees of M. (Here "remoteness" means that f is not itself computable from any real of M, in any time that is less than any ordinal which is itself computable from f.) The use of Σ^1_2-absoluteness is as follows: it is used to contract trees of M which have branches convergent in V with respect to a certain computation, to subtrees, still in M, with a uniform M-bound on the number of steps for convergence for all the branches (whether lying in V or M). This use of the Σ^1_2-absoluteness (or, eventually, of Σ_1-reflection) is in one sense a weakness: it means we can only prove the existence of minimal upper bounds for a restricted class of sets of degrees. However we have no idea how to contract trees otherwise.

2 Infinite Time Turing degrees

Hamkins and Lewis in [3] give the construction of these machines and develop the basic theory of this notion of computability. To summarise: an infinite time Turing machine has the hardware of an ordinary Turing machine, except that transfinite stages of computation are allowed (the action at successor stages being just that specified as usual by a finite program with a finite number of possible states). The action at limit stages ν of "time" is very definite: the machine head always returns to a specific cell on the tape, and enters a special limit state. The value in the cell C_n (thought of as a 0 or a 1) is specified as the limsup of the values at all previous "times". This specifies precisely a behaviour of a machine, and like ordinary Turing machines, such a device may "halt" with a specific output or may loop forever. The infinite amount of time available means however that such machines can accept an infinite string of integers as input (and can output the same). This implies that a machine can decide whether an input is a code of an wellfounded relation etc., etc. We list some of the basic facts that we shall need from [3], [12], and [13].

Fact 2.1 *([3] 2.5)(i) The relation $\phi_p^f(x)\downarrow$ (in α-steps) (the p'th program halts on input $x \in {}^\omega\omega$ using oracle $f \in {}^\omega\omega$ (additionally, in α-steps)) is a Δ^1_2 relation of f, p, x (and α).*
(ii) Similarly $\phi_p^f(x)\uparrow$ is also Δ^1_2. (It is only necessary to have a/all wellordered sequence(s) coding the course of a looping computation to see this.)

Fact 2.2 *([3] 3.7, 8.3) If $\mathcal{W} =_{df} \{x \in {}^\omega\omega \mid x$ is the output of $\phi_p(0)\downarrow\}$, then the "writable" ordinals, (those ordinals α so that there is $x \in \mathcal{W} \cap WO$ with $|x| = \alpha$) form an initial segment of the countable ordinals, λ. The class $H(\lambda)$ whose elements are coded by reals in $\mathcal{W}$ is a transitive admissible set, closed under hyperjump (and more).*

We use the notation $\mathcal{W}^f, \lambda^f \ldots$ for the notions relativised to a real. The following theorem states that the halting times of computations are themselves all realisable as outputs of such computations.

Fact 2.3 *([12]) Let γ be $\sup\{\alpha \mid \exists p\ \phi_p(0)\downarrow$ in precisely α steps$\}$. Then $\gamma = \lambda$ (and by relativisation $\forall f \in {}^\omega\omega \quad \lambda^f = \gamma^f$).*

The following is a corollary to the proof of the last fact:

Fact 2.4 *([12]) $H(\lambda) = L(\lambda)$, and for any $f \in {}^\omega\omega \quad H^f(\lambda^f) = L_{\lambda^f}[f]$.*

We shall need the following characterisation of the ordinal λ^f.

Fact 2.5 *([13] Thm 2.1) Let $f \in {}^\omega\omega$. Let (ζ^f, Σ^f) be the lexicographically least pair of ordinals with $L_\zeta[f] \prec_{\Sigma_2} L_\Sigma[f]$. Then $\lambda^f < \zeta^f$ and λ^f is the least $\overline{\lambda}$ with $L_{\overline{\lambda}}[f] \prec_{\Sigma_1} L_{\zeta^f}[f]$.*

Hamkins and Lewis define a notion of degree between reals:

Definition 2.1 *For* $f, g \in {}^\omega\omega$ *we write* $f \leq_\infty g \longleftrightarrow \exists p \in \omega\ \phi^g_p(0)\downarrow = f$.

We consider the following questions.

Question 1 Are there incomparable minimal ∞-degrees?

Hamkins has shown ([4]) that there are no degrees between that of 0 and the jump of 0. The following is a natural definition for a jump operator on reals alone (they call this the "weak jump" see [3]).

Definition 2.2 *For* $f \in {}^\omega\omega\ f^\nabla =_{df} \{n \in \omega \mid \phi^f_n(0)\downarrow\}$.

One may readily show that $f \to f^\nabla$ is a Δ^1_2 function. Clearly $\mathbb{R} \cap L$ is closed under this operation, as indeed is any Σ^1_2-correct model. Let M be a transitive rudimentary closed model, with $\omega \in M$. We define Σ^1_2-correctness for M in the usual way: if Φ is Σ^1_2 and $\vec{a} \in \mathbb{R} \cap M$, then $\Phi(\vec{a}) \iff \mathbb{R} \cap M \models \Phi(\vec{a})$.

Question 2 For what countable sets of ∞-degrees can one show the existence of minimal upper bounds?

Question 3 How long is the natural hierarchy of ∞-degrees? We define $\langle \boldsymbol{h_i} \mid i < \rho \rangle$ using as jump the operator $f \longrightarrow f^\nabla$, by $\boldsymbol{h_{i+1}} = \boldsymbol{h_i}^\nabla$ and taking $\boldsymbol{h_\lambda}$ as the least upper bound (if possible) of $\langle \boldsymbol{h_i} \mid i < \lambda \rangle$ at limit stages. If ρ is the least so that $\boldsymbol{h_\rho}$ is undefined, how large is ρ? Is $\rho > \omega$?

2.1 δ-degrees and minimality

The characterisation of $H^f(\lambda^f)$ in Fact 2.4 above can be reformulated as:

Fact 2.6 $x \leq_\infty y \longleftrightarrow x \in L_{\lambda^y}[y]$

Note then: $x \leq_\infty y \longrightarrow \lambda^x \leq \lambda^y$, and in fact one may easily see $x \leq_\infty y \longrightarrow (x^\nabla \leq_\infty y \longleftrightarrow \lambda^x < \lambda^y$. The assignment $x \longrightarrow \lambda^x$ then satisfies a Spector criterion.

Definition 2.3 *Let* $F = \{x \in {}^\omega\omega \mid x \in L_{\lambda^x}\}$

Then F, the set of "fast" reals x , consists of those x that can be computed as being in L, by an ordinal that is the length of time of an x-computation. This is by way of analogy with the largest thin Π^1_1 set of reals, $Q = \{x \mid x \in L_{\omega^x_{1ck}}\}$, the quickly constructible reals. It is easily seen that F is a strictly Δ^1_2 and thin set of reals. Those $x \notin F$ are correspondingly "slow".

Definition 2.4 *Let M be a Σ^1_2-correct model. Then f is* remote from M *if and only if, $\forall g \in {}^\omega\omega \cap M \; f \notin L_{\lambda^f}[g]$.*

It is the remote branches of our eventual tree that will provide minimal upper bounds. Clearly only slow reals can be remote. We define an auxiliary notion of degree to help establish minimality.

Definition 2.5 *Let δ be any p.r. closed ordinal. Then set $x \leq_\delta y \longleftrightarrow x \in L_\delta[y]$.*

Then again $\leq_\delta$ is a transitive partial ordering and we define the corresponding notion of "δ-degree" $\boldsymbol{x_\delta}$ as:

$$[x]_\delta = \{y \mid y \leq_\delta x \wedge x \leq_\delta y\}$$

Definition 2.6 *(i) If D is a set of δ-degrees, then f is* δ-minimal over *D, if $\forall g \in D \, g <_\delta f$, and for all h, if $(h \leq_\delta f \wedge \forall g \in D \, g \leq_\delta h)$ then either $h \leq_\delta g$ for some $g \in D$ or $f \leq_\delta h$.*
(ii) The notion of being ∞-minimal over a set of ∞-computability degrees is defined entirely similarly.

Lemma 2.1 *If f is λ-minimal (over $\mathbf{0}$), and $\lambda^f = \lambda$, then f is ∞-minimal over $\mathbf{0}$.*

Proof: Let f be λ-minimal. As $f \notin L_\lambda$, $f \not\leq_\infty 0$. Suppose $0 <_\infty h \leq_\infty f$. Then $h \in L_{\lambda^f}[f]$. As $\lambda^f = \lambda$, and $\lambda \leq \lambda^h \leq \lambda^f = \lambda$, by λ-minimality of f, $f \in L_{\lambda^h}[h]$ as required. **Q.E.D.**

This generalises straightforwardly to:

Lemma 2.2 *Let D be a countable set of (reals of) ∞-degrees. Let $\lambda^D =_{df} \sup_{f \in D} \lambda^f$. Let f be λ^D-minimal over D, with $\lambda^f = \lambda^D$. Then f is ∞-minimal over D.*

Lemma 2.3 *Let D be as above, and $\langle \delta_n \mid n < \omega \rangle$ an increasing sequence of p.r. closed ordinals with $\sup(\delta_n) = \lambda^D$. Suppose $\forall n \exists h_n \exists d \in D \, h_n \leq_\infty d \wedge (f \oplus h_n)$ is δ_n-minimal over h_n. If f satisfies $\lambda^f = \lambda^D$, and $\forall d \in D \, d <_\infty f$, then f is ∞-minimal over D.*

Proof: Show f is λ^D-minimal over D, and then apply the previous lemma. **Q.E.D.**

To fix notation we let $T, T^*, \ldots$ etc. range over sets of sequence numbers in $2^{<\omega}$. Actually we shall always assume that $T, T', \ldots$ range over perfect trees. We denote the set of branches of T, by $[T]$; $f, g, \ldots$ etc. will range over such branches. A tree T is then *∞-pointed* if for all $f \in [T]$, $T \leq_\infty f$.

By straightforward methods of Sacks ([8]) one has as usual:

Lemma 2.4 *Let T be ∞-pointed. Then:*
(i) $\{g \in {}^\omega\omega \,|\, T \leq_\infty g\} = \{g \in {}^\omega\omega \,|\, \exists h \in [T]$ *with* $h =_\infty g\}$.
(ii) Let $g \in {}^\omega\omega$ with $T \leq_\infty g$. Then there is an ∞-pointed tree $T^* \subseteq T$, $T^* =_\infty g$. In fact there is a uniform procedure for finding such a T^* recursively in T and g.

Lemma 2.5 *(i) Let $\delta > \omega$ be p.r.closed, and $g : \delta \longrightarrow \omega$ is (1-1). Then there is $T^* \subseteq T(T^* \in L_{\delta+1}[g, T] \wedge \forall f \in [T^*]((f \oplus T)$ is δ-minimal over $T)$.*
(ii) If T is ∞-pointed, and $\delta < \lambda^T$, then T^* can be taken ∞-pointed.

Proof: (i): This is just as in [7] 1.4 and 3.1. For (ii): just note then that we may take the $g : \delta \longrightarrow \omega$ as lying in $L_{\lambda^T}[T]$, as in the latter model, everything is countable. But then we may assume that T^* is there also. But then if $f \in [T^*]$, then $f \in [T]$, hence $T \leq_\infty f$ and $T^* \leq_\infty T$ as Fact 2.4 implies that $H^T(\lambda^T) = L_{\lambda^T}[T]$. **Q.E.D.**

3 Constructing minimal upper bounds

For x a set, let x^+ denotes the ordinal height of the smallest transitive admissible set containing $\{x\}$.

We first answer Question 3 of the Introduction.

Theorem 3.1 $\rho = \omega$

Proof: Set $\lambda_0 = 0$, and for $i \geq 1$, let $\lambda_i =_{df} \lambda^{0^{i\nabla}}$ where $\mathbf{0}^{n\nabla}$ is the n'th iterate of the jump operator on $\mathbf{0}$, and λ^f is defined at Fact 2.2 above. Let $\lambda_\omega = \sup_i \lambda_i$. Then each λ_i is a recursively inaccessible ordinal. We note that

$$L_{\lambda_1} \prec_{\Sigma_1} L_{\lambda_2} \cdots \prec_{\Sigma_1} L_{\lambda_\omega} \prec_{\Sigma_1} L_\zeta \prec_{\Sigma_2} L_\Sigma$$

by Theorem 2.1 and Lemma 2.2 of [13] (where ζ is least so that L_ζ has a Σ_2 transitive end extension L_Σ - this is just a natural generalisation of Fact 2.5). Consequently $L_{\lambda_n} \models$ "T is ∞-pointed" if and only if $L_{\lambda_\omega} \models$ "T is ∞-pointed" if and only if $L_\Sigma \models$ "T is ∞-pointed". It is also easy to see that "$\forall f \in [T]\, F(e, f) = \alpha$" is also absolute between these three models, for any $T, \alpha \in L_{\lambda_n}$.

Lemma 3.2 *Let $T \in L_{\lambda_\omega}$ be perfect, $f \in [T]$, $T \leq_\infty f$ and $F(p,f) \approx \beta < \zeta$, and $f \notin \Delta^1_1(h_\beta, T)$ where h_β is the L-least code for a wellorder of length β. Let n be least with $T \in L_{\lambda_n}$. Then there is $T^* \subseteq T$, $T^* \in L_{\lambda_n}$, and there is an $\alpha < \lambda_n \forall f \in [T^*]\ F(p,f) = \alpha$. If T is ∞-pointed, then T^* can be assumed ∞-pointed.*

Proof: Let $\beta = F(p,f)$. Note that for any $\mu < \zeta$, h_μ can be assumed definable over L_μ (by virtue of the smallness of ζ). Note:

(1) "$F(p,f) = \beta$" is Σ^1_1 in f and a code for β.
Define a set T' by:

(2) $T' =_{df} \{f : F(p,f) = \beta \wedge f \in [T]\}; \quad T' \in \Sigma^1_1(h_\beta, T)$.

Moreover T' contains elements not hyperarithmetic in (T, h_β) by assumption. By the "hyperarithmetic perfect set theorem", (see, *e.g.* [9] 6.2.III) T' contains a perfect set T^* recursive in $\mathcal{O}^{h_\beta,T}$, the hyperjump of (h_β, T), and so $T^* \in L_{(\mathcal{O}^{h_\beta,T})^+}[h_\beta, T] \subseteq L_{\lambda_\omega}$. Hence

(3) $L_\zeta \models \exists h_\beta \exists$ perfect $T^*, T^* \leq_T \mathcal{O}^{h_\beta,T}[\forall f \in [T^*]\, F(p,f) = |h_\beta|]$.

But $L_{\lambda_n} \prec_{\Sigma_1} L_\zeta$ so this reflects to L_{λ_n}. Consequently there are $\alpha, h_\alpha, T^* \in L_{\lambda_n}$ satisfying (3). (Note that λ_n is recursively inaccessible, hence L_{λ_n} is closed under hyperjump.)

To conclude the lemma now suppose T is ∞-pointed. Let $f \in [T^*]$; then $f \in [T]$, hence $T \leq_\infty f$, hence $\lambda^T = \lambda_n \leq \lambda^f$. But $T^* \in L_{\lambda_n} \subseteq L_{\lambda^f}[f]$. Hence $T^* \leq_\infty f$. Hence T^* is ∞-pointed. This finishes the lemma. **Q.E.D.**

Now let $\boldsymbol{d_i} = [0^{\nabla i}]$ for $i < \omega$ (so setting $\lambda_0 = 0$, $\boldsymbol{d_0} = \boldsymbol{0}$). Let $d_i = 0^{\nabla i}$, and $E = \{d_i\}_{i<\omega}$.

Lemma 3.3 *There exists a perfect tree T_0 definable over L_{λ_ω} such that $\forall f \in [T_0](f$ slow $\wedge \lambda^f \leq \zeta \Rightarrow \lambda^f = \lambda_\omega = \lambda^E \wedge f$ is ∞-minimal over $E)$.*

Proof: We define a binary system of ∞-pointed trees T_s for $s \in 2^{<\omega}$. Suppose $\langle \delta_n \mid n < \omega \rangle$ an increasing sequence of p.r. closed ordinals with $\delta_0 = \omega$, $\lambda_n < \delta_n < \lambda_{n+1}(= \lambda^{0^{\nabla n}})$. Set $T_\emptyset = 2^{<\omega}$; assume T_s is defined for any $s \in 2^{<\omega}$ with $lh(s) \leq i$ with the properties
(i) If $lh(s) = j \leq i$ then $d_j =_\infty T_s$ (hence $T_s \in L_{\lambda_{j+1}}$);
(ii) T_s is ∞-pointed.

For $lh(s) = i$ we define $T_{s\frown 0} =_\infty T_{s\frown 1} \in L_{\lambda_{i+2}}$, two subtrees of T_s preserving (i) and (ii), according to the following four step recipe.

1) First split T_s into two ∞-pointed subtrees $T_s'^0 =_\infty T_s'^1 =_\infty T_s$ with $[T_s'^0] \cap [T_s'^1] = \emptyset$;
2) Find (by appealing to Lemma 2.4), ∞-pointed $T_s^{*j} \subseteq T_s'^j$ in M, with $T_s^{*j} =_\infty d_{i+1}\,(j = 0, 1)$.
3) Find $T_s^j \subseteq T_s^{*j}$, $T_s^j =_\infty T_s^{*j}$ so that $\forall f \in T_s^j((f, T_s^{*j})$ is δ_{i+1}-minimal over $T_s^{*j})$. Such T_s^j can be chosen ∞-pointed, and in $L_{\lambda_{i+2}}$, by Lemma 2.5;
4) If there is any $\alpha_0 < \lambda_{i+2}$ and ∞-pointed $T_{s\frown 0} \subseteq T_s^0$ so that $T_{s\frown 0} \in L_{\lambda_{i+2}}$, and also so that $\forall f \in T_{s\frown 0}\ F(i, f) = \alpha_0$, then let $T_{s\frown 0}$ be the L-least such; otherwise set $T_{s\frown 0} = T_s^0$. Similarly define $\alpha_1, T_{s\frown 1} \subseteq T_s^1$.

Note that whatever happens at 4), $T_{s\frown 0} =_\infty T_{s\frown 1} =_\infty d_{i+1}$. Now let T_0 be the fusion of the $\langle T_s \,|\, s \in 2^{<\omega}\rangle$. Then $[T_0] = \bigcup_{g\in 2^\omega} \bigcap_n [T_{g\restriction n}]$. We now show that T_0 is as in the lemma. Let $f \in \bigcap_n [T_{g\restriction n}]$. Clearly $\forall i\, d_i \leq_\infty f$. As $\forall i \exists j\, d_i <_\infty d_j$, actually $\forall i\, d_i <_\infty f$. Hence f is a strict upper bound for E and $\lambda^E \leq \lambda^f$. Then if $s = g \restriction i$, (f, T_s^j) is δ_{i+1}-minimal over $T_s^{*j} =_\infty T_s^j$. If we set $h_n = T_s^{*j}$ we shall have by Lemma 2.3 that f is ∞-minimal over E if $\lambda^f = \lambda^E$. So suppose $F(i, f)$ is defined, where f is slow, and by assumption $\lambda^f \leq \zeta$. Thus $F(i, f) < \zeta$ and Lemma 3.2 is applicable. Thus we indeed could have found $T_{s\frown j} \subseteq T_s^j$ for $j \in \{0, 1\}$, again with both $T_{s\frown j}$ in $L_{\lambda_{i+2}}$, and ordinals $\alpha_j < \lambda_{i+2}$, with $\forall g \in [T_{s\frown j}]\quad F(i, g) = \alpha_j$. As this is true for all $i < \omega$, we conclude that $\gamma^f \leq \lambda^E$. But then by Fact 2.3 $\lambda^f \leq \lambda^E$. This finishes the Lemma. **Q.E.D.**

To conclude the theorem we must remark that there *are* slow reals f in $[T_0]$ with $\lambda^f \leq \zeta$; in fact, there are continuum many, as follows.

Let g be generic over L_Σ for the local Cohen forcing $\mathbb{P}_{T_0}$ whose conditions are finite initial segments of T_0. ("g is generic over L_α" means that g meets all predense sets of $\mathbb{P}_{T_0}$ in L_α.) If we suitably formulate forcing, using for definiteness the ranked language of [8], forcing for Σ_1-sentences is Σ_1 over any admissible L_α. By genericity, for any τ with $\lambda_\omega < \tau < \Sigma$, if τ is admissible, it is also g-admissible. Although Σ is not necessarily an admissible ordinal here, it is a limit of admissibles (in fact of Σ_2-admissibles) and the forcing relation for Σ_2 is Π_3 over L_Σ, and we get the reflection property that $L_\zeta[g] \prec_{\Sigma_2} L_\Sigma[g]$, (and are the least pair ζ, Σ with this property). By the Fact 2.5 $\lambda^g < \zeta$; further $L_{\lambda^g}[g] \prec_{\Sigma_1} L_\zeta[g]$. Clearly then g is slow. By construction $g \in [T_0]$. **Q.E.D.Theorem 3.1**

We now refine the above argument to construct a tree T_0 definable over L_λ with continuum many branches that are minimal over $\mathbf{0}$. Let now $\langle \delta_n \,|\, n < \omega\rangle$ be an increasing sequence of p.r. closed ordinals cofinal in λ. We construct trees T_s as before, but all T_s are in L_λ (and so are trivially ∞-pointed as $T_s =_\infty \mathbf{0}$). We perform steps 1),3), & 4) of the above recipe (2) now being redundant), choosing $T_s'^j, T_s^j \subseteq T_s'^j$ in L_λ, and $T_{s\frown j} \subseteq T_s^j$ with $T_{s\frown j}$, $\alpha_j \in L_\lambda$.

Again let T_0 be the fusion of the trees constructed, and we obtain:

Lemma 3.4 $\forall f \in [T_0](f$ *slow* $\wedge \lambda^f < \zeta \Rightarrow \lambda^f = \lambda \wedge f$ *is* ∞*-minimal over* $\mathbf{0})$.

Now let g be $\mathbb{P}_{T_0}$-generic over L_Σ. By the same argument as before $\lambda^g < \zeta$ so g satisfies the antecedent in the statement of the lemma. Hence:

Theorem 3.5 *There are continuum many reals g ∞-minimal over* $\mathbf{0}$.

This answers Question 1 of the Introduction. We turn now to consideration of Question 2.

Let M be a countable transitive KP-model. Let us call a real f *countably remote from M*, if it is remote from M and satisfies $L_{\lambda^f} \models$ "$V = HC$" (that is, every set is countable).

Remark: 1) The assumption of countable remoteness is used to ensure that we can extend ∞-pointed trees T to ∞-pointed subtrees in the argument of Lemma 3.7 below.

Theorem 3.6 *Let $M = L_\sigma$ be a Σ^1_2-correct KP-model with $\sigma < \omega_1^L$; assume also that M is projectible (that is there is some $\Sigma_n(M)$ map of ω onto M), and that $M \models$ "$V = HC$". Then:*
(i) There is a perfect tree $T_0 \subseteq {}^\omega\omega$, so that if $f \in [T_0]$ is any branch countably remote from M, then f is a minimal upper bound for the ∞-degrees of reals $g \in M$.
(ii) There are continuum many $f \in [T_0]$ which are countably remote from M branches.

Remark: 2) Countably remote from M is equivalent here to simply being "slow", *i.e.* $f \notin L_{\lambda^f}$, and satisfying $L_{\lambda^f} \models$ "$V = HC$".

Lemma 3.7 *Let $T \in M$ be perfect and suppose for some countably remote from M $f \in [T]$, $T \leq_\infty f$ and $\varphi^f_p(0)\downarrow$. Then there is $T^* \subseteq T$, $T^* \in M$, and there is an $\alpha < On \cap M \, \forall f \in [T^*]$ $F(p, f) = \alpha$. If T is ∞-pointed, then T^* can be assumed ∞-pointed.*

Proof: Let $\beta = F(p, f)$. Our assumption on f means we can find $h \in WO \cap Q \cap L_{\lambda^f}$ coding an ordinal $\mu > \beta$. An h satisfying this can be found definably over L_μ where $\mu \geq \beta$ is least with $L_{\mu+1} \models \mathrm{card}(\beta) = \omega$. Let $\Psi(h, \beta, n)$ stand for "h is the L-least $k \in WO$ with $|n|_k = \beta \wedge k \in L_{|k|+1}$"
Hence we may additionally assume:

(1) $\Psi(h, \beta, n)$ where $F(p, f) = \beta = |n|_h$.

Let $\Phi(f)$ stand for: "$\neg\exists\zeta'\exists\Sigma'[\zeta' < \Sigma' \wedge L_{\zeta'}[f] \prec_{\Sigma_2} L_{\Sigma'}[f]$". By the above and our choice of f, we thus have:

(2) $\exists h[h \in WO \wedge \exists f[F(p, f) = |n|_h \wedge L_{|h|+1}[f] \models$ "$\Psi(h, |n|_h, n) \wedge \Phi(f)$" $\wedge$ $\wedge f \in [T] \wedge f \notin \Delta^1_1(T, h)]]$ is a true $\Sigma^1_2(T, p, n)$ statement.

By the Σ^1_2-correctness of M, it is true in M. So pick $\tilde{h} \in M$, L-least, witnessing this with $\tilde{h} \in WO \cap L_{\delta+1}$ where $|\tilde{h}| = \delta$. Then:

(3) Let $T' =_{df} \{f : F(p, f) = |n|_{\tilde{h}} \wedge f \in [T] \wedge L_\delta[f] \models \Phi(f)\}$.

Then $T' \in \Sigma^1_1(\tilde{h}, T)$.

Moreover T' contains elements not hyperarithmetic in $(T, \tilde{h})$ by the last conjunct of (2). By the hyperarithmetic perfect set theorem already referred to above, T' contains a perfect set T^* recursive in $\mathcal{O}^{\tilde{h},T}$, the hyperjump of $(\tilde{h}, T)$, and so $T^* \in L_{(\mathcal{O}^{\tilde{h},T})^+}[\tilde{h}, T] \subseteq M$. Further $f \in T^* \longrightarrow F(p, f) = \alpha$ where $\alpha = |n|_{\tilde{h}}$ as required.

Suppose T is ∞-pointed. Let $f \in [T^*]$; then $f \in [T]$, hence $T \leq_\infty f$, and $F(p, f) = \alpha$. By the last conjunct of the defining clause of (3), $|\tilde{h}| < \Sigma^f$. But then $\tilde{h}$ has a unique Σ_1 definition in $L_{\Sigma^f}[f]$ (as the solution for k in the predicate $\Psi(k, \alpha, n)$); hence by Σ_1-elementarity, (Fact 2.5), $\tilde{h} \in L_{\lambda^f}$ and thus $\tilde{h} \leq_\infty f$. As $L_{\lambda^f}[f]$ is closed under hyperjump, (Facts 2.2, 2.4) $T^* \leq_\infty f$. Hence T^* is ∞-pointed. This finishes the lemma. **Q.E.D.**

Now let $\boldsymbol{d_0} \leq_\infty \boldsymbol{d_1} \leq_\infty \cdots \boldsymbol{d_n} \leq_\infty \cdots (n < \omega)$ be a cofinal (in $\leq_\infty \cap M \times M$) sequence of ∞-degrees of M. We assume that $\boldsymbol{d_0} = \boldsymbol{0}$. Let $d_i \in \boldsymbol{d_i}$. Let $E = \{d_i\}_{i\in\omega}$.

Lemma 3.8 *There exists a perfect tree T_0 such that* $\forall f \in [T_0]$
(f countably remote from M $\Rightarrow f$ is ∞-minimal over $E \wedge \lambda^f = \lambda^E$).

Proof: We define a binary system of ∞-pointed trees T_s for $s \in 2^{<\omega}$. Suppose $\langle \delta_n \mid n < \omega \rangle$ an increasing sequence of p.r. closed ordinals with $\delta_0 = \omega$, $\sup \delta_n = \lambda^E$ (which here is just σ). We may assume that $\delta_n < \lambda^{d_n}$. We adopt the same tactic as above. Set $T_\emptyset = 2^{<\omega}$; assume T_s is defined for any $s \in 2^{<\omega}$ with $lh(s) \leq i$ with the properties

(i) If $lh(s) = j \leq i$ then $d_j \leq_\infty T_s$;

(ii) T_s is ∞-pointed.

For $lh(s) = i$ we define $T_{s\frown 0} =_\infty T_{s\frown 1} \in M$, two disjoint subtrees of T_s preserving (i) and (ii), according to a suitable variation on the four step recipe.

1) As in 3.3 *verbatim.*
2) Let $k(i) \geq i+1$ be the least k so that $T_s <_\infty d_k$. Find (by appealing to Lemma 2.4), ∞-pointed $T^{*j}_s \subseteq T'^j_s$ in M, with $T^{*j}_s =_\infty d_{k(i)}$ $(j = 0,1)$.
3) Find $T^j_s \subseteq T^{*j}_s$, $T^j_s =_\infty T^{*j}_s$ so that $\forall f \in T^j_s((f, T^{*j}_s)$ is $\delta_{k(i)}$-minimal over $T^{*j}_s)$. Such T^j_s can be chosen ∞-pointed, and in M, by Lemma 2.5;
4) If there is any $\alpha_0 \in On \cap M$ and ∞-pointed $T_{s\frown 0} \subseteq T^0_s$ so that $T_{s\frown 0} \in M$, and also so that $\forall f \in T_{s\frown 0}$ $F(i,f) = \alpha_0$, then let $T_{s\frown 0}$ be the M-least such; otherwise set $T_{s\frown 0} = T^0_s$. Similarly define $\alpha_1, T_{s\frown 1} \subseteq T^1_s$.

If any of the $T_{s\frown j} \neq T^j_s$ (for s with $lh(s) = i$), then we may assume they are all of the same ∞-degree (by appealing to Lemma 2.4(ii) and shrinking our original choices of $T_{s\frown j}$).

Now let T_0 be the fusion of the $\langle T_s \mid s \in 2^{<\omega}\rangle$. Then $[T_0] = \bigcup_{g\in 2^\omega} \bigcap_n [T_{g\restriction n}]$. We now show that T_0 is as in the lemma. Let $f \in \bigcap_n [T_{g\restriction n}]$. Clearly $\forall i\, d_i <_\infty f$. Hence f is a strict upper bound for E and $\lambda^E \leq \lambda^f$. Then if $s = g \restriction i$, (f, T^j_s) is $\delta_{k(i)}$-minimal over $T^{*j}_s =_\infty T^j_s$. If we set $h_n = T^{*j}_s$ we shall have by Lemma 2.3 that f is ∞-minimal over E if we have that $\lambda^f = \lambda^E$. We now use our assumption that $f \in [T_0]$ is countably remote from M. So suppose $F(i,f)$ is defined. Then by Lemma 3.7 we indeed could have found $T_{s\frown j} \subseteq T^j_s$ for $j \in \{0,1\}$, again with both $T_{s\frown j}$ in M, and ordinals $\alpha_j \in M$, with $\forall g \in [T_{s\frown j}]$ $\;F(i,g) = \alpha_j$. As this is true for all $i < \omega$, we conclude that $\gamma^f \leq \lambda^E$. But then by Fact 2.3 $\lambda^f \leq \lambda^E$. This finishes the Lemma. **Q.E.D.**

Again we must show that there are continuum many countably remote reals in $[T_0]$. Again this is easily seen by a genericity argument. Note that such a T_0 can be defined over L_σ; let $\sigma < \zeta' < \Sigma' < \omega_1^L$ be the lexicographically least pair such that $L_{\zeta'} \prec_{\Sigma_2} L_{\Sigma'}$ Note that $L_{\Sigma'} \models$ "$V = HC$". As before, let g be generic over $L_{\Sigma'}$ for the local Cohen forcing $\mathbb{P}_{T_0}$. We shall have that $L_{\zeta'}[g] \prec_{\Sigma_2} L_{\Sigma'}[g]$, and that $\lambda^g < \zeta'$; clearly g is remote from $L_{\Sigma'}$ and so from the reals of $M = L_\sigma$; by construction $g \in [T_0]$. **Q.E.D.Theorem 3.8**

We note that this argument indeed requires some assumption about M to work: there are countable Σ_2-admissible models whose set of ∞-degrees have unique strict least upper bounds: L_ζ is such: the real $\tilde{0}$ coding the Σ_2-mastercode of L_ζ is the unique strict l.u.b. of this set (*cf* [13] Lemma 2.3). What is really being used is the extendibility of certain models to others that themselves have Σ_2-elementary end extensions. One can use the arguments above to show that more generally:

Theorem 3.9 *Let $\tau < \omega_1^L$ be such that there are $\overline{\zeta} < \overline{\Sigma}$ with $L_\tau \prec_{\Sigma_1} L_{\overline{\zeta}} \prec_{\Sigma_2} L_{\overline{\Sigma}}$ and $L_{\overline{\Sigma}} \models$ "$V = HC$". Then there is a perfect tree $\overline{T}_0 \in L_{\overline{\zeta}}$ so that the $\mathbb{P}_{\overline{T}_0}$ generic branches $g \in [\overline{T}_0]$ are ∞-minimal over the degrees of L_τ.*

However a full characterisation of which countable sets of degrees have minimal upper bounds is still wanting.

References

[1] H. FRIEDMAN, *Minimality in the Δ^1_2-degrees*, Fundamenta Mathematicae **81**, 1974, pp. 183-192.

[2] R.O.GANDY & G. SACKS, *A minimal hyperdegree*, Fundamenta Mathematicae, **61**, 1967, pp215-223.

[3] J.D. HAMKINS & A. LEWIS, *Infinite Time Turing Machines*, to appear in the J. for Symbolic Logic.

[4] ——, *Post's problem for Supertasks has both positive and negative solutions*, submitted to the Archive for Mathematical Logic.

[5] A. KECHRIS, *Forcing with Δ perfect trees and minimal Δ-degrees*, J. of Symbolic Logic, **46**, 1981, pp.803-816.

[6] ——, *Minimal upper bounds for Δ^1_{2n}-degrees*, J. of Symbolic Logic, **43**, 1978, pp.502-507.

[7] G. SACKS, *Forcing with perfect closed sets*, in **Axiomatic Set Theory**, Proceedings of Symposia in Pure Mathematics, **13** part I, (D.Scott, editor), AMS, Providence, Rhode Island, 1971,pp. 331-335.

[8] ——, *Countable admissible ordinals and hyperdegrees*, Advances in Mathematics, **20**, 1976, pp. 213-262.

[9] ——, *Higher Recursion Theory*, Perspectives in Mathematical Logic, 1990, Springer Verlag, Berlin, Heidelberg, New York.

[10] P.D.WELCH *Minimality in the Δ^1_3-degrees*, J. of Symbolic Logic, **52**, 1987, pp. 908-915.

[11] ——, *Determinacy and Δ^1_3-degrees*, to appear in Fundamenta Mathematicae.

[12] ——, *The Length of Infinite Time Turing degree Computations*, submitted.

[13] ——, *Eventually Infinite Turing Machine Degrees*, submitted to the J. Symbolic Logic.